POSTHUMAN FRONTIERS: Data, Designers, and Cognitive Machines:
Proceedings of the 36th Annual Conference of the Association for Computer Aided Design in Architecture (ACADIA)

EDITORS
Kathy Velikov, Sean Ahlquist, Matias del Campo, Geoffrey Thün

COPY EDITING
Pascal Massinon, Mary O'Malley

GRAPHIC DESIGN
Rebekka Kuhn

PRINTER
IngramSpark

COVER IMAGE
Matias del Campo

Conference hosted by the University of Michigan, Taubman College of Architecture and Urban Planning, Ann Arbor, Michigan.

Issued in print and electronic formats
ISBN 978-0-692-77095-5 (print)
ISBN 978-0-692-77096-2 (epub)

PROCEEDINGS

POSTHUMAN FRONTIERS:
    DATA, DESIGNERS, AND COGNITIVE MACHINES

Proceedings of the 36th Annual Conference of the
Association for Computer Aided Design in Architecture

University of Michigan Taubman College of Architecture
and Urban Planning, Ann Arbor

——

Edited by Kathy Velikov, Sean Ahlquist,
Matias del Campo, Geoffrey Thün

TAUBMAN COLLEGE
architecture + urban planning
University of Michigan

acadia

Proceedings

Foreword
Complex Entanglements

Jason Kelly Johnson
President, Association for Computer Aided Design in Architecture
Associate Professor, California College of the Arts
Founding Design Principal, Future Cities Lab

ACADIA was formed at a meeting 35 years ago on October 17, 1981 at Carnegie Mellon University in Pittsburgh. Since its inception, the conference has served as an incubator for emerging ideas in feedback loops between academia, industry, and professional practice. Over the years, ACADIA's members, leadership, and attendees have included some the most inventive and important figures in the fields of architectural education, design, computation, and engineering. While ACADIA is the most selective peer-reviewed conference of its kind in the world, it is also an open setting to discuss and debate experimental ideas no matter who you are or where you come from. The fruits of these debates can be found in influential schools and research centers around the globe; in award-winning software, hardware, products, furniture, and installations; and in much larger constructions that define cityscapes from California to New York, London, Dubai, Beijing, and beyond. Ideas percolate at the ACADIA conference, and are iterated, prototyped, questioned, refined, built, and further interrogated as time passes, and they continue to evolve.

At last year's event in Cincinnati, we organized a special session called *Pioneers of Computational Design* moderated by Robert Aish, featuring Don Greenberg, Tom Maver, and one of ACADIA's founding members, Chuck Eastman. This remarkable session revealed that our founding members' interests extended beyond "CAD" and included pioneering research in topics including virtual reality, computer graphics, and building information modeling. The session was also a reminder of how far ACADIA has come in 35 years, where computational and technical subjects are no longer partitioned from the complexities of the architecture studio. This year's conference sessions and publications will no doubt epitomize this transition. Presenters will describe emerging pedagogies and research models from schools, labs, shops, and offices around the globe, where computation and design are now pursued simultaneously, most often entangled with other unexpected disciplinary and non-disciplinary concerns and possibilities.

During a coffee break at one of my first ACADIA conferences, I recall finding myself in a conversation with the late Professor William J. Mitchell (also one of ACADIA's founding members). He had founded the MIT Media Lab's Smart Cities Program and his book *The Logic of Architecture: Design, Computation, and Cognition*, published in 1990, was credited by The New York Times as having "... profoundly changed the way architects approached building design." Bill had just listened to me present a project and asked me questions that surprised and inspired me: "It is a beautiful project, but what if a city was filled with projects like yours? What kind of world would it be?" While Bill was known as a technologist, he was also deeply interested in broader ideas about the role technology could play to positively shape cities and society. In many ways, his attitude thankfully lives on today. Just look at the range of this year's ACADIA papers, projects, participants, and speakers. In the words of this year's Conference Chairs, one of the defining features of this event is to explore the "complex entanglements" and feedback loops between a radically diverse set of design ecologies; what they call, "autonomous and semiautonomous states." Participants share a fascination with the interplay of these states where computation, artificial intelligence, and human ingenuity can yield radically new and innovative modes of designing, building, thinking, and interacting. In the spirit of William Mitchell, in the midst of our extraordinary experimentation and technological innovation, let's not forget to ask ourselves and our colleagues: "What kind of world would it be?"

On behalf of the ACADIA Board of Directors and its membership, I want to acknowledge the 2016 University of Michigan Taubman College of Architecture and Urban Planning team for their extraordinary organization, energy, and thoughtfulness. Special thanks to Conference Site Chair Geoffrey Thün, Conference Technical Co-Chairs Kathy Velikov, and Sean Alquist, and others members of the team, including Matias del Campo, Workshop Co-Chairs Wes McGee and Catie Newell, Exhibition Co-Chair Sandra Manninger, staff members Kate Grandfield and Deniz McGee, and many others. As they have now discovered, organizing an ACADIA conference can be a little like using your own backyard to host a wedding, a graduation and a funeral—all in one weekend. Each event requires the hosts to assume different personalities: the strategist, the enforcer, the MC, the inspirational speaker. It requires a thankless series of meetings and tasks that require vision, energy, a sense of humor, diplomacy, and above all patience. That being said, Geoffrey, Kathy, Sean, and the extraordinary team they assembled, have patiently and generously worked with us over two years to not only craft a thought-provoking conference, exhibition, and workshop series, but also produced some of the highest quality publications ACADIA has ever seen. We extend to you, and the entire Taubman College community, our sincerest admiration, respect, and appreciation.

I would also like to acknowledge ACADIA's many sponsors this year. Year after year, the support of sponsors allows us to host a world-class event with an unsurpassed roster of keynote speakers, awardees, exhibits, publications, workshops, special round-tables, events, and celebrations. Additional sponsorship from Autodesk allowed us to support more ACADIA Conference Student Travel Scholarships than ever before, and a new ACADIA Autodesk Awards Program will honor and financially support emerging paper and project research.

Lastly, I would like to thank the ACADIA Board of Directors and Officers. Through the leadership of this dedicated group of people, ACADIA's organization, finances, sponsorships, marketing, and other outreach efforts have never been stronger. We look forward to continuing to build upon and evolve these efforts in the coming year as ACADIA prepares to host its follow-up conference at MIT in Boston, Massachusetts in October 2017.

Introduction
Posthuman Frontiers

Kathy Velikov, Sean Ahlquist, Matias del Campo, Geoffrey Thün
University of Michigan, Taubman College of Architecture and Urban Planning

In the posthuman there are no essential differences or absolute demarcations between bodily existence and computer simulation, cybernetic mechanism and biological organism, robot teleology and human goals.

–N. Katherine Hayles, 1999[1]

If posthumanism can be defined as the condition in which humans are understood to be intertwined and co-evolutionary with technological, medical, informatic, and economic tools and networks, and exist within a continuum with biological, ecological, and machinic entities,[2] then computational design is by definition a posthumanist endeavor. In contrast to design paradigms dominant since the Renaissance, contemporary computational design is not "conceived in the mind" of the designer and then represented as instructions for construction,[3] but rather design emerges through informational procedures and feedbacks with one or more numerically controlled machines. Increasingly, feedback from environmental forces, other participants, and data streams becomes an active force shaping design outcomes, behaviors, and interactions.

The ACADIA 2016 Conference brings together new research and design work from practice and academia that explores the partnerships among human practices, procedural design methods, and autonomous machines. The papers presented in these proceedings identify and examine current tendencies in computational design to develop and use quasi-cognitive machines and to advance the integration of software, information, fabrication, and sensing in the generation of mechanisms capable of interfacing with the physical realm.

EMERGENT TOOL STREAMS

The conference proceedings extend architectural inquiry to incorporate fields that span material science, robotics, autonomous behavior, interaction, and data-driven approaches. The papers selected for the proceedings describe practices and techniques driven by procedural approaches and externalized knowledge that ultimately take on their own identity, instrumentality, and momentum. Design operates through complex entanglements and feedback with computational as well as physical matter. The processes and products of designers are now intertwined with an emergent tool stream that not only imbues us and our built environments with novel interfaces, senses, and sensibilities, but that also produces spontaneous material and informational excess as byproducts of its autocatalytic processes. Just one example of the manifold correlations demanded by contemporary methods of design—it is no longer a question of how to work with big data, but how to evaluate the quality of data, how to assess the politics of data, and how to consider the impact of design judgment, decision making, and aesthetics within data-driven processes.

POSTHUMAN DESIGN ECOLOGIES

Posthumanism does not entail a condition *after* the dominance of the human species or without humans, but rather emphasizes an alternative perspective on design that shifts focus away from an anthropocentric position of observation and control. Posthumanist design practices decentralize the role of human judgment and embrace the notion that creative agencies can be conferred to nonhuman entities such as objects, tools, materials, other species, and environmental forces. Externalized knowledge begins to take on identity and instrumentality, and participate in the process of generating novel design ecologies and alternative models for architectural design. The panoramic position of the (humanist) designer as sole genius is giving way to a design ecology that operates in autonomous and semiautonomous modes. While bottom-up design techniques have been discussed in architecture extensively, a primary difference in today´s conversation regards the amount of resolution that new technologies yield, the fine grain of the solutions, and the sophistication of their execution. All of these (technological as well as theoretical) aspects allow the integration of a certain level of codified information, and a level of detail that results in an architecture rich in expression and articulation. In a moment when reflective judgment, knowledge, and intent seem less and less understood as the basis of design professions, and when subjectivity and identity are increasingly augmented and fragmented, how can we consider the deep challenges posed to the future of design education, research, and professional practice?

POSTHUMAN FRONTIERS

The papers presented at the ACADIA 2016 Conference have been categorized into five sections that are intended to position the dominant conversations emerging from current work in the field.

Programmable Matter operates through the commonality between natural systems, architecture, and computational design. The papers describe methods for computation that translate the logic of biological systems into codified material behaviors, which allows formational and performative agency to be shared among designers, materials, computational procedures, and environmental forces.

Generative Robotics explores processes of design exploration that are developed through codified actions and procedures as opposed to constraints or predetermined instructions. Robotics are moving beyond instrumental tools to one that are fully immersed within the cyclical processes of design iteration. Robots and humans are becoming collaboratively engaged in the making of material and in the formation and assembly of architectural forms.

Procedural Design deploys collaborative and emergent protocols and processes to enable design exploration, ideation, form development, or the construction of physical architectural systems. Procedural designs do not result in singular solutions, but rather fields of possible outcomes emanating from protocols such as game engines, big data scanning, genetic algorithms, and self-assembling agents.

Posthuman Engagements explores awareness, interaction, and communication among humans, tools, and intelligent machines. From the use of learning algorithms that aim to achieve life-like

behavior in synthetic systems, to gesture-based drawing machines, the papers in this section experiment with material and digital languages that produce new relations and intimacies between humans, environments, and things.

Material Frontiers gathers two emerging areas of exploration: computational material agencies that extend beyond instrumentality and performance to engage aesthetics, ontology, and irregular formation, and design work with synthetic biologies, where architectural researchers deploy living matter crossbred with computational, biological, genetic, and electrochemical logics toward new species of architectural and landscape materialization.

The papers gathered in this volume represent some of the most innovative and exciting work currently occurring in the field. This conference helps to document a maturation of computational design into a discipline that embraces instrumental or formal sophistication while also expanding the potential fields of agents, matters, and environments in collaborative and co-evolutionary ways. With the increasingly effortless agility enabled by tools of computational design and digital fabrication, we see the ACADIA community not only address the synthesis of the human and the technological in the process of design, but also consider the participants of a posthuman architecture. We look forward to the conversations at the conference about the potential of such an expanded scope in theory, process, and practice, and anticipate future trajectories that build upon this work from the growing ACADIA community.

1. N. Kateherine Hayles, *How We Became Posthuman: Virtual Bodies in Cybernetics, Literature, and Informatics* (Chicago, University of Chicago Press, 1999), 3.

2. This definition is adapted from: Carey Wolfe, *What is Posthumanism?* (Minneapolis: University of Minnesota Press, 2009), xv; Ariane Lourie Harrison, "Charting Posthuman Territory," in *Architectural Theories of the Environment: Posthuman Territory*, ed. Ariane Lourie Harrison (New York: Routledge, 2013), 3; and Bruce Mazlish, *The Fourth Discontinuity: The Co-Evolutions of Humans and Machines* (New Haven, CT: Yale University Press, 1993), 5–7.

3. Leon Battista Alberti, *De re aedificatoria*, 2.1.4., quoted in Mario Carpo, *The Alphabet and the Algorithm* (Cambridge, MA: MIT Press, 2011), 21.

Procedural Design

Sean Ahlquist
University of Michigan, Taubman College of Architecture and Urban Planning

Procedural design is often classified as a computational approach relying upon a set of instructions that, when used in a particular sequence, are the generators of form. While within this framework certain methods may be iterative and cyclical, procedural design often denotes the construction, conceptually, of a linear solver. The work documented in this section, though, shows a significant evolution of this approach. Intelligent systems are formed in which computation is given the freedom to absorb, interpret, and respond within the sequential set of procedures, thus shifting from linear logics to networked ones. This is addressed through papers that discuss the language through which such processes are enacted and explore the emergence of a built architecture through dynamic logics of design computation.

Ludwig von Bertalanffy established the sequencing of a feedback system as a part of General Systems Theory.[1] This laid the groundwork for the semantics and structure of procedural design. In essence, it is a methodology that is used to test the relationships of parameters through iteration. Bertalanffy classified the components of a feedback system by *count*, *species*, and *association*. These have been superceded as the metrics of design space, since procedural operations allow for the exploration, testing, and refinement of ideal parametric relationships. In this application, the feedback system is an active agent of design exploration.

Traditionally, procedural design has offered means of testing the relationships of parameters, but the work shown in this section demonstrates an evolution of this approach. Procedural processes become an active agent for resolving the relationships of *systems*. In Gerber's "Multi-Agent System for Design" and Savov's utilization of gameplay, logics of fabrication shift from defining constraints to being exploratory agents for design ideation and the construction of architectural systems. Human inflection becomes an operational procedure in Johnson's work with SIFT algorithms and Sanchez's "Combinatorial Design." Both exploit the iterative facet of the feedback mechanism to scan massive datascapes while interjecting the transformational feature of human intuition.

Emergence is an innate function of a properly constructed procedural design process. Through works such as Andréen's "Large Swarms of Simple Robots" and Rusenova's "Aggregate Architectures," it is possible to see a shift from merely seeking emergence to enabling machine intelligence to learn from and respond to specific emergent behaviors. Koschitz, through the visual programming language "Beetle Blocks," proposes a platform that simplifies the construction of procedural processes and the conceptualization of emergent design. Davis's incorporation of building evaluation and Smith's method for building automation subsequently extends, on a grand scale, the scope of procedural design. Data is either a live agent orchestrating building systems or an encapsulation of the live agents—the building occupants—to re-inform successive design explorations.

Collectively, the research in this section brings a valuable ambiguity to the finality of the feedback system. This reverberates into processes and modes of design, where the work provides a clear indication that architectural form is the enmeshment of systems, not just a collection of geometric constructs.

1.   Ludwig von Bertalanffy, *General Systems Theory: Foundations, Development, Applications* (New York: George Braziller, 1969.

Procedural Design

A Multi-Agent System for Facade Design

A design methodology for Design Exploration, Analysis and
Simulated Robotic Fabrication

David Jason Gerber
Evangelos Pantazis
University of Southern California

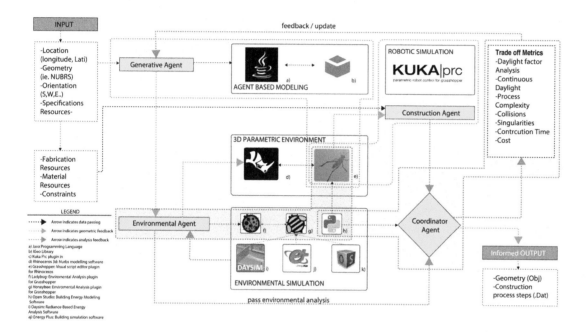

1

ABSTRACT

For contemporary design practices, there still remains a disconnect between design tools used
for early stage design exploration and performance analysis, and those used for fabrication and
construction of complex tectonic architectural systems. The research brings forward downstream
fabrication constraints into the up-stream design exploration and design decision making. This paper
addresses the issues of developing an integrated digital design work-flow and details a research
framework for the incorporation of environmental performance into a robotic fabrication for early
stage design exploration and generation of intricate and complex alternative façade designs.

The method allows the user to import a design surface, define design parameters, set a number
of environmental performance objectives, and then simulate and select a robotic construction
strategy. Based on these inputs, design alternatives are generated and evaluated in terms of their
performance criteria in consideration of their robotically simulated constructability. In order to
validate the proposed framework, an experimental case study of office building façade designs that
are generatively created from a multi-agent system for design methodology is design explored and
evaluated. Initial results define a heuristic function for improving simulated robotic constructability
and illustrate the functionality of our prototype. Project limitations and future research steps are
then discussed.

1   Diagram illustrating all the components of the Multi-Agent Systems for
Design framework and prototype.
The diagram shows the sequence
and data flow, system component
interactions and dependencies, the
agents, and feedback types and
loops.

INTRODUCTION

The Rise of the Robots in Architecture

The rapid evolution of digital design methodologies and technologies in the Architecture Engineering Construction (AEC) industry, and its coupling with new fabrication and construction technologies, have provided architects with opportunities to move away from Fordist standardization towards a post-Fordist realm of previously unattainable design possibilities. While symmetry and repetition brought about economical efficiencies for the AEC industry, these norms are now becoming antiquated and surpassed by an era of digital design, analysis, and fabrication. Parametric design and its evolution into Building Information

Modeling (BIM) has more efficiently facilitated the design exploration of new forms and multiple design alternatives for a broad spectrum of pre- and post-rationalization design approaches (Gerber 2012). Furthermore, simulations assist designers in making more informed design decisions by providing for higher degrees of certainty in the complex and synthetic design decision-making process (Roudsari, Pak, and Smith 2014).

While expert domains relied on the asynchronous handover of drawings, current practices seek the intelligent and rapid integration of environmental and structural analyses during the design exploration and development phases. By integrating these historically disparate and highly coupled architectural, engineering, and construction domains, the use of parametric models that accommodate design changes efficiently are now used to enhance environmentally conscious and cost-effective design solutions. With the explosion of new design technologies, there has also been an increase in the integration of novel digital fabrication techniques and industrial robots in architecture. These novel fabrication technologies have enabled the realization of complex geometric forms, and more generally, a paradigm shift towards mass customization, post-Fordism, and the non-standard (Racine et al. 2003).

Core to our research agenda is to continue to prove the importance of the architectural non-standard, and to use its intricacy and differentiation as an opportunity to not only provide higher performing design solutions in a multi-objective empirical fashion, but equally for the qualitative metrics of aesthetics and the communication of delight. In that regard, robotic fabrication offers a unique opportunity to reconsider the architectural design process as not just designing the building form to be constructed, but as designing the construction process itself, which equally infers a manifesting of its own aesthetic and inherent post-Fordist performance gains (Gramazio and Kohler 2008). The rapid integration of advanced digital and robotic simulation has

led to the reconsideration of the architect as both a contemporary master-builder and a digital toolmaker (Tamke and Thomsen 2009). This further highlights the need for more integrated approaches for the AEC industry, and an urgency to develop new design methodologies that support a more holistic approach towards the adoption of robotics, not just as tools, but as participants in the multi-objective, complex, and highly synthetic design decision-making process.

Despite the numerous advancements, there remains a gap in intuitive workflows for designers to incorporate simulation results as drivers and constraints early in the design process (Kilian 2006). In addition, disparate software is used for different design phases—design generation, simulation, manufacturing, and construction planning—and consequently continuous information exchange among the AEC disciplines still remains a challenge (Scheurer 2007; Schwinn and Menges 2015). Most critically, current design methods and computational tools remain limited in their foresight, and in registering design parameters and constraints from upstream through to downstream processes. Generally, these methods do not consider assembly and construction constraints, nor do they account for the adaptation of projects into local conditions and the reality of real world analogue noise. Currently, there are few robust and efficient programming strategies for controlling multiple robots in customized semi-automated construction conditions. Challenges for such programming tasks include: a) a constantly changing environment (construction sites); b) much smaller production volumes (buildings are one off products); and c) a much larger range of required tasks (Bechthold 2010). Our conjecture is that the next steps in digital design and robotic fabrication are to develop methodologies that consider computers and our robotics as collaborative partners in the design process, as having agency and the capacity to register contextual conditions, environmental analyses, and construction constraints in order to provide architects with design alternatives that fulfill complexly coupled criteria, such as environmental performance, structural efficiencies, and fabrication constraints (Shea, Aish, and Gourtovaia 2005).

This research presents our prototyping of an integrated design methodology that allows for validated simulations of the robotic assembly sequence to be integrated into the early phases of the design process. As a first step, we focus on coupling environmental performance simulations and robotic construction simulations for informing and optimizing a generative design process based on our multi-agent systems (MAS) for design work. We validate our framework by applying it to building envelopes (i.e., facades) of a set of commercial buildings representative of varying geometric forms. Building envelopes are among the most complex architectural components, combining aesthetic,

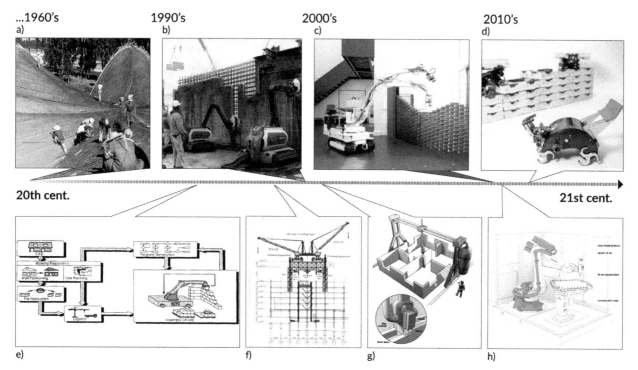

2   Timeline of Robotic Construction Applications: a) Felix Candela (1958), Los Manantiales Restaurant form works. Xochimilco; b) U.S. Construction Robotics (Robotworx (2012), Semi-Automated Masonry System; c) Gramazio Kohler Research (2014), Building Strategies for On-site Robotic Construction; d) Self Organizing Systems Research Group (2014), Termes Project. Harvard University; e) T. Bock (1998), Automatic generation of the controlling-system for a wall construction robot, Institute for Machinery in Construction, Karlsruhe; f) Obayashi Corporation (2006), Automated Building Construction System (ABCS), Japan; g) Zhang Jing (2013), Contour Crafting, USC; h) Achim Menges (2014), Landesgartenschau Exhibition Hall, ICD/ITKE Stuttgart.

structural, environmental, and construction concerns, and are a use case where robotic fabrications will be readily adopted (Bechthold et al., 2011).

BACKGROUND

Towards robotic agency, from design to construction

Architecture has increasingly become the materialization of digital information instead of the materialization of drawings (Mitchell 2005). Computational design, especially associative parametric Computer Aided Design (CAD), introduced in the 1990s, allowed for design variation through the use of geometric constraints and manipulation of free-form geometries. Associative geometry and parametric design enabled the establishment of a schema of geometric dependencies for the building components, and thus controls for the behavior of such objects under transformations within a topological constraint (Gerber 2012; Oxman 2008). However, a building can be described through multiple levels of abstraction, of which geometry is just one among many, such as program typology, performance, construction technologies, or physical elements. Oxman points out that there is great potential for combining parametric modeling with performance-based design techniques in a top down fashion to exploit simulations for the modification of geometrical form towards the objective of optimizing candidate designs. On the other hand, there are integrative approaches that situate themselves between bottom-up and top-down processes,

where based on nature and self-organization, agents are developed and fed with multiple types of input information (i.e., geometrical, structural, social), and are thus able to constantly adapt to required changes (Parascho et al. 2013).

Additionally, the utilization of large scale CAD/CAM manufacturing and robotics in the AEC industry is bridging the gap between what is digitally modeled and what could be physically realized. Such technologies allowed the production of non-standard components at almost the same price as standard (Kolarevic 2004). However, the AEC industry still remains a trade-oriented and labor-intensive industry with minimal automation of tasks. Research in construction-specific robots initiated in the 1980s in Japan, and focused on multiple construction-related activities. However, the unpredictability of the construction sites, hardware limitations, and high cost have resulted in few construction robots actually being used in construction sites (Maeda 2005). Related research has either focused on implementing different approaches for specific robotic construction subtasks, such as masonry placing robots (Bock et al. 1996; Steffani, Fliedner, and Gapp 1997), or for implementing large additive fabrication techniques, such as Contour Crafting (Khoshnevis 2004). These approaches use computer control to exploit the superior surface-forming capability of concrete and adobe structures, or to implement autonomous distributed robotic systems, which operate collectively in order to accomplish simple tasks

                                     A Multi-Agent System for Facade Design Gerber, Pantazis

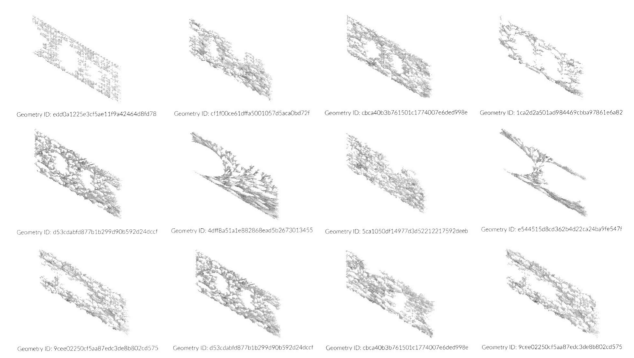

Geometry ID: edd0a1225e3cf5ae11f9a42464d8fd78   Geometry ID: cf1f00ce61dffa5001057d5aca0bd72f   Geometry ID: cbca40b3b761501c1774007e6ded998e   Geometry ID: 1ca2d2a501ad984469cbba97861e6a82

Geometry ID: d53cdabfd877b1b299d90b592d24dccf   Geometry ID: 4dff8a51a1e882868ead5b2673013455   Geometry ID: 5ca1050df14977d3d52212217592deeb   Geometry ID: e544515d8cd362b4d22ca24ba9fe547f

Geometry ID: 9cee02250cf5aa87edc3de8b802cd575   Geometry ID: d53cdabfd877b1b299d90b592d24dccf   Geometry ID: cbca40b3b761501c1774007e6ded998e   Geometry ID: 9cee02250cf5aa87edc3de8b802cd575

3   A subset of generated facade panel alternatives, with 2 and 3 openings on a planar input surface, which are high ranking in terms of environmental performance and therefore are tested for their constructability.

(Wawerla, Sukhatme, and Matarić 2002; Kube and Zhang 1993; Petersen 2014).

More recently in architecture, multi-functional industrial robots have become standard in automation precisely because, like personal computers, they have not been optimized for a single task but are suitable for a wide spectrum of applications (Gramazio and Kohler 2008). Perhaps most importantly, the widespread application of industrial robots has enabled the production of bespoke fabrications and mass customized manufacturing of architectural artifacts. This has been accompanied by a growing interest in reviving traditional techniques, or "digital craftsmanship," which leads to innovative yet regionally relevant architectural design that is formally intricate and complex, and often higher performing (Gramazio and Kohler 2008).

To date, digital design tools have mostly been utilized as a more efficient version of the drafting table, and not as an active aide in the design process (Oxman 2008). Current tools have incorporated performance feedback but only recently has research into integrating performance-based simulation for informing generative design become more accessible in practice (Malkawi et al. 2005). Although the tools for developing complex geometric forms are now commonplace, there is a lack of intuitive tools to enable designers to design and explore construction processes and particularly robotic solutions and their highly intricate and

specific geometric configurations (Braumann and Cokcan 2012; Helm et al. 2014). This is because of first, the open-endedness of design problems and their ill-defined nature, which often requiring hard-to-compute synthesis; second, a general lack of a deep understanding and abstraction of fundamental shape-forming processes in nature that allow us to create tools to support our design intuition (Shea, Aish, and Gourtovaia 2005); and third, the rapid evolution of digital fabrication and robotic technologies on one hand, and the emerging necessity for architects to develop programming skill-sets on the other hand. Combined, these causes have hindered the growth and role of computation and robotics to influence design cognition and to further contextualize and incorporate them into architectural design decision making (Daas 2014).

RESEARCH METHODOLOGY & OBJECTIVES

The objective of this research is to develop a new digital design methodology as an extension of our multi-agent systems (MAS) for design method. The extension of our MAS for design supports the generation of optimized context-specific building components (i.e., façade panels) by integrating a bottom-up evolutionary MAS design strategy with environmental analysis and construction process constraints. The following research questions are asked: One, how can we better enable our generative design process by coupling it with automated robotic fabrication constraints more efficiently? Two, how can

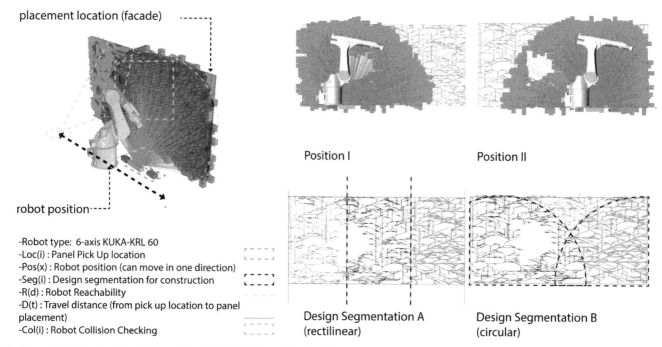

placement location (facade)

robot position

-Robot type: 6-axis KUKA-KRL 60
-Loc(i) : Panel Pick Up location
-Pos(x) : Robot position (can move in one direction)
-Seg(i) : Design segmentation for construction
-R(d) : Robot Reachability
-D(t) : Travel distance (from pick up location to panel placement)
-Col(i) : Robot Collision Checking

Position I

Position II

Design Segmentation A
(rectilinear)

Design Segmentation B
(circular)

4   Diagram illustrating the design of the experiment for the simulation of the robotic fabrication segmentation, positioning, and reachability, collision parameters of the construction agent.

we validate simulations that relate environmental and robotic constraints as performance drivers to improve the designer's decision-making process? Three, can we translate what we measure into a heuristic function for the design and incorporation of a robotic agency into the MAS for design approach of the research team.

It has been shown that the development of a bottom-up design approach, where robot-operating strategies drive the geometric design and assembly process right from the beginning, can help designers by providing them with design solutions that are already optimized for constructability issues and can provide feedback on how a specific design might behave in the assembly process (Schwinn and Menges 2015). Therefore, we explore the hypothesis that robotic simulations can become an integral part of the digital design workflow, integrating robotic agency with other objectives, including environmental analysis and a heuristic measurement of construction efficiency. The research methodology is a progression from the development of a bespoke MAS for design, where the starting point has been to use a bottom-up strategy to generate design alternatives constrained by building type, building system, environmental performance targets, and user preferences (Gerber, Pantazis, and Marcolino 2015). Here, the system is further extended to include the geometric constraints (Figure 1) of the complex façade fenestration or louvering pattern, as well as the downstream robotic fabrication and installation processes.

MAS for Design and Robotic Fabrication Agency

The research is based on an existing prototype and custom programming, but extends the existing workflow—which is described in detail here (Gerber et al. 2015a)—by addressing robotic assembly. In order to bridge the gap between design generation and construction, and in anticipation of physical proofs of concept of our MAS for design system, we develop a bottom-up design strategy that incorporates robotic fabrication constraints, and considers how local geometric rules and design parameters of building components are coupled with: a) global constraints (i.e., facade surface, openings); b) environmental analysis, such as Daylight Factor Analysis (DLA) and Continuous Daylight Autonomy (CDA); and, c) robotic simulation feedback that conditions the assembly process.

Our work implements an MAS for design framework for the intrinsic modularity and level of abstraction necessary for the development of the agencies and for their capacity to operate within distributed environments (Gerber et al. 2015a; Sycara 1998). The system evaluates the generated designs by utilizing two types of environmental analysis: DLA as the basic metric to measure the amount of light entering a space, and CDA as an environmental performance measure that calculates the percentage of occupied hours per year, when the minimum luminance level can be maintained by daylight alone (Reinhart 2010).

For the robotic simulation, we use a 6-axis robotic arm, the KUKA KR60H within the KUKA prc Grasshopper plug-in.

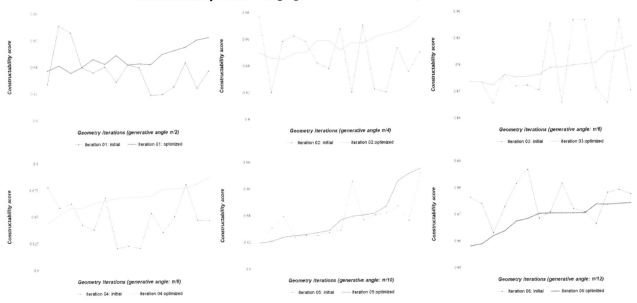

Constructability score/ Design generation for different generation angles

5   The graphs illustrate 6 different design generation cycles with variable angles with (bold, dark lines) and without (thin, light lines) the constructability score improvement over time per geometric iteration.

(Braumann 2011). MAS for design has four component agencies, which are modeled after specific design, analysis, and fabrication processes: a) a generative agent; b) a specialist environmental agent; c) a construction agent; and d), a coordinator agent (see Figure 1). The generative agent has a set of design parameters initially defined by the designer; for example, the design surface, orientation, panel length, generative angle, extrusion depth, and panel type preferences. This agent is tasked to iteratively create alternative geometries based on the input design parameters and context, which are updated on each run. The design parameters of the generative agent are defined by the designer, and are later conditioned by constraints imposed by the construction agent. The construction agent receives as inputs, the generative designs, which are evaluated to perform within a range defined in terms of the environmental analysis parameters described above. The system then performs the construction simulation and checks for constructability. Based on a constructability function and metric, the designs are ranked, and when necessary, the design parameters are updated and passed to the coordinator agent, forming one of the system's feedback loops. The objective of the construction agent is to find optimal positions for the robot that can facilitate and speed up the assembly process while decreasing collisions. This defines our ranking and heuristic function, further detailed below. The environmental agent performs two types of environmental analyses on the generated designs with the objective of combining the analyses and the passing of the analysis data—i.e., the lux values as CDA scores—to the coordinator agent along with a set of messages in the form of

a report. The goal for the environmental agent is to increase the natural daylight entering a typical office space (the DLA), while keeping the lux values within a threshold (CDA) defined by the space typology and dimensions. The coordinator agent establishes the communication among the agents, and ensures each process is being called, completed, and data is passed in a controlled fashion.

In order to test our MAS for design and robotic assembly methodology, we divide our experimental protocol into two phases. In the first phase, the designer develops; a) an initial design component of a façade and defines a sub set of alternative panel types (three in total for this experiment); b) provides the following information as inputs for the generative MAS for design system: length, angle, probability for each panel type, and depth or extrusion of panel; c) runs N number of iterations using a hill-climbing optimization algorithm; and d), generates a solution space of design alternatives that are then evaluated for their performance across the DLA and CDA metrics. In the second phase the best performing designs are; e) passed to a robotic simulation software where different construction strategies are explored; f) collisions and errors are registered as negative scores in the ranking equation; g) designs are ranked based on their constructability in terms of construction time; and, finally, h) these scores are then passed back to the generative system for further optimization. In the next section, we provide an overview of how we performed our data collection and analyzed the simulation results, and how these help the designer reconsider

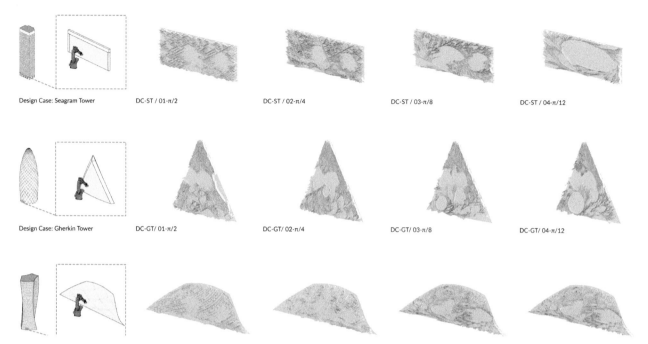

| Design Case: Seagram Tower | DC-ST / 01-π/2 | DC-ST / 02-π/4 | DC-ST / 03-π/8 | DC-ST / 04-π/12 |
| Design Case: Gherkin Tower | DC-GT/ 01-π/2 | DC-GT/ 02-π/4 | DC-GT/ 03-π/8 | DC-GT/ 04-π/12 |

6 Image showing design alternatives for 3 different building case studies, which represent different levels of complexity (flat surface, curvilinear surface, free-form surface).

the integration of the whole design cycle of exploration, analysis, and simulated fabrication processes.

To demonstrate the feasibility of the extension of our MAS for design framework, a prototypical design tool, in conjunction with an environmental performance and robotic simulation workflow, is set up in order to generate a highly varied, environmentally optimized shading system for a typical office environment. The system uses Rhinoceros as a platform that allows the designer to communicate across different Integrated Development Environments (IDE) that the integrated MAS platform utilizes. Python and Grasshopper visual scripting editor are used to call the different agent processes which have been implemented in Java using Processing libraries (Reas and Fry 2007). For the creation and management of geometry created by the agencies, the system uses the IGeo library (see Figure 1 components a) and b)), an open source NURBS based library, in order to output geometries that can be further used for fabrication purposes (Sugihara 2014). The specialist agent is developed in Python and uses Ladybug and Honeybee environmental simulation tools (see Figure 1 components g) and h)) in order to combine three different environmental analyses: Energy Plus, Daysim, and OpenStudio (Roudsari, Pak, and Smith 2014). For the construction agent, the team implemented KUKA prc, a simulation plug-in for Grasshopper, to simulate the robotic construction sequence and complex fenestration assembly process. Finally, the coordination agent developed in Python establishes communication between the generative process (generative agent), the analytical

process (specialist agent), and the simulation process (construction agent) by passing the analysis data and simulated results and messages back to the generative process, forming the culminating feedback loop in the system.

The work-flow is applied on the south facing façade of a typical academic office building, and is then further explored for other varied geometric forms. The following steps are followed in the design of our experiment:

1. The designer sets the input parameters and initializes a run of the generative system. The system implements a hill-climbing algorithm that searches for optimal window positions. At each iteration, the position of the windows is changed with a small increment in the surface domain of the facade. For each set of window positions, the system outputs one unique design, where the position of openings that provide more daylight to the interior are described as optimal. The data was collected by the system for 6 different generative angles: π/2, π/4, π/6, π/8, π/10, π/12. The generated designs are automatically passed to the specialist agent for lighting analysis. The system then performs two kinds of analysis, a) daylight factor analysis (DLA), and b) continuous daylight autonomy (CDA) for the generated designs.

2. Once the analysis is completed, the geometries with the highest ranking designs are sorted and passed to the robotic simulation.

A Multi-Agent System for Facade Design Gerber, Pantazis

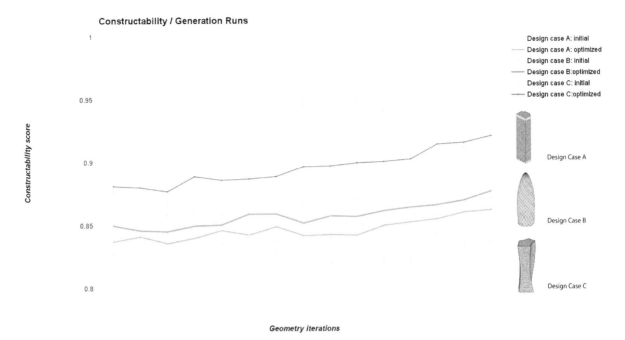

Constructability / Generation Runs

Design case A: initial
Design case A: optimized
Design case B: initial
Design case B:optimized
Design case C: initial
Design case C:optimized

Design Case A

Design Case B

Design Case C

Geometry iterations

7   Constructability metric in relation to design iterations for each of the different experimental case studies. Light curves represent the non-robotic-agent optimized constructability, bold curves represent the design generation inclusive of the robotic agency optimization.

3. The user selects the robot—in this case the 6-axis KUKA-KRL60—and defines the pick-up location and the robot axis of movement, here, parallel to the design surface.

4. A planning process is selected and the generated designs are segmented into groups based on the max reachability of the robotic arm. For the purpose of this research, only two different construction strategies are defined, thus dividing the generated design into rectangular and circular segmentations.

5. The robotic simulation is run and the system measures collisions, singularities, panels, and positions in order to develop the heuristic for the constructability agency. The following is the list of measurements and definitions in the heuristic function: a) Pos(x): the number of positions needed to assemble the whole façade and the number of max panels (Panels(n)) that the robot can place from a given robot position, where the best position is one in which the robot can place the maximum number of panels possible; b) Col(j): collisions between the robot and the already placed panels; c) Sing(k): singularity point positions that the robot can take while reaching a point in space that self-collide; d) D(t): and then calculates sum of travel distance from pick up location to panel placement (see Figure 2). Finally the constructability function is defined to be able to measure the performance of the construction process and rank the designs that depend on the parameters defined in steps a-d (see Equation below). Generally, more positions increase the construction time and

thus result in lower scores. Singularities (k) harm the robot and thus are defined as impacting the negative score further. Based on a user-defined number of maximum collisions when placing panels, we eliminate (delete) panels or alter the probability for each panel that is passed as input to the generative agent for the next iteration.

$$con(runID) = (Panels(n)/Pos(x))/(D(t)*\sum Col(j)) - \sum (Sin(k))$$

The objective of the MAS for design prototype is to initially search the generated solution space for the environmentally efficient facade design alternatives, based on different window configurations and variable design parameters—in particular, the angulation of each fenestration piece. The most efficient results are passed into the construction simulation for testing and for feasibility evaluation. Through this feedback loop, at each iteration, the generative process is informed by both simulations and tradeoffs between the agents established through each agency's utility functions. In this way, the system iteratively limits the solution space of possible better performing designs by constraining the design parameters to those that yield combined constructible and environmentally efficient outcomes.

EXPERIMENTAL RUNS

The team has completed a pilot study of the system and gathered initial results based on running the generative and analytical loops of the system for 450 cycles for three different case studies that vary in terms of geometry (Figure 6). The duration of each generation analysis cycle is approximately 5 min, including the geometry generation and the environmental analysis, while the robotic simulation cycle is approximately 10 min. Six different generative angles were tested to explore the capacity of our system to create an expanded and highly varied solution space of 450 unique geometries. The 10 highest-ranking designs from each cycle were selected and passed to the robotic construction agent. One pick-up location for all three types of panels was set and the range of possible segmentation positions was set from two to four. By segmentation position, we refer to the way we break up a generated design into parts, based on the work volume of a specified robot. In this study, we only looked into circular design segmentation based on the max reach of the available industrial arm.

Based on the simulation and the constructability heuristic function, the generated designs were improved for construction purposes either by eliminating panels or by changing the sequence of panel types. The updated parameters for each design were stored in XML files and were passed back to the generative agent. In Figure 5, graphs of the constructability score of the initial and improved geometries are presented for all the six different generation angles, while in Figure 7 we present the results for the three different case studies. It is observed that the constructability score is highly variable with the initial conditions, but becomes a line with zero or positive slope for the improved geometries across all the generation configurations. From the results on a flat façade surface, we can observe that geometries with generation angle $\pi/10$ performed better overall in terms of their constructability scores over time, as the deviation from initial to final score is the largest compared to the rest of the runs by a delta of 6%. In the initial runs, the constructability score from one design iteration to the next ranges from 0.1 to 0.6 in some cases, while in the optimized cases it drops below 0.15 and is constantly improving.

From the obtained results on different input geometries (flat surface: case A, curvilinear: case B, free form surface: case C), we can observe that for the planar geometry (design case A), the constructability score is higher, which means the system could find easier ways to assemble the façade panels, as expected, though the free form case seems to behave better than the curvilinear example (Figure 7), requiring further investigation. Thus in this work, we show that the system at this stage, with the integrated heuristic for constructability, is able to generate façade panels for different input geometries and output results that are performing well environmentally but are also optimized given a set of robotic fabrication constraints. Therefore, the designer can obtain feedback for multiple design objectives, based on both the environmental and construction simulation performance. As a result, designers can make more informed design decisions of one alternative over another, but also establish more seamless and efficient communication with façade engineers and fabricators. The system continues to be tested, validated, and further integrated.

In these experiments, a number of limitations are worth pointing out. Firstly, for reasons of simplicity, only one type of design segmentation was simulated and tested on the zero curvature (façade) surface domain and then the ruled curving and non-uniform domains. Moreover, at present the system is not fully automated and it still requires the designer to work across multiple interfaces. On occasion, during the generation of a design alternative, the criteria for creating a complete fenestration pattern across the domain of surface, while faced with a local minima problem within the hill-climbing algorithm, requires the manual deletion of collision points to enable the continuation of the pattern and design. Last but not least, in order to validate the results, we are planning to perform a physical experiment in order to be able to cross-reference the simulated with the physically obtained analogue and tectonic results.

DISCUSSION AND NEXT STEPS

At this stage of the research, we have successfully developed and tested an integrated MAS for design system. So far, the pilot study provided us with an intuition of how one design parameter, the angulation of the panels, can affect the design generation and the behavior of the system. We will continue to gather and analyze the data in order to further refine the heuristic function, as well as the definition of the agencies and their negotiation. As a next step, the full automation of the geometry and data passing across all agents will be further implemented. Another crucial enhancement to the MAS for design system is the further refinement of the generative agent to always output valid solutions given the feedback from the construction and simulation agent.

More work will also continue to revise and validate the results of our constructability utility function and to further test segmentation strategies. We will need to examine each parameter separately—doing so by running iterative simulations for each of the design parameters—and will check the behavior of the system, first in isolation, then in careful combination. Finally, the team will run simulations on a conventional existing façade design and compare our methodology in order to be able to benchmark the contribution of the work from design exploration, through multi-objective optimization, to efficient fabrication.

ACKNOWLEDGEMENTS
This material is partly based upon work supported by the National Science Foundation under Grant No. 1231001. Any opinions, findings,

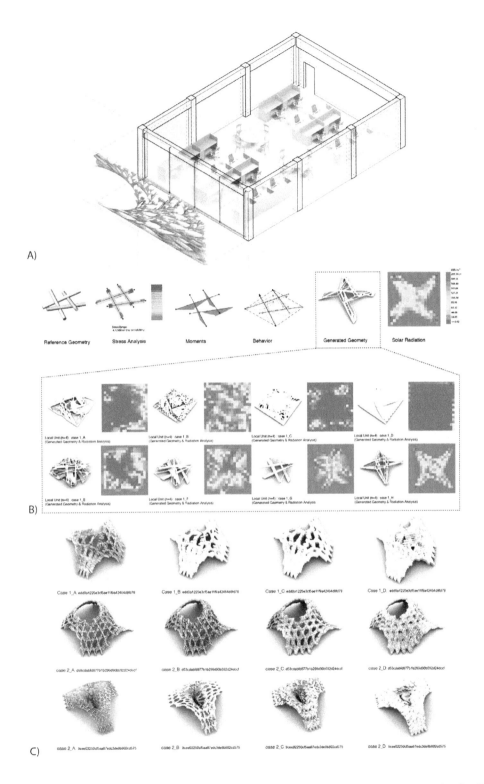

8   These images illustrate the Multi-Agent Systems for Design research thrusts and domains. This work presents research on the prototyping of multi-agent systems for architectural design of building envelopes, reciprocal frames, and form found shells. It presents a design exploration methodology at the intersection of architecture, engineering, and computer science, and focuses on bottom up generative methods coupled with multi-objective optimizing performance criteria; including for a) geometric complexity, b) objective functions for environmental and structural parameters, and c) robotic and digital fabrication. The Multi Agent System for design research has been developed, which looks at; A) building envelopes as a domain of interest for highlighting pro environmental capacities; B) local structural shell units in the form of reciprocal frames and in particular reciprocal frames as elements that can be used as form work for digitally deposited matrials that perform structurally as well as environmentally through voxelized porosity; and C) the combination of form found shells and the recirpocal frames to design explore and optimize the multi-objective aspects of these design space compelxities, specifically for structural matetrial minimzation in a trade off for lighing and temperature effects beneath the canopy. The work on enveloes and shells progress from non structural systems to structural systems and to the robotic implementation presented in this paper.

and conclusions or recommendations expressed in this material are those of the author(s) and do not necessarily reflect the views of the National Science Foundation. We would like to thank Professor Burcin Becerik-Gerber for her contributions in advising on the topic, and Alan Wang, an undergraduate researcher at USC, for his support in the development of the system and process.

REFERENCES

Bechthold, Martin. 2010. "The Return of the Future: A Second Go at Robotic Construction." *Architectural Design* 80 (4): 116–121.

Bechthold, Martin, Jonathan King, Anthony Kane, Jeffrey Niemasz, and Christoph Reinhart. 2011. "Integrated Environmental Design and Robotic Fabrication Workflow for Ceramic Shading Systems." In *Proceedings of the 28th International Association for Automation and Robotics in Construction.* Seoul, Korea: ISARC. 70–75.

Bock, T., D. Stricker, J. Fliedner, and T. Huynh. 1996. "Automatic Generation of the Controlling-System for a Wall Construction Robot." *Automation in Construction* 5 (1): 15–21.

Braumann, Johannes, and Sigrid-Brell Cokcan. 2012. "Digital and Physical Tools for Industrial Robots in Architecture: Robotic Interaction and Interfaces." *International Journal of Architectural Computing* 10 (4): 541–554.

Daas, Mahesh. 2014. "Toward a Taxonomy of Architectural Robotics." In *Proceedings of the 18th Conference of the Sociedad Iberoamericana Grafica Digital,* edited by Fernando García Amen. Montevideo, Uruguay: SIGRADI. 623–626.

Eastman, Charles M. 1994. "A Data Model for Design Knowledge." *Automation in Construction* 3 (2–3): 135–147.

Gerber, David J. 2012. "PARA-Typing Informing Form and the Making of Difference." *International Journal of Architectural Computing* 10 (4): 501–520.

Gerber, David J., Evangelos Pantazis, Leandro Marcolino, and Arsalan Heydarian. 2015a. "A Multi Agent Systems for Design Simulation Framework: Experiments with Virtual Physical Social Feedback for Architecture." In *Proceedings of the Symposium on Simulation for Architecture and Urban Design,* edited by Shajay Bhooshan and Holly Samuelson. Alexandria, VA: SimAUD. 1247–1254.

Gerber, David J., Evangelos Pantazis, and Leandro S. Marcolino. 2015. "Design Agency." In *Computer-Aided Architectural Design Futures. The Next City-New Technologies and the Future of the Built Environment, 16th International Conference, CAAD Futures 2015,* edited by Gabriela Celani, David Moreno Sperling, and Juarez Moara Santos Franco. Berlin: Springer. 213–235.

Gramazio, Fabio, and Matthias Kohler. 2008. *Digital Materiality in Architecture.* Baden: Lars Müller Publishers.

Helm, Volker, Jan Willmann, Fabio Gramazio, and Matthias Kohler. 2014. "In-Situ Robotic Fabrication: Advanced Digital Manufacturing Beyond the Laboratory." In *Gearing Up and Accelerating Cross-Fertilization Between Academic and Industrial Robotics Research in Europe.,* edited by Florian Röhrbein, Germano Veiga, and Ciro Natale. Vol. 94 of *Springer Tracts in Advanced Robotics.* Cham, Switzerland: Springer. 63–83.

Khoshnevis, Behrokh. 2004. "Automated Construction by Contour crafting—Related Robotics and Information Technologies." *Automation in Construction* 13 (1): 5–19.

Kilian, Axel. 2006. "Design Innovation Through Constraint Modeling." *International Journal of Architectural Computing* 4 (1): 87–105.

Kolarevic, Branko, ed. 2004. *Architecture in the Digital Age: Design and Manufacturing.* London: Taylor & Francis.

Kube, C. Ronald, and Hong Zhang. 1993. "Collective Robotics: From Social Insects to Robots." *Adaptive Behavior* 2 (2): 189–218.

Maeda, Junichiro. 2005. "Current Research and Development and Approach to Future Automated Construction in Japan." In *Construction Research Congress 2005: Broadening Perspectives,* edited by Iris D. Tommelein. San Diego, CA: CRC. 1–11.

Malkawi, Ali M., Ravi S. Srinivasan, Yun K. Yi, and Ruchi Choudhary. 2005. "Decision Support and Design Evolution: Integrating Genetic Algorithms, CFD and Visualization." *Automation in Construction* 14 (1): 33–44.

Mitchell, William J. 2005. "Constructing Complexity." In *Computer Aided Architectural Design Futures 2005, Proceedings of the 11th International CAAD Futures Conference,* edited by Bob Martens and Andre Brown. Vienna, Austria: CAAD. 41–50.

Oxman, Rivka. 2008. "Performance-Based design: Current Practices and Research Issues." *International Journal of Architectural Computing* 6 (1): 1–17.

Parascho, Stefana, Mark Baur, Ehsan Baharlou, Jan Knippers, and Achim Menges. 2013. "Agent-Based Model for the Development of Integrative Design Tools." In *Adaptive Architecture: Proceedings of the 33rd Annual Conference of the Association for Computer Aided Design in Architecture,* edited by Philip Beesley, Omar Khan, and Michael Stacey. Cambridge, ON: ACADIA. 429–430.

Petersen, Kirstin H. 2014. "Collective Construction by Termite-Inspired Robots." PhD Dissertation, Harvard University.

Racine, B., A. Pacquement, M. Burry, W. Prigge, F. Migayrou, and Z. Mennan. 2003. *Architectures non standard.* Paris: Editions du Centre Pompidou.

Reas, Casey, and Ben Fry. 2007. *Processing: A Programming Handbook for Visual Designers and Artists.* Cambridge, MA: MIT Press.

Roudsari, Mostapha S., Michelle Pak, and Adrian Smith. 2014. "Ladybug: A Parametric Environmental Plugin for Grasshopper to Help Designers

Create an Environmentally-Conscious Design." In *Proceedings of BS2013: 13th Conference of International Building Performance Simulation Association*. Chambéry, France: IBPSA. 3128–3135.

Scheurer, Fabian. 2007. Getting Complexity Organised Using Self-Organisation in Architectural Construction." *Automation in Construction* 16 (1): 78–85.

Schwinn, Tobias, and Achim Menges. 2015. "Fabrication Agency: Landesgartenschau Exhibition Hall." *Architectural Design* 85 (5): 92–99.

Shea, Kristina, Robert Aish, and Marina Gourtovaia. 2005. "Towards Integrated Performance-Driven Generative Design Tools." *Automation in Construction* 14 (2): 253–264.

Steffani, H. F., J. Fliedner, and R. Gapp. 1997. "A Vehicle For a Mobile Masonry Robot." In Proceedings of the *23rd International Conference on Industrial Electronics, Control, and Instrumentation*, vol 4. New Orleans, LA: IECON. 1337–1342.

Sugihara, Satoru. 2014. "iGeo: Algorithm Development Environment for Computational Design Coders with Integration of NURBS Geometry Modeling and Agent Based Modeling." In *ACADIA 14: Design Agency— Proceedings of the 34th Annual Conference of the Associa on for Computer Aided Design in Architecture*, edited by David Gerber, Alvin Huang, and Jose Sanchez. Los Angeles: ACADIA. 23–32.

Sycara, Katia P. 1998. "Multiagent Systems." *AI Magazine* 19 (2): 79–92.

Tamke, Martin, and Mette Ramsgard Thomsen. 2009. "Digital Wood Craft." In *Joining Languages, Culturse, and Visions–Proceedings of the 13th International CAAD Futures Conference*, edited by Temy Tifadi and Tomás Dorta. Montreal, QC: CAAD.

Wawerla, Jens, Gaurav S. Sukhatme, and Maja J. Matarić. 2002. "Collective Construction With Multiple Robots." In *Proceedings of the IEEE/RSJ International Conference on Intelligent Robots and Systems*, vol. 3. Lausanne, Switzerland: IROS. 2696–2701.

IMAGE CREDITS

Figures 1, 3–8: © Gerber and Pantazis, 2015
Figure 2: Timeline of Manual&Robotic Construction Applications. All images retrieved in July of 2016 from:
a) https://gr.pinterest.com/pin/294141419383708208/;
b) http://www.bdcnetwork.com/
robots-drones-and-printed-buildings-promise-automated-constructionc);
c) http://gramaziokohler.arch.ethz.ch/web/e/forschung/273.html;
d) http://www.eecs.harvard.edu/ssr/projects/cons/termes.html;
e) T.Bock, et.al. "Automatic generation of the controlling-system for a wall construction robot", Automation and Construction Journal 5 (1996) 15–21;
f) https://www.obayashi.co.jp/;
g) J. Zhang, B Khoshnevis, "Optimal machine operation planning for construction by Contour Crafting", Automation in Construction 29 (2013) 50–67; h) http://icd.uni-stuttgart.de/?p=11173

Dr. David Jason Gerber is an assistant professor at USC's School of Architecture with a courtesy joint appointment in the Sonny Astani Department of Civil and Environmental Engineering. He conducts NSF and corporate funded research at the intersection of design, science, and computing for Architecture. From 2008 to 2010 Dr. Gerber was a Vice President at Gehry Technologies Inc., a leading innovator in Building Information Modeling and building industry technology consulting. Dr. Gerber has worked as an architectural and urban designer in the US, Europe and Asia, for the Steinberg Group, Moshe Safdie, and as a project architect for Zaha Hadid Architects. He has been awarded research fellowships at MIT's Media Lab, as well as numerous teaching and research fellowships at Harvard University Graduate School of Design and as Harvard University's Frederick Sheldon Fellow. Dr. Gerber studied archtiecture at the University of California Berkeley (BA 1996), the Architectural Association in London (MArch 2000) and completed his design compotuing research and doctoral studies at Harvard University (MDesS 2003, DDes 2007).

Evangelos Pantazis is currently pursuing a PhD at the Viterbi School of Engineering at the University of Southern California. His research lies on the intersection of architecture and engineering and computational modelling and focuses on the use of Multi Agent Systems for building design . Evangelos holds a Masters of Advanced studies in the field of Computer Aided Architectural Design from the ETH in Zurich (2012). He received his Diploma in Architecture with honors from Aristotle's University of Thessaloniki (2010), and has also graduated from the MOKUME jewelry design school in 2007, where he was trained as a jeweler.

20,000 Blocks

Can gameplay be used to guide non-expert groups in
creating architecture?

Anton Savov
Digital Design Unit (DDU),
Technical University Darmstadt

Ben Buckton
Invent the World

Oliver Tessmann
Digital Design Unit (DDU),
Technical University Darmstadt

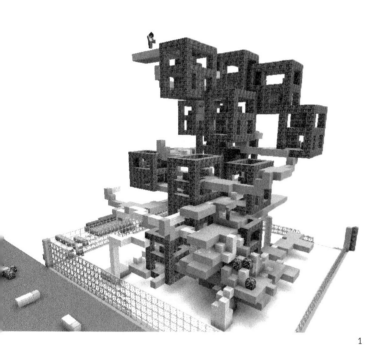

1

2

ABSTRACT

The paper follows research in engaging groups of non-trained individuals in the creation of architectural designs using games and crowdsourcing for human-directed problem-solving. With the
proposed method, architectural experts can encode their design knowledge into custom-developed
multiplayer gameplay in Minecraft. Non-expert players then are constrained by this gameplay which
guides them to create unique architectural results. We describe a method with three components:
guiding rules, verification routines and fast feedback. The method employs a real-time link between
the game and structural analysis in Grasshopper to verify the designs. To prove the viability of these
results, we use robotic fabrication, where the digital results are brought to reality at scale. A major
finding of the work is the suite of tools for calibrating the balance of influence on the resulting
designs between the Experts and the Players. We believe that this process can create designs
which are not limited to parametrically optimal solutions but could also solve real-world problems in
new and unexpected ways.

1 A 20,000 Blocks game map titled
HOW HIGH – created by Andrea
Quartara, Marc Emmanuelli and
Ulrik Montnemery during our
cluster at SmartGeometry 2016.

2 A 20,000 Blocks game map titled
WOULD YOU MIND YONA
FRIEDMAN? – created by Marios
Messios and Max Rudolph during
our cluster at SmartGeometry
2016.

3 The Project 20,000 Blocks links
participatory design, automated
structural analysis and robotic
fabrication.

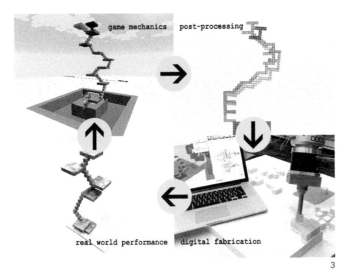

game mechanics post-processing

real world performance digital fabrication

3

PROBLEM STATEMENT

The authors pose the hypothesis that a better built environment could come into being if digital technology is used to include the non-experts, together with the experts, in the architectural design process. Arguments in favor of participatory design methods are not new in architecture and vary from increasing speed of design and construction, through fulfilling the preferences of inhabitants, to building cohesive urban communities (Friedman 1975, Parker 2014).

For a meaningful inclusion of the non-architect in the design process, the architectural expert knowledge needs to be partially encoded into an agile, open and generative rule-based system. For this we needed a modeling environment that combines the ease and popularity of computer games with the control and evaluation tools of Computer-Aided Architectural Design (CAAD) software. In this paper, we present such a system with three key components:

1) *Guiding rules* employed to direct the participants towards feasible design solutions.
2) *Verification routines* to automatically process the resulting designs on key parameters and performances.
3) *Fast feedback* to keep the players aware of what is happening, allowing for decision corrections to happen in short cycles.

Our aim is to let online communities use a game to prototype and test various building designs in short cycles. Our approach can be categorized as crowdsourcing. Daren C. Brabham defines crowdsourcing as "an online, distributed problem-solving and production model that leverages the collective intelligence of online communities to serve specific organizational goals. Online communities, also called crowds, are given the opportunity to respond to crowdsourcing activities promoted by the

organization, and they are motivated to respond for a variety of reasons." (Brabham 2013).

Games are accessible to people of all ages and backgrounds and can process large amounts of data input from many players. Hence our approach seeks to crowdsource within a game environment, particularly focusing on online multiplayer games. CAAD software, on the other hand, takes input from specialists and is capable of advanced performance analysis as well as controlling robotic fabrication. Verification routines based on CAAD software are needed to test the reliability of a multitude of architectural designs created by the non-trained crowd.

Crowdsourcing through gaming, seen as a means of user engagement and problem solving, is gaining ground in the sciences with projects such as FOLD.IT and Eve Online's Project Discovery, and recently in architecture with the work of Jose Sanchez, in particular the game Block'hood (Cooper et al. 2010, Szantner 2016, Sanchez 2015). At the same time, Minecraft (www.minecraft.net) has turned from a game into a phenomenon, being used in initiatives such as "Block by Block" to help the United Nations Habitat involve local communities in the refurbishment of their public spaces (BBC News 2012).

To assess our hypothesis, we set up an experiment, under the project *20,000 Blocks Above the Ground* (www.20000blocks.wordpress.com), which links multiplayer gaming, participatory digital design tools and robotic fabrication. (Figure 3)

The research project is run at the *Digital Design Unit (DDU)* at Technische Universität Darmstadt and started with the question: "Can gameplay mechanics be used to guide groups of non-experts throughout the collaborative creation of architectural designs?" Anyone can design a building in *20,000 Blocks Above the Ground* and have it 3D printed by our robotic arm Ginger. We use a game called *Minecraft* that is played with simple graphics and rules. It is a game where you build and demolish cubes of 1x1x1 meters. Currently more than 40 million people play it (Warren 2016).

Minecraft is a sandbox game where players can choose on their own which goals and adventures to pursue. The game consists of a 3D procedurally generated world where materials are spread for the players to mine. Players can combine various materials to "craft" new objects such as pick-axes, wooden planks, buckets, doors etc. Each Minecraft world can be loaded on a server and made available for online access to other players offering multiplayer functionality. Minecraft's rich catalogue of materials and objects plus the ability for in-game scripts allows the creation of custom maps where a goal or an adventure can be defined by

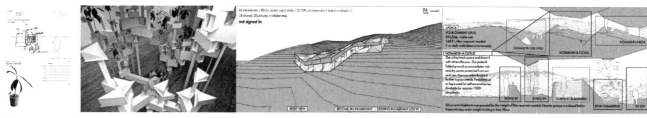

4   Project Avocado (first two from left) and Box in a Cloud.

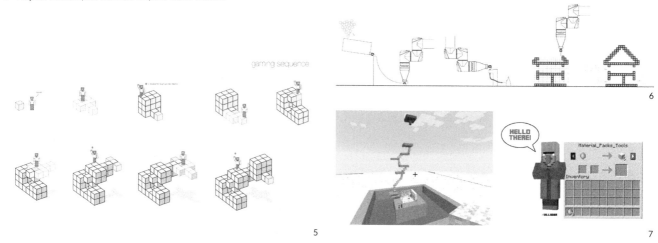

the map creator. The project **20,000 Blocks** uses such a custom-made map to define the game rules leading groups of players through the creation of architectural designs.

Minecraft's abstract, non-photorealistic, voxel world makes it a good choice for our research, based on Fröst and Warren's conclusion that "low-detail of sketch-like real-time 3D models often promotes creativity and discussion"(Fröst and Warren 2000). The make up of a Minecraft world out of 1 meter large blocks offers a suitable resolution for mass-modelling of architectural concepts from the scale of a room up to urban planning. Smaller scale objects such as the details fenestration or furniture are not easily possible in Minecraft but are also irrelevant to our research. Our focus is rather on crowd motivation and guiding techniques for the production of initial, schematic designs.

Both Peter Fröst and the team of Achille Segard have conducted similar experiments in architecture using the game engine of Half-Life (Fröst 2003, Segard, Moloney, and Moleta 2013). Their work is however targeted mostly at architects and focuses entirely on the design's shape. Claudia Otten has conceptualized a house-making game tool for single player use, targeted at non-architects (Otten 2014).

In our previous work, particularly *Sensitive Assembly* (2015), we encountered the power of the players as topology optimization agents (Savov, Tessmann, and Nielsen 2016). *Project Avocado* (2012) and *Box in a Cloud* (2013), on the other hand, have

successfully allowed participants to customize a collectively designed building to their individual preferences (Figure 4) (Savov 2012, 2013).

The project **20,000 Blocks above the Ground** builds upon these references and spans the digital and the material worlds. It combines a user-friendly game environment as a front-end with the precise performance feedback from specialized parametric design tools.

METHOD

Our method places importance on constraining the player choices through game mechanics and immersing players in the gameplay to increase motivation and investment. We consider there to be three key roles involved in this process:

The first role is of project **Organizers**. As researchers this is the role we take, bringing together the architects, the game designers, the players and the roboticists. We define the technology so others can use what we make to generate designs.

The second role is the **Experts**, whose knowledge will be encoded. These are the people setting and modifying the vocabulary and defining the goal. The game industry analogy would be of a Level designer, who is in charge of creating the challenges within a game.

Players are the third role. The experience of the players is tightly controlled — the Organizers decide which materials they can

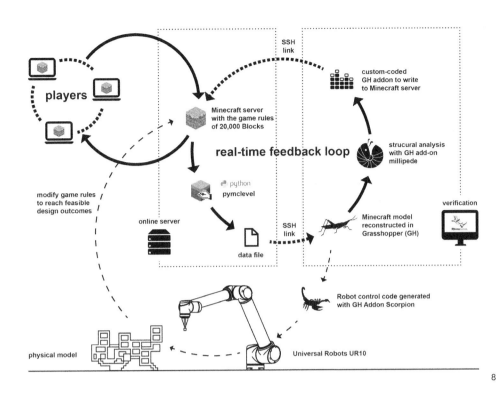

5   The vocabulary-based rewarding leads to a player-controlled growth process of the design (drawing by Andrea Quartara).

6   The robotic fabrication process in 20,000 Blocks (drawing by Jörg Hartmann).

7   Players need to build in such a way so they can walk up and down to buy more materials from the villagers. The main goal is modeled in red and when reached the game is saved and reset.

8   Real-time link between the Minecraft game of 20,000 Blocks and structural verification done in Grasshopper/Millipede.

8

place and break, and where they can walk. The players are given a goal but insufficient resources to achieve it. To progress, players build shapes out of Minecraft blocks, choosing from an architectural vocabulary defined by the Experts. Players are rewarded with resources for building one of the shapes. While players compete to reach the goal, a building emerges out of the shapes that they have built.

To steer the players, we can modify the three components of the method: the guiding rules, the verification routines and the feedback.

Guiding rules

It is of interest to us how we can steer the resulting typology while keeping the player focused only on the game mission and not on the artefact they are creating. The concept of vocabulary was introduced early on in the process and has evolved over several implementations and test iterations (Figure 5).

An example of a goal given to the players is to build up to a certain height, let's say 30 blocks. At the same time, their initial resources allow them to build only 12 blocks high. Each player can create a shape out of the vocabulary for which they get rewarded with additional 12 blocks of material. However, the shape is designed to consume 10 of their blocks yet is just one block high. That means that the maximum height the player can reach now is 15 blocks. Players can combine the shapes from the vocabulary to form the best strategy to reach the height of 30

blocks and win. They can also interact with the other players by building together or stealing each other's achievements. Another type of goal, for example, could be to build a certain number of a given vocabulary shape with the constraint that, again, the material needed for them is initially insufficient. Completing a mission takes approximately 20 minutes, after which the result is saved and the game world is reset.

Verification routines

At the heart of the project lies a real-time link between the online Minecraft server and analysis routines in parametric design tool Grasshopper (www.grasshopper3d.com) (Figure 8). We use the open-source Python library *pymclevel* by David Vierra to get geometry from Minecraft to Rhino/Grasshopper, which is then reconstructed in a format suitable for structural performance evaluation via the add-on *Millipede* (Vierra 2014, Sawako and Panagiotis 2014). The results from this analysis are fed back to the Minecraft server, either by creating new blocks or changing the material/color of existing ones. For this we developed our own Grasshopper components that can send commands to the server over SSH. Along with the limited vocabulary, this information is used to help to guide the players' actions in the game world so that they create 'viable' structures. In section *Fast Feedback,* we discuss how this is done.

We use a robot fabrication process to verify the stability and constructability of a game result. We feed the point information for each Minecraft block to our robot arm, Ginger, who builds

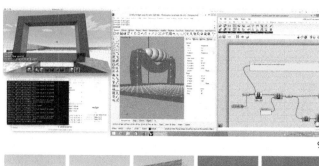

9

10

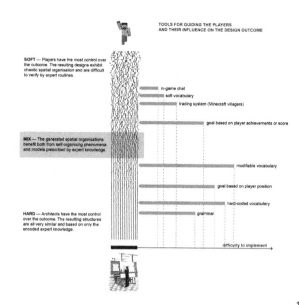

TOOLS FOR GUIDING THE PLAYERS
AND THEIR INFLUENCE ON THE DESIGN OUTCOME

SOFT — Players have the most control over the outcome. The resulting designs exhibit chaotic spatial organisation and are difficult to verify by expert routines.

in-game chat
soft vocabulary
tracing system (Minecraft villagers)
goal based on player achievements or score

MIX — The generated spatial organisations benefit both from self-organising phenomena and models prescribed by expert knowledge.

modifiable vocabulary
goal based on player position
hard-coded vocabulary
grammar

HARD — Architects have the most control over the outcome. The resulting structures are all very similar and based on only the encoded expert knowledge.

difficulty to implement

11

the model in a scale of 1:100. It grabs wooden cubes with a vacuum gripper, pushes them into the glue and positions them on the correct spot according to the digital model. If a cube is cantilevering, Ginger places free-standing cubes underneath it as a support. These are removed after the model is finished. This additive manufacturing process is similar to selective fusing of materials in a granular bed such as *binder jetting*, with the difference that the granules are much larger (Figure 6).

The second purpose of the models is to visually and materially represent the player-created designs as a form of feedback for both Players and Experts.

Fast feedback
We provide on-the-fly feedback to the players in several ways: posting messages to the in-game chat, showing particle effects and changing the material of blocks in the game world.

In the chat console, we display messages to inform players what to do when they join the server. While playing, if a player successfully builds a new element out of the vocabulary, we inform everyone. Goal achievements are also broadcast in this fashion.

Particle effects are used to visually attract players to important positions in the game world. The most important of these is used to inform a player that the system is verifying a newly built vocabulary element.

At the game world level, a structurally weak spot computed in Millipede would be marked in Minecraft and turned into an optional, secondary mission for the players. This would award them extra points for resolving the structural problem (Figure 9).

RESULTS AND DISCUSSION
We aimed to learn how to calibrate design sessions of the project *20,000 Blocks* so that while playing, groups of non-experts could generate architectural designs with prescribed features. One important design decision which needs to be made is the degree of freedom given to players. In other words, who should have the control over the design: the Expert or the Player.

As Brabham states, for a well-functioning crowdsourcing model it is important that the locus of control regarding the creative production of goods, exists between the organization and the crowd. If the locus of control is closer or larger in the community, such as in the case of open-source software or Wikipedia, or if the control is largely in the organization's hands, such as when a company wants the community to simply vote for the color of a product, we are not seeing a true crowdsourcing model (Brabham 2013).

Therefore, we took into account in our analysis of the results the balance of control that players and experts had. We imagined it on a gradient scale of soft to hard, where the soft end of the spectrum gives more control to *Players* and the hard end of the spectrum, more control to the *Experts* (Figure 11).

Soft vocabulary
We conducted the first test with soft vocabulary. The Players were presented with a catalogue of architectural elements such as rooms, terraces, etc. (Figure 10). A player would build any one of those, anywhere within the confined building site in Minecraft. They would then come to the Organizers, who would visually verify if the built element matches the catalogue and award the player the corresponding points and resources. This happened

9   Structural verification in real-time. The block marked in magenta is a new secondary mission for the players.

10 The initial soft vocabulary is a visual catalogue, including three variations of platforms, three bridges, two gardens and two pools.

11 Gradient from soft to hard. Each tool empowers players or transfers an expert's will differently. It also comes with a certain cost in effort to implement.

12 Top: Three game results from a hard vocabulary and one main goal. Bottom left: A game played with soft vocabulary. Bottom middle, right: Two screenshots from gameplay with hard vocabulary and secondary goals marked in gold.

12

verbally in the game chat. The goal was also soft — get the highest number of points.

The outcomes were rather fuzzy designs as all the aspects of the system — the guiding rules, the verification process and the feedback — were negotiable between the player and the organizers. Another hindrance, we noticed, was that the elements in the vocabulary were too numerous and too complex to be remembered by players easily. This required the players to refer to the vocabulary too often, hence breaking their flow of play. Furthermore, it was also slow — a game was marked as completed when the confined building site was filled up, which took around two hours (Figure 12 bottom left).

Therefore, we looked for ways to make the rules stricter by automating them.

Hard vocabulary

The guiding rules of the hard vocabulary method we tested were based on three automated routines:

1) *Detection routine*: The elements that players build to gain rewards are recognized automatically by the game engine. We described one element — a platform of 5x5 blocks — in code using Minecraft's programming language and command blocks (Figure 14). The player places a trigger block (blue diamond) and positions their character on top of it to activate the detection routine (Figure 17). Only exact copies of the catalogue structures will be recognized and rewarded, thus reducing the fuzziness of the design solutions.

2) *Trading*: The resources for building need to be purchased from Non-Player Characters (NPCs) — a Minecraft villager

— in exchange for emeralds, gained when successfully building an element from the vocabulary. Villagers can be summoned only at the ground level. This ensures that the structure being created can be walked up and down (Figure 7).

3) *Goal detection*: We defined a main goal, achievable within 15–20 minutes that gives a clear end to each session (Figure 7). This limits the time for generating a design and allows for many solutions to be created under the same conditions. Reaching the goal automatically triggers a save of the built structure, and the game is reset. When we noticed that game plays resulted in self-similar linear structures (Figure 12 top row) we introduced a set of secondary goals as incentives to break the linearity of the gameplay and create spatially richer designs (Figure 12 bottom middle and right).

At this stage of the project, many structures built by the players had cantilevers which would be not be able to sustain the force of gravity. To unlock the true architectural potential of mass-participation, we need a set of evaluations that relate, and possibly rate, a shape on its performance as an architectural structure in the real world. Verification was done post-factum for most of the game outcomes using Grasshopper/Millipede. The results prompted us to change the elements in the vocabulary until the play results had less problematic overhangs (Figure 15).

The feedback from the game actions was printed automatically to the game chat (Figure 13) and kept the players informed of what to do in the game, who scored a new point by building a 5x5 platform and when the game was over, being saved and reset. This proved very useful and successful in keeping the players aware.

13

14

Modifiable vocabulary

The latest iteration of the guiding gameplay mechanics implements a vocabulary of predefined design elements that follows combinatorial rules (Figure 16). This is somewhat similar to an L-System (Dapper 2003).

It has all the three main principles of the **Hard Vocabulary** variation with the difference being that detectable structures are not described in code but are built on dedicated slots next to the building platform as a visual catalogue. This allows us to modify and prototype the vocabulary much faster. It furthermore allowed us to separate the roles of Organizer and Expert without needing to teach Minecraft scripting.

The fact that the vocabulary consisted of more than one element proved to soften the play outcomes. Therefore, we tried a system where the trigger blocks were placed automatically with the vocabulary element and not by the players. We called this new notion Grammar because it meant Experts could define which of the elements could be built upon each other thus opening or limiting choices for the players.

We didn't use the game chat as extensively as in the **hard-vocabulary** approach and relied on players orienting themselves in the game world. This proved confusing for most people.

Findings

The project has been in continuous development and testing for one year at the time of this writing. Each of the three vocabulary types (soft, hard and modifiable) was put through online play-testing. The hard vocabulary was additionally tested at events at the **Digital Design Unit (DDU)** at TU Darmstadt with university students, as well as at **Invent the World (ITW)** with 7–10-year-old kids (Figure 18). The modifiable hard vocabulary was tested at the **SmartGeometry 2016 Conference** (Figure 19). At sg2016 we also tested how other architects, acting as Experts, could use the modifiable vocabulary to define their own purpose and use-case for the designs (Figure 1 and 2).

With the current iteration, we as Organizers find that the systems in place allow a great range of creativity from both Experts and Players but feel that the results are still too hard in terms of spatial organization.

So far, 8 people, split in three teams, have participated in the project as Experts. With the game levels contributed by the Experts and ourselves, we tested a total of 7 game maps, resulting in 57 player-created structures. The Experts were very happy with their experience and were able to both generate unique use cases, as well as the needed gameplay to create suitable results.

Around 70 players have taken part in the various on-site and online play sessions. It takes 5-10 minutes to explain to a new player the basic rules, and then they enter the tutorial part of the game map which we have designed to introduce them to the principle of vocabulary element detection. Currently, the Players are still too overwhelmed when they begin participation. The project requires more guidance and feedback for players, and the skill barrier to entry must be lowered. Much too much time was spent dealing with simple issues such as the players accidentally wasting their rewards or quitting in the middle of a game.

From all 57 game results, we selected, post-processed and robotically manufactured five models. The robotic process is currently too slow to keep up with the digital iterations. The speed is 200 blocks per hour, and an average design requires 1200 to 2000 placed blocks to build, i.e. 6–10 hours per model. This includes both the model blocks as well as the supporting blocks. As the focus of the research is on game design as a guiding instrument, the analysis of the models built is secondary and will be subject to further development and publications.

Key findings:
1) Harder rules are better than soft ones in delivering a feasible architectural design.
2) If the rules become too hard, the players no longer feel part of the design and as such are unmotivated to play.

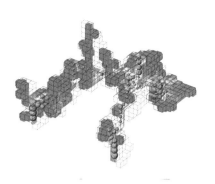

13 The game chat is used to feed current information to players.

14 The interface for coding in Minecraft using command blocks.

15 Millipede screenshot (left) and wooden robotically fabricated model (right) based on a vocabulary designed by Max Rudoph and Marios Messios.

15

3) If the rules are too soft, the resulting structures become too chaotic. This makes them difficult to verify and construct. In addition, players find those designs confusing to navigate and play through.
4) Robotic fabrication is possible but currently too slow to provide meaningful feedback.
5) An easily modifiable vocabulary allows other architects to participate as Experts and that opens up our method to more possible applications.
6) Clear, non-architectural, time-bound goals make participation easier and more entertaining than tasks where players need to understand the spatial and architectural qualities of the game elements they are building.

The work thus far reveals that the calibration process — the positioning between hard and soft — is an ongoing challenge and main focus of our research. Finding the right balance between hard and soft requires constant testing. With every project iteration and every game played, we are able to calibrate the balance better.

CONCLUSION

The paper explained how the use of Minecraft, in connection with Grasshopper, can allow for non-expert players to generate and evaluate architectural design concepts. We presented a methodology that utilizes a mixture of expert and non-expert participation, crowdsourcing for design creation with real-time structural analysis feedback on design decisions, and robotic fabrication.

The two main hurdles to overcome in our research in the future development of *20,000 Blocks Above the Ground* are:

1) *Fool proofing* the game world so that players are able to play unsupervised at any time. We can achieve this by making game rounds shorter (10–15 minutes) and with an automated start and end. If a player joins while a game is in progress, they are put in observer mode and added to the queue for the next round.

2) *Diverse and continuous feedback* using the game chat console for detailed progress reports as well a continuous structural performance evaluation. We are implementing a new message system able to display full screen large text messages to the players. We also intend to introduce a third type of feedback as an online catalogue of player-created designs. A web-gallery is in the making to display screenshots of all played games immediately after they have been completed.

Possible applications we see for our method of game-to-CAD-to-Robot transfer of geometry are:

- In the research field, new forms of architecture could be explored that transgress established typologies. This is helped by engaging the unbiased minds of the non-experts.
- In the practice, a specific design task could be crowdsourced, tapping into the decision-making power of inhabitants, neighbors and investors by defining the corresponding in-game rules and providing the suitable background evaluation algorithms.

We consider the ideal scenario to test the method in reality is the massing out of an architectural concept, such as defining the rough placement and orientation of rooms in a building or determining the location and infrastructure of buildings in an urban design scheme (Figure 20). The ability to involve a team of experts in the making and testing of the architectural vocabulary holds the potential for feasible and well-performing solutions. To make this viable, a main hurdle of the architectural field to overcome is developing post-processing routines that can quickly turn player-generated designs into CAD models for further specification.

ACKNOWLEDGEMENTS

The authors would like to thank: all the Players of 20,000 Blocks; Jörg Hartmann and Sebastian Kotterer for assistance with the robotic fabrication process; the participants at our workshop at SmartGeometry 2016 who entered into the role of Experts — Andrea Quartara, Marc Emmanuelli, Ulrik Montnemery, Marios

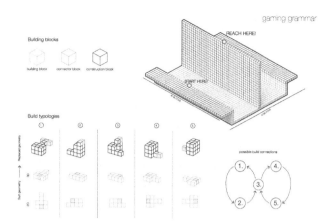

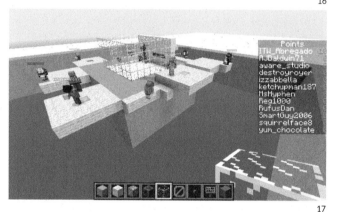

16 The vocabulary shown here, designed by Alexander Gösta, Samuel Eliasson and Ashris Choudhury results in bridge-like building designs.

17 Screenshot from the game play at Invent the World. The trigger blocks are in blue.

18 Play session at Invent the World, Australia.

19 Our setup with 4-way game terminal and a robot at SmartGeometry 2016.

Messios, Max Rudolph, Alexander Gösta, Samuel Eliasson, Ashris Choudhury; the organizers of SmartGeometry 2016; Oskar Gerspach for shooting and cutting the project's video; Christian Schwamborn from TU Darmstadt's IT department for setting up our server; and our colleagues at the Digital Design Unit (DDU) and Invent the World (ITW).

REFERENCES

BBC. 2012. "Minecraft to Aid UN Regeneration Projects." BBC News, November 26. Accessed June 22, 2016. http://www.bbc.com/news/technology-20492908.

Brabham, Daren C. 2013. Crowdsourcing. Cambridge, MA: MIT Press.

Cooper, Seth, Adrien Treuille, Janos Barbero, Andrew Leaver-Fay, Kathleen Tuite, Firas Khatib, Alex Cho Snyder, Michael Beenen, David Salesin, David Baker, Zoran Popovic, and >57000 Foldit players. 2010. "The Challenge of Designing Scientific Discovery Games." In Proceedings of the Fifth International Conference on the Foundations of Digital Games. Monterey, CA: FDG. 40–47.

Dapper, Timm. 2003. "Practical Procedural Modeling of Plants." Accessed March 25, 2016. http://www.td-grafik.de/artic/talk20030122/overview.html.

Friedman, Yona. 1975. Toward a Scientific Architecture. Translated by Cynthia Lang. Cambridge MA: MIT Press.

Fröst, Peter, and Peter Warren. 2000. "Virtual Reality Used in a Collaborative Architectural Design Process." In Proceedings of the IEEE International Conference on Information Visualization. London: IV. 568–573.

Fröst, Peter. 2003. "A Real Time 3D Environment for Collaborative Design." In Proceedings of the 10th International Conference on Computer Aided Architectural Design Futures, edited by Mao-Lin Chiu, Jin-Yeu Tsou, Thomas Kvan, Mitsuo Morozumi, and Tay-Sheng Jeng. Tainan, Taiwan: CAAD. 203–212.

Otten, Claudia W. 2014. "Everyone is an Architect." In ACADIA 2014: Design Agency—Proceedings of the 34th Annual Conference of the Association for Computer Aided Design in Architecture, edited by David Gerber, Alvin Huang, and Jose Sanchez. Los Angeles: ACADIA. 81–90.

Parker, Laura. 2014. "Not Just Playing Around Anymore: Games for Change Uses Video Games for Social Projects." New York Times, April 21, 2014. Accessed May 10, 2016. http://www.nytimes.com/2014/04/22/arts/video-games/games-for-change-uses-video-games-for-social-projects.html?_r=2.

Sanchez, Jose. 2015. "Block'hood, Developing an Architectural Simulation Video Game." In Real Time: Proceedings of the 33rd eCAADe Conference, vol. 1, edited by B. Martens, G. Wurzer, T. Grasl, W. E. Lorenz, and R. Scharanek. Vienna, Austria: eCAADe. 89–97.

Savov, Anton, Olivier Tessmann, and Stig A. Nielsen. 2016. "Sensitive Assembly: Gamifying the Design and Assembly of Façade Wall Prototypes." International Journal of Architectural Computing 14 (1): 30–48.

Savov, Anton. 2012. "Project Avocado." Accessed March 3, 2016. http://www.aware-studio.com/progeny/project-avocado/.

———. 2013. "Box in a Cloud." Accessed March 3, 2016. http://www.aware-studio.com/progeny/box-in-a-cloud/.

20 A rendering showing an architect's interpretation of a schematic design created by players in the 20,000 Blocks' game map WOULD YOU MIND YONA FRIEDMAN? by Marios Messios and Max Rudolph (rendering Max Rudolph).

Sawako, Kaijima, and Michalatos Panagiotis. 2014. "Millipede." Accessed September 14, 2015. http://sawapan.eu/.

Szantner, Attila. 2016. "What Did the Easter Bunny Bring Us." MMOS, March 28. Accessed April 15, 2016. http://mmos.ch/news/2016/03/28/what-did-the-easter-bunny-bring-us.html.

Segard, Achille, Jules Moloney, and Tane Moleta. 2013. "Open Communitition: Competitive Design in a Collaborative Virtual Environment." In Open Systems: Proceedings of the 18th International Conference on Computer-Aided Architectural Design Research in Asia, edited by R. Stouffs, P. Janssen, S. Roudavski, and B. Tunçer. Singapore: CADRIA. 231–240.

Vierra, D. 2014. "pymclevel — Minecraft Levels for Python." Accessed November 10, 2015. https://github.com/mcedit/pymclevel.

Warren, Tom. 2016. "Minecraft Sales Top 100 Million." The Verge, June 2, 2016. Accessed July 2, 2016. http://www.theverge.com/2016/6/2/11838036/minecraft-sales-100-million.

IMAGE CREDITS

Figures 1, 5: Quartara, 2016, ©

Figure 2: Messios, 2016, ©

Figures 3, 8, 9, 10, 11, 12, 13, 14, 17, 19: Savov, 2016, ©

Figure 4: Savov, 2013, ©

Figure 6: Hartmann, 2016, ©

Figure 7: Savov/Messios, 2016, ©

Figure 15: Rudolph/Messios, 2016, ©

Figure 16: Eliasson, 2016, ©

Figure 18: Buckton, 2016, ©

Figure 20: Rudoplh, 2016, ©

Anton Savov works as a research associate at the Digital Design Unit (DDU), Faculty of Architecture at the Technical University Darmstadt and is the founder of AWARE — an architectural design studio using crowdsourcing and game design to engage the non-experts in the creative process. Previously, Anton has worked at Bollinger+Grohmann Engineers and taught at the Städelschule Architecture Class, Frankfurt. In 2012 Anton received the Artists and Architects in Residence Program Scholarship of the MAK Center in Los Angeles.

Ben Buckton is a Freelance Game designer, consultant and educator. As a completely self-taught developer he has a somewhat unorthodox way of approaching problems which has gotten him involved in many interesting projects. As a partner at Invent the World, Ben designs and runs Holiday programs for 'nerdy' children in Australia. This sees him traveling back and forth across the globe with great regularity. Ben is also the event organizer for the Frankfurt Independent Developer scene, helping to bring together unique and talented people.

Oliver Tessmann is an architect and Professor for Digital Design at TU Darmstadt. His research and teaching is located in the field of computational design and digital fabrication. After receiving his doctorate from the University of Kassel in 2008, he served as the director of the performativeBuildingGroup at Bollinger+Grohmann. From 2010 to 2012 he was Visiting Professor at the Städelschule Architecture Class in Frankfurt and, from 2012 to 2015, held the position of Assistant Professor at KTH Stockholm.

Architectural Heat Maps: A Workflow for Synthesizing Data

Jason S. Johnson
Matthew Parker
University of Calgary

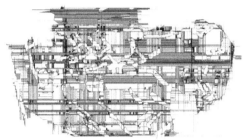

SIFT composite plan

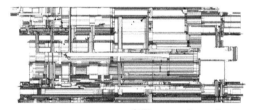

SIFT composite section

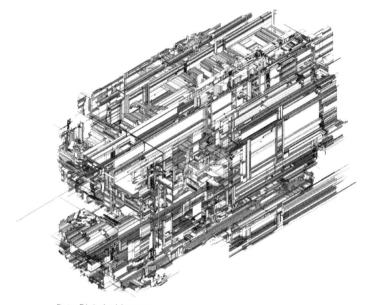

Data-Rich Architecture

1

ABSTRACT

Over the last 5 years, large-scale 'data dumps' of architectural production have been made available online through project-specific websites (mainly competitions) and architectural aggregation/dissemination sites like Architizer, Suckerpunch, and Archinect. This reinforces the broader context of Ubiquitous Simultaneity, in which large amounts of data are continuously updated and easily accessed through a dizzying array of mobile devices. This condition is being exploited by sports leagues and financial speculators through the development of tools that collect, visualize, and analyze historical data for the purpose of producing speculative predictive simulations that could lead to strategies for enhanced performance.

We explore the development of a workflow for deploying computer vision, SIFT algorithms, image aggregation, and heteromorphic deformation as a design strategy. These techniques have all been developed separately for various applications and here we combine them in such a way as to allow for the embedding of the historical and speculative artifacts of architectural production into newly formed three-dimensional architectural bodies. This work builds on past research, which resulted in a more two-dimensional image-based mapping and translation process found in existing imaging protocols for projects like Google Earth, and transitions towards the production of data-rich formal assemblies. Outliers and concentrations of visual data are exploited as a means to encourage innovation within the production of architecture.

1   SIFT composite plans and sections are mobilized within the workflow described in this paper to produce an architecture that operates through a condition we describe as ubiquitous simultaneity.

INTRODUCTION

Outliers are often the generative fuel for innovation within a wide variety of fields. Something or someone breaks away from the collective pack due to an aberrational quality or glitch that was not fully understood or accepted by the larger system. Innovators seek to both find and describe the aspects of these outliers as a way of exploiting their potential to produce emergent models of differentiated productivity. The cumulative effect of this continual process of the collective either destroying or moving towards these mutations is perhaps best illustrated in the world of sports, where the previously niche field of data analytics has become deeply ingrained across disparate sports at various levels. The success of this model depends on the ability to aggregate and analyze large pools of information in order to understand and leverage the relationships between the individual attributes (singular performance metrics) and the corresponding performance of the team. The National Basketball Association has been using tracking systems to monitor a wide range of statistics relative to a player's positioning, time on the court, and performance. Similar systems that pinpoint the location of players relative to the location of both moving and static objects on the field of play have been deployed by football, baseball, and a number of other sports, and are now being tied to biometric sensors attached to a player's body. These tools of measurement are being used to both analyze past performance and simulate future outcomes in ways that are impacting the strategies deployed in real-time game situations (Goldsberry 2012). The visualizations for these tools as deployed in baseball (Figure 2) and basketball (Figure 3) hint at the potential for overlaying historical data onto a speculative or strategic performance image.

The potential for adapting these approaches for the production of architecture is one that could begin to re-situate the conversation about parametricism and generative design away from a coalescing visual style and towards a more fulsome engagement of historical and speculative formal and functional relationships within designed artifacts of all scales. Schumacher's manifesto for parametricism as a style speculated that the move towards the "inter-articulation of sub-systems" would be critical to the production of differentiated field conditions vs. the production of bounded space (Schumacher 2008). What is perhaps missed in the ways in which generative design projects are currently being developed is the integration of historical and speculative inter-articulation into the parametric models. These models could become more like the data-driven approaches that are redefining the ways in which sports are being visualized in both historical and speculative ways. Here, we propose an approach that leverages computational tools for vision, sensing and recombining the data associated with architectural production, towards a workflow that might begin to engage the age

of Ubiquitous Simultaneity (Johnson and Parker 2016), which is allowing for an unprecedented amount of visual, environmental, and regulatory data to be gathered and processed. This research builds on previous explorations into the potential for computer vision and Scale-Invariant Feature Transform (SIFT) algorithms to be deployed for the production of two-dimensional images and introduces into the workflow techniques for three-dimensional form finding and the embedding of bias for the purpose of innovation (Johnson and Parker 2014). It builds out an approach that speculates about the nature of design as a profession that must begin to leverage massive repositories of historical and speculative design strategies and artifacts in order to better understand both past tendencies and future directions. For the purpose of this research we have focused this case study on the design competition held for the Helsinki Guggenheim. With over 1700 entries that share a site as well as programmatic criteria all formatted in two-dimensional images, this dataset allowed us to set up a series of protocols that could form an approach towards a new generative production technique.

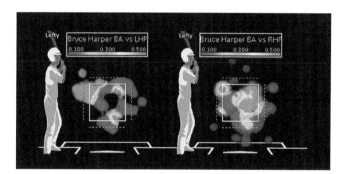

2

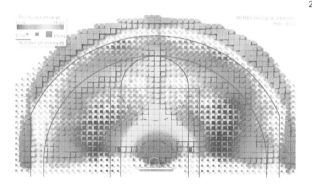

3

2   Heat Map with two variables. Map charts the relative productivity by the specific player when he swings at a pitch in relation to the location of the pitch and handedness of pitcher. Image courtesy of Kirk Goldsberry.

3   This map reveals league-wide tendencies in both shot attempts and points per attempt. Larger squares indicate areas where many field goals were attempted; smaller squares indicate fewer attempts. The color of the squares is determined by a spectral color scheme and indicates the average points per attempt for each location. Image courtesy of Kirk Goldsberry.

BACKGROUND

Computer Vision

Architecture and the city are increasingly experienced through what this paper refers to as algorithmic observation, a type of machine vision that 'sees' despite a lack of eyeballs, rods, cones, and a visual cortex; instead, algorithmic observers perceive the city through the use of sensors capable of detecting light, heat, motion, and color data to produce composite 'images' that describe the physical world. By further abstracting these images into their geometric components, algorithmic observers possess an inhuman ability to identify, sort, and catalog image-based information, making them ideal for deployment within areas of security, surveillance, military observation, feature identification, and a wide array of other uses that are building up a continuously updated catalog of 'seen' data. Within this research, algorithmic observers are utilized for their ability to identify and organize deep catalogs of preexisting speculative projects. These catalogs allow for the indexing of a body of work that not only embeds physical characteristics typical of all buildings, but also the theoretical tendencies that are typically amplified in projects developed for design competitions.

The design competition has long been a vehicle for the avant garde in architecture to push for new formal, social, and functional agendas within the context of a given typology. As such, they typically form a body of work that exists at the edges of architectural practice as a whole. The ability of algorithmic observers to dissect and make sense of the prolific amount of free architectural content disseminated through architectural competitions and the emergence of project aggregation sites like Architizer, Archinect, and Suckerpunch allows for previously 'invisible' insights to be collected and analyzed against each other and other data sets.

The use of computer vision to identify and organize deep catalogs of information originates in the work of Lev Manovich, who produces software in response to a growing desire to track and map the "global digital cultures with their billions of cultural objects and hundreds of millions of contributors." This is the result of the "exponential growth of a number of [...] media producers over the last decade [which] has created a fundamentally new cultural situation and a challenge to our normal ways of tracking and studying culture" (Manovich 2012). Manovich claims that the rise of social media and globalization leaves no other choice than to rely on computational tools for the organization and mapping of the visual artifacts that describe and construct culture. By developing techniques for automatic digital image analysis, Manovich is capable of generating numerical descriptions of various visual characteristics of an image, which allows him to identify tendencies and outliers that aggregate to form a

cultural reading (Figure 4). Whereas Manovich's work relies on information codified at the level of a pixel (its hue, saturation, brightness) to catalog images, architecture is faced with a problem of needing to identify specific characteristics within traditional architectural drawings of plans and sections. To accomplish this, we rely on the algorithmic observers' ability to computationally reconstruct images into collections of unique features that can be identified, organized, and matched within and across extensive image sets, through the use of SIFT algorithms.

SIFTs (Scale-Invariant Feature Transform)

SIFTs, first developed by David Lowe, enable algorithmic observers to identify specific images features that are invariant to scaling, rotation, changes in illumination, and 3D camera viewpoint (Lowe 1999; 2004). SIFT descriptors are extracted from each pixel of an input image and encoded with contextual information through processes that reduce images to a large number of highly distinctive features, facilitating the filtration of visual clutter/noise within the image, while providing a high probability of feature matching and correlation across images. SIFTs, with their strong matching capabilities and computational stability, are mobilized for

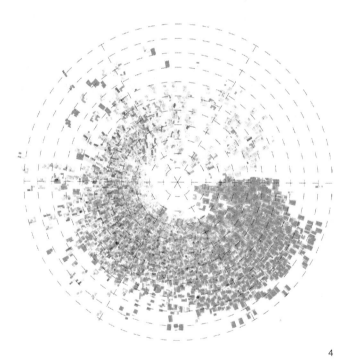

4

4   Identifying outliers: A technique for identifying outliers based off their SIFT-generated heat maps (description to follow in section 3.2), This graph illustrates the outliers within an initial sample of 1700 heat maps of Casa de Fascio when viewed from varying viewpoint perspectives.

5   SIFT production of Warped Images: By identifying correlate SIFTs between 2 input images, the workflow maps and superimposes corresponding data onto other bodies of soft data, tasking SIFT algorithms to exert agency within the production of 'warped images' by speculating towards which data might 'fill-in-the-blanks'.

the purposes of image retrieval, image stitching, machine vision, object recognition, gesture recognition, match moving, and the digital construction of three-dimensional virtual environments (McClendon 2012; Wu et al. 2013; Yang et al. 2011).

Previous research (discussed in Johnson and Parker 2014; Parker 2014) explored the computational protocols of SIFT algorithms and their ability to facilitate computer vision towards the digital (re)construction of the physical environment. These papers draw attention to the ability of algorithmic observation to shape our experiences and relationships with the city, challenging architecture to engage these technologies by intentionally assuming multiple identities within the superimposed virtual and physical layers that compose the contemporary city. The overarching theme of this previous research is that architecture, by and large, retains a level of ambivalence towards algorithmic observation that stems from a traditional desire to cling to certain disciplinary claims, specifically architecture's role in shaping and defining our spatial relationships to the physical materiality of the city. Whereas previous research attempted to hack the virtual layers of the city by augmenting architecture's exterior envelope towards direct interface and communication with algorithmic observers, this research explores a SIFT-based workflow that challenges architects to engage a form of practice that no longer situates the physical act of making buildings as their primary role, but instead challenges them to take part in the shaping and designing of the attributes and actions that redefine the ecologies of the city.

In order to reframe the relationship between algorithmic observers and architecture, it is necessary for architects to address algorithmic observers and the space they produce as valuable, despite their lack of traditional materiality and familiar formal manifestations. This research builds off a previously developed understanding of the computational logics of SIFT algorithms in order to produce a workflow for the production of three-dimensional inter-articulated architecture that can be biased towards historical or speculative tendencies within pre-defined bodies of architectural production. These bodies could be defined as specific sites and programmatic contexts, typologies, or a combination of the two constrained by one another.

Within the context of SIFT algorithms and their role in computer vision, this research claims that at their most basic level, SIFTs convert 2D image data into soft data for the purposes of recombination and superimposition. Soft data possesses the ability to elastically deform in response to external forces while retaining its original topology. With respect to computer vision, SIFT algorithms convert digital images to recognizable geometric descriptors for the production of data-rich territories (DRT) (Figure 5). Initial research into the production of ubiquitous simultaneity utilized a homeomorphic relationship between images, allowing each image to exert an equal force toward the production of DRT. By contrast, this workflow explores techniques that bias particular characteristics within a body of soft data, allowing said characteristics to exert greater influence within the production of DRT.

Importantly, the conversion of images to soft data does not alter the unique characteristics of a particular body of data. To accomplish this, SIFT keypoints—which are unique codified features of an image—are utilized for their capability to transform

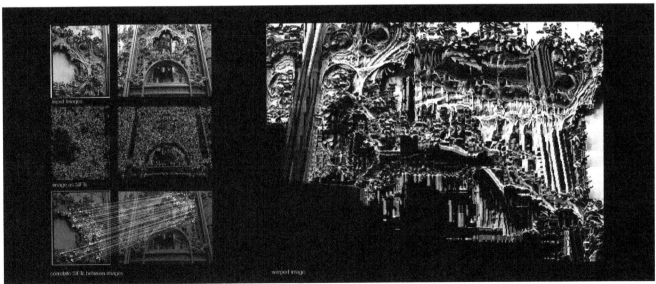

and receive information from correlate keypoints present across dynamic data sets. Simply put, keypoints allow specific geometric compositions of an image to map and accept multi-dimensional information produced from similar keypoints within a dataset. This process results in an augmentation of the keypoint but not the production of a new keypoint. In this sense, keypoints allow for the codification of multiple features to be contained within a single unique image feature, resulting in what we dub a feature heat map. In the same way that the individual data points within a shot chart (figure 3) both accept a number of characteristics (number of shots, percentages made) and assign a location within a two-dimensional representation of physical space, these feature heat maps allow for a densely calibrated visual dataset to emerge.

By deploying heteromorpholgic deformation within the production of DRT, this research deviates from previous explorations that exploited a one-to-one mapping between bodies of soft data. As such, we are currently invested in developing a workflow for the production of unequal influence in the production of said bodies, allowing for a more speculative generative approach for the establishment of a trajectory towards or away from specific characteristics present in any given image dataset. This notion of bias allows for design decisions to be made within a framework that promotes a gradient approach to projected outcomes.

METHODS AND RESULTS

In their seminal essay "Transparency: Literal and Phenomenal" (1964), Colin Rowe and Robert Slutzky champion a definition of 'transparency' within architecture that supersedes an object's mere ability to transmit light through it without optical destruction. Instead, Rowe and Slutzky suggest 'phenomenal transparency,' the ability to concurrently perceive different spatial locations of an object, allowing space to not only recede but to fluctuate in a continuous activity that allows objects to be understood simultaneously as closer and further away. What this definition does is transform our understanding of the 'transparent' from that which is perfectly clear to that 'which is clearly ambiguous.' This understanding allows for the superimposition of objects/data within a framework that does not necessitate spatial privileging, allowing us to ask if the multiple datasets that (potentially) define a physical object can be compressed into single data-fields for the production of the speculative new. And are there ways in which this transparency between the physical (historical) and the image (speculative) allow for new formulations of space that leverage the latent potentials found within each dataset.

By compositing sets of soft data into a single manifestation of an architectural object, this investigation marks a radical departure from the previous experiments that sought to add information into existing objects. Whereas past research focused on mapping surfaces of existing buildings and producing combinations that resulted in hybrid conditions, this research builds off notions of recycled intentionality or the compression of multiple design-intentions in order to "transpose insignificant singularities into meaningful complexities" (Maholy-Nagy in Rowe and Slutzky 1964). In order to accomplish this, large aggregate collections of architectural drawings (plans, sections, elevations) obtained from the Guggenheim Helsinki Design Competition are reinserted into the workflow to produce a series of volumes that embed the functional aspects of the projects inside of the speculative proposals. These volumes can be read as heat maps of functional and aesthetic intentions. No single instance is as legible as the general tendencies to reinforce the outliers at their perimeters while at the same time producing aggregations that do not lose definition at any one point within the larger assembly.

Producing Bias: Soft Drawings

In order to bias outliers, this workflow relies on protocols that allow a particular instance of soft data to exert varying degrees of influence within the compositional processes involved in the production of DRTs. Within these experiments, we have identified plans that read as clusters and sections that exhibit a relative tallness as outliers within a general population of over 1700 entries. Clustering can be considered an outlier, as the majority of the projects rely on a linear or grid based plan, with tallness being any building that exhibits sectional heights in excess of four stories (Figure 6). This is not to say these are the only two outlier conditions within the submission, or even the most prevalent, but

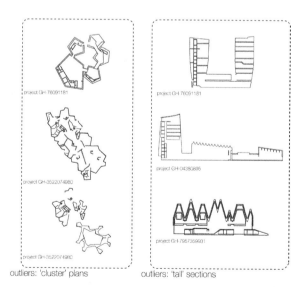

outliers: 'cluster' plans   outliers: 'tall' sections

6

6   Outliers: examples of 'cluster' plans and 'tall' sections that we have identified as outliers for the purposes of this investigation.

merely two outliers identified to situate as the biasing characteristics within these investigations.

Bias is produced by converting all available drawings to soft data for the initial purposes of superimposition within a system that allows each body of data to exert an equal force within the process of (re)composition (Figure 7). This process results in a large body of agile data in the form of speculative plans and sections that are ready to be deformed relative to the identified outliers. Bias is introduced once the architectural drawings have been converted to soft data through a computationally codified bias slider that operates similar to an alpha channel or image mask in Photoshop. The bias slider controls how much influence a particular dataset will exert within the composite whole, with a bias value of zero producing no effect within the composite whole and a value of 1.0 exerting maximum influence. If soft dataset A has a bias of 0.8 it will exert a force 200% greater than soft dataset B that exhibits a bias of 0.4, thus the resulting DRT will exhibit a biased towards the characteristics inherent in soft dataset A (figure 8). It follows that if one wishes to bias the characteristics included within the plans identified as 'clusters,' one would assign a higher bias value closer to these image-driven datasets within the processes of producing data-rich territories.

Soft-Data: Superimpostion, Composition, and 3D Form

Once SIFT algorithms have converted drawings to soft data and superimposed them towards the production of DRT, the data must be rendered into architectural drawings legible for human observation. These new drawings expose speculative architectural drawing(s) that have elastically deformed in response to a body of soft data's ascribed bias. Previous research has demonstrated that the process of superimposition that renders soft data as images produces sets of n-dimensional vectors that correspond to the uniquely codified entities of a particular body of soft data. Whereas previous research deployed these vectors towards the production of speculative images, this research remaps these vectors to their correlate SIFT keypoints for extrusion and the production of three-dimensional form (Figure 9),

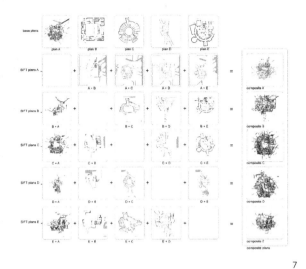

7

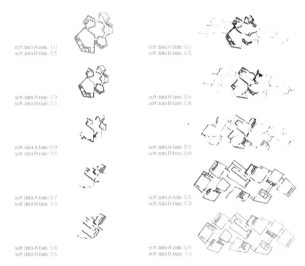

8

7   Drawings are converted to soft data for SIFT processing. Within this initial process, bias is negated and all drawings are a result of equal forces between bodies of soft data.

8   Gradient of biased transformation: Illustration of the bias slider enabling a particular soft dataset to exert more or less influence within the production of data-rich territories.

9   Identified keypoints of SIFTed plans and sections are extruded along their corresponding n-dimensional vectors for intersection and meshing in accordance with the computational protocols of the workflow.

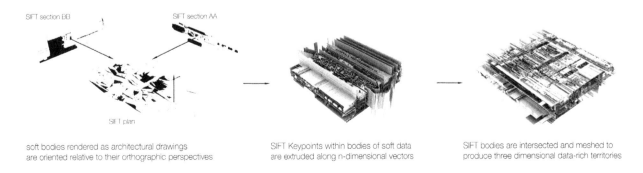

soft bodies rendered as architectural drawings are oriented relative to their orthographic perspectives

SIFT Keypoints within bodies of soft data are extruded along n-dimensional vectors

SIFT bodies are intersected and meshed to produce three dimensional data-rich territories

9

producing forms that read as a compression of recycled intentions as opposed to singular ideals.

As this workflow relies on extensive datasets towards the production of extensive collections of data-rich territories, we have started to produce a catalog of deep intentionality that seeks to produce novel forms by intentionally biasing specific characteristics in previously developed architectural artifacts (Figure 10). The characteristics we might choose to bias could be many, of course, but they would share an identification as being instances that fall outside of the formal qualities found in most of the dataset. In the case of the Guggenheim Helsinki Competition, the fact that these projects share a site and space programming allowed us to focus first on the physical characteristics that would be distinct in some projects when compared to the larger group of projects.

Within this workflow, bias is not limited to the production of soft drawings, but is extended to the processes of three-dimensional constructions. Just as any particular body of soft data can exert more influence within the process of image production, the workflow allows for the biasing of a particular plan or section to exert greater force within the resulting formal manifestation. Figure 11 demonstrates how different morphologic manifestations result from the same plan and section simply by biasing the plan over section and vice versa.

Program Identification and Extraction

Just as program is identified through architectural plans and sections this workflow allows for the identification of program within DRTs. By assigning a specific color to designated programmatic areas (public, exhibition, circulation, commercial, service, private), composite image color maps are produced that correlate to a corresponding soft drawing (Figure 12). These color maps produce programmatic heat maps that allow for program identification and organization within an otherwise chaotic environment. Following similar protocols to those developed to produce and extract n-dimensional vectors from soft drawings, programmatic heat maps are SIFTed, producing new sets of vectors that can be mapped to their corresponding image keypoints. These vectors

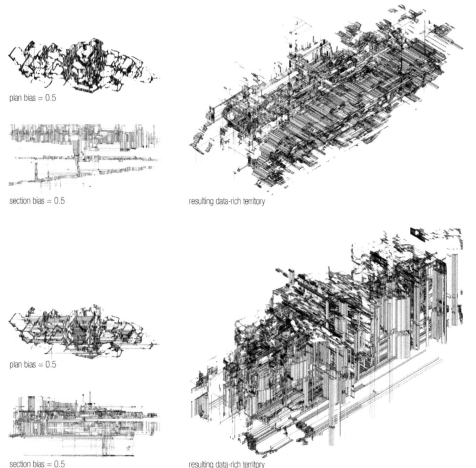

plan bias = 0.5

section bias = 0.5

resulting data-rich territory

plan bias = 0.5

section bias = 0.5

resulting data-rich territory

10 Selected instances from a growing catalog of deep intentionality. The bias slider allows for the manipulation of certain characteristics that belong to singular plans or sections within the dataset.

10

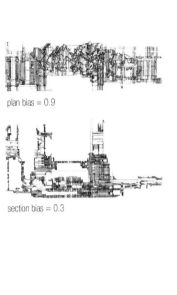

plan bias = 0.9

section bias = 0.3

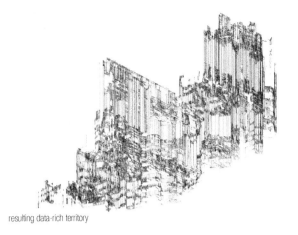

resulting data-rich territory

11 The initial plan and section that generated these geometries is the same: simply by biasing one drawing set over the other, unique instances can be produced from similar drawings.

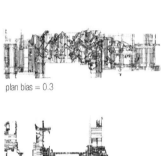

plan bias = 0.3

section bias = 0.9

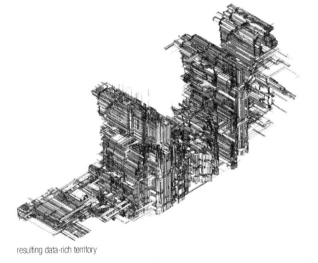

resulting data-rich territory

11

are intersected with the DRT, allowing for the identification of program within 3D forms (figure 12). The use of coloration allows for a performative reading of the newly formed territories that is both associative (related to the initial relationships as described in any given project) and hybridized (capable of producing program gradients in three-dimensional space).

CONCLUSION

Generative design strategies have until now predominantly focused on integrating data specific to the environmental responsiveness and formal articulation of buildings within their given regulatory contexts and stylistic/theoretical positions. What we propose here is the exploitation of an ever-increasing catalog of architectural production that is pre-loaded with many of these aspects through a process of computational sensing and evaluation that leads to recombinant data-rich territories. These territories are underutilized resources that represent millions of hours of architectural production and this workflow looks to exploit that resource in ways that are only possible through a computationally mediated framework.

While there has been some success here in creating a workflow that begins to scratch at the potential for mining the available information and processing it iteratively towards new architectural proposals, there are still a number of issues we see as critical to building out a more robust tool. Currently, we are relying on manually identifying outliers for biasing the workflow and an inefficient labor-intensive process for gathering, scaling, and converting architectural image data into a usable format. The ability to automate this function would greatly increase our ability to produce robust catalogs of biased outputs.

Lastly, while the introduction of visual and spatial 'heat mapping' programs has led to increased legibility into the potential for innovation and evaluation of outcomes, they would benefit from the introduction of a voiding function which could begin to define spatial relationships within what are currently effectively dense mass structures. This workflow has successfully leveraged existing techniques of data visualization that relate to strategic spatial organization used in sports, and in our view could be used as much as a strategic tool for creating spatial and material

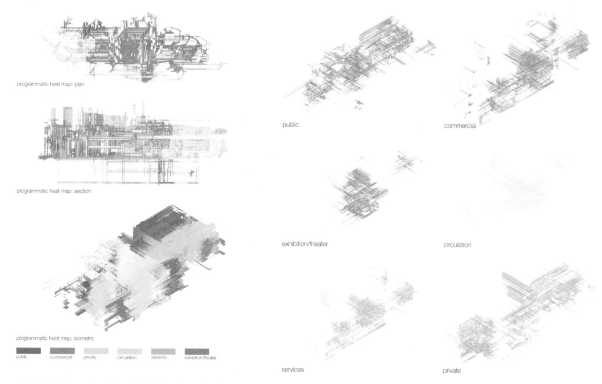

public

commercial

exhibition/theater

circulation

services

private

programmatic heat map: plan

programmatic heat map: section

programmatic heat map: isometric

public   commercial   private   circulation   services   exhibition/theater

12 Programmatic Heat Maps produced through 2D drawings and mapped to 3D assemblies with the identified programs extracted on the left.

relationships as it could for the production of physical assemblies. The potential exists then for a protocol that is scalable and capable of accessing and processing the ever-increasing body built and unbuilt design proposals whose current value is only that of dead content.

NOTES

1. Ubiquitous Simultaneity is a term coined by the author that identifies a condition which is characterized by the prevalence of massive amounts of data continuously updated and easily accessible through a number of digital interfaces.

REFERENCES

Franks, Alexander, Andrew Miller, Luke Bornn, and Kirk Goldsberry. 2015. "Counterpoints: Advanced Defensive Metrics for NBA Basketball." In *Proceedings of the 9th Annual MIT Sloan Sports Analytics Conference*. Boston: SSAC.

Goldsberry, Kirk. 2012. "CourtVision: New Visual and Spatial Analytics for the NBA." In *Proceedings of the 6th Annual MIT Sloan Sports Analytics Conference*. Boston: SSAC. 1–7.

Google. 2012. *The Next Dimension of Google Maps*. YouTube video, 51:18. Posted by Google, June 6, 2012. Accessed November 23, 2014 from https://www.youtube.com/watch?v=HMBJ2Hu0NLw

Lowe, David G. 1999. "Object Recognition from Local Scale-Invariant Features." In *Proceedings of the Seventh IEEE International Conference on Computer Vision*, vol. 2. Kerkyra, Greece: ICCV. 1150–1157. doi:10.1109/ICCV.1999.790410.

———. 2004. "Distinctive Image Features from Scale-Invariant Keypoints." *International Journal of Computer Vision* 60 (2): 91–110. doi:10.1023/B:VISI.0000029664.99615.94.

Johnson, Jason S., and Matthew Parker. 2014. "This Is Not a Glitch: Algorithms and Anomalies in Google Architecture." In *ACADIA 14: Design Agency—Proceedings of the 34th Annual Conference of the Association for Computer Aided Design in Architecture*, edited by David Gerber, Alvin Huang, and Jose Sanchez. Los Angeles: ACADIA. 389–398.

———. 2016. "Ubiquitous Simultaneity: A Design Workflow for an Information Rich Environment." In *CROSS-AMERICAS: Probing Disglobal Networks—The 2016 International Conference of the Association of Collegiate Schools of Architecture*. Santiago, Chile: ACSA.

Manovich, Lev. 2012. "How to Compare One Million Images?" In *Understanding Digital Humanities*, edited by David M. Berry. New York: Palgrave MacMillan. 249–278.

Parker, Matthew. 2014. "SIFT Materiality: Indeterminacy and Communication between the Physical and the Virtual." In *What's the Matter: Materiality and Materialism at the Age of Computation*, edited by Maria Voyatzaki. Barcelona: ENHSA-EAAE. 313–326.

Rowe, Colin, and Robert Slutzky. 1963. "Transparency: Literal and Phenomenal." *Perspecta: The Yale Architectural Journal* 8: 45–54.

Schumacher, Patrik. 2008. "Parametricism as Style - Parametricist Manifesto." Delivered at *Darkside Club, 11th Venice Architecture Biennale*. Venice: Darkside Club: 1–5.

Wu, Jian, Zhiming Cui, Victor S. Sheng, Pengpeng Zhao, Dongliang Su, and Shengrong Gong. 2013. "A Comparative Study of SIFT and its Variants." *Measurement Science Review* 13 (3): 122–131.

Yang, Donglei, Lili Liu, Feiwen Zhu, and Weihua Zhang. 2011. "A Parallel Analysis on Scale Invariant Feature Transform (SIFT) Algorithm." In *Proceedings of the 9th International Conference on Advanced Parallel Processing Technology*, edited by Olivier Temam, Pen-Chung Yew, and Binyu Zang. Shanghai: APPT. 98–111.

IMAGE CREDITS

Figures 1, 6, 8, 10-12: Johnson and Parker, 2016
Figures 2–3: Goldsberry, 2012
Figures 4, 7, 9: Parker, 2015
Figure 5: Parker, 2014

Jason S. Johnson has practiced and taught architecture in North and South America and Europe. He is a co-director of the Laboratory for Integrative Design, where he holds the position of Associate Professor of architecture. He has exhibited and published his work internationally. His most recent book, co-edited with Joshua Vermillion, *Digital Design Exercises for Architecture Students*, was published in 2016. Mr. Johnson is the founder of Minus Architecture Studio, a design research practice in Calgary, where he lives with his wife and three sons.

Matthew Parker completed his Master of Architecture from the University of Calgary's Faculty of Environmental Design, where he received honors recognition and the AIA Gold Medal. Currently he is completing a Post Professional Masters with his current research focusing on the ability of algorithmic observation to transform, mediate and re-animate architectures' image. Matthew is also a researcher with the Laboratory for Integrative Design (LID), and a studio designer and parametric consultant with Minus Architecture Studio and Synthetiques / Research + Design + Build.

Combinatorial design

Non-parametric computational design strategies

Jose Sanchez
University of Southern California

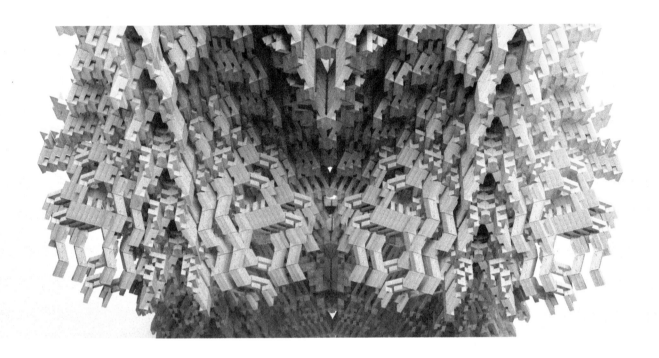

1

ABSTRACT

This paper outlines a framework and conceptualization of combinatorial design. Combinatorial design is a term coined to describe non-parametric design strategies that focus on the permutation, combination and patterning of discrete units. These design strategies differ substantially from parametric design strategies as they do not operate under continuous numerical evaluations, intervals or ratios but rather finite discrete sets. The conceptualization of this term and the differences with other design strategies are portrayed by the work done in the last 3 years of research at University of Southern California under the Polyomino agenda. The work, conducted together with students, has studied the use of discrete sets and combinatorial strategies within virtual reality environments to allow for an enhanced decision making process, one in which human intuition is coupled to algorithmic intelligence. The work of the research unit has been sponsored and tested by the company Stratays for ongoing research on crowd-sourced design.

1   Design – Year 3 – Pattern of one
    unit in two scales. Developed by
    combinatorial design within a game
    engine.

INTRODUCTION—OUTSIDE THE PARAMETRIC UMBRELLA

To start, it is important that we understand that the use of the terms parametric and combinatorial that will be used in this paper will come from an architectural and design background, as the association of these terms in mathematics and statistics might have a different connotation. There are certainly lessons and a direct relation between the term 'combinatorial' as used in this paper and the field of combinatorics and permutations in mathematics. As it will be explained further, the term combinatorial design encapsulates notions of both permutation and combinatorics and uses the studies of discrete finite sets of units and their possible arrangements by an algorithmic or intuitive process.

The term parametric, as understood for this paper, will be demarcated by the series of geometrical operations that concern the continuous variation and differentiation of form by the manipulation of ratios, intervals and intensities as a form to increase or reduce detail. It is in this description in which the term parametric, with its association of infinite continuous numbers, can be opposed to the term combinatorial, associated with finite discrete numbers.

Parametric design has received a great deal of attention since Patrik Schumacher elevated the design techniques to a form of architectural style (Schumacher, 2011). From a formal perspective, it is only natural that a particular set of mathematics would provide a path of least resistance for certain forms, and with a massive adoption for tools that work in such a paradigm, the yield of the claimed style would grow.

The tools of parametric design have flourished within a NURBs geometry environment, where as opposed to vertices, faces and edges from a mesh environment, the notions of curvature and parameters describe the building blocks for geometric constructions. A point in a curve represents an infinitesimal location along a domain which can be easily mapped to a parametric slider. This simple operation can be understood as a ratio or intensity described by an infinitesimal number. These notions do not only apply to NURBs geometry, as meshes as well rely on the positioning of points in a continuous space. The location of vertices can be described by a parametric function, and the resolution of a mesh geometry can also be described by a parameter.

Architecture and design have developed a large repertoire of tools and operations that operate in such a framework, creating and detailing geometric forms out of parametric functions. But the design discipline has quickly denominated under the same umbrella, doing a disservice to the further research and study of

2

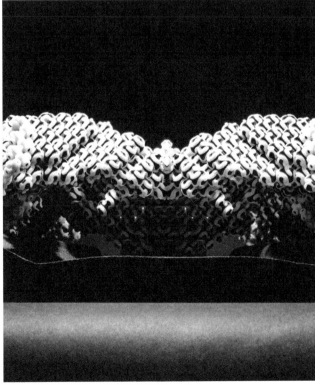

3

2   Project's game simulation developed in Unity3D.

3   Pavilion design – Year 1 – Aggregation printed out a pattern developed within the game simulation.

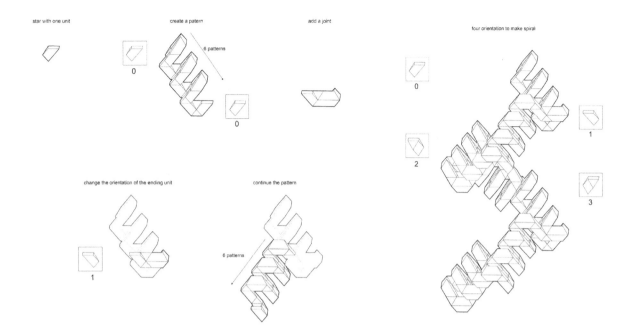

star with one unit create a patern add a joint four orientation to make spiral

4

Combinatorial design

A different set of techniques will be described here as combinatorial design: the use of a finite set of units or operations that can be arranged in different configurations. It's important to be clear that combinatorial design is not just the study of possible permutations of parameters, what Kostas Terzidis calls 'Permutation design' (Terzidis, 2014), where any given variable of a design problem establishes a degree of freedom that can be catalogued and cross referenced to other variables, yielding the solution space of a given system. In fact, Terzidis rejects how intuition and experience can play a role in the design process, favoring a framework of optimization performed by a deep search of algorithms over the permutation space. He explains:

"Traditionally, such arrangements are done by human designers who base their decision making either on intuition (from the point of view of the designer) or on random sampling until a valid solution is found. However, in both cases the solution found may be an acceptable one but cannot be labeled as "the best possible solution" due to the subjective or arbitrary nature of the selection process. In contrast, an exhaustive list of permutation-based arrangements will eventually reveal the "best solution" since it will exclude all other possible solutions." (Terzidis, 2014)

Combinatorial design is a design strategy that starts from the definition and individuation of parts, describing an open ended series of relations with one another. These parts will be coupled and aggregated to generate larger assemblies, describing meaning, performance and function at different scales of configuration. The system will always remain open ended and malleable, allowing for the replacement of parts within it.

The open-endedness of the system implies, in contrast to Terzidis ideas, that there is no possible optimization, as the solution space of permutations grows with each unit added at an exponential rate, becoming computationally impossible to search for an optimum.

The malleability of the system implies a temporal relevance, a contingent emergence of a pattern out of contextual conditions. No pattern might be the optimal one for a long period of time, but rather an ad-hoc solution harvested by designers or as we will see, a crowd.

GRANULAR ASSEMBLIES

What is described here as combinatorial design could be associated with what Ian Bogost calls 'Unit Operations'. Bogost explains:

"I will suggest that any medium – poetic, literary, cinematic, computational – can be read as a configurative system, an arrangement of discrete, interlocking units of expressive meaning. I call these general instances of procedural expression unit operations." (Bogost, 2006)

Combinatorial Design Sanchez

4   Diagrams – Year 2 – Diagrams of unit
     orientations and geometric growth.

5   Design – Year 2 – Larger pattern
     design using combinatorial catalog.

5

For Bogost, meaning emerges from the coupling of units without belonging to a larger holistic system. He is able to describe units in their autonomy to a larger structure rather than parts of a whole. His distinction between wholes and multitudes allows for the existence of units without any overarching structure. He explains:

> "A world of unit operations hardly means the end of systems. Systems seem to play an even more crucial role now more than ever, but they are a new kind of system: the spontaneous and complex result of multitudes rather than singular and absolute holisms" (Bogost, 2006)

In the studies of combinatorial strategies, the research team has addressed these multitudes in design as granular assemblies. While the granularity could be understood as a tectonic condition, it encapsulates the autonomy and spontaneous interactions of units. The idea of assemblies addresses the properties of re-configurability of parts and potentially their reversibility. It describes a temporal condition, one of a contingent, almost convenient, performative configuration.

It is Bogost who also explains the struggle that units need to maintain their individuation:

> "Unit-operational systems are only systems in the sense they describe collections of units, structured in relation to one another. However, as Heidegger's suggestion advises, such operational structures must struggle to maintain their openness, to avoid collapsing into totalizing systems" (Bogost, 2006)

This is a further distinction between the formal expression of combinatorial systems and parametric ones. While the former uses the language of patterns and arrangements, the latter uses the language of gradients and flows that transition into one another. Parametric systems are designed via articulation of flows while combinatorial ones are designed by the patterning of units.

Three years of combinatorial research, Polyomino studio.
The Polyomino agenda started in 2013 at University of Southern California, working with students from the Master in Advanced Architectural Studies. The research brief for the studio is that of defining and designing a combinatorial building system, one that could produce a series of architectural conditions. The projects would be developed both physically and digitally, allowing material experiments to inform a simulation within a game engine.

By studying games and interactive experiences, the project was able to develop a digital platform with all the material constraints to explore the combinatorial possibilities. Often, many of the tools of aligning, orienting and snapping complex geometric shapes to one another require a series of commands in any software platform. By pre-writing code that could execute quickly and effectively these alignment calculations, the team is able to create user experiences for intuitive combinatorial design.

The games, developed as simulations, would instantly offer alternatives of growth for the designer and provide the data to make decisions every step of the way.

6   Design – Year 3 – Combinatorial patterning of one identical unit.

Evolution of Polyomino

The Polyomino agenda was developed over 3 years. In each iteration of the studio, the team would analyze the developments of the previous years and innovate and experiment testing different assumptions. The first iteration of the studio focused on the symmetry breaking operations of a truncated octahedron, which became the boundary geometry for more complex forms which would establish structural connections with a specific number of neighbors. Each unit was designed in great detail, attempting to communicate and guide the assembly and valid topologies through the geometry and curvature of each unit. This is the model that was carried forward for the further development with the company Stratasys in the Polyomino series released in 2016.

The second iteration of the studio allowed for a broader study of perfect packing, in the interest of exploring granular assemblies that could host new topological networks. One of the projects decided to abandon the perfect packing framework altogether to allow for the unit assemblies to possess a larger functional difference. This strategy proved to be highly productive as the combinatorial space grew, offering a larger possibility of neighbors and rotations. The project required a deep study of permutations defining the building blocks of larger tectonic assemblies. The project reduced the complexity and information of the unit itself, but increased the differentiation of assembles. The combinatorial strategy was exacerbated and became a central advancement for the 3rd iteration of the studio.

The third iteration of the studio was an attempt to interrogate in a more radical form the framework of combinatorial design. Three projects were developed, each with an attempt to question what had become a familiar setup.

The first project was developed as a study of 'systems and exceptions', where a perfect packing arrangement allowed a small polyhedron to propagate and define large granular tectonic assemblies. This operation was repeated in different areas of a test project but without an underlying structure that would connect these different 'islands' together. The idea here was to allow for a small percentage of bespoke units to work as the connecting tissue of the perfectly packed parts. The implications of these strategy is far reaching, as it suggests a possibility to couple the power of serially repeated units working in a combinatorial framework with unique units machined and detailed for tolerance and to operate as a system of exceptions for a formal rigid system.

The second project from this series attempted to introduce analog connections to operate as an alternative to discrete rotations. This operation certainly starts blurring the boundary

What is described here as a game draws concepts from gaming culture, but mainly refers to a guided simulation where the decisions of, in this case, the player start generating a branching narrative. The narrative is the geometric development and the decisions of function and performance embedded by a designer. These game simulations generate a strong feedback loop between the computational constraints and the intuition and experience of a human designer.

The role of the designer is to play and design meaningful patterns. The relevance of a pattern is defined collectively by the studio analyzing structural concerns, porosity, light scattering or delineation, among other criteria, but it's up to any designer to define the criteria for design. Often a team would try to design diverse patterns with different performances, in order to later allow for the combinatorial game to operate at a different scale between patterns rather than units.

7   Model – Year 3 – Material patterning of one identical unit.

between combinatorial and parametric design, but was a proof of concept to allow for gradient intensity arrangements to coexist within a combinatorial framework. This was achieved by designing a sliding joint that would allow two units to describe a continuous set of positions to one another. This breaks the notion of a finite set, as the analog connection offers an infinitesimal set of positions between two units. Nevertheless, the combinatorial strategy prevailed within the design assembly. The sliding connection became an opportunity for differentiation without overtaking the main framework of the project agenda. The units, as expressed in the figure below, offer two discrete face to face connections, each with 4 possible rotations, and a 2 sliding faces, to allow for the bundling of units parallel to each other.

The third project of this series stayed within a purely combinatorial space growing upon the lessons of previous years. The project defined a smart simple unit that could connect to itself in a large arrange of permutations. In the traditional Polyomino methodology, a large cataloging of permutations defined the 'vocabulary' of arrangements that created meaningful patterns. These arrangements were analyzed for their recursive capabilities, structural capacity, porosity and even formal delineation - all design criteria that could be deployed as a granular field condition for a given project. This project proved that a simple combinatorial system can yield a great deal of differentiation providing the basis for the development of open ended building systems for use in architecture.

These operations could not have been performed by a permutation algorithm, as what became meaningful in terms of configurations was at all times a human synthesis mediating among diverse criteria. The framework is indeed computational, but has been designed to allow for a human-machine symbiosis, allowing human designers to interact with generative aggregates in an intuitive way.

Evolution of Polyomino

For the collaboration with the company Stratays, the research unit envisioned the use of gaming technology and virtual reality to generate a frictionless combinatorial environment where users could immerse themselves into a design experience. The deployment of a gaming platform implies a real-time feedback with a simulation where tools provide enough information to generate a feedback loop with the designer. As explained in the ACADIA paper of 2014 (Sanchez, 2014), the use of gaming platforms has a great potential to allow for new designers to emerge from crowdsourced operations. As Jeremy Rifkin describes it, the 'prosumers' (Rifkin, 2014), consumers that become producers of their own goods, can potentially drive a new explosion of manufacturing.

This was the objective of the series of projects developed with Stratasys. Framed in the notion of 'from gaming to making', the Polyomino prototypes establish a bridge between a massive entertainment platform such as video games and a growing maker movement empowered by tools like 3D printing.

8

9

10

The challenges of the virtual reality platform have to do with the framerate requirements while dealing with complex geometry and the user experience at the moment of design. The team makes the distinction between a 3D modeling software in the form that we know them in computers, and a game experience that inserts a player into a constrained space with particular combinatorial operations available. This is an area that is still in early development as virtual reality is a very recent technology and the research group sees great potential in future development.

One of the most radical innovations of 3D printing has nothing to do with form but rather with forms of distribution. By digitizing objects to digital files, transferred through the web and manufactured on local machines, we bypass the whole distribution chain that is highly inefficient and has a great environmental impact. The team envisions how users would be able to use gaming platforms to adapt and create new objects that they could print locally.

The modular design of the Polyomino print is a reflection on how modular combinatorics will play a role into the customization of products, where a user can decide to exchange modules to adapt to diverse conditions. Polyomino uses magnetic connections to allow an intuitive snapping into place of units, allowing the user to decide diverse patterns and arrangements. The project uses color and the new technology developed by Stratasys with the J750 to print a large range of units in different digital materials defined by their color and flexibility. The model establishes an

analogy of how materials work at an atomic scale, allowing for the bonding of certain molecules to one another and generating larger assemblies.

CONCLUSION

The work developed over the past three years in the framework of combinatorial design has developed a rich repository of strategies and a clear framework to continue further research. There are indications that industry will continue pursuing systems and products with a strong modular open-ended architecture, as the business model coming from a serialized repetition of units is still the strongest and most economical form of fabrication. The addition of combinatorial strategies adds a new layer of heterogeneity and customization where families and sets of products could be combined and re-organized by an interconnected community.

The tools and infrastructure developed for achieving these design strategies also resonate with the zeitgeist of the manufacturing industry, where the issue of distribution of goods is one of the largest industries that could be radically transformed by the digitalization of physical goods with the advent of local and cheaper 3D printing centers. Gaming platforms could take an educational role and help to minimize the friction between creativity and production.

This future is currently in the making and pushed by those institutions and individuals who are building the tools to allow its implementation.

8 Elevation – Year 3– Pavilion design using perfect packing system with exceptions. 20% of units breach between different orientation planes that do not align to each other.

9 Section – Pavilion design.

10 Plan – Pavilion design.

11 Detail – Year 3 – Assembly of wooden members with sliding connection.

11

ACKNOWLEDGEMENTS

The design work was developed within University of Southern California since 2013 with the participation of:

Year 1: Yuchen Cai, Setareh Ordoobadi, Can Jiang, Jingjing Li, Yanping Chen,
Year 2: Hanze Yu, Kaining Li, Siyu Cui
Year 3: Team 1: Jingbo Yan, Dechen Zeng, Wansu Zhang, Xinran Ding, Team 2: Aditya Gulati, Ashley Kasuyama, Zhixue Zheng, Qianqian Xu
Team 3: Jiawen Ge, Yuchuan Chen, Yu Zhao, Yu Zhao, Qian Liu

The work could not have been realized without a collaborative group environment where both students and tutors work together towards a common goal.

REFERENCES

Bogost, Ian. 2006. *Unit Operations: An Approach to Videogame Criticism.* Cambridge, MA: The MIT Press.

Rifkin, Jeremy. 2014. *The Zero Marginal Cost Society: The Internet of Things, The Collaborative Commons, and the Eclipse of Capitalism.* New York: Palgrave Macmillan.

Sanchez, Jose. 2014. "Polyomino: Reconsidering Serial Repetition in Combinatorics." In *ACADIA 2014: Design Agency—Proceedings of the 34th Annual Conference of the Association for Computer Aided Design in Architecture,* edited by David Gerber, Alvin Huang, and Jose Sanchez. Los Angeles: ACADIA. 91–100.

Schumacher, Patrik. 2011. *The Autopoiesis of Architecture: A New Framework for Architecture.* Chichester, UK: Wiley.

Terzidis, Kostas. 2014. *Permutation Design: Buildings, Texts and Contexts.* London: Routledge.

IMAGE CREDITS

Figure 1: Polyomino Studio year 3 – Team 1: Jingbo Yan, Dechen Zeng, Wansu Zhang, Xinran Ding.
Figure 2: Polyomino Research – Plethora Project with Yuchen Cai, Setareh Ordoobadi.
Figure 3: Polyomino Research – Plethora Project with Yuchen Cai, Setareh Ordoobadi.
Figure 4: Polyomino Studio year 2 – Hanze Yu, Kaining Li, Siyu Cui
Figure 5: Polyomino Studio year 2 – Hanze Yu, Kaining Li, Siyu Cui
Figure 6: Polyomino Studio year 3 – Team 1: Jingbo Yan, Dechen Zeng, Wansu Zhang, Xinran Ding.
Figure 7: Polyomino Studio year 3 – Team 1: Jingbo Yan, Dechen Zeng, Wansu Zhang, Xinran Ding.
Figure 8: Polyomino Studio year 3 – Team 3: Jiawen Ge, Yuchuan Chen, Yu Zhao, Yu Zhao, Qian Liu
Figure 9: Polyomino Studio year 3 – Team 3: Jiawen Ge, Yuchuan Chen, Yu Zhao, Yu Zhao, Qian Liu
Figure 10: Polyomino Studio year 3 – Team 3: Jiawen Ge, Yuchuan Chen, Yu Zhao, Yu Zhao, Qian Liu
Figure 11: Polyomino Studio year 3 – Team 2: Aditya Gulati, Ashley Kasuyama, Zhixue Zheng, Qianqian Xu
Figure 12: Polyomino Research – Plethora Project with Yuchen Cai, Setareh Ordoobadi.
Figure 13: Polyomino Research – Plethora Project with Yuchen Cai, Setareh Ordoobadi.

12 3D Printed piece developed out of the gaming platform. The units have been printed separately and magnets have been added to the face connections to allow the growth of the voxel structure.

Combinatorial Design Sanchez

13 Virtual reality gaming software – Player is able to place different kinds of units out a pre-established inventory, pick color and rotate them to define the necessary connectivity with other units.

Jose Sanchez is an Architect / Programmer / Game Designer based in Los Angeles, California. He is the director of the Plethora Project, a research and learning project investing in the future of on-line open-source knowledge. He is also the creator of Block'hood, an award winning city building video game exploring notions of crowdsourced urbanism named by the Guardian one of the most anticipated games of 2016.

He has taught and guest lectured in several renowned institutions across the world, including the Architectural Association in London, the University of Applied Arts (Angewandte) in Vienna, ETH Zurich, The Bartlett School of Architecture, University College London, and the Ecole Nationale Supérieure D'Architecture in Paris.

Today, he is an Assistant Professor at USC School of Architecture in Los Angeles. His research 'Gamescapes', explores generative interfaces in the form of video games, speculating in modes of intelligence augmentation, combinatorics and open systems as a design medium.

Emergent Structures Assembled by Large Swarms of Simple Robots

David Andréen
Lund University

Petra Jenning
FOJAB arkitekter AB

Nils Napp
University at Buffalo

Kirstin Petersen
Max Planck Institute for
Intelligent Systems

1

ABSTRACT

Traditional architecture relies on construction processes that require careful planning and strictly defined outcomes at every stage; yet in nature, millions of relatively simple social insects collectively build large complex nests without any global coordination or blueprint. Here, we present a testbed designed to explore how emergent structures can be assembled using swarms of active robots manipulating passive building blocks in two dimensions. The robot swarm is based on the toy "bristlebot"; a simple vibrating motor mounted on top of bristles to propel the body forward. Since shape largely determines the details of physical interactions, the robot behavior is altered by carefully designing its geometry instead of uploading a digital program. Through this mechanical programming, we plan to investigate how to tune emergent structural properties such as the size and temporal stability of assemblies. Alongside a physical testbed with 200 robots, this work involves comprehensive simulation and analysis tools. This simple, reliable platform will help provide better insight on how to coordinate large swarms of robots to construct functional structures.

1   View of testbed with robots (orange) and passive blocks (white) at the SmartGeometry conference in Göteborg 2016.

INTRODUCTION

Driven by new technologies and increasingly complex and challenging construction environments, the field of multi-robot systems with distributed control is gathering interest. The allure lies in the potential for rapid, in-situ construction, continuously responsive and evolving buildings, and the colonization of human-hostile environments such as areas of disaster, the deep sea, or space.

In such situations, swarms—coordinated groups of autonomous agents—exhibit a number of advantageous traits: *Redundancy*, where robustness to failure is achieved through the sheer number of agents; *parallelism*, where many agents can work efficiently on the same structure at once; *scalability*, where additional agents can be added to perform a larger task or complete one more rapidly; and *adaptability*, where the system has the potential to respond to external disturbances. In other words, swarms of robots are potentially robust, responsive, and adaptable constructors, lacking many of the limitations found in automated centralized processes.

However, building, maintaining, and coordinating large robot swarms is a challenging problem, especially as the complexity of individual robots increases. Furthermore, achieving predictable complex emergent outcomes from bottom-up programming is difficult. Instead, it may be beneficial to find simpler methods for collaborative organization, such as the mechanical programming explored here. Minimally complex agents physically interact, and outcomes are controlled by carefully designing the geometry of the agents and their environment.

In pursuit of this goal, we have set up a testbed and organized a workshop to explore emergent self-organizing behavior of a large number of simple robots. The testbed is composed of a physical arena which includes active robots (bristlebots) and passive building blocks, a digital simulation of the physical testbed, and a kit of analysis tools which allow for a performative evaluation of the emerging structures and behaviors. In the future, we hope to use this testbed to understand and formalize methods of achieving emergent structures with predictable properties, based solely on manipulation of the geometry of the robots and building blocks.

Inspiration

Architects have long been intrigued by the emergent properties of swarms. Miranda and Coates (2000) outline some of the system characteristics, albeit in a virtual environment, highlighting the combination of simple mechanics and complex emergent phenomena. Tibbits (2012) takes the investigation a step further by moving to physical contexts and omitting programmed control, instead relying on geometric properties and

2   No two termite mounds look the same; the structure emerges from the local interactions between millions of relatively simple insects always resulting in a unique, but functional, output.

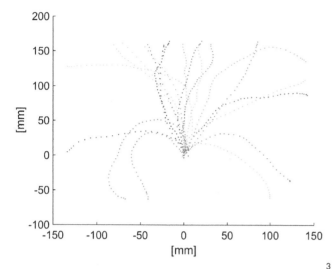

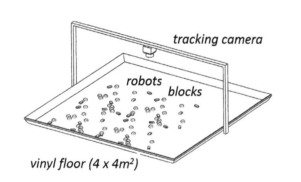

3 Tracked paths of 20 bristlebots started from (0,0) heading along the vertical axes; although most move somewhat straight, some are predispositioned to circular motion patterns.

4 Photos showing the original bristlebot, bristlebot with Styrofoam cover, and final bristlebot with a colored cardboard top for tracking purposes.

5 Sketch of testbed area.

mechanical coding to achieve predetermined outcomes.

Leaving the field of architecture, the natural and biological precedent is strong and forms the clearest inspiration for the work documented here. Self-organization is found in many places in nature, often as a performative survival strategy. One example, as outlined by Ben-Jacob (2010), is in single cell bacteria, where the brainless organisms form complex spatio-temporal patterns. Another example is the construction of functional nests and mounds by termite colonies without any centralized coordination

or sensing (Turner and Soar 2008; Figure 2). The termites have served as a model for previous efforts to design collective construction robots (Werfel, Petersen, and Nagpal 2014; Napp and Nagpal 2014), but these systems quickly escalate in design complexity due to the requirements of executing specific bio-inspired algorithms.

In general, large collectives of robots are rarely presented, because of implications regarding maintenance, cost, and reliability. The single largest swarm presented was 1,024 robots (Rubenstein, Cornejo, and Nagpal 2014). Using distributed control and local communication, these robots could aggregate to form loosely connected 2D shapes. In contrast to the hardware platform presented in that research, we use inexpensive commercially available robots, which are very robust and able to move over a much larger variety of surfaces. In the analysis section, we show how system properties such as temporal stability, structural stability, and propensity to form clusters are greatly affected by geometry and can be tuned by changing the degree and type of mechanical interlocking. Through this project, we seek to explore how more fundamental mechanisms of emergent form can be found and navigated; this may later serve as a foundation for more complex templates.

METHODOLOGY
Robot Testbed
The robot testbed is based on a large two-dimensional arena containing passive building blocks and up to 200 bristlebots. Input variables such as the geometric shape and number of robots and blocks, initial configuration, and permanently mounted mechanical shapes determine the properties of the output structure. A camera (Logitech C920) mounted overhead records the experiments for subsequent analysis.

The robots are commercially available (Hexbug Nano ver. 1), propelled forward by a simple vibrating motor on top of angled soft legs, and have no means of programming other than changes to their body geometry. They all exhibit some degree of randomness in their movement; furthermore, some robots tend to move fairly straight, while others have a circular motion pattern of different radii (see Figure 3). By bending the legs, these patterns can be altered. The robots operate on an AG13 battery, with a battery life of approximately 45 minutes, holding an average speed of approximately 11 mm/s.

The robot shape is altered by adding a customizable cover (Figure 4). Although the movement pattern of the robot is not strongly influenced by the covers tested here, covers that are too large or front heavy will severely alter the movement/decrease the velocity of the robot.

Emergent Structures Andréen, Jenning, Napp, Petersen

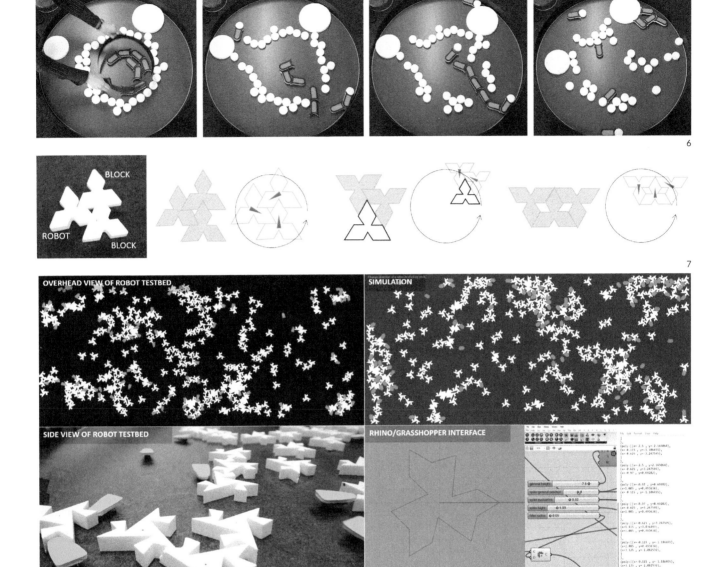

The robots operate in an arena of variable size, up to 4 x 4 m², covered with a dark vinyl rug. Because the robots tend to get stuck along vertical walls, the edges of the vinyl are bent up to form a soft boundary that provides a passive incentive for the robots to turn back into the arena (Figure 5).

In addition to the bristlebots, passive building blocks of CNC-cut styrofoam are placed in the arena. The tracking of these blocks is the primary mode to evaluate the outcome of the experiment, along with behavioral analysis of the robots. The photo sequence in Figure 6 gives an example of the type of complex outcome one can achieve despite the simplicity of the system. By matching the robot gripper to the robot rear they can form long emergent chains that can move objects too heavy for a single robot. Figure 7 shows a system where the same shape is used for

6   Photo sequence of small-scale experiment: the geometric shape of the robots prompts them to form emergent chains for collective transport of heavy objects.

7   Spatio-temporal patterns formed using passive and active triangular shapes, depicted with motion patterns in decreasing order of stability.

8   Upper left: frame from the recorded video where each pixel is classified as background, building block, or robot, and colored accordingly. Upper right: snapshot of the simulation. Lower left: close-up photo of the robots and building blocks. Lower right: Rhinoceros 3D/Grasshopper interface allows users to quickly modify and test new geometric shapes in both simulation and real life.

robots and building blocks, providing interesting metastable and stable shapes for spatio-temporal outputs.

Simulation

A simulation environment was implemented with Processing (Reas and Fry 2006) using Box2D (Catto 2011) as a physics

9

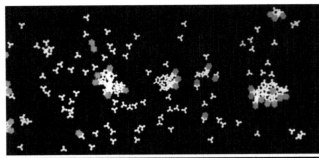

10

9   Experiment with triangular shaped robots (from Figure 4) without passive
building blocks. The figure shows a screenshot from the digital simulation and
an analysis of the cluster sizes and migration between two timestamps.

10   Top: Still image from video recording of experiment. Building blocks are white,
and robots orange. Bottom: Analysis of video, robots with fading trails are shown
in green. Single robots pushing only one block tend to get stuck in a circular
pattern of a set size, while the bigger clusters are moved by the collective effort
of several robots.

engine; it consists of a 2D world with both active agents and
passive building blocks. This simulation tool complements the
robot testbed by enabling fast iterations of geometric design,
easy replication of experiments, and the ability to explore the
population scale at which the rules, which are hard-coded into
the robots, no longer produce meaningful structural properties.

All simulated objects have ground friction (modeled as velocity
damping), object-object friction, density, and a coefficient
of restitution; their shape is defined by one or more convex
polygons and circles. The agents differ from the passive blocks
by an applied force vector along with a certain degree of noise
to create movement patterns which match the physical robots.
The object parameters were adjusted to achieve behavior
similar to that observed in the physical experiments. To mimic
the soft boundaries of the physical arena, a repellent force has
been added to the world boundaries. As the simulation runs, it
automatically produces video output and log files for subsequent
analysis.

Finally, a small script was added in Grasshopper (www.grasshop-
per3d.com) and Rhinoceros 3D (www.rhino3d.com) to enable
quick iterations of geometric shapes, simulation, and production
of physical blocks and robots (Figure 8).

Analysis

A set of analysis tools has been developed to aid in understanding the emergent behavior arising from the mechanical
program, as well as understanding the similarities between the
physical experiments and the digital simulations. They look at
both the passive building blocks and the robots themselves. The
primary focus is on visualizing the assembly process by analyzing
the evolution of building block clusters and their robot induced
motion.

The video data is analyzed using the Python interface to the
computer vision library OpenCV (Bradski 2000). Through the use
of color information, the video is segmented into background,
passive building blocks, and robots, and can be passed on for
further analysis. In the case shown in Figure 9, the assemblies
are analyzed using Grasshopper 3D for the number and size
of discrete clusters over time by counting the number of parts
associated with each cluster.

The ability to process robots and passive building blocks separately also makes it possible to visualize robot motion; in Figure
10, for instance, the location of the robots is plotted over time.
This analysis can be done both on recorded videos of the physical experiments as well as the digital simulations to visualize
how collective motion patterns emerge when several robots
form clusters linked via passive blocks. This collective transport
behavior emerges even though there is no central control, and
the robots are programmed using mechanical shape only. A single
robot can push a maximum of approximately 3 building blocks,
depending on size and shape of the blocks, while larger clusters
of blocks can be pushed when several robots work together.

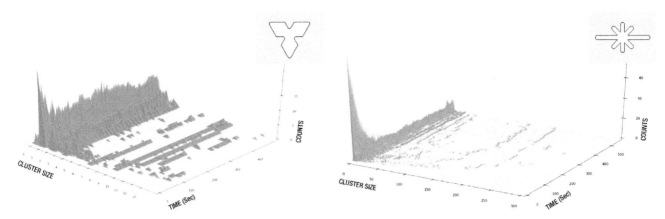

11 Cluster sizes of building blocks plotted over time. Peak shows initial state where blocks are randomly dispersed, after a while stable clusters form indicated by "lines" along the time axis. Different physical shapes lead to fundamentally different behaviors. The plot to the left shows clusters of triangular robots, which can quickly form stable small clusters (Figure 4), while the plot to the right shows clusters of star-shaped blocks, which tend to aggregate into larger stable clusters too large to move.

Similarly, the assembly process can be observed by plotting cluster properties over time. Figure 11, for instance, shows how semi-stable clusters of passive blocks form over time depending on their geometry.

Observations

The intention of this article is primarily to describe an experimental set-up, rather than to establish definitive links between the agents' geometry and the large scale patterns that emerge, which is left for forthcoming research. However, some observations were made during the course of the workshop that may serve as initial guidelines or hypotheses for such research. These observations are somewhat intuitive and qualitative in nature and would require further confirmation from quantitative data.

1. *Density*. The overall density of blocks and robots greatly affect the outcome. When the robots or blocks are sparse, random movements tend to dominate, and the blocks and robots aggregate along the outer perimeter of the testbed. Conversely, with too many blocks or robots, the system locks up due to congestion and little change over time is observed.

2. *Cluster stability*. In order to form stable clusters that do not disperse over time, the robots and/or blocks must be able to interlock with one another. Figure 4 shows an example where the blocks aggregate to form stable clusters over time, and Figure 3 shows how robots can interlock to make them move in conjunction with each other.

3. *Directed motion*. Blocks may counteract the random movement of a robot through negative feedback. An example is the triangular blocks shown in Figure 4; a passive block in front of a robot will dampen the random movements of the robot, making it move mostly straight.

4. *Collective transport*. Combined, interlocking features and the ability of blocks to alter motion patterns of the robots lead to interesting swarm behaviors, such as robots jointly moving large clusters of blocks along straight paths, as shown in Figure 7. In Figure 3, robots sporadically interlock and are thereby able to push objects that would be too heavy for an individual robot.

CONCLUSION

In this article we explore 2D construction processes with emergent outcomes by using swarms of robots and passive building blocks; the swarm behavior is programmed by changing the geometric shape of both robots and blocks. We develop and present the necessary tools for designing and analyzing such systems, including a physical robot testbed and a detailed simulation environment. The behavior of each individual robot is simple, but the shape-mediated interactions with other robots and passive blocks can lead to complex swarming patterns, for example, forming stable clusters of specific size or collective transport.

By focusing on a simplified 2D environment with commercially available, cheap, and robust robots, we can physically implement the swarm on a scale much larger (200 robots) than is commonly seen in literature. Having full control over all input parameters, including population size and geometric shape, as well as the world perimeter, allows efficient exploration of which parameters have the greatest effect on structure outcome.

The physical testbed is complemented by a customized simulation platform and design tools allowing more rapid iterations and automated tests of new shapes than is possible in real life, which is essential due to the complexity of agent interaction.

The presented analysis tools are general and scale well with respect to swarm size. By working directly on video data, we can use the same analysis toolchain on both simulated and real-world data, and establish that the two systems behave similarly. Here, we have focused on evaluating the evolution of cluster sizes, but more specialized measures are possible. For example, cluster shapes and locations could be influenced by designing building blocks that induce preferential growth directions or initiate cluster formation based on arena geometry.

Looking forward, we aim to use this testbed to improve our understanding and develop formalized theories of how to adapt local rules (geometry) according to desired global outcomes. Other extensions of this work might focus on closing the loop in the design process by optimizing shapes for specific properties, either through stochastic optimization schemes like genetic algorithms (Davis 1991) or recent work on Bayesian optimization that aims to efficiently use simulations to form priors on measurements in physical experiments. These methods have been successfully used for optimizing large design spaces, and coupling our design and analysis tools to run these algorithms might enable us to find shapes that produce strong and reliable emergent behavior in a bottom-up fashion.

The advantages of the simple robot swarms presented here—their robustness, adaptability, and scalability—can provide significant benefits in real world construction scenarios. By providing this framework where questions of control and predictability of the emergent structures can be explored and tested, we hope to contribute to the field of construction by robot swarms.

ACKNOWLEDGEMENTS
We would like to thank all the participants of the workshop "Swarmbot Assemblage" at Smartgeometry 2016: Judyta Cihocka, Jie-Eun Hwang, Erik Larsen, Grzegorz Lochnicki, Martin Nygren, Evangelos Pantazis, Jasmi Sadegh, Mohammad Soorati, Emmanouil Vermisso.

Furthermore, we thank Forbo Flooring for providing the vinyl rug used as testbed floor, and the GETTYLAB Program for partially sponsoring travel costs related to the project.

REFERENCES
Ben-Jacob, Eshel, 1997. "From Snowflake Formation to Growth of Bacterial Colonies II: Cooperative Formation of Complex Colonial Patterns." *Contemporary Physics* 38 (3): 205–241.

Bradski, Gary. 2000. "The OpenCV Library." *Doctor Dobbs Journal* 25 (11): 120–126.

Catto, Erin. 2011. "Box2D: A 2D Physics Engine for Games." http://box2d.org/

Davis, Lawrence, editor. 1991. *Handbook of Genetic Algorithms*. New York: Van Nostrand Reinhold.

Miranda, Pablo, and Coates, Paul. 2000. "Swarm Modelling: The Use of Swarm Intelligence to Generate Architectural Form." In *Proceedings of the 3rd International Conference on Generative Art*. Milan: GA.

Napp, Nils, and Nagpal, Radhika. 2014. "Distributed Amorphous Ramp Construction in Unstructured Environments." *Robotica* 32 (2): 279–290.

Reas, Casey, and Ben Fry. 2006. "Processing: Programming for the Media Arts." *AI & Society* 20 (4): 526–538.

Rubenstein, Michael, Alejandro Cornejo, and Radhika Nagpal. 2014. "Programmable Self-Assembly in a Thousand Robot Swarm." *Science* 345 (6198): 795–799.

Tibbits, Skylar. 2012. "The Self-Assembly Line." In *ACADIA 12: Synthetic Digital Ecologies—Proceedings of the 32nd Annual Conference of the Association for Computer Aided Design in Architecture*, edited by Jason Kelly Johnson, Mark Cabrinha, and Kyle Steinfeld. San Francisco: ACADIA. 365–372.

Turner, J. Scott, and Rupert C. Soar. 2008. "Beyond Biomimicry: What Termites Can Tell Us About Realizing the Living Building." In *Proceedings of the First International Conference on Industrialized, Intelligent Construction*, edited by Tarek Hassan and Jilin Ye. Loughborough, UK: I3CON. 221–237.

Werfel, Justin, Kirstin Petersen, and Radhika Nagpal. 2014. "Designing Collective Behavior in a Termite-Inspired Robot Construction Team." *Science* 343 (6172): 754–758.

IMAGE CREDITS
Figure 8: Lochnicki, Hwang, and Nygren, 2016
Figures 9–10: Cihocka, Larsen, and Soorati, 2016
All other figures: Andréen, Jenning, Napp, and Petersen, 2016

Emergent Structures Andréen, Jenning, Napp, Petersen

David Andréen is currently a lecturer at Lund University in Sweden, where he also received his Master of Architecture. He was awarded an Engineering Doctorate (Eng.D) from University College London along with a Master of Research in Adaptive Architecture and Computation. David has taught architecture at institutions such as Greenwich University, the Bartlett School of Architecture and Lund University, as well as holding workshops and serving as examiner and critic in numerous places around the world. His research interest lies in the intersection of computation, architecture and biology, with a particular emphasis on complex geometries and termites.

Petra Jenning (SAR/MSA) recieved her MArch from Lund University in Sweden in 2007. She has since practiced architecure in Shanghai, Paris, London and Malmö. She has taught at University of Greenwich, been an invited critic at the AA, CITA, and LTH, and run workshops in several universities and companies. Currently she is head of computational design at FOJAB architects in Sweden.

Nils Napp received his Bachelor of Science in Engineering and Mathematics from Harvey Mudd College, Claremont CA, in 2003 and his MS and PhD in Electrical Engineering from the University of Washington, Seattle WA, in 2006 and 2011 respectively. Before joining the University at Buffalo, Buffalo NY, as an Assistant Professor of Computer Science and Engineering in 2014 he was a post-doctoral fellow at the Wyss institute for Biologically Inspired engineering at Harvard University, Cambridge MA. His research focuses on robotics and biologically inspired algorithm design that enables engineered systems to robustly interact with their environment.

Kirstin Petersen received a BSc in electro-technical engineering from the University of Southern Denmark, and a PhD in Computer Science from Harvard University in 2014. Her thesis involved design of autonomous robots for collective construction inspired by African mound-building termites. She did a 2-year postdoc on novel soft actuator mechanisms at the Max Planck Institute for Intelligent Systems in Germany, and is currently an assistant professor in the ECE department at Cornell University, USA. Her research interests involve bio-inspired robot collectives able to interact with and manipulate their environment according to user-specified goals.

Feedback- and Data-driven Design for Aggregate Architectures

Analyses of Data Collections for Physical and Numerical Prototypes of Designed Granular Materials

Gergana Rusenova
Karola Dierichs
Ehsan Baharlou
Achim Menges
ICD/ University of Stuttgart

1

ABSTRACT

This project contributes to the investigations in the field of aggregate architectures by linking two research areas: the numerical simulation of aggregate formations, and a concept for an online-controlled pneumatic formwork system.

This paper introduces a novel approach for constructing with designed particles based on a feedback process. The overall aim was to investigate the capacity of aggregates as an architectural material system, which create emergent spatial formations. Initially the particles´ micro-mechanical behavior and the fragile stability of the formations were analyzed using numerical simulations. Based on this, an online-controlled inflatable formwork system was developed. The formwork was designed to react to the actual stability state of an aggregate formation; for this, a statistical set of simulation data was gathered, which directly informed the physical system. This overall concept was proven and verified in a one-to-one scaled physical model.

The methods developed within this research provide a first set of baselines for comparison between the behavior of simulated and physical designed granular materials.

1 Comparison between physical and numerical prototype constructed with synthetically produced non-convex particles and using inflatable formwork system.

INTRODUCTION

Aggregate Architecture

In the context of architecture, aggregates are a material system defined as a large amount of natural or designed particles in unbound contact. Based on the ability to change their state from one of a fragile solid to one of a dry liquid, they form a separate branch in the context of material systems (Rivier and Fortin 2011; Hensel and Menges 2006a). As such, aggregates necessitate the development of a feedback-driven design approach, where the fabrication of a formation consists of iterative steps and is based on the actual state of matter. The field of designed particles offers a wide range of research topics, such as the development of observation techniques for better understanding of the material behavior (Dierichs and Menges 2010); studies about possible element shapes (Dierichs and Menges 2016; Miskin 2016; Murphy et al. 2016; Athanassiadis et al. 2014); and robotic-driven design (Angelova et al. 2015; Dierichs et al 2013).

This research project seeks to build upon research based on a previously developed material system consisting of non-convex synthetic particles. The focus is on two aspects of the material performance: first, the fragile stability (Cates et al. 1998) of aggregate formations relying on friction, and second, the self-organizational capacity of granular materials to demonstrate emergence (Wolf and Holvoet 2004).

Fragile Stability

In the context of this research, the concept of stability should not be viewed from the perspective of continuum mechanics where the behavior of materials is modeled and analyzed as a continuous mass rather than as discrete particles. This paper suggests that the principles of aggregates´ behavior are aligned with investigations in the field of soft matter physics (Jaeger et al. 1996) and more precisely with the concept of fragile matter (Cates et al. 1999), which in materials science is described as a granular matter conditionally prevented from moving. The marginal stability of the formations constructed with such matter is related to the force chain network within the material (Cates et al. 1998). In this context, one of the core questions is how fragile stability can be understood and analyzed. One approach recently published in the context of physics and nonlinear complex systems suggests to quantify stability of columns constructed with non-convex aggregates by measuring the mass of particles which drop off when a column collapses (Zhao et al. 2016). The points of stability and of collapse then become constitutive parameters of scientific material analysis; they can, however, also become active design drivers.

Therefore, this paper proposes a rather architectural strategy for using fragile stability of aggregate formations. The research

project focused on investigations on the aggregates´ micro-mechanical behavior, and data collected through analyses of both numerical simulations and physical prototypes. The aim was to extract parameters that describe the marginal stability of an aggregate composition. These parameters were used as lever points to form an IF/THEN conditional statement in a complex adaptive system (CAS) (Holland 1995, 2006).

Emergence

The behavior of the aggregates as a material system can be considered fundamentally emergent, since the global shape and the state of matter of the overall aggregate formation depend on the interactions and the movement of the individual particles on a local level (Wolf and Holvoet 2004). The performance of a single aggregate is contingent on two factors: first, the environmental conditions, such as pouring methodologies, boundary conditions, and external impacts, and second, the interactions with other particles. Thus, the collective self-organizational capacity of the material system is the cause for changes in the state of the overall composition from fluid to static and backwards. This dynamic behavior predefines the ability of aggregates to exhibit emergence.

Furthermore, this paper presents the concept of an online-controlled pneumatic formwork system, whose behaviors can also be described as emergent. The logic behind the system´s performance is based on statistical series of digital simulations, where the performance of the aggregates was traced and analyzed, and clear parameters for the stable state of the digital models were established.

Both systems, formwork and material, worked together in a process described as a continuous feedback-loop. The behavior of one system is the cause for the performance of the other one, and vice-versa.

STATE OF THE ART

The state of the art for this project has two primary trajectories: first, experimental setups with aggregates, which focus on the implementation of formwork, and second, Discrete Element Method (DEM) simulations of discontinuous materials.

Experimental Setup

A common construction technique for small-scale experiments with designed particles is the casting of material in predefined molds. Non-convex designed particles take the shape of the confining space they are initially poured into. Moreover, because of the particle geometry and static friction, aggregate formations exhibit a stable state even after the supporting structure is removed. Thus, the concept of simple and reusable formwork is a

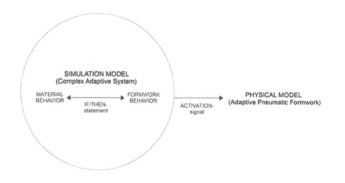

2

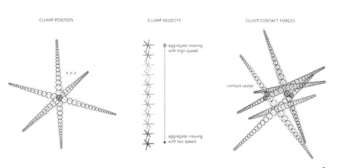

3

2   Complex adaptive system and formwork activation diagram.

3   Simulation parameters - clump position, velocity and number of contact forces.

4   Analyses of simulation parameters.

5   Velocity-contact forces diagram.

key aspect of the experimental setup. In the context of architecture, such flexible systems were first proposed by Eiichi Matsuda (Hensel and Menges 2006b). In his research project, a series of experiments was conducted where aggregates were poured into cubic test containers filled with balloons. Once the balloons were deflated, a self-stabilizing process took place. A new fragile stable state of the aggregate formation was achieved, and the final result was the formation of spatial voids inside of the material. The project described within this paper also concentrated on the implementation of inflatables as a relevant experimental setup. However, the focus was shifted toward the development of a complex adaptive system (CAS) where the inflatables' performance was synchronized with the behavior of the digitalized material system.

DEM Simulations

The DEM simulations are mathematical models for simulating the dynamic micro-mechanical behavior of large numbers of small particles. The theoretical background is based upon the laws of motion by Sir Isaac Newton. Later, several scientists, such as Cundall and Strack, created the foundations of the method when solving problems related to rock mechanics (Cundall

1971; Cundall and Strack 1979). The simulations are based on algorithms for calculating the nearest neighbor, which makes it possible to observe the interactions inside the material and the motion of each individual particle.

DEM simulations are widely used for solving engineering problems in discontinuous materials like powders, grains, rocks etc. Thus, for the purposes of this research, their implementation was included as a relevant mathematical method for numerical modeling of aggregate formations. The simulations were executed with a state of the art DEM-software. In its first part, this research used already developed algorithms for observing the behavior of non-convex particles (Dierichs and Menges 2015). Based on the statistical information gathered, the digital behavior of the adaptive pneumatic formwork system was added to the already existing algorithms. Figure 2 illustrates the design process.

METHODS

The computational and analytical methods described in this paper demonstrate a mixture of observation techniques from both material and machine computation (Dierichs and Menges 2012). These two categories are inseparable when investigating the behavior of aggregates. The main goal was to collect information about the material system through numerical and physical prototyping. As speculated in this paper, the data gathered can be later used to inform the design process.

Machine Computation Method 01: DEM Simulations

In the DEM-software, each individual aggregate (clump) is represented by a number of bonded spheres (pebbles), which allows for time-saving calculation. A confining space is defined and a formwork mesh-geometry is imported. Aggregates are randomly distributed and gravitational force is applied to each of them, simulating their fall. The model state is time-based where series of calculations cycles are conducted. Each cycle consists of a sequence of operations finding the solution to Newton´s laws, which updates the position of the particles and consequently the next cycle is introduced (Pöschel and Schwager 2005). The calculations run until the particles find an equilibrium state. After that the formwork is removed and a second self-stabilizing process takes place until final settlement. The variables in the simulations executed are material friction, damping, particle surface geometry, pebble number per model and cycling. The calibration of the material properties for the simulations is described in an earlier publication (Dierichs and Menges 2015).

The parameters observed from the simulations were the position of each aggregate as x, y and z values; the clump velocity represented by colors ranging from red to blue indicating high to low velocity respectively; and the number of clump contact forces

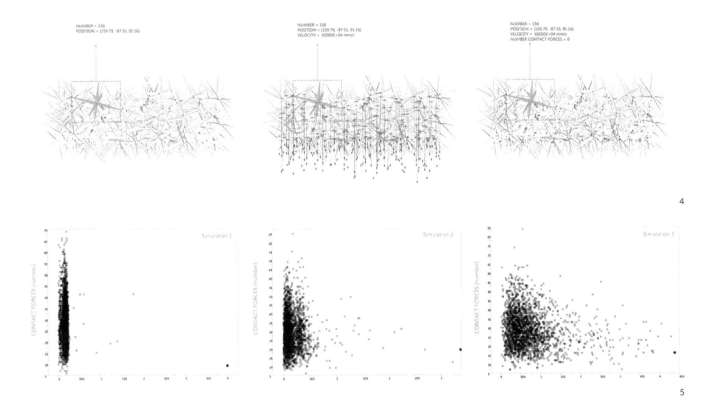

4

5

(see Figure 3). In the context of this project, it is speculated that a simulated aggregate formation exhibits a marginal stable state when the majority of aggregates have short velocity vectors and high number of contacts with their neighbors.

Machine Computation Method 02: Analyses

Two different software platforms were used during this research project—a 3D-modeling platform and a DEM-simulation program. A computational tool for analysis of the simulation results was developed to link them. First, the formwork mesh-structures were created and exported to the DEM-software. After the completion of the simulations, the output values were exported as *.csv-files to the 3D modeling environment where the behavior of the formations was traced and documented. Additionally, the simulated clumps with their final position and rotation in space were also imported as mesh-objects in the 3D modeling software. Figure 4 illustrates the results achieved with the tool developed.

Machine Computation Method 03: Statistical Analyses

Because of the stochastic behavior of the material system, statistical analyses are a key aspect of the observation technique. Figure 5 shows the relationship between numbers of contact forces and lengths of velocity vectors for three numerical studies of the same simulation setup. In the algorithm, a generate-function was used for defining the initial clumps' positions before they start falling. These were different for both cases, simulating

the fact that in reality no aggregate composition can be built twice exactly the same way. Thus, despite the same bounding conditions for both tests, the simulation values after the same settling time were different. However, statistics enable the opportunity for estimating patterns of behavior after analyzing data sets. Therefore, large information collections of experiments were managed using a statistical software. The mean values extracted from the data formed the IF/THEN statement which activated a feedback-loop in the simulation process.

Material Computation Method 01: Physical Models

The second branch of methods used within this research is material computation. Physical models were built and the number of inflatables and their position in the confining space were varied for design and stability purposes. Two types of non-convex particles were used throughout the conduction of experiments: ten-armed with a diameter of 120 mm, and six-armed with diameter of 300 mm. The high degree of friction due to the particle geometry and the large amount of resources available were the main reason for this particular material choice. The physical prototypes built with both material systems are considered crucial to preparing the project, since they are essential to illustrating the material's behavior. Additionally, physical experiments were necessary in order to verify the reliability of the simulation and computational methods used. Figure 6 illustrates an example of a prototype built with 2500 ten-armed aggregates. The space formed after removing the inflatable remained stable even

6

7

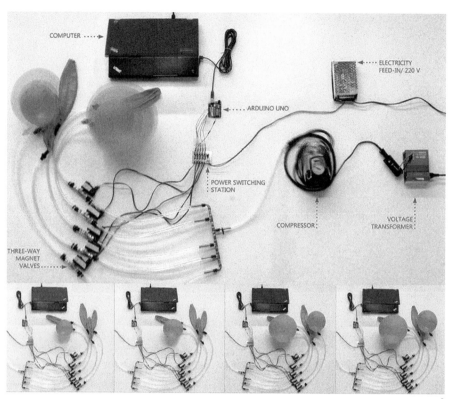

8

after destroying half of the model. The stability of formations built with non-convex particles is based upon their interlocking behavior under compression forces. Therefore, enclosed spaces, for example dome-like structures are preferred for the global design of the formations. The design strategy of all experiments, including the final prototype, follows this logic.

Material Computation Method 02: Photogrammetry

A set of physical prototypes were documented using photogrammetry techniques which allowed for more precise analyses of physical models. The formations were photographed from all sides, and sets of 80 pictures for each were imported in a photogrammetry software. The program generated a dense point-cloud for each experiment with approximately one million points. A close view is illustrated in Figure 7. Each point holds the values of its x, y and z coordinates. This information was used for quantitative and numerical comparison of physical and digital prototypes.

Material and Machine Computation: Adaptive Pneumatic Formwork System

A direct connection between machine and material computation was established with the development of an adaptive pneumatic formwork system, following the concept illustrated in Figure 8. The tool was designed to be activated via computer.

The balloons were attached on the one side to plastic pipes and

an electrically powered compressor that constantly pumped air in one direction, and on the other side to three-way magnetic valves. The three outputs of the valves allowed for both inflation and deflation of balloons individually. The system was controlled via an Arduino Uno board connected with a USB-cable to a computer. A power-switching station was designed for converting the 5 V output signal of the Arduino board to 12 V for each valve.

The system developed allowed for adaptation of the physical formwork based on the behavior of its digital representation during on-going simulations.

CASE STUDIES

Case Study 01: First Experiments with DEM Simulations

A series of simulation tests were needed to gain knowledge about the numerical representation of both particle-particle, and particle-formwork interactions. These first studies were of great importance for the development of further methods.

One investigation was dedicated to the comparison between two types of physical particles: six-armed and ten-armed, both with diameter of 110 mm. Four hundred and fifty particles from both material systems were separately poured in two identical boxes. The same settling cycle number was set for the two samples (see Figure 9). The first expected outcome was that the

Feedback- and Data-driven Design for Aggregate Architectures
Rusenova, Dierichs, Baharlou, Menges

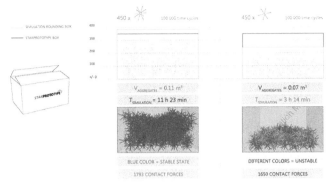

6 Physical prototype constructed with ten-armed aggregates.

7 3D-scan of an aggregate composition.

8 Adaptive pneumatic formwork system and sequence of picture demonstrating the individual inflation of different balloons.

9 Simulation results of case study 01: ten-armed vs. six-armed particles.

10 Section through formations with different amount of aggregates.

9

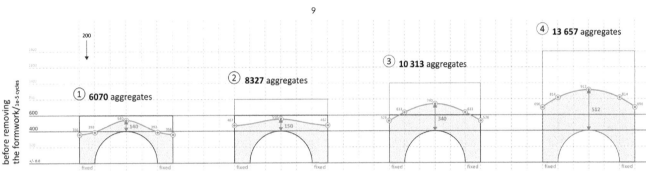

10

volume occupied by the six-armed particles would be almost the half of the volume occupied by the ten-armed particles. The second conclusion was that after the same settling duration, the ten-armed particles would exhibit more static and stable behavior. However, a crucial parameter was also the time needed for both calculations. This case study proved that the simulations of six-armed particles are less time-consuming than those of ten-armed particles (3 hours 14 minutes vs. 11 hours 23 min on a PC equipped with a 4th generation Intel® Core™ processor). Since a regular computer was used in this project, it was decided to continue all experiments with the six-armed aggregates.

Another important examination is documented in Figure 10. It is a case study on the amount of particles needed to achieve a stable aggregate formation with a hemispherical void in the center. Before starting a simulation, one has to estimate the number of aggregates needed to achieve a stable configuration after their fall. Based on a number of experiments, the count was defined as a function of both the formwork´s and the particles´ volumes.

Case Study 02: Statistical Sets of DEM Simulations
The next two sets of simulations focused on observing the stochastic behavior of the material system. Two different setups were defined (see Figure 11). The first one consisted of a test container and two spheres, which allowed the formation of a column. One thousand six-armed aggregates with diameter of 300 mm were poured into the test container. The cycle time

before and after removing the formwork was set to 1e+5. The second setup was similar to the first one. The two formwork spheres were designed to intersect each other and to allow a larger span in the middle of the formation. The fall of 800 six-armed particles was simulated with the same duration conditions as in the first setup.

To analyze the behavior of the material system, each setup was tested ten times with different start positions for the aggregates. The output values for the clump velocity and the number of contact forces were first analyzed with a computational tool. Images of the top, right and front view were generated for each formation showing the last movement of the aggregates in the simulation (see Figure 12). The color and length of the arrows indicate the clump velocity. The critical parts of the formations are marked in red. Each formation, even within the same simulation setup, exhibits unstable behavior in a different place. This conclusion is of great importance for the feedback-driven design approach suggested.

The information gathered from the two simulation setups formed two separate statistical data sets (see Figure 13). Mean values were extracted for both parameters, the clump velocity and the number of contact forces. As expected and according to these parameters, the second simulation set showed more unstable behavior after the same amount of settling time. This is primarily due to the absence of a column in the middle, which requires the

11 Simulation setup 1 and 2.

12 Collection of front, side, and top views from simulation setup 1.

13 Comparison of the results from setup 1 and 2 after statistical analyses.

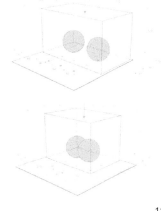

11

aggregates to span two dome volumes instead of one as in the first simulation set.

The second set of simulations was also quantitatively compared to a statistical set of ten physical prototypes. Each model was 3D-scanned using photogrammetry. The generated 3D point-clouds of the physical prototypes were compared with the mesh-geometries exported from the simulations. Figure 14 shows that from a visual perspective, the results were identical. Additionally, the x-, y- and z-coordinates of the points for one point cloud were extracted in mm. The z-values of the points situated in the middle of the span were numerically compared to a mesh-model from one of the simulations (see Figure 15). The value deviation is considered acceptable for the stochastic nature of the material system.

Case Study 03: Feedback-driven Design Approach

The proposal for a design approach can be described as a feed-back-driven looping process consisting of four individual steps:

- Simulation of initial aggregate formation.
- Analyses of the results according to the two parameters.
- Adaptation of the digital formwork according to the actual stability state of the simulated formation.
- Activation of the physical formwork system.

The setup for this last case study for both the simulation and the physical prototype consisted of four spherical inflatables, two intersecting each other and two with a column between them. The initial aggregate formation was simulated. Aggregates were designed to fall in the predefined boundary space and to settle on top of the inflatables. The initial formwork was removed and the second self-stabilizing phase started. This point of the process was crucial, since it was the appropriate moment for the feedback-loop to take place and for the adaptive inflatable system to be triggered. The digital balloons were designed to

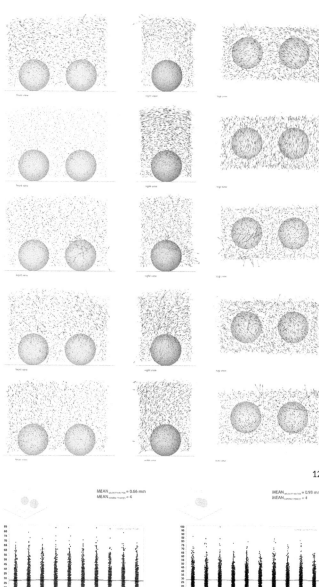

12

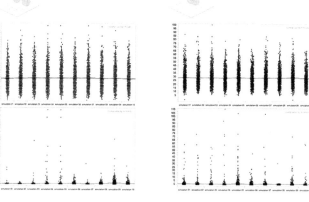

13

observe the aggregates´ behavior during the simulation and to react to it by inflating or deflating in the simulation. The behavior of the individual balloons was calculated directly in the simulation software. The feedback-algorithm was based on the actual state of matter of the numerical aggregate formation. It was designed to measure the length of the velocity vector for each aggregate in the system and to count the number of contacts with its neighbors. The mean values extracted in case study 02

Feedback- and Data-driven Design for Aggregate Architectures
Rusenova, Dierichs, Baharlou, Menges

formed an IF/THEN statement that triggered the inflation of the balloons needed. The code calculated the x and y centroid coordinates of the aggregates moving with higher velocity and having least number of contacts. If an instable group of particles was found, the nearest digital balloon was activated to support the structure (see Figure 16). The information about the position of the next inflation point was saved as a *.txt file and forwarded to the according balloon in the physical world. This process was designed to be iterative until a final stable state of the overall aggregate formation was achieved (see Figure 17).

RESULTS AND REFLECTION

This study builds upon the concept of discrete designed particles as a building material system and contributes to the investigations in the field by developing a simulation model which works as a complex adaptive system (CAS). This digital representation consists of two parts: the behavior of the aggregates and the behavior of the formwork. The adaptation occurs online and during the simulation. The feedback-loop is regulated through an IF/THEN statement based on mean values estimated via statistical simulation sets. Once the adaptation takes place in the digital environment, a signal is sent to the physical formwork (see Figure 2). Thus, the methods described here, lead to two main results.

First, a set of tools for direct connection between physical aggregate formations and their simulated representations was developed. This allows for a simultaneous observation of both digital and physical prototypes and, therefore, enables a straightforward comparison between them. However, due to the material's stochastic behavior, a one-to-one recreation of the physical model in the simulation, and vice-versa, is clearly impossible. As a further development, a series of experiments could be conducted and analyzed using statistical methods. In this way behavior patterns for both the simulated and the real aggregates could be estimated with a focus on better understanding the material system.

Second, the project combined the self-organization capacity of aggregates to demonstrate emergence with the ability of the pneumatic formwork system to react according to the material state of matter. The outcome is the development of a CAS. This design strategy builds upon one of the most important advantages of aggregates as a building material system, namely their ability to reconfigure. The designer is driven to work with a design tool and material system, both of which remain outside of his/her complete control. Given this context, a consequential question needs to be answered: should the formations discussed find a final equilibrium state, or will they remain in a constantly self-adjusting mode, shifting periodically between static and flowing states?

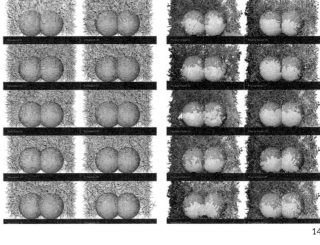

14

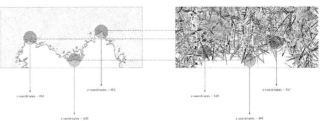

15

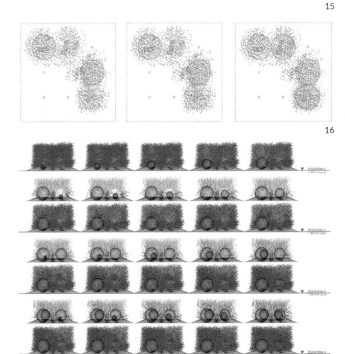

16

17

14 Physical vs. numerical prototypes.

15 Numerical comparison.

16 Diagram for the calculation of the next inflation point.

17 Simulation results from case study 03.

FURTHER RESEARCH

The findings discussed here present opportunities for further development in two directions.

On the technical side of the project, the methods suggested could be improved. For example, the current version of the algorithm developed for the formwork actuation does not take the physical properties of the inflatables into account. A possible line of investigation could include measurements of pressure on the surface of the balloons in the physical model during construction and incorporation of this variable into the code. Otherwise, improving synchronization between the digital and physical models may also be of benefit, since the whole process is set online.

The second project branch with possibility for further investigation is the design process. The work presented here concentrates on the micro-mechanical behavior of the material and on the fragile stability of the formations as a design generator. All investigations were based on the self-supporting capacity of the structures. In a following step, static and dynamic loads, such as wind or directed air flows could be introduced into the system. This would allow for a design strategy based on the material properties and their interaction with the surrounding environment.

ACKNOWLEDGEMENTS

The authors would like to thank Sacha Emam, Lothar te Kamp, Mattew Purvance and Reza Taghavi as well as Benedikt Wörl of ITASCA Consulting Group, Inc. for their support in the simulation model development, as well as the ITASCA Education Partnership Program for providing the software. They would like to express their gratitude to Desislava Angelova, Alexander Wolkow, Jasmin Sadegh, Martin Alvarez, and Emil Rusenov for their support during the execution of the project and the preparation for this publication.

REFERENCES

Angelova, Desislava, Karola Dierichs, and Achim Menges. 2015. "Graded Light in Aggregate Structures - Modeling the Daylight in Designed Granular Systems Using Online Controlled Robotic Processes." In *Real Time- Proceedings of the 33rd eCAADe - Volume 2*, edited by B. Martens, G. Wurzer, T. Grasl, W. Lorenz and R. Schaffranek. Vienna. 399–406.

Athanassiadis, Athanassiadis G., Marc Z. Miskin, Paul Kaplan, Nicolas Rodenberg, Seung H. Lee, Jason Merritt, Eric Brown, John Amend, , Hod. Lipson, and Heinrich M Jaeger. 2014. "Particle Shape Effects on the Stress Response of Granular Packings." *Soft Matter* 10 (1): 48–59.

Cates, Michael E., Joachim P. Wittmer, Jean P. Bouchaud, and Philippe Claudin. 1999. "Jamming and Static Stress Transmission in Granular Materials." *Chaos* 9 (3): 511–22.

———. 1998. "Jamming, Force Chains, and Fragile Matter." *Physical Review Letters* 81 (9): 1841–44.

Cundall, Peter A. 1971. "A computer model for simulating progressive large scale movements in blocky rock systems." Proceedings from *Symposium International Society of Rock Mechanics*. vol 1: Paper II– 8.

Cundall, Peter A., and Otto D. L. Strack. 1979. "A Discrete Numerical-Model for Granular Assemblies - Reply." *Geotechnique* 29: 47–65.

Dierichs, Karola, and Achim Menges. 2010. "Natural Aggregation Processes as Models for Architectural Material Systems." *Design and Nature V: Comparing Design in Nature with Science and Engineering* 138: 17–27.

———. 2012. "Aggregate Structures: Material and Machine Computation of Designed Granular Substances." *Architectural Design* 82 (2): 74–81.

———. 2013. "Material and Machine Computation of Designed Granular Matter. Rigid-Body Dynamics Simulations as a Design Tool for Robotically-poured Aggregate Structures Consisting of Polygonal Concave Particles" In *Simulation, Prediction and Evaluation- Proceedings of the 30rd eCAADe - Volume 2*, edited by R. Stouffs and S. Sariyldiz. Delft. 701–9.

———. 2015. "Simulation of Aggregate Structures in Architecture: Distinct-Element Modeling of Synthetic Non-Convex Granulates." *Advances in Architectural Geometry 2014*. Berlin/ Heidelberg Springer-Verlag. 1–13.

———. 2016. "Towards an Aggregate Architecture: Designed Granular Systems as Programmable Matter in Architecture." *Granular Matter* 18 (25).

Hensel, Michael and Achim Menges. 2006a. "Material Systems - Proto-Architectures (Part 2)." In *Morpho-Ecologies*, edited M. Hensel and A. Menges. London: AA Publications. 62–67.

———. 2006b. "Eiichi Matsuda - Aggregates 01 2003-2004." In *Morpho-Ecologies*, edited M. Hensel and A. Menges. London: AA Publications. 262–272.

Holland, John H. 1995. *Hidden Order: How Adaptation Builds Compexity*. Boston: Addison-Wesley.

———. 2006. "Studying Complex Adaptive Systems." *Journal of Systems Science and Complexity* 19 (1): 1–8.

Jaeger, Heinrich M., Sidney R. Nagel, and Robert P. Behringer. 1996. "The Physics of Granular Materials." *Physics Today* 49 (4): 32–38.

Miskin, Marc Z. 2016. *The Automated Design of Materials Far From Equilibrium*. Switzerland: Springler International Publishing.

Murphy, Kieran A., Nicolaj Reiser, Darius Choksy, Clare E. Singer. and Heinrich M. Jaeger. 2016. "Freestanding Loadbearing Structures with Z-shaped Particles." *Granular Matter* 18 (26).

Pöschel, Thorsten and Thomas Schwager. 2005. *Computational Granular Dynamics: Models and Algorithms*. Berlin/Heidelberg: Springer.

Feedback- and Data-driven Design for Aggregate Architectures
Rusenova, Dierichs, Baharlou, Menges

18 Spatial aggregate formation.

Rivier, Nicolas, and Jean-Yves Fortin. 2011. "Unjamming in Dry Granular Matter : Second-Order Phase Transition between Fragile Solid and Dry Fluid (Bearing) by Intermittency." *Solid-Solid Phase Transformations in Inorganic Materials*, Pts 1–2: 1106–11.

Wolf, Tom D. and Tom Holvoet. 2004. "Emergence Versus Self-Organization: Different Concepts but Promising when Combined." In *Engineering Self-Organizing Systems*, edited S.A. Bruekner, G. Di Marzo Serugendo, A. Karageorgos, R. Nagpal. Berlin Heidelberg: Springler-Verlag. 1–15.

Zhao, Yuchen, Kevin Liu, Mattew Zheng, Jonathan Barés, Karola Dierichs, Achim Menges, and Robert P. Behringer. 2016. "Packings of 3D stars: stability and structure." *Granular Matter* 18 (24)

IMAGE CREDITS

Figures 1,6,8,18 : Gergana Rusenova, ITECH 2015, 21.10.2015

Gergana Rusenova is an architect and PhD researcher at the Chair for Architecture and Digital Fabrication (Prof. F. Gramazio, Prof. M. Kohler), ETH Zurich. Her research focuses on the development of design strategies for a novel material system composed of low-cost bulk material and tensile fiber-reinforcement. Gergana Rusenova holds a bachelor degree in Architecture and Urban Planning and a master degree as part of the program Integrative Technologies and Architectural Design (ITECH), both from the University of Stuttgart and under the lead of Prof. A. Menges and Prof. J. Knippers. She has worked as student employee and junior architect for several architectural and engineering offices, including Knippers Helbig Advanced Engineering.

Karola Dierichs is an architect, researcher and tutor at the Institute for Computational Design (ICD) with Prof. Menges, University of Stuttgart. She has been educated at the Technical University of Braunschweig, the ETH Zurich and the Architectural Association (AA) in London and graduated from the Emergent Technologies and Design Program with Distinction in 2009. She has taught at the Architectural Association in London, the Städelschule Architecture Class in Frankfurt and the Institute for Computational Design in Stuttgart. At the Institute for Computational Design (ICD), Karola Dierichs is leading the research field of aggregate architectures, where she is developing synthetic granular materials as designed matter in architecture. Among others her recent work has been recognized with the Holcim Acknowledgement Award Europe 2014.

Ehsan Baharlou is a doctoral candidate at the Institute for Computational Design (ICD) at University of Stuttgart. He holds a Master of Science in Architecture with distinction from the Islamic Azad University of Tehran. Along with pursuing his doctoral research, he has taught seminars on computational design techniques and design thinking at the ICD since 2010. His research interest is currently focused on the integration of fabrication and construction constraints into computational design for form generation via agent-based modeling and simulation.

Achim Menges is a registered architect and professor at the University of Stuttgart, where he is the founding director of the Institute for Computational Design. He also is Visiting Professor in Architecture at Harvard University's Graduate School of Design. He graduated with honours from the AA School of Architecture in London where he subsequently taught as Studio Master of the Emergent Technologies and Design Graduate Program, as visiting professor and as Unit Master of Diploma Unit 4. Achim Menges' practice and research focuses on the development of integral design processes at the intersection of morphogenetic design computation, biomimetic engineering and computer aided manufacturing. His projects and design research have received many international awards, has been published and exhibited worldwide, and form parts of several renowned museum collections, among others, the permanent collection of the Centre Pompidou in Paris.

What Bricks Want:
Machine Learning and Iterative Ruin

Paul Harrison
University of Toronto

1

ABSTRACT

Ruin has a bad name. Despite the obvious complications, failure provides a rich opportunity—how better to understand a building's physicality than to watch it collapse? This paper offers a novel method to exploit failure through physical simulation and iterative machine learning. Using technology traditionally relegated to special effects, we can now understand collapse on a granular level: since modern-day physics engines track object-object collisions, they enable a close reading of the spatial preferences that underpin ruin. In the case of bricks, that preference is relatively simple—to fall.

By idealizing bricks as rigid bodies, one can understand the effects of gravitational force on each individual brick in a masonry structure. These structures are sometimes able to 'settle,' resulting in a stable equilibrium state; in many cases, it means that they will simply collapse. Analyzing ruin in this way is informative, to be sure, but it proves most useful when applied in series. The evolutionary solver described in this paper closely monitors the performance of constituent bricks and ensures that the most successful structures are emulated by later generations. The tool consists of two parts: a user interface for design and the solver itself. Once the architect produces a potential design, the solver performs an evolutionary optimization; after a few hundred iterations, the end result is a structurally sound version of the unstable original. It is hoped that this hybrid of top-down and bottom-up design strategies offers an architecture that is ultimately strengthened by its contingencies.

1   Ruined brick towers in various stages of evolutionary growth. The visible 'capital' that terminates each tower is an artifact of the rigid-body physics simulation.

INTRODUCTION

Bricks don't aspire to be in arches—they want to be on the ground. As a kind of fundamental antagonist of architecture, gravity is sadly under-represented in the digital realm: consider the surfaces that weightlessly float across your screen each day. What's needed now is a kind of realism about materiality, an acknowledgement that buildings can and do fall down.

Though the losses associated with failure can often be devastating, the ruined building sometimes provides an opportunity for realignment: look no further than the cathedral at Beauvais. Intended as a structurally audacious successor to the cathedral at Chartres, the ambitious 144 ft vaulting collapsed in 1284, causing later additions to be scaled down significantly. Contemporary builders certainly took note; Grodecki (1976) characterizes the collapse as "signaling the end of the gigantic Gothic undertakings of the thirteenth century." As it stands, the cathedral has been the site of ongoing structural modification, playing host to a variety of ad-hoc supports in an effort to keep the daring structure together.

Architecture's recent symbiosis with structural engineering now largely mitigates our risks of collapse. A deep understanding of a building's statics is now *de rigueur*; while this development has certainly resulted in safer buildings, the opportunity to learn from structural failure is almost entirely absent from contemporary discourse. Building to failure allows for limit states to be reached and for analysis to move beyond the statically indeterminate. In short, a ruined building offers a richer diagram than that achieved by analysis alone (Figure 2).

This new approach aims to both exploit and iterate that diagram. Recent developments in rigid-body physics engines allow us to simulate the spatial preferences of vast numbers of independent bodies; the collapse of an entire masonry building can be simulated over the course of a few minutes. New analysis tools can perform a close reading of collapse and provide a detailed understanding of each brick's stability over time.

The use of rigid-body physics in architecture is hardly without precedent (Dierichs and Menges 2013; Dierichs and Menges 2012; Meredith and Sample 2013). Though previous studies have focused on the simulation of relatively unstructured aggregates, the application of rigid-body physics to traditional masonry construction has pedigree in the world of structural engineering. Jacques Heyman has argued for an emphasis on stability over strength in the analysis of structural masonry, noting that the maximum stresses encountered in a typical dry-stone cathedral are insignificant compared to the crushing strength of the material. The cathedral at Beauvais, for instance,

2

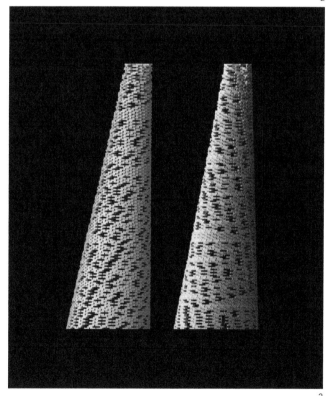

3

2   Partial collapse reveals underlying corbelling "diagram."

3   Towers with bricks removed using cellular automata rules.

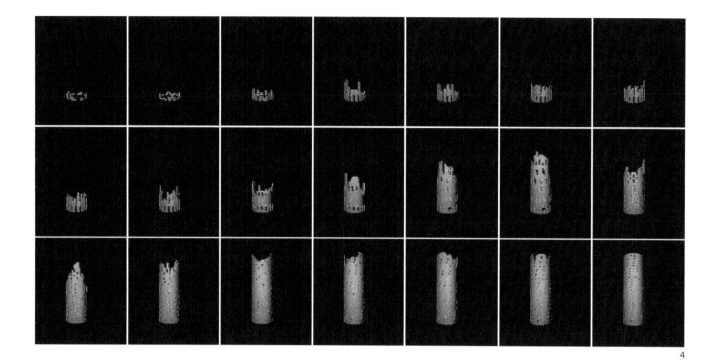

was found by Benouville to have a maximum stress of only 1.3 N/mm², only 1/30th of the stone's compressive strength of 40 N/mm² (Heyman 1995). Clearly, the structural issues at play were failures of stability; in most dry masonry applications, abstracting each brick as a rigid body should accurately model the failure mechanism at play.

This approach stands in contrast to the pioneering computational masonry work undertaken by the Block Research Group, which adapts the traditional graphical method of funicular design to a three-dimensional interactive design tool. Not unlike the hanging-chain method employed by Gaudi, Otto, and others, the success of this approach lies in its abstraction: funicular vaults are accurately modeled through 'Thrust Network Analysis,' a form-finding method that models the structures as "spatial representations of compression forces in equilibrium with the applied loads" (Rippmann and Block 2012).

The algorithm described in this paper is modeled on a considerably less elegant method: the centuries-long process of trial-and-error structural optimization. As Heyman notes, the historical evolution of masonry was led mostly by the study of precedent—in the absence of static analysis, builders were encouraged to follow rules of thumb and expand on passed-down experience. The machine-learning techniques described herein fulfill a similar purpose, as successful iterations are copied and mutated until new, better precedents are established. By automating the process of collapse and analysis, this new algorithm effectively short-circuits the historical 'brute force'

approach to establish a novel general method for masonry structural optimization.

METHOD

Before the evolutionary solver begins its iterative collapse, a user interface allows the designer to specify a pre-simulation genotype, encoding a template within which the physics engine can begin to place its bricks. After simulation occurs, the evolutionary solver mutates the initial genotype and passes the resulting template back to the physics engine; the process is then iterated and successful mutations are stored within the evolved genotype.

Beyond simply controlling the tower's shape, the user interface also allows for bricks to be removed from its overall form. This has the effect of both increasing the solver's rate of attrition and making the resultant collapse less computationally expensive. The first version of the solver removed these bricks using cellular automata rules (Figure 3) (Wolfram 2002), encoding a degree of spatial preference in each brick and resulting in a truly object-oriented outcome. Later versions of the solver implemented a top-down approach, controlling on/off states with an 18-parameter sinusoidal equation.

The net effect of both methods is that of aesthetic porosity. Although the cellular automata technique was conceptually preferable, initial results showed that slight variations in the solver's Wolfram rule genotype had systemic rather than incremental effects on the phenotype. Further, the equation method allows deep customization and a fine-grained approach to incremental

4 Sample generations in the evolutionary structural optimization.

5 Detail of sinusoidal plan variation. The relative overlap of each brick is controlled by the solver's genotype; greater overlaps sometimes allow for high degrees of corbelling but often introduce instability into the system.

5

change that dramatically increases the performance of the solver.

Computational efficiency lies behind the simulation strategy as well. Vaulted masonry structures have traditionally relied on temporary centering for support during construction. Early versions of this algorithm followed suit: instead of resting on wooden scaffolding, the bricks were allowed to propagate before the physics engine was enabled. This created a kind of bottleneck, as each simulation started with every brick in play. In order to ease the computational load, later versions of the algorithm adopted a brick-by-brick approach; after each course is laid, the physics engine is enabled for long enough to detect and remove any unstable bricks.

Aside from the obvious correlation with physical bricklaying, this approach has historical precedent: Brunelleschi's dome in Florence was famously constructed without interior centering, though this meant that the dome needed to be far sturdier than those supported in the traditional way (King 2000). This approach may be stricter than real-life construction, but it stands to reason that an incrementally simulated masonry structure will be easier to construct than one conceived with virtual 'centering.'

A rigid-body physics engine known as Bullet performs the simulation itself. Originally written for C++ by Erwin Coumans, it was later ported to Java by Martin Dvorak; the current implementation of my own algorithm uses bRigid, an excellent Processing.org port written by Daniel Kohler. This engine works in close conjunction with the evolutionary solver, which was written from scratch using typical strategies for evolutionary computation (Dillenburger, Braach, and Hovestadt 2009; Rosenman 1996; von Buelow 2002). The genotype of the solver is an array of eighteen parameters which control both the aforementioned sinusoidal equation and the initial rotation and separation of each brick (Figure 6). The fitness criterion is a user-weighted factor of brick placement success (i.e., the percentage of intended bricks that remain standing) and height, divided by the overall number of bricks required to reach that height. The obvious criteria here might be the total height divided by number of bricks standing, but this almost always results in the generation of a single-brick-width tower. Factoring in the brick placement success rate encourages the solver to build complete courses and enforces the "design intent" of the sinusoidal equations.

A single parameter of the genotype is mutated for each iteration; as each course is laid, unstable bricks are removed until an entire course proceeds unplaced, at which point the overall fitness is determined. If the fitness is greater than the previous generation, the genotype mutation is preserved. A lower fitness means that the mutation is discarded and the previous highest-fitness generation is mutated again. The cycle continues until the brick placement ratio reaches a specified amount; the net result is a stable, structurally sound brick tower (Figure 4).

RESULTS

Over the course of 205 generations, the tower depicted in Figure 6 grew from an initial height of 18 courses to a final, stable height of 120 courses, shown in Figure 7. Using an Intel

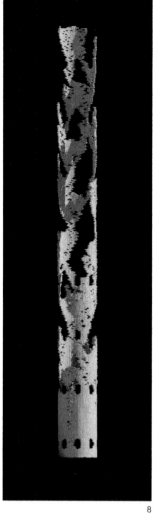

6   The first generation of an evolutionary sequence, initially using 418 bricks.

7   The 205th; over 8 minutes, the number of bricks has reached 3,517.

8   An early test produced while the algorithm was still undergoing calibration.

Core i7 processor running at 3.6 GHz, the structural optimization took roughly 8 minutes. More porous structures typically take longer; the evolution sequence depicted in Figure 4 took place over 2647 generations and roughly three and a half hours.

Although the solver is able to complete a tower with nearly every attempt, the net result is often somewhat under-built by physical standards—though the bricks are able to resist the vertical force enacted upon them by the solver, the algorithm does not yet simulate lateral loading such as wind. Since forces are applied on a per-object basis, this change will likely be relatively easy to implement.

Resisting this force may prove more difficult. The most obvious structural shortcoming of the solver's current output is that the tower's walls consist of only a single wythe of bricks. Adding two

or more wythes may slow things down considerably: doubling the number of bricks significantly increases the computational overhead of each simulation.

A new technique for computational optimization may provide a way forward. Rather than simulate the entirety of the structure at all times, a set number of courses are passed to the physics engine; the remainder are stored as traditional Processing objects and are locked in space. At the time of writing, simulating the ten most recent courses seems to produce results indistinguishable from a fully simulated model, though this option will need to undergo sensitivity testing before being fully implemented. It is hoped that greater efficiencies will allow for further increases in realism and complexity.

CONCLUSION

Somewhat predictably, this evolutionary solver has the tendency to make towers more structurally conservative: porosity and formal variation are typically decreased over time. While the algorithm is correctly optimizing for structural stability, the results are relatively normative. Future explorations will focus on evolution towards more novel ends—how porous can a stable brick tower possibly be? An early trial may point the way forward. The tower depicted in Figure 8 was the unexpected outcome of an incorrect gravity value, but its radical porosity serves as a kind of working target for the next version of this solver.

Postscript

As it stands today, the cathedral at Beauvais is somewhat compromised—the vaulting, once ambitiously broad, is cluttered with a bird's nest of ad-hoc structural steel reinforcement. Despite its aesthetic shortcomings, there is something poetic about this provisional arrangement; the intended cathedral was mutated and adapted over centuries, adopting the revisions of a host of designers. With its plurality of authors, one might say that Beauvais is more bazaar than cathedral (Raymond 1999). This is not a standalone case—architecture's claim to authorship is inevitably muddled by the messy realities of construction. Any built project is an embodiment of concession: an authored work, sure, but one inflected (and often strengthened) by the unintended.

Ruin is perhaps the most extreme form of this concession. Though we must grapple with the demands of client, budget, or program, the architect's most elemental struggle lies in resisting collapse. This paints materiality itself as a kind of stakeholder— bricks want something, after all, which is to fall to the ground. By incorporating the bricks themselves as secondary authors, it is hoped that this new approach helps to celebrate the gap between intent and execution. Maybe Kahn was right after all?

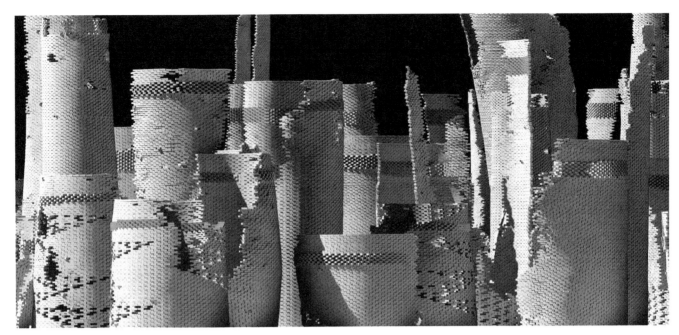

9   Failed iterations of the cellular automata solver.

ACKNOWLEDGEMENTS

This project owes a great deal to the guidance of Benjamin Dillenburger, who provided valuable advice throughout its beginning stages.

REFERENCES

Dierichs, Karola, and Achim Menges. 2012. "Material and Machine Computation of Designed Granular Matter: Rigid-Body Dynamics Simulations as a Design Tool for Robotically-Poured Aggregate Structures Consisting of Polygonal Concave Particles." In Digital Physicality: Proceedings of the 30th eCAADe Conference, vol. 2, edited by Henri Achten, Jiri Pavlicek, Jaroslav Hulin, and Dana Matejovska. Prague: eCAADe. 711–719.

———. 2013. "Aggregate Architecture: Simulation Models for Synthetic Non-convex Granulates." In Adaptive Architecture: Proceedings of the 33rd Annual Conference of the Association for Computer Aided Design in Architecture, edited by Philip Beesley, Omar Kahn, and Michael Stacey. Cambridge, ON: ACADIA. 301–310.

Dillenburger, Benjamin, Markus Braach, and Ludger Hovestadt. 2009. "Building Design as Individual Compromise Between Qualities and Costs: A General Approach for Automated Building Generation Under Permanent Cost and Quality Control." In Joining Languages, Cultures, and Visions–Proceedings of the 13th International CAAD Futures Conference, edited by Temy Tifadi and Tomás Dorta. Montreal, QC: CAAD. 458–471.

Grodecki, Louis. 1978. Gothic Architecture. New York: Harry N. Abrams.

Heyman, Jacques. 1995. The Stone Skeleton: Structural Engineering of Masonry Architecture. Cambridge, UK: Cambridge University Press.

King, Ross. 2000. Brunelleschi's Dome: The Story of the Great Cathedral in Florence. London: Chatto & Windus.

Meredith, Michael, and Hilary Sample. 2013. Everything All at Once:

The Software, Videos and Architecture of MOS. New York: Princeton Architectural Press.

Raymond, Eric S. 1999. The Cathedral and the Bazaar: Musings on Linux and Open Source by an Accidental Revolutionary. Sebastopol, CA: O'Reilly Media.

Rippmann, Matthias, Lorenz Lachauer, and Phillipe Block. 2012. "Interactive Vault Design." International Journal of Space Structures 27 (4): 219–230.

Rosenman, M. A. 1996. "The Generation of Form Using an Evolutionary Approach." In Artificial Intelligence in Design '96, edited by John S. Gero and Fay Sudweeks. Boston, MA: Kluwer Academic Publishers. 643–662.

von Buelow, Peter. 2002. "Using Evolutionary Algorithms to Aid Designers of Architectural Structures." In Creative Evolutionary Systems, edited by Peter J. Bentley and David W. Corne. San Francisco, CA: Morgan Kaufmann. 315–336.

Wolfram, Steven. 2002. A New Kind of Science. Champaign, IL: Wolfram Media.

IMAGE CREDITS

All figures © Paul Harrison, 2016

Paul Harrison received his Master of Architecture from the University of Toronto, where he was awarded the Faculty Design Prize in 2015. He also holds a professional degree in mechanical engineering from Queen's University; prior to his studies in architecture, he worked as a parametric design consultant for Rand Worldwide. Paul currently works as a project designer for Williamson Chong Architects.

Form-Making in SIFT Imaged Environments

Matthew Parker
Joshua M. Taron
University of Calgary

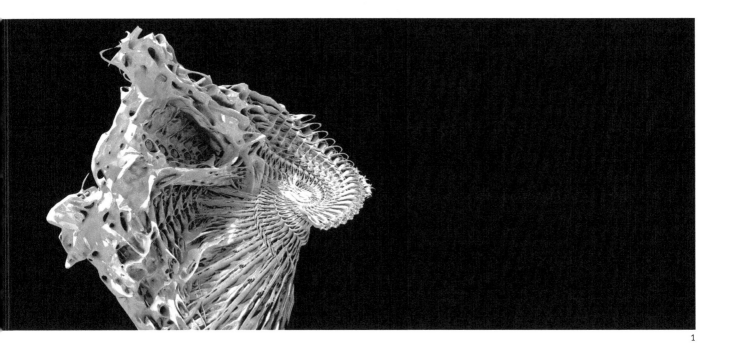

1

ABSTRACT

Within the contemporary condition, turbulence that confronts architecture is no longer unpredictable weather patterns or wild beasts, but the unintended forces of a constantly connected digital infrastructure that demands constant attention. If, as Mark Wigley puts it, "architecture is always constructed in and against a storm" it is time for architecture to reevaluate its ability to separate us from a new storm—one that situates technology, global connectivity, human, non-human and composite users, and algorithmic architecture itself as new weather systems. Toward this end, this paper explores architecture's ability to mediate and produce algorithmic turbulence generated through image-based sensing of the built environment. Through a close reading of Le Corbusier's *Urbanisme*, we argue that for much of the 20th and the early part of the 21st century, cities have been designed to produce diagrams of smooth and homogeneous flows. However, distributed personal technologies produce virtual layers that unevenly map onto the city, resulting in turbulent forces that computational platforms aim to conceal behind a visual narrative of accuracy, cohesion, anticipation, and order. By focusing on *SIFT algorithms* and their ability to extract n-dimensional vectors from two-dimensional images, this research explores computational workflows that mobilize turbulence towards the production of indeterminate form. These forms demarcate a new kind of challenge for both architecture and the city, whereby a cultural appetite to deploy algorithms that produce a smooth and seamless image of the world comes hand in hand with the turbulent and disruptive autonomy of those very same algorithms. By revisiting *Urbanisme*, a new set of architectural objectives are established that contextualize SIFTs within an urban agenda.

1   Indeterminate n-dimensional form that results from turbulence produced by algorithmic observation.

INTRODUCTION

"Ambiguity is absorbed when it becomes an unambiguous goal. Architects keep turbulence as a secret, secreting it inside to produce the effect of a clear line between the wild and the calm, such that this line, the facade of the building pressed up against the ever-changing exterior and exposed to unpredictable forces, is seen to be unchanging and therefore the calmest line of all." Mark Wigley

Western architectural theory contends that all buildings descend from a primitive hut, a small structure consisting of four poles, four beams, and a roof. The parable of the primitive hut demonstrates a desire to order and stabilize an unruly and chaotic world through the production of interiority that shelters us from a turbulent exterior. It wasn't until Le Corbusier's embrace of technology in the 1920s that an alternative to this framing was produced. For Le Corbusier, shelter from the outside was not only an insufficient task, it was a mode of framing that would have doomed architecture to obsolescence within an increasingly technological city. Le Corbusier recognized the architectural interior not as a shelter from the exterior, but rather as a point of interface and engagement with an ordered city. In *Urbanisme* (1924), the city is framed as a landscape of flows that relies on architecture to mediate the exchange between interior and exterior. But this mode of framing complicates the viability of the architectural interior-exterior distinction and is only exacerbated by the ubiquity of ever-intensifying orders and ever-expanding scales of algorithmic computation. Perhaps the distinction between turbulent and laminar flows—terms borrowed from fluid dynamics—provides a more useful reading of *Urbanisme* that allows algorithmic processes to be layered on top of and throughout Le Corbusier's figure of architecture. In particular, this fluid distinction provides a scalable way to address complex territories that simultaneously blur and articulate interiors.

Within the contemporary condition, turbulence that confronts architecture is no longer unpredictable weather patterns or wild beasts, but the unintended forces of a constantly connected digital infrastructure that demands constant attention. If, as Mark Wigley puts it, "architecture is always constructed in and against a storm" (2007), it is time for architecture to reevaluate its ability to separate us from a new storm—one that situates technology, global connectivity, human, non-human and composite users, and algorithmic architecture itself as new weather systems.

But where exactly do these stormy flows lie? What algorithms do the work of imaging them? How do these algorithms see the world and how do they both smooth and agitate the turbulent and laminar images captured within their gaze? What are the effects of their sensing and making sense of architectural images and images of the environment—and do they even recognize

architecture as a thing within the built environment? If so, how might architecture take advantage of its privileged position? If not, does this compromise the architectural object as an agent of environmental mediation; and is it possible to extend the architectural objectives of *Urbanisme* to agents more capable of the task?

This paper explores the ability of Scale-Invariant Feature Transform (SIFT) algorithms to mediate and produce algorithmic turbulence generated through image-based sensing of the built environment while evaluating them in terms specified in *Urbanisme*.

LE CORBUSIER'S TECHNO-DEMATERIALIZATION

Urbanisme expresses a desire for architecture to embrace technologies that flatten spatial and temporal dimensions of the city. 'Intense agitations' of technology would reshape the city by destroying traditional habits of urban planning and force a 'mutation' of both the concept and form of the city itself. Le Corbusier's tactic was to dematerialize architecture such that it would integrate technology into the interior by "domesticat[ing]" all the lines of communication in the city [...] turning the interior into a landscape in which nothing is ever still" (Wigley 2007). He would dissolve walls in order to 'minimize friction' and rooms to give way to 'trajectories.' The transition from a Cartesian system of extensively defined spaces and boundaries to an intensive logic of vectoral trajectories not only marks the obsolescence of the interior-exterior distinction, it also spells out the agenda for a techno-architecture concerned with mediating gradients of informational friction and flow. So for Le Corbusier, the project for architecture was very much one of the city—a utopian agenda of frictionless architecture that would both allow and enable technological integration rather than opposing or resisting it. What Le Corbusier recognized is that for architecture to be most effective, it would have to radically open itself up (thus subjecting its inhabitants) to technological force through the annihilation of interior-exterior disruptions via laminar boundaries.

Corbusier's framing also belied the delicacy of technological performance manifesting in a tabula rasa approach, whereby architecture would strive to produce only the most minor instances of turbulence amidst an otherwise smooth and laminar context—cities and architecture designed free of resistance. He writes, "a city should be treated by its planner as a blank piece of paper, a clean table-cloth, upon which a single, integrated composition is imposed" (Le Corbusier 1971). Within this smooth environment, Le Corbusier called for a centrally located core that was to perform all the 'higher functions' of society; a core consisting of skyscrapers that form the 'brains' not only of the city, but of the entire country. These brains "embody the work of elaboration and command on which all activities depend.

Everything is concentrated there: the tools that conquer time and space; telephones, telegraphs, radios, the banks, trading houses, the organs of decisions for the factories: finance, technology, commerce" (Le Corbusier 1971). Architecture therefore is responsible for turbulent agitations that act as controlling signals. The center of the city does not consult with the citizens of the city, instead, it issues commands that are to be followed towards the production of homogeneity and smoothness—able to register the most insignificant perturbations within the neural network of the technological city. This metaphorical model is as viable now as it was then. However, what Le Corbusier could not possibly have comprehended is the scale and intensity that this model would assume, nor the radically ambiguous and interchangeable positions that urban objects themselves would be able to occupy.

What we arrive with from *Urbanisme* is the privileged territory of the architectural envelope defined as a disruption between some interior and its environment. This territory can be evaluated as architecture by asking the following questions: How is the boundary produced and/or imaged? What is the logic of the envelope? What qualities do its signals produce? What new architectural forms emerge from such an investigation?

The Agitated Relationship Between Technology and the City

Facilitated by advancements in tele-communication, planetary scale computation, and global exchange networks, the city of the early 21st century has become a *Digital City*, a city "composed of multiple 'intelligent' layers, based on 'real-time' interaction, communication and location based content, [existing] beyond the physical buildings and urban environment" (Handlykken 2011). Within the Digital City, architecture is forced to look beyond the physical materiality of buildings and toward software as means of imaging, occupying, and using the city. Furthermore, interior/exterior distinctions also fall away within a digital paradigm. Thus,

Form-Making in SIFT Imaged Environments Parker, Taron

2 Visual anomalies of Google Earth: These images illustrate a selection of (mis)representations that result from architecture's agitated relationship to algorithmic observers.

3 SIFT cathedral study: digital images are converted to SIFT keypoints for the purpose of matching, stitching, etc., within the computational protocols of SIFT algorithms.

3

Le Corbusier's model demonstrates a kind of extension from the modern city into the digital city, where buildings have been replaced by code as the agents of turbulent disruption. But digital layers comprised of virtual mapping platforms and the ubiquitous production and dissemination of images produce a type of virtual tourism that destabilizes and distorts architecture's ability to even render an outside. Instead, what we have is a seamless image of an interior space that scales from our living rooms to the entire surface of the planet. But how is such a turbulent and inaccessible environment rendered in such a flattened and accessible way? And what is architecture to do if it is in the least way concerned with agitating or augmenting the smooth seamlessness of that space? Rather than beginning with the objects captured by this particular algorithmic observer (we will return to that later), the answer to these questions perhaps lies in the algorithms that produce the meta-image of the earth itself.

Despite its best efforts to "iron out the creases," turbulence is no more evident than in the visual anomalies of Google Earth (figure 2). As Johnson and Parker (2014) demonstrate, the visual anomalies of Google Earth are not glitches but the algorithms exposing themselves from behind a veil of anonymity. These digital artifacts are produced through architecture's attempt to present its outside (façade) to a class of *algorithmic observers* that have been tasked with sensing and making sense of the built environment. Algorithmic observers are entities that possess a type of sight despite a lack of eyeballs, rods, cones, and visual cortex. Instead they 'see' through the use of sensors to detect light, heat, motion, and color data, which are computationally

processed to produce 'images' of the physical world. To digitally reconstruct the built environment, algorithmic observers reduce images of architectural façades to geometric data that is superimposed, composited, stitched, and texture mapped to produce virtual three-dimensional environments.

Algorithmic observers are interesting precisely because they lack "humanlike or human-level perceptual and aesthetic capacities, but rather [demonstrate] something that is uncanny and interesting because it does not possess those things" (Bratton 2015). It is their lack of humanlike perception and cognition that destabilizes our relationship to the built environment and produces new modes, scales, and patterns of turbulence within systems designed to produce and maintain smooth flows. The (in)ability of algorithmic observers to digitally reconstruct architecture produces turbulent flows previously concealed from human perception. These flows illustrate a new excess that has yet to be folded into the purview of human—or even architectural—perception. This research is not necessarily interested in either absolutely smoothing or agitating these possible flows, but rather focuses on possible ways in which turbulence might be modeled and imaged by misusing a tool that is generally biased toward smooth or laminar flows.

METHODOLOGY: HACKING SIFTS

Algorithmic observers rely on algorithmic protocols of machine vision to deconstruct images into collections of unique Scale-Invariant Feature Transform (SIFT) features that can be identified, organized, and matched across dynamic image sets. Of particular importance within these computational protocols are the algorithms associated with SIFT descriptors. SIFTs, developed by David Lowe (1999), enable algorithmic observers to identify specific image features that are invariant to scaling, rotation, changes in illumination and 3D camera viewpoint (Lowe 1999; 2004). SIFT descriptors are extracted from each pixel of an input image and encoded with contextual information through processes that reduce images to a large number of highly distinctive features, facilitating the filtration of visual clutter/noise within the image, while providing a high probability of feature-matching and correlation across images (figure 3). SIFTs, with their strong matching capabilities and computational stability, are mobilized for the purposes of image retrieval, image stitching, machine vision, object recognition, gesture recognition, and the digital reconstruction of cities (McClendon 2012; Wu et al. 2013; Yang et al. 2011).

Due to the inherent ability of SIFTs to identify invariant geometric features within an image, an accurate 3D digital model with high-resolution renderings produces nearly identical SIFT data-fields to those of an actual image (figure 4). Additionally, the use of computer modeling and rendered outputs allows for the control of other image characteristics such as camera aperture size, focal length, white balance, global illumination, and lighting levels, along with rigorous control over viewpoint perspective, ambient noise, and visual clutter (trees, clouds, cars, people, garbage, etc.).

The use of 3D modeling and rendering allows for the production of an initial input dataset, where the camera viewpoint with respect to the object is the only dependent variable, with all other variables remaining constant (figure 5). By controlling all variables except for the camera's position to an object, the resulting dataset is defined primarily through an object's ability to produce data through its geometric composition, outside of variable lighting, camera, or environmental impacts. This methodology allows the object of observation to become transferable, as the control over image variables provides a framework for replicable and repeatable investigation with respect to any physical/virtual object.

Earlier research—presented in "This is Not a Glitch: Algorithms and Anomalies in Google Architecture" (Johnson and Parker 2014)—illustrates the internal protocols of SIFT algorithms and a methodology to extract n-dimensional pixel data from two-dimensional images. Whereas this previous research engaged SIFTs and their ability to reveal concealed vectors as a two-dimensional image problem, this paper extends this methodology to explore how these n-dimensional datasets can be mobilized towards the production of 3D form.

Voxels present a unique opportunity for working with turbulent systems, as they possess the ability to contain and respond to a multiplicity of datasets within a single volumetric pixel.

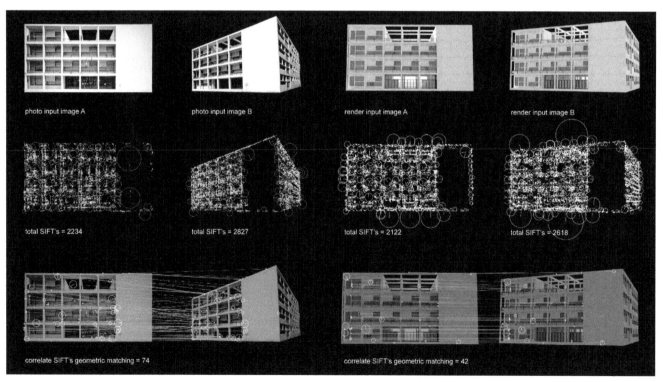

4   SIFTs in Photos vs. Renders: This diagram illustrates how high fidelity renderings produce similar SIFT data fields to those of an actual object.

Form-Making in SIFT Imaged Environments Parker, Taron

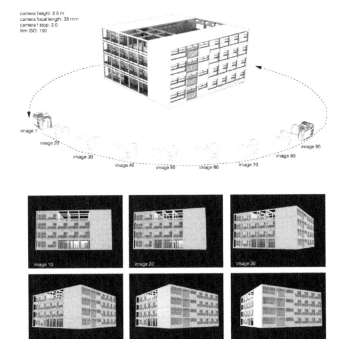

camera height: 2.5 m
camera focal length: 35 mm
camera f stop: 2.0
film ISO: 100

image 1
image 20
image 30
image 40
image 50
image 60
image 70
image 80
image 90

image 10
image 20
image 30
image 40
image 50
image 60
image 70
image 80
image 90

5

5   Virtual Camera Path and Outputs: By using virtual objects to simulate the phys-
     ical, image variables can be constrained allowing the objects relationship to the
     observer to be the only dynamic variable.

Due to this ability, voxels are often used within medical image reconstruction, terrain scanning, and computational fluid dynamics. Whereas particles and points are defined by their relationship to coordinate space and their position relative to other particles (thus making them good for swarm logics, etc.), voxels need not respond to their neighbors or community as they can be programmed to operate as independent unique entities. The ability to uniquely codify each voxel is valuable as it pertains to SIFT algorithms, because once a SIFT keypoint is identified and codified, it becomes an autonomous entity, for which SIFT algorithms search for correlate SIFTs regardless of their neighbors and surrounding image context. It is the autonomy of SIFT keypoints that produce the warped images previously discussed in "SIFT Materiality: Indeterminacy and Communication between the Physical and Virtual" (Parker 2014), where correlate SIFTs are superimposed and mapped onto each other, with the surrounding context resulting from degrees of algorithmic interpolation, or what Luciana Parisi refers to as 'soft thought' (2013). By mapping the n-dimensional vectors produced through processes of superimposition directly to voxels within

programmed gas-solver environments, turbulent systems are mobilized towards the production of dynamic form.

RESULTS AND REFLECTIONS: UNPACKING FLATTENED DIMENSIONS

Early investigations only looked at how voxel fields would deform when the only forces within the system were those internally produced through algorithmic observation. The resultant objects demonstrate the autonomous ability to produce turbulence without context (figures 6–9).

As images are abstracted to SIFTs and composited to produce smooth yet warped images, an excess of n-dimensional vectors are revealed through the production of seamless and 'accurate' composite images that have been algorithmically curated to represent a physical object. By mapping these n-dimensional vectors to a corresponding voxel field, voxels assume the attributes of a particular pixel as it transforms within the process of re-composition. By allowing these charged voxels to organize within a computationally turbulent gas-solver, voxels organize, not in relation to their neighbors or a specific coordinate system, but in relation to the dynamic data on a voxel-by-voxel basis. These experiments provide a technique to produce 3D form from 2D images, as each pixel of a warped image generates n-dimensional vectors that are mapped to unique voxels within the simulation (a warped image of 640 X 480 pixels produces vectors mapped to a two-dimensional voxel field of 640 X 480 voxels). Despite the initial field being two-dimensional, the voxels, free to interact within the smooth constraints of a three-dimensional computational environment, result in indeterminate n-dimensional form (figure 1). This is the structural logic of SIFT weather systems.

The workflow allows for images to be mined for their n-dimensional vectors, producing dynamic datasets that aggregate to form increasingly resolved, multi-dimensional 'images' of a particular object. Similar to the manner in which algorithmic observers sense an object in the built environment through a series of images or scans from multiple viewpoints over a specified duration, each new set of n-dimensional vectors produces unique magnitudes and directions that are introduced to their corresponding voxel on a per-frame basis. This process creates a turbulent system that negotiates data-intensities over time. By establishing a one-to-one relationship between datasets and frames, the simulation is capable of producing a distinct morphologic manifestation on a per-frame basis, each one a reflection of data excess and the turbulence produced as algorithmic observers attempt to process and composite a multitude of discrete images into a cohesive whole (figure 12).

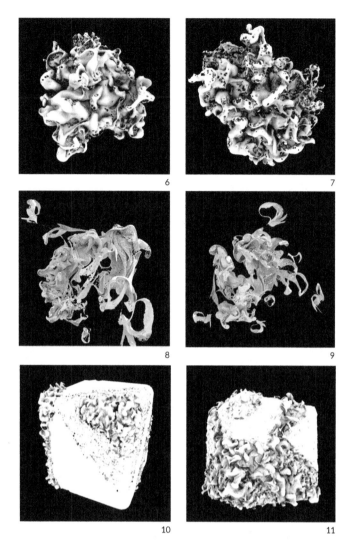

6

7

8

9

10

11

Turbulent Façades

From this investigation emerges an interesting discussion about the ability of the familiar to produce the speculative, and a symbiotic relationship between specific smoothing agents and their turbulent byproducts becomes a bit clearer. This set of investigations explores how the previously developed turbulent systems can exert a force back onto and into an existing object towards the production of the *uncanny*. By defamiliarizing the familiar, the uncanny has the ability to mobilize a dialogue around what else an object might be, and what other bodies of information may be concealed from human observation (figures 10 and 11).

Within this research, the process of destabilization relies on computational protocols to expand the intelligence of a mesh, allowing it to operate indeterminately within an environment. To accomplish this, high fidelity computational objects are converted to voxels through a process that reduces an existing poly-mesh into a number of spatial voxels that correspond to the location and number of pixels an object occupies within a digital image. The technical problem emerging from this conversion process is determining exactly how many vectors a 3D mesh is responsible for producing within a 2D image. As the visual noise and clutter are controlled within this workflow through the production of computation and renders, objects exist within the image outside of their context. Whereas an image may be 640 X 480 pixels, the object may only occupy a small number of these pixels, leaving the rest of the image devoid of information and thus incapable of producing n-dimensional vectors. The challenge is to then

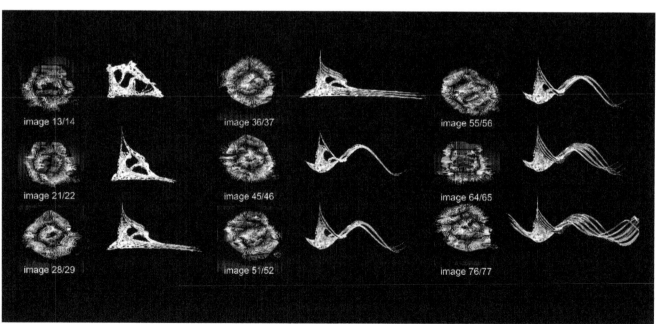

image 13/14

image 36/37

image 55/56

image 21/22

image 45/46

image 64/65

image 28/29

image 51/52

image 76/77

12

Form-Making in SIFT Imaged Environments Parker, Taron

SIFT geometry

13

6-9 Turbulent Object: produced as vector-flow-fields operate indeterminately across a set of voxels in accordance with the governing logics of a codified computational environment.

10-11 Primitive deformation (Cube A): the faces of a volumetric cube are charged relative to their correlate SIFTs in a 2D image plane. The charged surface exerts a force within a voxel system describing the geometry and the output is deformed relative to the resulting vector-flow-fields produced through algorithmic observation.

12 N-dimensional workflow for extracting and mobilizing the output vector-flow-fields produced through algorithmic observation. Extracted n-dimensional vectors deform voxel based geometry within a computational environment.

13 De-familiar façade: This diagram illustrates two input images warping to produce a composite data-rich territory and resulting n-dimensional vector field. This vector field is extracted from the 2D image and mapped to its 3D form where turbulent thresholds are identified and deformed within the predefined logics of the system.

identify the location and number of pixels an object occupies and convert the computational object to a corresponding number of voxels. To further compound this problem, an object may occupy a different number of pixels within different images, dependent on camera view-point perspective and orientation to the object. To account for these kinds of problems a process of *Taylorian Projection* (Cohen 2001) is utilized to convert a 3D object to its corresponding number of voxels relative to the number of pixels in an image, and to accurately remap the n-dimensional vectors onto their corresponding voxels.

As this process maps a specific vector to each voxel that describes an object, it is possible to deform an object in its entirety; however, this research has explored the potentials of only deforming areas that exhibit charges over a predetermined threshold. By constraining deformation to specified thresholds of intensities, a legibility becomes apparent within the destabilization of an object. This legibility allows for the production of a narrative that supports the uncanny. Through a process of curation that allows the indeterminate nature of the protocols to operate in accordance with predetermined thresholds, the output remains legible. However, the object exhibits a speculative deformation in the areas of high complexity in accordance with the perceptive abilities of algorithmic observers (figure 13). The resulting deformation illustrates the agitated relationship between architecture and its algorithmic observers, and distorts our perception of an object by mobilizing a concealed technologic turbulence.

CONCLUSION: NEW OBJECTIVES

Le Corbusier argues that old Paris subjected its citizens to chaotic, terrifying, and confusing conditions through its 'disturbing,' 'twisted,' 'mis-shapened,' and 'abnormal' patterns of urbanization. To rectify these problems, Le Corbusier proposed an architecture that sheltered its inhabitants from the chaos of the city, producing order by facilitating co-existence. It can be argued that for much of the 20th and the early part of the 21st century, cities have been designed to produce diagrams of smooth and homogeneous flows, with the increasing prevalence of distributed personal technologies allowing us to navigate the urban fabric no matter how twisted, hostile, and misshapen a city's form may be. However, the virtual layers that unevenly map onto the terrestrial constraints of the city produce turbulent forces that computational platforms tasked with sensing and making sense of the built environment aim to conceal behind a visual narrative of accuracy, cohesion, anticipation, and order. What is at stake with the geometries produced within this research is not that the forms are necessarily novel or unique, but that the methods through which they are produced are born from both the image of the world, as well as the algorithms that observe and produce those images. This restlessness demarcates a new kind of challenge for both architecture and the city, whereby a cultural appetite to deploy algorithms that produce a smooth and seamless image of the world comes hand in hand with the turbulent and disruptive autonomy of those very same algorithms.

Whether we accept it or not, algorithmic observation shapes our experiences and relationships to the city. And so looking toward Le Corbusier's *Urbanisme* through a different lens, we are presented with a variety of opportunities to re-frame the problem. Instead of eliminating boundaries, we are now tasked with identifying latent boundaries and querying them for the signals they might be sending. This demands that architecture address more than buildings—allowing Google Earth into our houses, our bedrooms, and our bodies and training them to produce new turbulent patterns—life as art in the eyes of a SIFT observer. In a less conventional sense, it could also mean that architecture treats SIFT networks as architectural sites themselves—exploring the way in which the algorithm can be inhabited, occupied and, instrumentalized for user-based, programmatic purposes. Algorithmic observation challenges architecture to engage a form of practice that no longer situates the physical act of making buildings as the primary role of the architect, but to take part in shaping and designing the technologies that increasingly redefine new ecologies of the city. In order to formally establish algorithmic observation as an architectural thing, it is necessary for architects to address it and the spaces produced as valuable, despite their lack of traditional materiality and familiar formal manifestations. It also demands that we come to understand the nature of architectural boundaries in the context of algorithmic observation as being both laminar and turbulent at the same time. This should not be seen as compromising Corbusier's laminar agenda, but rather as an expression of the ability for architectural flows and objects to simultaneously occupy multiple dimensions within contested fields of algorithmic observation.

This research is merely a first step towards investigating the agitated relationship that exists between architecture and its virtual processes, but it shows that the new forms of urban inhabitation challenge the smoothness of the city and call into question architecture's desire to accommodate heterogeneous forces towards the production of laminar flows. Instead, algorithmic observation challenges architecture to exploit the fluid and vectoral aspects of imaged buildings, to explore turbulence as an agent of potentially catastrophic change, and to investigate the technologies that are obsessed with squeezing new dimensions out of flattened environmental images.

REFERENCES

Bratton, Benjamin. 2015. "Machine Vision: Benjamin Bratton in Conversation with Mike Pepi and Marvin Jordan." *DIS Magazine*. Accessed August 14, 2015. http://dismagazine.com/discussion/73272/benjamin-bratton-machine-vision/

Cohen, Preston S. 2001. *Contested Symmetries and Other Predicaments in Architecture*. New York: Princeton Architectural Press.

Google. 2012. *The Next Dimension of Google Maps*. YouTube video, 51:18. Posted by Google, June 6, 2012. Accessed November 23, 2014 from https://www.youtube.com/watch?v=HMBJ2Hu0NLw

Handlykken, Asne K. 2011. "Digital Cities in the Making: Exploring Perceptions of Space, Agency of Actors and Heterotopia." *Ciber Legenda* 25 (2): 22–37.

Johnson, Jason S. and Matthew Parker. 2014. "This Is Not a Glitch: Algorithms and Anomalies in Google Architecture." In *ACADIA 14: Design Agency—Proceedings of the 34th Annual Conference of the Association for Computer Aided Design in Architecture*, edited by David Gerber, Alvin Huang, and Jose Sanchez. Los Angeles: ACADIA. 389–398.

Le Corbusier. 1971. *The City of To-morrow and its Planning*. Translated by Frederick Etchells. Cambridge, MA: MIT Press. Originally published as *Urbanisme* (Paris: Éditions Crès, 1925).

Lowe, David G. 1999. "Object Recognition from Local Scale-Invariant Features." In *Proceedings of the Seventh IEEE International Conference on Computer Vision* vol. 2. Kerkyra, Greece: ICCV. 1150–1157.

———. 2004. "Distinctive Image Features from Scale-Invariant Keypoints." *International Journal of Computer Vision* 60 (2): 91–110.

Parisi, Luciana. 2013. *Contagious Architecture Computation, Aesthetics, and Space*. Cambridge, MA: MIT Press.

Parker, Matthew. 2014. "SIFT Materiality: Indeterminacy and Communication between the Physical and the Virtual." In *What's the Matter: Materiality and Materialism at the Age of Computation*, edited by Maria Voyatzaki. Barcelona: ENHSA-EAAE. 313–326.

Wigley, Mark. 2007. "Towards Turbulence." *Volume* 10. Accessed April 19, 2016 from http://c-lab.columbia.edu/0135.html

Wu, Jian, Zhiming Cui, Victor S. Sheng, Pengpeng Zhao, Dongliang Su, and Shengrong Gong. 2013. "A Comparative Study of SIFT and its Variants." *Measurement Science Review* 13 (3): 122–131.

Yang, Donglei, Lili Liu, Feiwen Zhu, and Weihua Zhang. 2011. "A Parallel Analysis on Scale Invariant Feature Transform (SIFT) Algorithm." In *Proceedings of the 9th International Conference on Advanced Parallel Processing Technology*, edited by Olivier Temam, Pen-Chung Yew, and Binyu Zang. Shanghai: APPT. 98–111.

IMAGE CREDITS

Figure 1: Parker, 2014
Figure 2: Google Earth, accessed 2014
Figure 3–4, 6–11: Parker, 2015
Figures 5, 12–13: Parker, 2016

Matthew Parker completed his Master of Architecture from the University of Calgary's Faculty of Environmental Design, where he received honors recognition and the AIA Gold Medal. Currently he is completing a Post Professional Masters with his current research focusing on the ability of algorithmic observation to transform, mediate and re-animate architectures' image. Matthew is also a researcher with the Laboratory for Integrative Design (LID), and a studio designer and parametric consultant with Minus Architecture Studio and Synthetiques / Research + Design + Build.

Joshua M. Taron is an Associate Professor of architecture at the University of Calgary Faculty of Environmental Design where he also co-directs the Laboratory for Integrative Design. His current research focuses on designing for divertability as an alternative to conventional wholesale building demolition through the use of advanced fabrication techniques. Taron is also Principal of Synthetiques Research & Design, Inc, a consultancy that specializes in complex architectural form-making. He earned his undergraduate degree in architecture from the University of California, Berkeley and holds a Master of Architecture degree from the Southern California Institute of Architecture.

Load Responsive Angiogenesis Networks

Structural Growth Simulations of Discrete Members
using Variable Topology Spring Systems

Christoph Klemmt
University of Cincinnati
University of Applied Arts Vienna
Orproject

Klaus Bollinger
B+G Ingenieure Bollinger und
Grohmann GmbH
University of Applied Arts Vienna

1

ABSTRACT

Venation systems in leaves, which form their structural support, always connect back to one seed
point, the petiole of the leaf. In order to develop similar structural networks for architectural use
which connect to more seed points on the ground, an algorithm has been developed which can
develop from two or three seed points, inspired by angiogenesis, the process through which the
vascular system grows. This allows for the generation of structurally suitable topologies based
on discrete members, which can be evaluated using Finite Element Analysis and which can be
constructed from linear structural members without an additional interpretation of the results.

The networks have been developed as load bearing spring systems above the support points.
Different structures have been compared and tested using Finite Element Analysis. Compared
to traditional column and beam structures, the angiogenesis networks as well as the venation
networks are shown to perform well under load.

1   Angiogenesis Network.

INTRODUCTION

The venation systems in leaves fulfill both circulatory as well as structural functions (Roth-Nebelsick et al. 2001), and their generation has been shown to relate to physical stress (Laguna et al. 2008). The systems have been used by various designers to create formations for architecture and product design (Andraos 2015, Tamke et al. 2014, Seepersad 2014), often digitally simulated by algorithms similar to the one developed at the University of Calgary (Runions et al. 2005, Runions et al. 2007, Runions 2008). Following this algorithm, a network is grown from seed points towards a set of target points which need to be reached (Runions et al. 2005, Runions et al. 2007, Runions 2008). Digitally generated venation systems have been shown to perform well as architectural load-bearing structures (Klemmt 2014).

Venation systems always connect back to a single support, the petiole of the leaf, which makes their use as architectural load-bearing structures difficult. Structural beams and arches, which form connections between two different support points, cannot be generated using the venation algorithm.

Previous work attempted to solve this by growing networks from different support points towards the same target points. This led to separate structures, each supported above a single seed point, which were leaning against each other, connected by a set of the thinnest members along their common edges (Klemmt 2014).

In order to grow similar networks which can generate beam-like formations on top of two or three support points, an algorithm based on angiogenesis has been developed. Every target point in the network is connected to at least two support points and therefore lies on a connection between those supports (Figure 2).

While the growth following the venation algorithm starts from the seed point, the growth of the angiogenesis algorithm starts from an initial line which connects two seed points. This connection then splits up repeatedly in order to reach the set of target points which are to be supported (Figure 3).

The proposed methodology allows for the generation of structurally acting topologies. Unlike the ESO/BESO topology generation (Querin et al. 1998, Huang et al. 2007), the proposed system does not use a voxel grid but instead uses discrete members, which means that the outcomes can be evaluated using Finite Element Analysis (FEA), and the resulting networks can be constructed from linear structural members without an additional interpretation of the results.

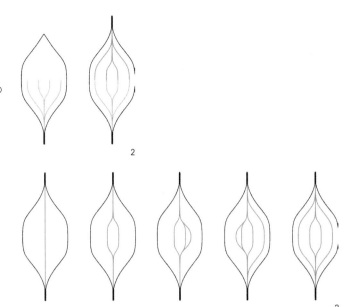

2   Left: Venation network connecting to one seed point.
Right: Angiogenesis network connecting to two seed points.

3   The network is developed through a division and refinement of one initial connection between the two seed points.

ANGIOGENESIS

Angiogenesis is the process by which new blood vessels grow from existing ones. The veins of leaf venation systems contain both xylem and phloem cells which transport water and sap towards and away from the leaf cells. By contrast, vascular systems are directional and the blood always flows in a defined direction through its arteries and veins. Therefore, a venation system has a single starting point, while a vascular system has a starting as well as an end point, the atria and ventricles of the heart (Birbrair et al. 2014).

A growth model based on angiogenesis can therefore be used to generate networks which connect two support points together. The vascular network attempts to reach proximity to every cell in the organism. This behavior is translated to the requirement of the structural system to reach proximity of every load point which is to be supported. Both the support points as well as the load points are given, and the algorithm generates a network between the support points through a splitting of connections, with the aim of supporting the load points.

As the load points are placed at a higher level than the support points, the network is pulled upwards from the support points. It is expected that the networks form shell-like formations with possible similarities to the rib systems below Gothic domes.

Two types of angiogenesis are known: sprouting angiogenesis, the formation of a new vein between two existing veins, and intussusceptive angiogenesis, the splitting of a vein into two parallel veins. Vasculogenesis, the initial embryonic formation of the first veins, is a different process (Risau and Flamme 1995). This paper explores simulations using intussusceptive angiogenesis.

The aim of the research has not been to realistically simulate the development of vascular systems as they develop in nature, but rather to explore the biological precedent of angiogenesis in the development of iteratively refining spring systems.

RELATED WORK
Related Work in Design and Engineering
In structural engineering, algorithms are used to develop efficient structural systems, as with the use of genetic algorithms (Byrne et al. 2011, Clune et al. 2012).

Tools are available for the generation of topology, such as the voxel-based SIMP method (Bendsoe and Sigmund 2013) and the ESO and BESO algorithms (Querin et al. 1998, Huang et al. 2007). However, as those methods are voxel-based, the outcomes need to be interpreted in order to be constructed as systems with discreet structural members (Huang and Xie 2009).

In parametric architectural design, structural performance is often used as a driver to influence the parametric or algorithmic models (Makris et al. 2013, Turrin et al. 2011).

Research into a growth or additive development of structural systems has been explored using physical components (Dierichs & Menges 2012a, Dierichs and Menges 2012b, Zhao et al. 2015) or algorithmic simulations (Andraos 2015, Sugihara 2015, Richards et al. 2012, Richards and Amos 2014, Kicinger et al. 2005a, Kicinger et al. 2005b).

Related Work in Biology and Medicine
In developmental biology, the growth of organisms is studied on a cellular level (Wolpert et al. 2011). The growth of angiogenesis and the factors influencing it are of special interest in the field of cancer research, as an inhibition of angiogenesis can stop the growth of a tumor. Simulations of angiogenesis in this field are used as a research as well as visualization tool (Shirinifard et al. 2009, Szczerba & Székely 2002).

ALGORITHM FOR TWO SUPPORT POINTS
In the computational model, the vein network is represented by nodes which are connected by the vein segments. Similar to the previous experiments with venation systems (Klemmt 2014) that also aim to support loads, support points are placed at the bottom and the load points are placed at a higher level above.

In relation to the biological precedent of a vascular system and its angiogenesis, the blood flow leads from one of the two support points through the veins to the other.

Set-up
The inputs for the simulation are two support points, S_A and S_B, and a set of load points. The load points are to be supported by the vein structure on top of the support points.

In an initial step, the two support points are connected in a straight line by a set of evenly spaced nodes with spring connections (the veins) between them.

In relation to the biologic precedent, this can be seen as establishing an initial vein from S_A to S_B, and it serves to define a flow axis along which, in the vascular vein model, the liquid would flow from S_A to S_B. This flow axis can be defined at every node in the network. It can be calculated as a unitized vector which points from the neighboring node in the direction of S_A (upstream) towards the neighboring node in the direction of S_B (downstream). This flow axis is used in the simulation to control the splitting of the veins in a sideways direction.

After a vein has split, the node at the junction will have two or more neighbors in the direction of either S_A or S_B. For the purposes of calculation, the flow axis is defined as the unitized vector which points from the midpoint of all neighboring nodes in the direction of S_A towards the midpoint of all neighboring nodes in the direction of S_B.

Every load point establishes a spring connection towards the node closest to it. Those load point springs have a smaller strength than the vein springs. There is a fixed amount of load point springs, one per load point, while the amount of vein springs in the network grows when the veins split.

Iterative Calculations
During each iteration, the springs of the network are relaxed. All springs are calculated as having a rest length of 0. The new positions of all nodes are updated, and the flow axis and the strength of forces acting on the nodes is recalculated.

The node with the strongest forces acting on it will split into two. The two new nodes are moved apart from each other sideways, orthogonal to the flow axis.

The new nodes are re-establishing spring connections to the neighbors of the previous node, separately for the upstream and downstream neighbors. If there is only one neighbor in a direction before splitting, both of the new nodes will connect to it. If

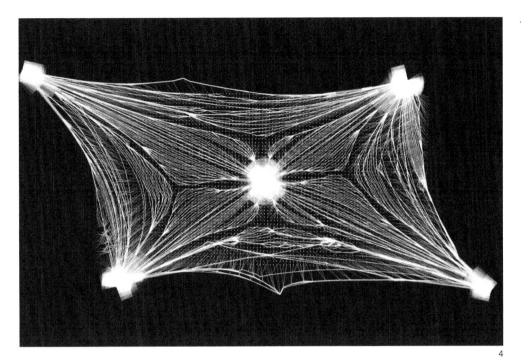

4 Set of eight adjacent two-support-point systems, plan. Each two-support-point system starts as a straight line in plan. The distribution of the target springs (in blue) result in the fanning and the deformations.

4

there is more than one neighbor in a direction, the neighbors are distributed evenly between the two new nodes.

Also, the load points which had been connected to the splitting node become evenly distributed between the two new nodes. Therefore, if a node only has one load point connection, it is not allowed to split, and the node with the next strongest forces acting on it will split instead (Figure 4).

ALGORITHM FOR THREE SUPPORT POINTS
In order to generate a three-point support system, the algorithm has been adjusted to use three support points, S_A, S_B and S_C. Initially, those points are connected to their midpoint (resulting in a Y-formation) and those connections are evenly subdivided by nodes.

Flow Directionality
As in the algorithm with two support points, it is necessary to identify a main flow axis at every node, so that a splitting of the node can be programmed to happen sideways to this flow axis. In the version with two support points, the flow axis can be identified at every node from one support point towards the other. This is not possible in the version with three support points, but it is still possible to identify which support point is closest to any node. This can be used to trace a path from a node to each of the support points. Every neighbor of a node is marked towards which support point it is leading, and the acting forces are traced towards each of the three supports. This can be used to describe a main flow axis at the node along which the forces are acting.

Angiogenesis
The splitting of the cells happens as in the algorithm with two support points, however, the reassignment of the neighbors becomes more disparate, especially in order to avoid diagonally crossing connections. The neighbors are reassigned to the two new nodes while at the same time being marked as leading towards one or two of the support points. The neighbors are therefore distributed between the new nodes along the flow axis, while ensuring that each node still has at least one neighbor leading towards each support point.

FINITE ELEMENT ANALYSIS
Analysis Setup
The generated networks have been tested using Finite Element Analysis (FEA) and have been compared to a column and beam structure and to a venation structure as developed previously (Klemmt 2014).

The support points have been placed at the corners of a right triangle of 10 m width and 15 m length. The load points have been placed in the same triangle 5 m above the supports, in a square grid of 250 mm. A slab has been simulated by connecting the load points with structural members, which have been tested in three different sizes. Steel has been used as the material of all members. In order to compare the performances, all member sizes have been scaled so that the overall mass of each model was 10 t.

A load of 1 kN/m² has been applied at the load points, as separate load cases in the X, the Y and the Z direction.

Networks

The following networks have been tested:

- A column and beam structure, with vertical columns above the three support points and three horizontal beams connecting the columns
- A reticulate venation system as described by Klemmt 2014 (Figure 5)
- A two-support-point angiogenesis network, consisting of three systems which each form a connection between two of the three support points (Figure 6)
- A three-support-point angiogenesis network with one load point spring per load point as structural member
- A three-support-point angiogenesis network with every load point spring of the simulation as structural member (Figure 7)

In the venation network, every member has a value according to the amount of load points it supports, which is used as the member size (Klemmt 2014). In the angiogenesis models, the force of every spring connection equals its length, which defines the member size.

DEFLECTION RESULTS

Low Slab Strength

Slab member length:	629.14 m
Member diameter:	1 mm
Slab mass:	4.938 kg

Table 1. Deflection with low slab strength, in mm.

	Load Direction		
	X	Y	Z
Column Beam	238.88	167.88	2147483
Venation	91.15	83.76	76.89
Angiogenesis 2 Supports	6751.71	781.30	211.04
Angiogenesis 3 Supports 1 member per load point	1613.02	374.42	875.94
Angiogenesis 3 Supports all members	65.88	53.79	11.03

Medium Slab Strength

Slab member length:	629.14 m
Member diameter:	10 mm
Slab mass:	493.877 kg

Table 2. Deflection with medium slab strength, in mm.

	Load Direction		
	X	Y	Z
Column Beam	168.19	136.28	6008.83
Venation	72.98	66.14	43.74
Angiogenesis 2 Supports	207.11	111.31	58.17
Angiogenesis 3 Supports 1 member per load point	109.70	60.21	20.73
Angiogenesis 3 Supports all members	37.03	27.41	6.17

High Slab Strength

Slab member length:	629.14 m
Member diameter:	100 mm
Slab mass:	49387.65 kg

Table 3. Deflection with high slab strength, in mm.

	Load Direction		
	X	Y	Z
Column Beam	79.68	67.63	49.07
Venation	21.84	11.98	5.20
Angiogenesis 2 Supports	16.68	12.60	5.92
Angiogenesis 3 Supports 1 member per load point	35.30	30.65	3.38
Angiogenesis 3 Supports all members	28.90	19.46	1.01

Table 4. Topological comparison of the structural networks.

	Member length (m)	Diameter small (mm)	Diameter large (mm)
Column Beam	74.84	130.46	130.46
Venation	625.21	3.98	275.69
Angiogenesis 2 Supports	1689.55	20.65	81.86
Angiogenesis 3 Supports 1 member per load point	1311.12	17.33	117.82
Angiogenesis 3 Supports all members	13817.4	8.72	59.29

EVALUATION

Amongst the tested structures, the column and slab structure performs least well, especially the central areas of the slab which are furthest away from the beams and deform the most.

Similarly, the central slab area of the two-point angiogenesis structure deforms significantly as it is relatively far away from any supporting beam members. However, this deficiency can be made up for by a stronger slab. This model then also works well

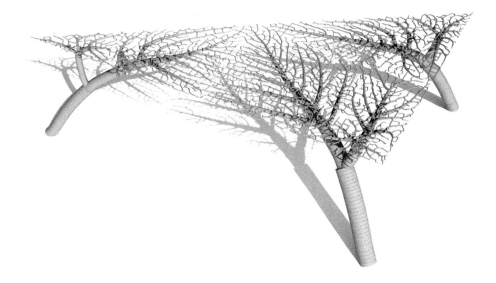

5 Venation System.

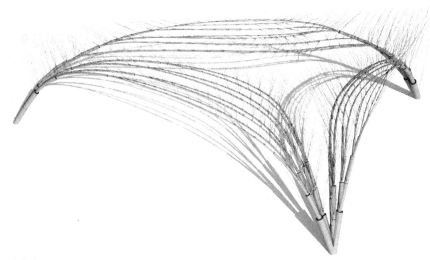

6 Two-Support-Point Angiogenesis System.

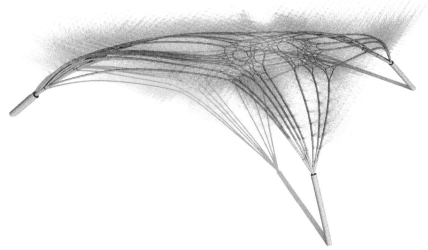

7 Three-Support-Point Angiogenesis System.

for the lateral load cases. The arches which have been generated between the support points appear to perform well in taking the lateral loads.

The three-support-point angiogenesis structures perform very well, especially the one with all load point springs used as structural members. The dense load point springs then form a space frame geometry below the slab. However, this network consequently has a large amount of individual members which may make it less feasible for construction.

As the algorithm used to develop the angiogenesis networks has a structural logic, it was expected that those systems would act better as load-bearing networks than the venation structure. However, the venation structure performs very well in comparison. Also the venation structure has a shorter cumulative length of the members, which makes it more economically efficient to construct physically (Figure 8).

CONCLUSIONS AND FUTURE WORK

By using the process of angiogenesis as a precedent, an algorithm has been developed that can successfully generate branching networks that develop from two or three seed points.

The algorithm has been used to generate load-bearing networks which have been tested using Finite Element Analysis. The generated structures were shown to perform very well. However, structures generated with the venation algorithm (Klemmt 2014) perform similarly well, even though the algorithm to generate those does not have a structural logic.

The proposed methodology is a first step towards the algorithmic generation of structural topology based on discreet members. Currently, the simulations generate shell-like formations, which should be generalized to various other topologies.

The algorithm which has been developed uses springs to simulate forces. However, this does not reflect the bending moments which occur in structural members. Future developments will provide a more accurate simulation by taking those forces into account.

ACKNOWLEDGMENTS

The paper was written by Christoph Klemmt as part of a doctorate under the supervision of Prof. Dr. Klaus Bollinger at the University of Applied Arts Vienna.

REFERENCES

Andraos, Sebastian. 2015. "Generating Adaptable Structural Systems Using Natural Growth Algorithms." Master's thesis, ENSA Paris Malaquais, http://andraos.co.uk/GeneratinganAdaptableStructuralSystem_LR.pdf

Birbrair, Alexander, Tan Zhang, Zhong-Min Wang, Maria Laura Messi, John D. Olson, Akiva Mintz, and Osvaldo Delbono. 2014. "Type-2 Pericytes Participate in Normal and Tumoral Angiogenesis." *American Journal of Physiology - Cell Physiology* 307 (1): C25–38. doi:10.1152/ajpcell.00084.2014.

Bendsøe, Martin P. and Ole Sigmund. 2013. *Topology Optimization: Theory, Methods, and Applications*. Berlin: Springer Science & Business Media.

Byrne, Jonathan, Michael Fenton, Erik Hemberg, James McDermott, Michael O'Neill, Elizabeth Shotton, and Ciaran Nally. 2011. "Combining Structural Analysis and Multi-Objective Criteria for Evolutionary Architectural Design." In *Proceedings of the 2011 International Conference on Applications of Evolutionary Computation - Volume Part II*, edited by Cecilia Di Chio et al. Berlin, Heidelberg: Springer-Verlag. 204–213

Clune, Rory, Jerome J. Connor, John A. Ochsendorf, and Denis Kelliher. 2012. "An Object-Oriented Architecture for Extensible Structural Design Software." *Computers & Structures* 100–101 (June): 1–17. doi:10.1016/j.compstruc.2012.02.002.

Dierichs, Karola, and Achim Menges. 2012. "Functionally Graded Aggregate Structures: Digital Additive Manufacturing with Designed Granulates." In *Proceedings of the 32nd Annual Conference of the Association for Computer Aided Design in Architecture (ACADIA)*, edited by M Cabrinha, J Johnson, and K Steinfeld. San Francisco: ACADIA. 295–304.

Dierichs, Karola, and Achim Menges. 2012. "Aggregate Architectures, Observing and Designing with Changeable Material Systems in Architecture." In *Change, Architecture, Education, Practices – Proceedings of the International ACSA Conference*, edited by X Costa and M Thorne. Barcelona: ACSA. 463–468.

Huang, Xiaodong, and Yi Min Xie. 2009. "Bi-Directional Evolutionary Topology Optimization of Continuum Structures with One or Multiple Materials." *Computational Mechanics* 43 (3): 393–401.

Huang, Xiaodong, Yi Min Xie, and Mark C. Burry. 2007. "Advantages of Bi-Directional Evolutionary Structural Optimization (BESO) Over Evolutionary Structural Optimization (ESO)." *Advances in Structural Engineering* 10 (6): 727–737.

Kicinger, Rafal, Tomasz Arciszewski, and Kenneth De Jong. 2005. "Evolutionary Computation and Structural Design: A Survey of the State-of-the-Art." *Computers & Structures* 83 (23–24): 1943–1978.

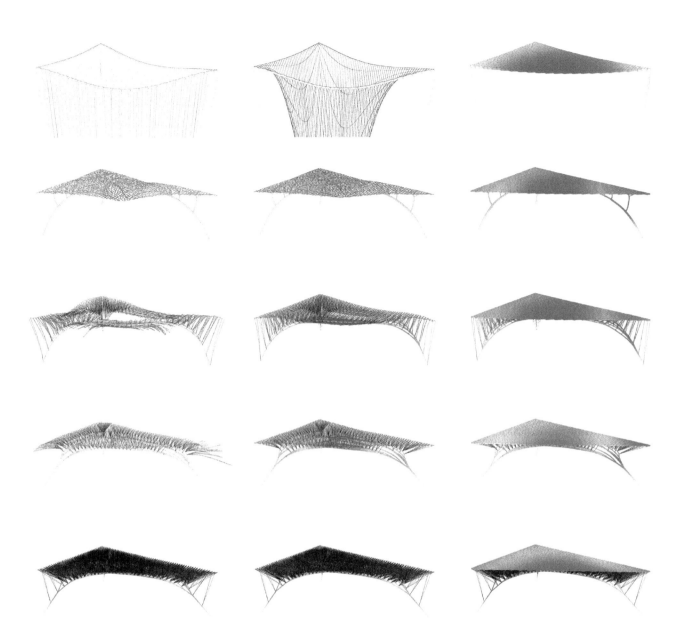

8   Deflection under loading in the Z-direction. Columns from left to right show low, medium and high slab strength. Rows show from top to bottom: Column and Beam structure, Venation system, Angiogenesis system with two supports, Angiogenesis system with three supports with one connecting member per load point, Angiogenesis system with three supports with all members connecting to the load points.

Kicinger, Rafal, Tomasz Arciszewski, and Kenneth De Jong. 2005. "Parameterized Versus Generative Representations in Structural Design: An Empirical Comparison." In *Proceedings of the 7th Annual Conference on Genetic and Evolutionary Computation (GECCO '05)*, edited by HG Beyer. New York: ACM. 2007–2014.

Klemmt, Christoph. 2014. "Compression Based Growth Modelling." In *Proceedings of the 34th Annual Conference of the Association for Computer Aided Design in Architecture (ACADIA)*, edited by David Gerber, Alvin Huang, and Jose Sanchez. Los Angeles: ACADIA. 565–572.

Makris, Michael, David Gerber, Anders Carlson, and Doug Noble. 2013. "Informing Design through Parametric Integrated Structural Simulation: Iterative Structural Feedback For Design Decision Support of Complex Trusses." In *eCAADe 2013: Computation and Performance–Proceedings of the 31st International Conference on Education and research in Computer Aided Architectural Design in Europe*. Delft, The Netherlands: eCAADe.

Querin, Osvaldo M., Grant P. Steven, and Yi Min Xie. 1998. "Evolutionary Structural Optimisation (ESO) Using a Bidirectional Algorithm." *Engineering Computations* 15 (8): 1031–1048.

Richards, Daniel, Nick Dunn, and Martyn Amos. 2012. "An Evo-Devo Approach to Architectural Design." In *Proceedings of the 14th Annual Conference on Genetic and Evolutionary Computation*. Philadelphia: ACM. 569–576.

Richards, Daniel, and Martyn Amos. 2014. Evolving "Morphologies with CPPN-NEAT and a Dynamic Substrate." In *ALIFE 14: The Fourteenth Conference on the Synthesis and Simulation of Living Systems*, edited by Hiroki Sayama, John Rieffel, Sebastian Risi, René Doursat, and Hod Lipson. New York: ALIFE. 255–262.

Risau, Werner, and Ingo Flamme. 1995. "Vasculogenesis." *Annual Review of Cell and Developmental Biology* 11 (1): 73–91.

Roth-Nebelsick, Anita, Dieter Uhl, Volker Mosbugger, and Hans Kerp. 2001. "Evolution and Function of Leaf Venation Architecture: A Review." *Annals of Botany* 87 (5):553–566.

Runions, Adam, Brendan Lane, and Przemyslaw Prusinkiewicz. 2007. "Modeling Trees With a Space Colonization Algorithm." In *Proceedings of the Eurographics Workshop on Natural Phenomena*. Aire-la-Villa, Switzerland: Eurographics. 63–70.

Runions, Adam, Martin Fuhrer, Brendan Lane, Pavol Federl, Anne-Gaëlle Rolland-Lagan, and Przemyslaw Prusinkiewicz. 2005. "Modeling and Visualization of Leaf Venation Patterns." *ACM Transactions on Graphics* 24 (3): 702–711.

Runions, Adam. 2008. Modeling Biological Patterns Using the Space Colonization Algorithm. M.Sc. Thesis, University of Calgary.

Seepersad, Carolyn C. 2014. "Challenges and Opportunities in Design for Additive Manufacturing." *3D Printing and Additive Manufacturing* 1 (1): 10–13.

Shirinifard, Abbas, J. Scott Gens, Benjamin L. Zaitlen, Nikodem J. Popławski, Maceij Swat, and James A. Glazier. 2009. "3D Multi-Cell Simulation of Tumor Growth and Angiogenesis." PLoS ONE 4 (10): e7190.

Sugihara, Satoru. 2015. "Guidelines for the Agent Based Design Methodology for Structural Integration: Inanimate Material Physics Simulation and Animate Generative Rules." In *Paradigms in Computing: Making, Machines, and Models for Design Agency in Architecture*, edited by David Jason Gerber and Mariana Ibanez. eVolo Press. 92–15.

Szczerba, Dominik, and Gábor Székely. 2002. "Macroscopic Modeling of Vascular Systems". In *Proceedings of the 5th International Conference on Medical Image Computing and Computer-Assisted Intervention-Part II (MICCAI '02)*, edited by Takeyoshi Dohi and Ron Kikins. London, UK: Springer-Verlag. 284–292.

Tamke, Martin, Paul Nicholas, and Jacob Riiber. 2014. "The Agency of Event: Event based Simulation for Architectural Design." In *ACADIA 2014 Design Agency: Proceedings of the 34th Annual Conference of the Association for Computer Aided Design in Architecture*, edited by David Gerber, Alvin Huang, and Jose Sanchez. Los Angeles: ACADIA. 63–74.

Turrin, Michela, Peter von Buelow, and Rudi Stouffs. 2011. "Design Explorations of Performance Driven Geometry in Architectural Design Using Parametric Modeling and Genetic Algorithms." *Advanced Engineering Informatics* 25 (4): 656–675.

Wolpert, Lewis, Cheryll Tickle, Thomas Jessell, Peter Lawrence, Elliot Meyerowitz, Elizabeth Robertson, and Jim Smith. 2011. *Principles of Development*. Oxford, UK: Oxford University Press.

Zhao, Yuchen, Kevin Liu, Matthew Zheng, Jonathan Barés, Karola Dierichs, Achim Menges, and Robert P. Behringer. 2015. "Packings of 3D Stars: Stability and Structure." *arXiv*:1511.06026

IMAGE CREDITS

Figures 1–8: Christoph Klemmt, 2016.

Christoph Klemmt, AA dipl, graduated from the Architectural Association in London in 2004. He is Assistant Professor at the University of Cincinnati and a doctoral candidate at the University of Applied Arts Vienna.

He has worked amongst others for Zaha Hadid Architects. He has lectured and given workshops at the Architectural Association, Nottingham University, the University of Wuppertal, AA Visiting School, Tsinghua University and Tongji University.

In 2008 he co-founded Orproject, an architect's office specializing in advanced geometries with an ecologic agenda. Orproject has exhibited at the Palais De Tokyo in Paris, the China National Museum in Beijing and the Venice Biennale. The work of Orproject has been featured world-wide in magazines and books and the practice has won several international Awards.

www.orproject.com

Klaus Bollinger, o.Univ.-Prof. Dipl.Ing. Dr.techn., has studied Civil Engineering at the Technical University Darmstadt and taught at Dortmund University.

Since 1994 he has been assigned Professor for Structural Engineering at the School of Architecture/University of Applied Arts at Vienna and since 2000 guest professor at the Städelschule in Frankfurt.

In 1983 Klaus Bollinger and Manfred Grohmann established the practice Bollinger + Grohmann, now located in Frankfurt am Main, Vienna, Paris, Oslo and Melbourne with around 100 employees. The office provides a complete range of structural design services for clients and projects worldwide. For years they have been collaborating successfully with numerous internationally recognized architects and strive to always provide the best solution through their creativity and technical excellence. Their scope of work includes building structures, facade design and building performance for commercial, retail and exhibition facilities as well as classic civil engineering structures such as bridges, roofs and towers.

www.bollinger-grohmann.de

Machine Learning Integration for Adaptive Building Envelopes

An Experimental Framework for Intelligent Adaptive Control

Shane Ida Smith, Ph.D.
Chris Lasch
University of Arizona

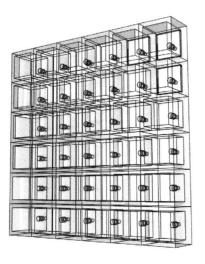 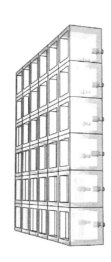 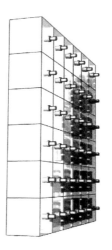 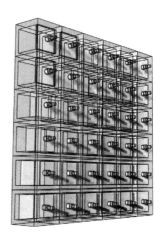

1

ABSTRACT

This paper describes the development of an Intelligent Adaptive Control (IAC) framework that uses machine learning to integrate responsive passive conditioning at the envelope into a building's comprehensive conventional environmental control system. Initial results show that by leveraging adaptive computational control to orchestrate the building's mechanical and passive systems together, there exists a demonstrably greater potential to maximize energy efficiency than can be gained by focusing on either system individually, while the addition of more passive conditioning strategies significantly increases human comfort, health and wellness building-wide.

Implicitly, this project suggests that, given the development and ever increasing adoption of building automation systems, a significant new site for computational design in architecture is expanding within the post-occupancy operation of a building, in contrast to architects' traditional focus on the building's initial design. Through the development of an experimental framework that includes physical material testing linked to computational simulation, this project begins to describe a set of tools and procedures by which architects might better conceptualize, visualize, and experiment with the design of adaptive building envelopes. This process allows designers to ultimately engage in the opportunities presented by active systems that govern the daily interactions between a building, its inhabitants, and their environment long after construction is completed. Adaptive material assemblies at the envelope are given special attention since it is here that a building's performance and urban expression are most closely intertwined.

1 Responsive facade modules with distributed control actuation points.

INTRODUCTION

Parallel advances are occurring in the fields of dynamic building facades and building automation control systems, exposing an increasingly complex terrain of dynamic systems' theory between exterior and interior built environments. Global trends in computational optimization strategies for automated control systems include the addition of intelligent control schemes, such as adaptive neuro-fuzzy inference systems, and optimization algorithms, such as multi-objective genetic algorithms, simulated annealing, meta-analysis, and others (Shaikh et al. 2014). In addition, efficiencies of conventional HVAC controllers are greatly improving, with emerging studies of applied reinforcement learning techniques indicating 4%–11% energy conservation over conventional control for heat-pumps (Ruelens et al. 2015). At the same time, emphasis on adaptive building envelope performance in response to dynamic environments is gaining heightened interest (Erickson 2013; Kolarevic and Parlac 2015; Zamella and Faraguna 2014). The ever-expanding portfolio of dynamic facade technologies exposes great promise to reduce a building's reliance on fossil-fuel based mechanical air conditioning in favor of natural, passive mechanisms that consume significantly less energy and simultaneously improve occupant well-being.[1] While each of these fields is receiving significant interest, there is not yet an explicit effort to link the two areas together for reciprocal benefits between proactive automated control systems and responsive envelope actuation functions.[2]

It is not possible to design in advance a system with a fixed control policy capable of anticipating dynamic outdoor and indoor conditions while also capitalizing on the qualitative and quantitative benefits that are possible in the synergistic inter-actions of these different socio-environmental control systems. In order to maximize the energy efficiency potential of these technologies, in addition to the qualitative potential for occupant experience and wellbeing, a building's environmental control operations must be considered holistically within an intelligent and adaptive framework. Such a framework shall be capable of orchestrating all of the building's systems in concert and adapt to simultaneous changes in internal and external conditions. The Intelligent Adaptive Control (IAC) architecture that we are developing is able to synthesize and adapt an integrated suite of control policies to coordinate building-wide active and passive environmental conditioning systems. IAC learns over time from sensors and history of control actions made during its operation. IAC policies constantly evolve so that its response becomes more finely tuned to the idiosyncrasies of each building's particular environmental landscape.

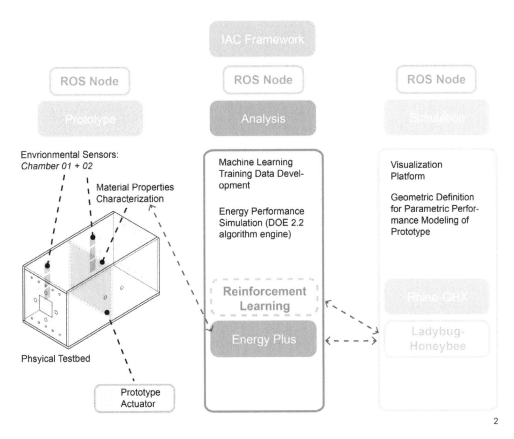

2   Intelligent Adaptive Control (IAC) experimental framework for building envelope integration.

2

In this project, our collaborators in Information Science and Electrical and Computer Engineering have identified two relevant control frameworks: Model Predictive Control and an area of machine learning known as Reinforcement Learning. Model Predictive Control (MPC) has become the dominant popular approach to HVAC control (Morari and Lee 1999; Maciejowski 2001; Ernst et al. 2009). Because MPC incorporates an accurate model of its task environment, it can anticipate future events and adjust accordingly based on decision point or fixed-horizon algorithms. MPC controllers require accurate knowledge of the operating environment conditions and become ineffective in unknown and changing operating environments. Reinforcement Learning (RL) is an area of machine learning concerned with how software agents learn to perform a series of sequential actions within an environment in order to maximize some notion of a long-term reward (Sutton and Barto 1998). A reinforcement learner does not rely on an a-priori model of its operating environment like an MPC does; it learns its optimal policy from its history of interaction with the environment. Reinforcement learning has been proposed as one approach to regulate controls within an environment as dynamic and complex as a building interior (Dalamagkidis et al. 2007).

For the conception of linking an intelligent automation system with the building envelope functions, an adaptive controller synthesis paradigm is preferred because the task environment dynamics are more uncertain. For this particular integration, our team has established a direction towards a hybrid approach to the computational control system, blending the benefits of RL with those of MPC (Peng and Morrison 2016). Comprehensively, the IAC framework engages concurrent development of physical dynamic envelope prototypes, simulation of digital design concepts, and analysis of building energy performance.

METHODS

Our experimental framework is an ecosystem consisting of a physical testing apparatus linked to both a digital simulation and analysis environment [Fig. 2]. A range of adaptive facade material assemblies can be inserted within the physical environmental test chamber. Digital configurations of these assemblies are simultaneously developed within a simulation environment for design purposes and in order to apply our experiments to the building scale for energy performance analysis. The bridge between these three environments is the IAC computational control framework. The IAC framework is an autonomous adaptive control architecture based on an adaptive machine-learning methodology. The IAC regulates the electronic controls within the physical testbed as well within the digital simulation. Over time, the data generated within these two experimental arenas train the IAC's control algorithms toward adaptive performance improvements.

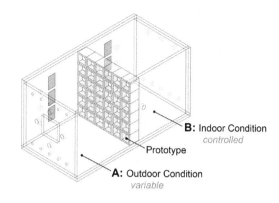

B: Indoor Condition *controlled*
Prototype
A: Outdoor Condition *variable*

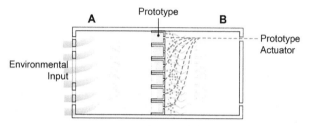

A   Prototype   B

Environmental Input

Prototype Actuator

3

4

Physical Testbed and Prototypes

The physical environmental test bed is an acrylic chamber designed as a modular kit of parts for testing a range of materials, control mechanisms and data processing models. The chamber is divided into two equal volumes separated by a slot for a removable prototype. Different active facade assembly prototypes can be inserted and tested between chamber A, representing exposure to an external environment, and chamber B, which represents a controlled internal environment [Fig. 3]. This test chamber permits experimental control of environmental conditions (humidity, temperature, light, heat flow) on each side of the testing facade and the monitoring of the response and adaptability of the apparatus to variations in conditions. The testbed is modular by design to enable experimental evaluation of the adequacy of various categories of adaptable materials and data-driven adaptive control policies within the IAC.

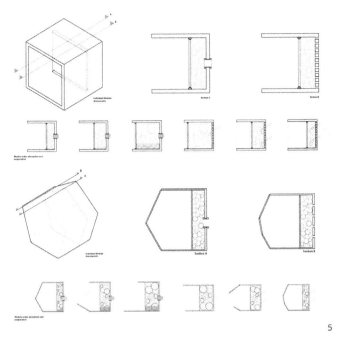

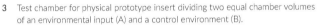

5

6

3   Test chamber for physical prototype insert dividing two equal chamber volumes of an environmental input (A) and a control environment (B).

4   Physical testbed construction with initial baseline dynamic glazing film and halogen dimmable lamp array input.

5   Digital design studies of dynamic responsive passive envelope conditioning hydrogel modules for evaporative cooling, heat capacitance, and natural daylighting functions.

6   Digital design studies of dynamic responsive building envelope modules in various pattern configurations.

Multi-sensory device arrays (thermocouples, photodiode, humidity sensors, infrared camera, etc.) embedded throughout the interior chamber, the facade itself, and the external space produce large-scale data flows used to generate responsive behaviors through adaptive learning. Initial physical prototype baseline studies are being prepared with electroactive photochromic dynamic glazing film technology, which responds to photometric measures in graduated increments of opacity and transparency based on a dimmable halogen lamp array input and photosensor data collection [Fig. 4]. The anticipated result of the combination of advanced facade materials with adaptive control is an autonomously responsive envelope that can maintain internal environmental conditions with appropriate performance levels compared to conventional methods. Specific targets in next-generation building energy management systems indicate the merging of sensor data and predictive statistical models to allow for more proactive modulation as signals are changing (Zavala et al. 2011). Future predictive modeling may also be linked with online sources provisioning communication from urban microclimate data from external sensory networks and utility providers (Pang, Hong, and Piette 2013).

Digital Simulation Environment

The work within the design development process of machine learning integration with adaptive building envelope and reciprocal building energy performance is conducted in the Rhino 3D – Grasshopper platform with Ladybug-Honeybee plug-ins and EnergyPlus simulations. In the current work, Python scripts access reinforcement learning algorithms that, along with weather data input and energy simulation output through base building analyses, inform dynamic changes in building envelope properties. Dynamic envelope design concepts developed with the parametric visualization tools [Figs. 5 and 6] can be correlated to dynamic properties for analysis engine input.

The framework developed for this project serves as a design process tool, in addition to informing potential building envelope technologies. There are three primary facets to the simulation process: a) defining the building envelope system and the influence of external environmental stimuli, b) determining the interior building environmental performance through dynamic envelope properties, and c) defining the learning algorithm to actuate change in properties or functionality of the building envelope system. Current building energy simulation models have two drawbacks in these areas—limitation on dynamic envelope analysis and limitation on reinforcement learning algorithm integration (Magoules and Zhao 2016; Sanyal et al. 2014).

The simulation process is more complex than current standards for energy performance models because real-time building performance results are continuously analyzed through algorithmic comparison with concurrent external stimuli to actuate

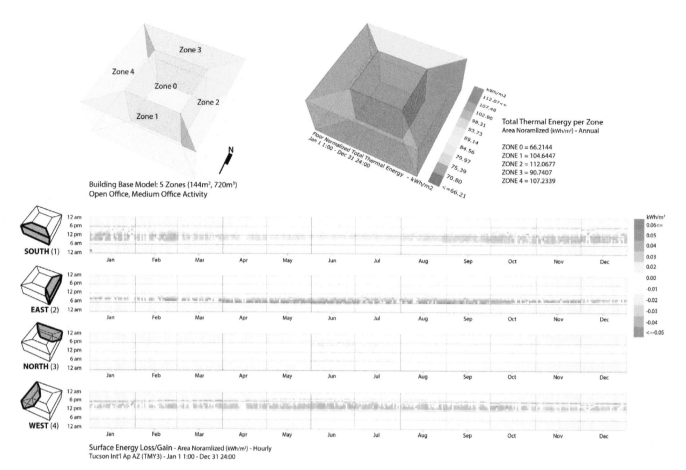

7 Base building energy analysis model establishing open office floor plan with four perimeter zones and specific analysis data for total thermal energy per zone (upper right) and surface energy losses and gains through envelope system (graphs) for each facade in cardinal orientations at hourly resolution for an annual timeframe.

change in the building envelope properties. The process is dynamic rather than static, and intends for adaptability of an envelope system beyond a two-state control process. Our current analysis framework includes a baseline building model with selective data processing for the surface energy losses and gains at the building envelope for each cardinal orientation [Fig. 7]. The data provides the MPC learning sets, which are utilized for initial building envelope property changes in response to the algorithm actuation. Further development is required to model the dynamic behaviors of envelope response stimulated through the IAC with the EnergyPlus interface for predictive environmental performance results.

Computational Control Framework

The multifaceted nature or our testing environment, containing physical as well as virtual components, supports our hybrid approach to the development of the IAC framework. The project methodology allows for two parallel sets of learning data to be developed - one in the test chamber with physical prototypes and sensors, and another in the simulation environment with analysis tools. Our project is also developing a compromise

between planning and learning, where planning is represented by the framework of an MPC and learning is represented by a model-free reinforcement learning technique (Morari and Lee 1999; Maciejowski 2001; Afram and Janabi-Sharifi 2014). Both approaches seek to define a series of policies for state-dependent actions to maximize cumulative long term reward.

In planning, it is assumed that a complete model of the task environment is available and the planner induces a policy for choosing the action in each state that achieves optimal performance in terms of total long term reward. The RL approach, on the other hand, does not assume the environment is known ahead of time. Instead, the learning agent has to interact directly with the environment to gather data about the effects of its actions on the world and their reward value, and while doing so searches for an optimal policy for action.

The a-priori model of the task environment is both the strength and weakness of the planning approach. In an environment as complex as a building interior, these conditions are unlikely to be completely known in advance and may change over time.

The learning framework provides a general approach to solving sequential decision-making problems without relying on a pre-existing model of the task environment, but incurs the prohibitively high cost of real-time interaction with the environment that would require multiple parallel processors (Magoules and Zhao 2016). We can potentially get the advantages of both frameworks through a hybrid approach in which we use a suitable platform for offline training of an adaptive learning system through our complementary methodology of physical and simulation environments. In this case, the policy learned in simulation is used to initiate the learner with a reasonable performance that is then transferred to and fine-tuned in real-world interaction. This approach reduces the amount of costly real-world experience required to achieve high performance (Liu and Henze 2006; Cutler et al. 2015).

By constructing a feedback loop between actual and simulated environments, we streamline the development of the learner. At the same time, we iteratively increase the accuracy of our simulation by recalibrating it each cycle based on results recorded from the physical test chamber. The result is a prototyping environment where we can develop a novel environmental control framework that adapts to its ever changing context and continually improves its performance over time. This experimental setup is also designed to anticipate how our IAC might be employed in the field. While in operation as a building control system, a parallel simulation driven by real-time data collected from building sensors would provide an environment where alternative control policies may continuously be explored and evolve.

RESULTS

Initial work on the development of the IAC has focused on its integration with a conventional HVAC control system. Subsequent development will confront the more complex prospect of passive conditioning through an adaptive facade. In simulations of our initial model, the thermostat controller tuned via a reinforcement machine learning algorithm performed approximately 5% more efficiently than a simulated conventional automatic thermostat. We expect that this performance will improve over the course of our work towards single system performance improvements of 8–12%, and our target for accumulated efficiency between a passive facade system and a conventional HVAC system centrally controlled by the IAC technology is 25–35% (Jacobs 2003). The integration of the IAC with adaptive building envelope actuation could provide up to 50% reduction in energy demands.

CONCLUSION

Current work to date has demonstrated performance benefits from the application of an early version of the IAC. The time series of environmental conditions and the state of the facade will be analyzed to create models of system dynamics at multiple time scales. These dynamic models shall serve as the instrument for developing control algorithms to maintain desired chamber-internal environment states while optimizing for low energy consumption. This includes using adaptive learning techniques that explore control strategies under different optimizing constraints.

Our efforts have occurred through interdisciplinary collaborations with Information Sciences, Electrical and Computer Engineering, and Material Science Engineering. In order to develop robust IAC policies, further interdisciplinary connections are warranted that will enhance the possibilities for holistic evaluative frameworks in the reinforcement learning algorithms. The primary focus in near-term work required specific attention to linking the RL with the dynamic facade for parametric development of environmental data feedback informing actuation signals. The simplified physical model for the electrochromic dynamic glazing film and photometric analysis will be analyzed with concurrent development of dynamic simulation scripting in the EnergyPlus platform.

This work ultimately pursues the systematic evaluation of a range of adaptive envelope technologies with regard to environmental performance. When complete, the resulting comparative study will undoubtedly have value beyond the bounds of this project and be useful to architectural designers at large. Furthermore, our experimental testbed serves as a generalizable process and kit of parts. It may be adopted and improved upon by future designers for use as tool kit for the study and design of adaptive facades in all of their aspects.

ACKNOWLEDGEMENTS

The authors acknowledge a Renewable Energy Network (REN) Faculty Exploratory Grant award (#5825479) from the University of Arizona (UA) to make this research possible. We also thank our collaborators on the framework development: Clayton Morrison, Associate Professor in the School of Information Science, Technology, and Arts at the UA; Pierre Lucas, Professor in Material Science Engineering at the UA; and Kuo Peng, Doctoral Student in Electrical and Computer Engineering at the UA. We also thank all of the students involved on the project to date, including Alice Wilsey, Tyler Connel, Long Long, Nikki Hernandez, and Di Le.

NOTES

1. Design integration of high performance envelope systems has shown 30%–70% energy savings. "Federal R&D Agenda for Net-Zero Energy, High-Performance Green Buildings Report" National Science and Technology Council, Report of the Subcommittee on Buildings Technology Research and Development, Oct 2008, p.21.

2. The only identifiable emerging market technology in this area is the Provolta Energy OS patent pending machine learning platform by Rengen supported in part by R&D partner Lawrence Berkeley National Laboratory; however, this platform does not explicitly focus on integration with dynamic building envelopes, but rather on improving performance of HVAC control systems through building integration feedback.

REFERENCES

Afram, A. and F.Janabi-Shari. 2014. "Theory and applications of HVAC control systems-a review of model predictive control (MPC)." *Building and Environment* 72: 343–355.

Cutler, M., T.J. Walsh, and J.P. How. 2015. "Real-world reinforcement learning via multifidelity simulators." *IEEE Transactions on Robotics* 31 (3): 655–671.

Dalamagkidis, K., D. Kolokotsa, K. Kalaitzakis, and G.S. Stavrakakis. 2007. "Reinforcement learning for energy conservation and comfort in buildings." *Building and Environment* 42 (7): 2686–2698.

Erickson, James. 2013. *Envelope as Climate Negotiator: Evaluating adaptive building envelope's capacity to moderate indoor climate and energy.* Dissertation, Arizona State University.

Ernst, D., M. Glavic, F. Capitanescu, and L. Wehenkel. 2009. "Reinforcement learning versus model predictive control: A comparison on a power system problem." *IEEE Transactions on Systems, Man, and Cybernetics, PartB (Cybernetics)* 39 (2): 517–529.

Jacobs, Pete. 2003. "Small HVAC System Design Guide: Design Guidelines." *California Energy Commission* (October).

Kolarevic, Branko and Vera Parlac. 2015. *Building Dynamics: Exploring Architecture of Change.* United Kingdom: Taylor and Francis.

Liu, S. and G.P. Henze. 2006. "Experimental analysis of simulated reinforcement learning control for active and passive building thermal storage inventory: Part1 – theoretical foundations." *Energy and Buildings* 38 (2): 142–147.

Maciejowski, J. 2001. *Predictive Control with Constraints.* Englewood Cliff, NJ: Prentice Hall.

Magoulès, Frederic and Hai-Xiang Zhao. 2016. *Data mining and machine learning in building energy analysis.* Hoboken, NJ: John Wiley & Sons.

Morari, M. and J.H. Lee. 1999. "Model predictive control: Past, present and future." *Computers & Chemical Engineering* 23 (4): 667–682.

Pang, Xiufeng, Tianzhen Hong, and Mary Ann Piette. 2013. "Improving Building Performance at Urban Scale with a Framework for Real-time Data Sharing" *Lawrence Berkeley National Laboratory* LBNL-6303E (June).

Peng, Kuo and Clayton Morrison. 2016. "Model Predictive Prior Reinforcement Leaning for a Heat Pump Thermostat." *IEEE International Conference on Automatic Computing: Feedback Computing* '16 (July).

Ruelens, F., Iacovella, S., Claessens, B., and Belmans, R. 2015. "Learning Agent for a Heat-Pump Thermostat with a Set-Back Strategy Using Model-Free Reinforcement Learning." *Energies* 8 (8): 8300–8318.

Shaikh, P. H., N.B.M. Nor, P. Nallagownden, I. Elamvazuthi, and T. Ibrahim. 2014. "A review on optimized control systems for building energy and comfort management of smart sustainable buildings." *Renewable and Sustainable Energy Reviews* 34: 409–429.

Sanyal, J., J. New, R.E. Edwards, and L. Parker. 2014. "Calibrating building energy models using supercomputer trained machine learning agents." *Concurrency and Computation: Practice and Experience* 26 (13): 2122–2133.

Sutton, R.S. and A.G. Barto. 1998. *Reinforcement Learning: An Introduction.* Cambridge: The MIT Press.

Zamella, Giovanni and Andrea Faraguna. 2014. *Evolutionary Optimisation of Facade Design: A New Approach for the Design of Building Envelopes.* New York, NY: Springer.

Zavala, V.M., C. Thomas, M. Zimmerman, and A. Ott. 2011. "Next-Generation Building Energy Management Systems and Implications for Electricity Markets" *Argonne National Laboratory* ANL/MCS-TM-315 (July).

IMAGE CREDITS

Figure 1: Responsive facade modules (Smith, 2016)
Figure 2: IAC Framework Methodology (Smith and Lasch, 2015)
Figure 3: Testbed diagram (Smith and Le, 2015)
Figure 4: Testbed photograph (Smith, 2016)
Figure 5: Adaptive module designs (Smith and Le, 2015)
Figure 6: Adaptive module patterns (Smith and Le, 2015)
Figure 7: Building energy analysis model results (Smith, 2016)

Shane Ida Smith is an architect, designer, educator, and philosopher. She has over fifteen years experience in professional architecture practice and is fluent with building performance analytics. She holds a Doctorate of Philosophy in Architectural Sciences from the Center for Architecture, Science, and Ecology at Rensselaer Polytechnic Institute. She is currently an Assistant Professor in Architecture at the University of Arizona conducting design research on emerging environmental building technologies.

Chris Lasch is a principal and co-founder of Aranda\Lasch, a design studio dedicated to experimental research and innovative building through a deep investigation of structure and materials. The studio was awarded the United States Artists Award and Young Architects + Designers Award in 2007, the Architectural Record Design Vanguard Award in 2014, the Architectural League Emerging Voices Award in 2015, and was named one of Architectural Digest's 2014 AD Innovators. Their early projects are the subject of the best-selling book, *Tooling*. Aranda\ Lasch has exhibited its work internationally in galleries, museums, design fairs and architecture biennials.

Space Plan Generator

Rapid Generation & Evaluation of Floor Plan Design
Options to Inform Decision Making

Subhajit Das
Autodesk / Georgia Tech

Colin Day

Anthony Hauck
Autodesk

John Haymaker

Diana Davis
Perkins+Will

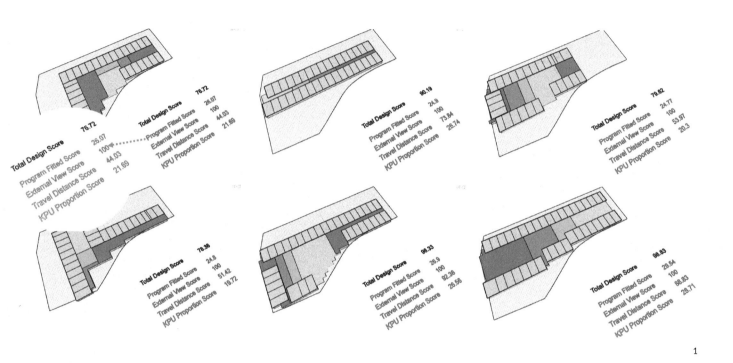

1

ABSTRACT

Design exploration in architectural space planning is often constrained by tight deadlines and
a need to apply necessary expertise at the right time. We hypothesize that a system that can
computationally generate vast numbers of design options, respect project constraints, and analyze
for client goals, can assist the design team and client to make better decisions. This paper explains
a research venture built from insights into space planning from senior planners, architects, and
experts in the field, coupled with algorithms for evolutionary systems and computational geometry,
to develop an automated computational framework that enables rapid generation and analysis
of space plan layouts. The system described below automatically generates hundreds of design
options from inputs typically provided by an architect, including a site outline and program document with desired spaces, areas, quantities, and adjacencies to be satisfied. We envision that this
workflow can clarify project goals early in the design process, save time, enable better resource allocation, and assist key stakeholders to make informed decisions and deliver better designs. Further,
the system is tested on a case study healthcare design project with set goals and objectives.

1   Generated space plan layout
options with design score for
reference.

INTRODUCTION

Design teams work under schedule and resource constraints that limit the range of design solutions they can generate and analyze, impeding informed design and optimal decision making. We have surveyed building guidelines, codes, facility-planning rules, literature, and case studies, and conducted interviews with senior professional medical planners and architects to understand industry best practices and acceptable design methodologies (Das, Haymaker, and Eastman 2015). This work clarified the implicit and explicit domain knowledge and processes in space planning and identified opportunities to leverage computation for repetitive design generation and analysis tasks, liberating architects to invest time in problem formulation and decision making.

In this paper, we present Space Plan Generator (SPG), an emerging methodology and tool to automate aspects of architectural design. We first briefly describe foundational work in data structures and space-planning methodologies, then explain the working methodology of SPG (see Figure 2) by describing the Hierarchical Space Assignment strategy, a top-down approach from whole to part, and the K-dimensional (K-d) Tree Data Structure, which showcases an efficient data storage, retrieval, and traversal technique. Next, we discuss the splitting strategy to assign departments and programs to the site, and the implementation of a cell grid that enables circulation computation with analysis and scoring of the generated space plans. We then describe the implementation of the methodology in Autodesk Dynamo, and the testing of its the validity and efficacy within a healthcare facility case study from a large architectural practice.

RELATED WORKS
TreeMap Data Structure

Architectural design is a process necessitating multiple iterations until design goals are achieved. Consequently, we have implemented a data structure which can rapidly access identical data multiple times and store spatial data in a manner supporting nearest neighbor search, which is necessary to build department and program topology maps for design fitness appraisal. Owing to its spatial partitioning structure, K-d tree serves our purposes by using nested elements to store data. Nested elements provide a means to retrieve data through bidirectional traversal, top down and bottom up.

K-d trees store and organize data as a set of 'n' points in a multi-dimensional space, in a structure of a binary search and partitioning trees. Due to their efficiency and speed, they are primarily applied to nearest-neighbor queries, search algorithms, database applications, and ray-trace methods. Originating in computational geometry, their efficiency arises from the space partitioning algorithm organizing objects in K-d space (Knecht and König 2010). Average running time of an 'n' data point database for:

- Insertion – $O(\log n)$;
- Deletion of root – $O(n(k-1)/k)$
- Deletion of any random node – $O(\log n)$ (Bentley 1975).

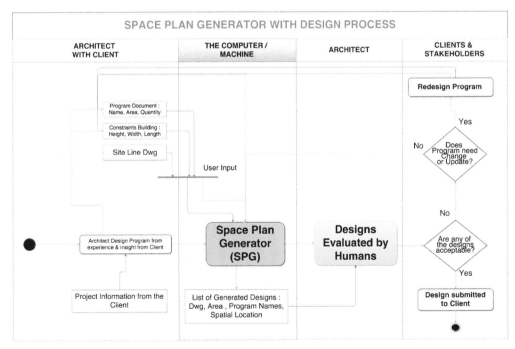

2 Activity diagram showing SPG and architect together in a design process.

Space Plan Generation

Eastman (1972) automated space generation in two dimensions by implementing decision rules to guide subsequent placement and arrangement of design units. These rules were driven by operators that transformed the state of the design units iteratively to satisfy a set of relationships between them. Jo and Gero (1998) highlighted an evolutionary-design model describing a schema to represent design knowledge capable of providing design solutions for the given problem requirement. Their work highlighted topological and geometrical arrangements of spatial elements tested on a large office layout problem. Though not many variations of the spatial arrangement were highlighted, their work showed the robustness of coupling genetic algorithm-based searches with design workflows to produce good results in space planning. Michalek, Choudhury, and Papalambrosa (2002) presented an optimization model integrating optimization and subjective decision making during conceptual design. Coupling gradient-based algorithms and evolutionary algorithms, they innovated to include human decision making in the workflow. They implemented topology optimization algorithms on top of geometry optimization algorithms. The results were automated space plans but limited in variation. Nassar (2010) presented new findings in graph theory with direct implications in space planning problems. He described architectural space plans as simple, connected, labeled planar graphs, and elaborated on the relevance of finding a rectangle dual for every planar graph to increase solution space. This work outlined a tool for architects to generate space plans. Realizing spatial relationships as planar graphs with nodes as rooms and edges as adjacencies, it claimed that the proposed model could provide a truly exhaustive set of potential designs. However, this model was limited to two-dimensional space plans only.

Boon et al. (2015) employed genetic algorithms to generate 3D space stacking, respecting input adjacency requirements by the user. Their algorithm evaluates space plans based on adjacency constraints to minimize the total distance of all interconnected programmatic elements. They prioritized the program spaces based on practice expertise and user input, which helps discard unrealistic design solutions. The algorithm stacks spaces in three dimensions, distributing program elements over multiple floors and sometimes unnecessarily complicating architectural space layouts.

Some relevant research in automated space plan generation comes from game design, which requires extensive 3D environments at architectural scale. Lopes et.al (2010) generate floor plans for different classes of buildings, many with connected floor levels, offering limited control to the designer for functional constraints. One of the salient features of this system is the grid-based strategy for placing and growing rooms, which generates

building zones, followed by room areas, constrained by adjacency and connectivity. Marson and Musse (2010) implement a squarified treemap algorithm (previously used to represent hierarchical information graphically) to compartmentalize the input space into different zones or regions. These zones are organized into a hierarchy that satisfies design goals and site constraints and are visualized into a square tree map to generate various floor layouts. Their work excels at building sophisticated circulation networks. After the rooms are site located, a circulation network graph is built to understand which rooms are connected or disconnected from each other. They use an A* algorithm to traverse the connectivity graph and find shortest paths to access spaces from the lobby.

METHODOLOGY

The Space Plan Generator has distinct components orchestrated to generate and analyze space plans. It generates layouts as closed polylines representing each space type (either the department or the program element), with the circulation network represented as colored poly surfaces for any generic architectural design problem.

Hierarchical Space Assignment

Our approach is hierarchical (see Figure 3). Program elements are spaces to fit on the site. Certain types of program elements can be clustered to form departments. The recursive algorithm first places the department on site and then programs within departments, finally placing circulation within and between departments.

K-d Tree Data Structure

Our space-assignment algorithm uses a K-d data structure dividing the K-d space by partition planes perpendicular to one of the coordinate axes (see Figure 4). Conventionally, the median or average of the point coordinates in the database is calculated in the split dimension. Site space is the root and the first node after the spatial split, dividing the point set in two. All points whose coordinates are smaller than the split value reside in the tree's left branch, while the other points are relegated to the right branch. This step is repeated until reaching a threshold depth of the K-d tree. Each generated node is assigned a space or region, each of which can contain child nodes (Bentley 1990). K-d Trees efficiently search and traverse data sets to create spatial plan partitions (Bentley 1975).

Space data trees implement a modified version of K-d data structures, where the point set is split by a line dividing a monolithic space into two, using an algorithm to allocate program elements at particular site locations. For example, a programmatic requirement to locate patient rooms on the site periphery or the need of an entry lobby on the site's west side. Any region on the site

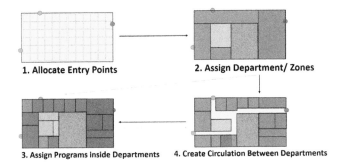

1. Allocate Entry Points

2. Assign Department/ Zones

3. Assign Programs inside Departments

4. Create Circulation Between Departments

3 Hierarchical space planning approach.

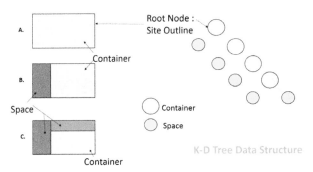

Root Node :
Site Outline

Container

Space

Container

K-D Tree Data Structure

Container

Space

4 Space Data Tree construction based on K-d data structure binary tree.

unassigned to a department or program element is termed a 'container,' while any region allocated to a department or program is termed a 'space.' The root of the data tree is the container (the site), with its left child the region of the site to the left of the splitting line. The remaining region becomes the right child of the root and the current container. The algorithm recursively repeats the operation for each node until all spaces are stored in the tree. After each split of the container, the left node becomes a space node, and the right node becomes a container node. The direction of the split can be flipped from horizontal to vertical alignment at every iteration, similar to slice and dice techniques (Shneiderman 1992), depending upon the aspect ratio of the container and the program elements awaiting allocation.

Space and Container Node Data

After the split to represent site location, every space node contains data including: Space Node ID, Parent Node ID, Left Child Node ID, Right Child Node ID, Split Line, Department ID Assigned, Cells Allocated, and Polyline representation.

Every container node retains data above except for Department ID. As each node stores links to its parent and child nodes, the space data tree can be traversed upwards and downwards (see Figure 4).

Benefits of using Space Data Tree:

- Provides fast and efficient data storage to represent spatial data.
- Supports nearest neighbor searches, which help build department topology maps and appraise the efficiency of a layout by representing department neighbors and adjacency.
- Finds neighbors for departments and program elements and shared edges between them, with each Space and Container Node storing the splitting line. Thus allows the computation of circulation networks between and within departments.
- Can randomly ascertain circulation paths between any two spatial locations in the space plan, aiding in space plan appraisal through key metrics such as nurse travel routes and distances, patient flow paths, fire egress routes, etc., in a healthcare facility.

Splitting Strategies

We deployed several custom methods to split a polyline into two or more by a dividing line, including:

- *Split by Distance*: Creates a splitting line from a given point on a polyline by a specified distance or by a specified line orientation. Returns two polylines on a successful split.
- *Split by Ratio*: Splits a polyline into two polylines by a supplied ratio and direction.
- *Split by Area*: Ensures split polylines meet area requirements as specified by the user or the algorithm, employing a brute force splitting strategy by iteratively applying the split by distance method, altering the distance variable after each iteration.
- *Split by Line*: Requires the user or the algorithm to provide a splitting line when there is an existing splitting line for further splits.
- *Split Recursively to meet Minimum Dimension*: A recursive split strategy adapted to split a single polyline into multiple polylines until the minimum dimension of each polyline meets an acceptable width or length constraint specified by the user. Employing a slice and dice technique, after each split the split direction is toggled between horizontal and vertical. This strategy arrives at interesting spatial patterns with an acceptable level of architectural rationality. The listed methods divide the input site polyline into individual departments containing program elements. Method outputs coupled with department and program information are assigned to the space data tree to organize them for further computation.
- *Split by Offsetting Polyline Points*: A computationally faster strategy generates two polylines by shifting the points of the parent polyline, avoiding more expensive algorithms such as line-poly intersection or ordering points in a list to rebuild a polyline as used in other split strategies.

Cell Grid Underneath the Spaces Assigned

From an input site outline, the system builds a bounding object containing the site polyline (see Figure 5b). The algorithm

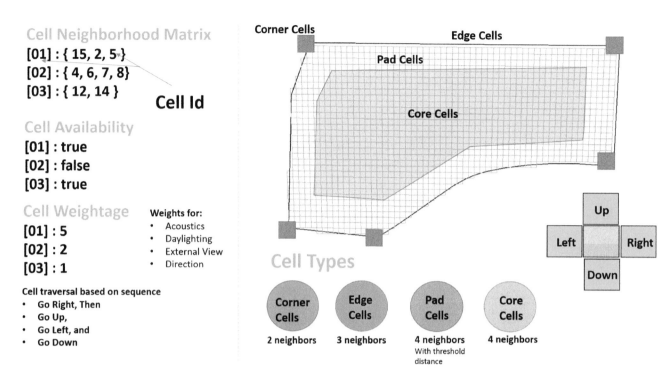

Cell Neighborhood Matrix

[01] : { 15, 2, 5 }
[02] : { 4, 6, 7, 8}
[03] : { 12, 14 }

Cell Id

Cell Availability

[01] : true
[02] : false
[03] : true

Cell Weightage

[01] : 5
[02] : 2
[03] : 1

Weights for:
- Acoustics
- Daylighting
- External View
- Direction

Cell traversal based on sequence
- Go Right, Then
- Go Up,
- Go Left, and
- Go Down

Corner Cells   Edge Cells   Pad Cells   Core Cells

Up / Left / Right / Down

Cell Types

Corner Cells — 2 neighbors
Edge Cells — 3 neighbors
Pad Cells — 4 neighbors With threshold distance
Core Cells — 4 neighbors

5   (a) Shows Cell storage based on ID, Cell neighbor matrix, Cell weights to account for specific objectives. (b) Shows types of cells based on the number of neighbors they have.

supplies the bounding box with points on an orthogonal grid, with spacing specified by the user. A site outline test determines the contained points used to build each cell. A collection of cells is termed a grid object, employed to compute the shortest path results between program elements, placement of doors and windows, and similar problems. Grid objects are used to build the cell neighbor matrix, which tracks the neighbors of every cell in three dimensions. The benefit of such a matrix is in its capability to efficiently traverse the site.

Cell Neighbor Matrix

Each cell in the grid object possesses a location ID. The neighbor matrix of a cell is a list of lists, which stores identifiers of all neighboring cells. Neighboring cells are traversed in a fixed order of right, up, left, and down. A cell identified as 0 might have cells 05, 08, 02, 15 stored in its neighbor matrix row, with 05 identifying the right-side cell, 08 denoting the upside cell, cell 02 on the left side, with 15 identifying the downside cell. -1 indicates a cell lacking any neighbors. The matrix provides a means to traverse the site from one corner to another to find specific locations, such as doors and exit routes (see Figure 5a).

Cell Weighting for Specific Metrics

Each cell is given a custom weight to gauge different aspects of the project, such as acoustic performance, daylighting, and site constraints. Weighting influences the allocation of spatial locations for a certain program. For example, if the northwest side

of the site is conducive for daylighting, then cell weights can be increased for that zone to arrive at an appropriate distribution bias.

Shortest Path to Find Doors / Access Points

The cell neighbor matrix also helps find the shortest path between site regions by implementing Dijkstra's Algorithm (Wang 2012). Specific cells are marked as doors or windows, with the objective of pathfinding to discover the shortest path to those cells from a given program element.

Polyline Boundary

This algorithm finds the maximum sized orthogonal polyline boundary of any input polyline having both orthogonal and non-orthogonal angles between its sides (see Figure 6). To locate the bounding polyline, the algorithm first identifies border cells by traversing the cell neighbor matrix. A cell is identified as a border cell if it's either a corner cell with two neighbors or an edge cell with three neighbors. The algorithm rebuilds the cell neighbor matrix from the border cell, finds the lowest leftmost cell in the border cell list, and subsequently traversesg the cells. Initially, traversal attempts a rightward search, proceeding on failure through successive tries up, left, and down. The centroid of each visited cell is stored in a point list, and the cell is marked as unavailable for further visits, preventing infinite search loops. Upon traversing every cell, the complete centroid list will be used to build the polyline boundary.

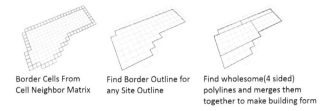

Border Cells From Cell Neighbor Matrix

Find Border Outline for any Site Outline

Find wholesome(4 sided) polylines and merges them together to make building form

6   Polygon Boundary algorithm showing the orthogonal boundary for any input arbitrary site outline.

Analyzer

In addition to the design generator, an analyzer component validates the efficacy of the generated space plan with respect to the original project goals. Some analysis types implemented include gauging the percentage of program elements fitted to the site outline in comparison to specified requirements, determining the quantity of day-lit rooms with external views, and determining circulation efficiency. Please see Figure 7 to understand the working methodology of SPG further.

IMPLEMENTATION

The prototype employs an Autodesk Dynamo Package written in C# as a library of 'zero touch' custom nodes (Helsberg et al. 2010), which helps construct the space plan generator as explained below (see Figure 8). The Dynamo graph is composed of two components, a 'Generator' and an 'Analyzer.' After every graph evaluation, the 'Generator' yields distinct space plan layouts rendered as a list of polyline geometries. The 'Analyzer'

gauges the fitness of each design option to input goals and constraints. The prototype proceeds as follows:

Step 1: User Inputs

The graph needs requirements in the form of a site outline '.sat' file and program document '.csv' file (see Figure 9). The '.csv' should contain information for each program element to be fit onto the site, with each element possessing the attributes, namely: ID, name, department, quantity, area or dimension, program, preference value, and adjacency list.

Preference value for each program element specifies the order in which the element will be placed onto the site. The adjacency list is a comparison chart to which the program and department topology map will adhere while allocating program elements on the site. Other user inputs include corridor width, the aspect ratio for circulation and program elements, grid cell dimension, etc.

Step 2: Cell Grid Information

The graph builds the cell grid on the site and the initial cell neighbor matrix, which are used repeatedly in the workflow explained above.

Step 3: Data Stack

This step extracts input information from the program document and builds a Department Data and Program Data class object. The data stack node sorts and stores the department and program data object based on the preference values from the

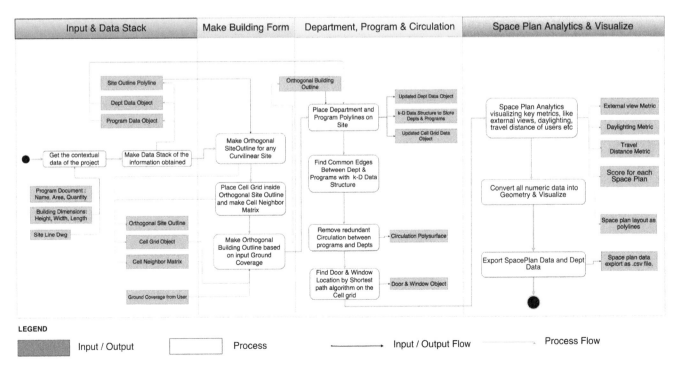

7   Activity diagram showing Space Plan Generator's working methodology.

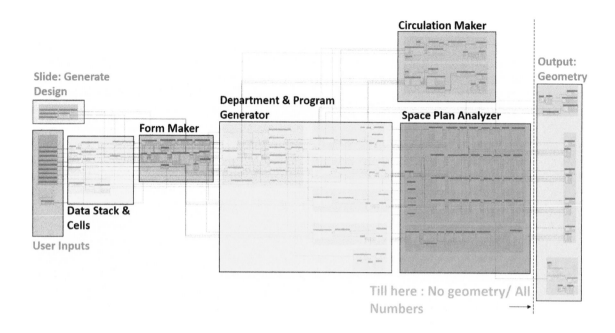

program document. Any computation or analysis regarding the spatial assignment of program elements is done with the aid of the data stack of departments and programs.

Step 4: Form Generation

The form maker node constructs the orthogonal building outline for an arbitrary site outline (see Figure 10 and 11) traversing the cell grid via the cell neighbor matrix, implementing two strategies to pack programs.

Strategy 1: Function Follows Form

This strategy splits the orthogonal building outline into sub-polylines until each polyline has exactly four sides, and subsequently merges a random number of such four-sided orthogonal polylines together to get the building form. One advantage of this approach is that the user can set a percentage of total site area for the building to occupy, resulting in building perimeters of specific site coverage. The merge operation employs the Polyline Boundary algorithm as explained above.

Strategy 2: Form Follows Function

Unlike Strategy 1, Strategy 2 develops the building outline while placing department and program elements. After each depart- ment polyline placement, the algorithm evaluates adherence to site constraints and design requirements. The algorithm will only proceed to subsequent departments after constraints and requirements are satisfied or until reaching an allowed quantity of trials. When area requirements are satisfied for programs and departments, the algorithm discards any leftover waste space,

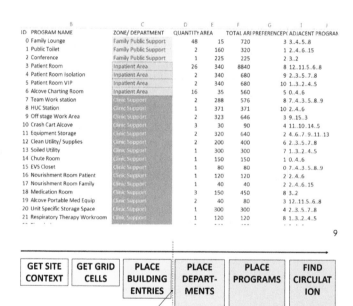

ID	PROGRAM NAME	ZONE/ DEPARTMENT	QUANTITY	AREA	TOTAL AR	PREFERENCE	ADJACENT PROGRAM
0	Family Lounge	Family Public Support	48	15	720	3	3..4..5..8
1	Public Toilet	Family Public Support	2	160	320	1	2..4..6..15
2	Conference	Family Public Support	1	225	225	2	3..2
3	Patient Room	Inpatient Area	26	340	8840	8	12..11.5..6..8
4	Patient Room Isolation	Inpatient Area	2	340	680	9	2..3..5..7..8
5	Patient Room VIP	Inpatient Area	2	340	680	10	1..3..2..4..5
6	Alcove Charting Room	Inpatient Area	16	35	560	5	0..4..6
7	Team Work station	Clinic Support	2	288	576	8	7..4..3..5..8..9
8	HUC Station	Clinic Support	1	371	371	10	2..4..6
9	Off stage Work Area	Clinic Support	2	323	646	3	9..15..3
10	Crash Cart Alcove	Clinic Support	3	30	90	4	11..10..14..5
11	Equipment Storage	Clinic Support	2	320	640	2	4..6..7..9..11..13
12	Clean Utility/ Supplies	Clinic Support	2	200	400	6	2..3..5..7..8
13	Soiled Utility	Clinic Support	1	300	300	7	1..3..2..4..5
14	Chute Room	Clinic Support	1	150	150	1	0..4..6
15	EVS Closet	Clinic Support	1	80	80	0	7..4..3..5..8..9
16	Nourishment Room Patient	Clinic Support	1	120	120	2	2..4..6
17	Nourishment Room Family	Clinic Support	1	40	40	2	2..4..6..15
18	Medication Room	Clinic Support	3	150	450	8	3..2
19	Alcove Portable Med Equip	Clinic Support	2	40	80	3	12..11.5..6..8
20	Unit Specific Storage Space	Clinic Support	1	300	300	4	2..3..5..7..8
21	Respiratory Therapy Workroom	Clinic Support	1	120	120	8	1..3..2..4..5

9

GET SITE CONTEXT	GET GRID CELLS	PLACE BUILDING ENTRIES	PLACE DEPART- MENTS	PLACE PROGRAMS	FIND CIRCULAT ION

FORM MAKER

10

8   Current State of the Autodesk Dynamo Graph. Every node is a custom node, written in C#.

9   Requirements supplied to the graph via .csv program document.

10 'Form Maker' in the sequence of hierarchical space planning approach to generate architecturally rationale space plans.

11 Form maker in action placing colored poly surfaces as individual departments.

12 (a) Shows the Dept. Analytics node highlighting area, cells, and programs for all four departments. (b) Shows the area percentage desired and achieved in relation to other departments for all four departments.

merging the outer lines of the department polylines to arrive at the final building outline.

For both of the strategies, custom functions remove single or multiple notches from the building outline using minimum edge distance, thus refining the building form and increasing the number of design choices for the user.

Step 5: Department Placement and K-d Space Data Tree for Department

This step assigns departments to the site based on preference values from the user and simultaneously builds the space data tree. The iterative process is appropriately integrated with the form maker from the previous step, depending upon which form-making strategies are adopted and how well design goals and constraints are satisfied.

Step 6: Program Placement

This step assigns program elements inside each department, as prioritized by user goals, and updates the grid object by assigning each cell a certain program type and updating each cell's weight.

Step 7: Circulation Computation

Circulation computation discovers circulation networks between departments and subsequently finds circulation networks

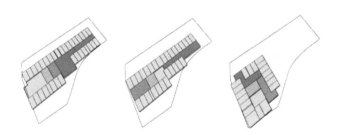

11

Dept Analytics

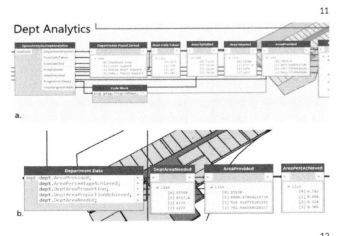

a.

b.

12

between program elements by using K-d data structure referenced above. Shared edges between departments and program elements form the initial circulation network. Next, a circulation redundancy check, with the aid of the grid object, is deployed to select only those edges in the network needed to access all spaces. Further, it employs pathfinding algorithms with the aid of cell neighbor matrix to discover the shortest paths within available circulation pathways between points, and places doors and windows along the discovered path. Grid Object also accounts for those spaces which do not get any access and places additional network lines to render them accessible from the main public space in the layout (Mirahmadi and Shami 2012).

Step 8: Space Plan Analytics

Evaluates the generated design based on metrics as summarized above (see Figure 12).

Step 9: Space Plan Output / Geometry Rendering

One of the salient features of the SPG is that it maintains a distinction between geometry and computation, with geometry only rendered at the process conclusion for visualization (see Figures 13a and 13b). Until step 8, no geometry is rendered on screen or temporarily saved in memory, which significantly speeds the production of results. Since geometry is not used before this step, custom methods are included in determining line intersections and line/polygon intersections, removal of duplicate geometries, determine point containment, merge polylines, etc., and made available as 'zero touch' Dynamo nodes.

Step 10: Storing and Scoring Generated Designs

Generated designs are stored as a collection of cells, where each cell stores information about its assigned program or department. Cells store space plan analytic information, such as distance to external windows, visibility of the cell from circulation areas, etc., and these metrics help determine the overall scored success of a space plan. Scoring conveys to the user the success of design options relative to input goals and constraints (see Figure 1).

USE CASE AND RESULTS

The prototype is being tested on a new hospital bed tower to be built on an existing healthcare facility site which includes an existing bed tower and hospital facility (see Figure 14). With site constraints and specific client goals, such as maximizing patient beds per floor, employing new hospital facility design guidelines and building codes, minimizing nurse travel distance (Rechel, James, and Martin 2009), maximizing connectivity to the existing hospital, and minimizing view impedance from the existing bed tower, this project is a valuable test case to understand how a goal-driven design workflow can be automated using generative design strategies. The current state of the Dynamo package

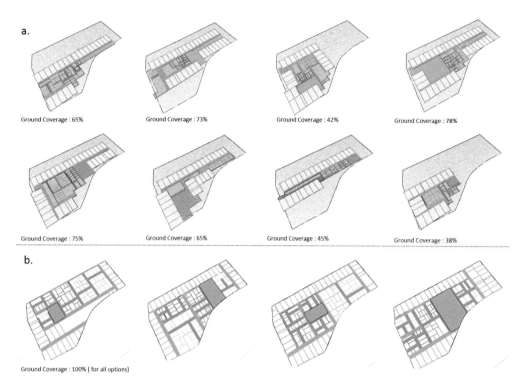

a.

Ground Coverage : 65%   Ground Coverage : 73%   Ground Coverage : 42%   Ground Coverage : 78%

Ground Coverage : 75%   Ground Coverage : 65%   Ground Coverage : 45%   Ground Coverage : 38%

b.

Ground Coverage : 100% (for all options)

13 (a) Generated Space Plan Layouts after using 'Form Maker' based on set ground coverage. (b) Space Plan Layouts without using 'Form Maker' filling the whole site space with program elements.

restricts designs to a single floor as a limited case to ensure stability and reliability before being implemented for multiple floors.

At present, the system is capable of generating, scoring, and analyzing design options each time the user adjusts a slider within the graph. Currently, each space plan is scored with respect to the percentage of program elements placed in comparison to program document requirements, the number of patient rooms with access to external views and daylighting, nurse travel distance to all patient rooms, and percentage number of Key Planning Units (inpatient patient beds in this case). Metrics are user-weighted, allowing architects to seek design solutions for project goals. We plan to couple the

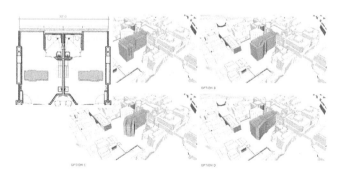

14 Concept forms generated manually by the design team. Top left shows modular patient room for the case study Hospital Bed Tower.

generator with Genetic Algorithm (GA) optimization to improve design candidates by learning from each iteration to reach optimal floor layouts, as driven by project goals.

CONCLUSION

Though this is not the first attempt to use generative design strategies to develop space plans, this research leverages efficient data structures, coupled with robust, scalable algorithms from computational geometry and generative design, to deliver rational architectural space plans. Building the system on Autodesk Dynamo allows a large number of users to benefit from the system. Separating geometry and computation significantly improved system performance, allowing iterative solution searches to arrive at satisfactory results. Nevertheless, we recognize current system limitations such as an inability to handle non-orthogonal or curved spaces, as well as an inability to distribute spaces on multiple levels, but we plan to address these shortcomings. Currently, the system generates design options without learning from its previous iterations, sometimes leading to architecturally inadequate proposals. We envision surpassing this limitation when the generator is coupled with genetic algorithm optimization in future work.

REFERENCES

Becker, S., M. Peter, D. Fritsch, D. Phillip, P. Baier, and C. Dibak. 2013. "Combined Grammar for the Modeling of Building Interiors." *ISPRS Annals* of the Photogrammetry, Remote Sensing and Spatial Information Sciences II-4 W1: 1-6.

Bentley, Jon Louis, and Jerome H. Friedman. 1979. "Data Structures for Range Searching." ACM Computing Surveys 11 (4): 397-409.

Bentley, Jon Louis. 1975. "Multidimensional Binary Search Trees Used for Associative Searching." Communications of the ACM 18 (9): 509-517.

———. 1990. "K-d trees for semidynamic point sets." In Proceedings of the Sixth Annual Symposium on Computational Geometry. Berkeley, CA: SOCG. 187-197.

Boon, Christopher, Corey Griffin, Nicholas Papaefthimious, Jonah Ross, and Kip Storey. 2015. "Optimizing Spatial Adjacencies using Evolutionary Parametric Tools: Using Grasshopper and Galapagos to Analyze, Visualize, and Improve Complex Architectural Programming." Perkins + Will Research Journal 7 (2): 25-37.

Das, Subhajit, John Haymaker, and Chuck Eastman. 2015. "Data Model and Processes for Building Programming." Atlanta: Georgia Tech Digital Building Lab. http://www.dbl.gatech.edu/node/15402

Eastman, Charles. 1970. "Automated Space Planning." Communications of the ACM 13 (4): 242-250.

Helsberg, Anders, Mads Torgersen, Scott Wiltamuth, and Peter Golde. 2010. C# Programming Language. Boston, MA: Addison-Wesley Professional.

Hicks, Chris, Tom McGovern, Gary Prior, and Iain Smith. 2015. "Applying Lean Principles to the Design of Healthcare Facilities." International Journal of Production Economics 170 (Part B): 677-686.

Jo, Jun H., and John S. Gero. 1998. "Space Layout Planning Using an Evolutionary Approach." Artificial Intelligence in Engineering 12 (3): 149-162.

Knecht, Katja, and Reinhard König. 2010. "Generating Floor Plan Layouts with K-d Trees and Evolutionary Algorithms." Proceedings of the 13th International Conference on Generative Art. Milan: GA. 238-253.

Liggett, Robert S. 2000. "Automated Facilities Layout: Past, Present and Future." Automation in Construction 9 (2): 197-215.

Lopes, Ricardo, Tim Tutenel, Ruben M. Smelik, Klaas Jan de Kraker, and Rafael Bidarra. 2010. "A Constrained Growth Method for Procedural Floor Plan Generation." In Proceedings of GAMEON. Leicester, UK: EUROSIS. 13-22.

Marson, Fernando, and Soraia Raupp Musse. 2010. "Automatic Real-Time Generation of Floor Plans Based on Squarified Treemaps Algorithm." International Journal of Computer Games Technology 2010 (7): 10.

Michaleka, Jeremy, Ruchi Choudhury, and Panos Papalambrosa. 2002. "Architectural Layout Design Optimization." Engineering Optimization 34 (5): 461-484.

Mirahmadi, Maysam, and Abdallah Shami. 2012. "A Novel Algorithm for Real-time Procedural Generation of Building Floor Plans." arXiv:1211.5842v1

Nassar, Khaled. 2010. "New Advances in the Automated Architectural Space Plan Layout Problem." In Proceedings of the International Conference In Computing in Civil and Building Engineering, edited by Walid Tizani. Nottingham, UK: ICCBE.

Rechel, Bernd, Buchen James, and Mckee Martin. 2009. "The Impact of Health Facilities on Healthcare Workers' Well-Being and Performance." International Journal of Nursing Studies 46 (7): 1025-1034.

Shneiderman, Ben. 1992. "Tree visualization with tree maps : 2-d space-filling approach." ACM Transactions on Graphics (TOG) 11 (1): 92-99.

Wang, Shu-Xi. 2012. "The Improved Dijkstra's Shortest Path Algorithm and Its Application." Procedia Engineering 29: 1186-1190.

Subhajit Das has a diverse background in Architecture and Computer Science, with degrees in M Arch Design Computing (University of Pennsylvania) and MS in Computer Science (Georgia Tech). Being part of the Autodesk Generative Design group, Subhajit was instrumental to collaborating with Perkins+Will Healthcare Design team to steer this research forward. Currently, he is pursuing a PhD in Computer Science at Georgia Tech with core interests in Machine Learning, Graphics, and Visual Analytics.

Colin Day is a Principal Engineer in the Autodesk Generative Design Group, with years of experience in the development and implementation of algorithms in Computational Design and Computer Graphics.

Anthony Hauck joined the Autodesk Revit team in 2007, holding a succession of product management positions in the group until joining Autodesk AEC Generative Design in 2015 as its Director of Product Strategy, where he is responsible for helping define the next generation of building software products and services for the AEC industry.

John Haymaker, PhD, AIA, LEED AP serves as Perkins+Will's Director of Research, overseeing areas of inquiry including materials, design process, building technology, and healthcare practices. Previously a Professor of Civil Engineering at Stanford University, and at Georgia Tech, John has contributed over 80 articles to professional literature in the areas of design process communication, optimization, and decision-making.

Diana Davis has over 18 years of experience in the design, planning, and delivery of healthcare projects. She brings a special interest in Lean planning and evidence-based design to her work, paired with a commitment to improving the healthcare environment for caregivers, patients, and families. She is mentoring graduate level researchers at the Georgia Institute of Technology and Autodesk in the development of new digital tools to aid healthcare planning and design.

Evaluating Buildings with Computation and Machine Learning

Daniel Davis
WeWork

1

ABSTRACT

Although computers have significantly impacted the way we design buildings, they have yet to meaningfully impact the way we evaluate buildings. In this paper we detail two case studies where computation and machine learning were used to analyze data produced by building inhabitants. We find that a building's 'data exhaust' provides a rich source of information for longitudinally analyzing people's architectural preferences. We argue that computation-driven evaluation could supplement traditional post occupancy evaluations.

1   WeWork's Wonderbread office in Washington, DC is one of seventy offices studied as part of this paper.

INTRODUCTION

Although academia and practice have found many novel ways to improve the design process with computation, almost none of this work has filtered through to impact how designers evaluate buildings. Searching for 'post occupancy' in CuminCAD, an index of over 12,000 papers from ACADIA, CAADRIA, eCAADe, SIGraDi, ASCAAD, and CAAD Futures, returns 22 results, only 9 of which were published in the past 10 years. While computers are used in some aspects of modern building evaluation, such as emailing surveys and taking temperature readings, the use of computation in building evaluations hasn't had nearly the same impact, or warranted the same degree of attention, as the use of computation in the design process.

In some ways, the focus on 'computation for design' reflects a larger preference in the industry for doing design instead of evaluating design. In a survey of 29 mid-sized American architecture firms in 2015, Julie Hiromoto found that "post occupancy evaluation is currently rare" because of the "design team time and cost required to produce meaningful results" (Hiromoto 2015). And when firms find the budget, it generally only allows for a short-term study, often at the start of the building's life, making long-term longitudinal studies of building performance exceptionally rare. Bordass, Leaman, and Ruyssevelt (1999), who have spent their careers evaluating buildings, conclude that "the sad fact is that hardly any architectural or engineering design practices consistently collect information on whether or not their buildings work."

Although designers seldom evaluate the performance of their designs, they are under increasing pressure to create designs that perform. In particular, there seems to be increasing pressure from clients, governments, and society to attain environmental, sociological, and economic performance targets. But since architects rarely study their completed projects, they don't necessarily understand the impact of design decisions on previous projects, putting them in a difficult position to forecast the impact of future decisions. Frank Duffy says that "because our heuristic seems to be 'Never look back,' we are unable to predict the longer term consequences in use of what we design" (Duffy 2008).

In this paper, we explore new potentials for applying computation to the evaluation of buildings. Using two longitudinal case studies based on real data collected from seventy buildings, we consider whether metadata from a room booking application can be used to evaluate meeting rooms and whether machine learning can be used to identify building maintenance patterns. Throughout this paper, we will make the case that people's behavior online often carries with it latent information about how they perceive their physical environment, and that this data can be a valuable tool in the long-term analysis of physical environments.

EXISTING METHODS OF BUILDING EVALUATION

The process of building evaluation can vary dramatically. In some instances it involves a quick survey sent to the client, while in other instances it comprises = an in-depth study involving extensive data collection. Although there is variety in the scope of an evaluation, for the most part all evaluations rely on a few established research methods. In the book, *Learning from Our Buildings: The State of Practice Summary of Post Occupancy Evaluation*, the Federal Facilities Council writes that "traditionally post occupancy evaluations are conducted using questionnaires, interviews, site-visits, and observations of building users" (2001).

These methods of building evaluation have been in development since the 1960s (Preiser 2002), and as such, they tend to focus on aspects of the building that were readily available five decades ago. This is evidenced in Hiromoto's (2015) survey of 29 architecture firms, which found that building evaluations tended to rely on a few well established techniques:

- Quantitative measurements primarily of daylight, acoustics, and the thermal environment.
- Observational studies of density, utilization, efficiency, and differences between plans and occupation.
- Data collected from facility managers on energy and water usage.
- User surveys focused primarily on happiness, energy levels, perceived health benefits, and personal perceptions of the space.

In recent years there has been significant commercial interest in the possibility of adding sensors to the built environment in order to automate the process of taking quantitative measurements. Currently there are many vendors selling varieties of sensors that enable building owners to monitor everything from electricity usage to air quality, occupation rates, and light levels. While these systems are often an effective means of collecting high-fidelity, longitudinal data, the cost of installing and maintaining a sensor network means that this type of analysis is still the exception rather than the norm. Furthermore, the sensors are designed to capture only the most readily measurable aspects of the built environment, which means that more subjective interpretations of the space are not captured. For example, sensors can tell a researcher about a room's mean temperature and CO_2 levels, but this data only gives the researcher a rudimentary understanding of whether the room is successful and gives almost no insight into whether the user likes it.

2 Two meeting rooms at WeWork's City Hall location in New York.

In this paper we explore ways of longitudinally collecting data about people's spatial preferences. In other fields, such as advertising, traditional methods of capturing a person's opinion, such as surveys and interviews, have been supplemented with modern methods of inferring people's preferences through their online actions (Neef 2014). Google, for instance, has developed sophisticated machine-learning techniques to predict which advertisements people will prefer, not by surveying them about their preferences, but by observing which websites they visit (McMahan et al. 2013).

There is reason to suspect that a similar analysis could help evaluate buildings by using people's online behavior as a means to understand their preferences in the physical environment. To date, there seem to be no previous studies in this specific area, with prior research tending to focus on traditional post-occupancy evaluation methods coupled with electronic sensors (such as the research from Berkley's Centre for the Built Environment, Stanford University's Space-Mate program, and from Humanyze). One reason for the lack of previous investigation may be that only a few companies currently have access to the quantity and quality of data used in this paper (which captures how tens of thousands of people interacted with over seventy buildings during a span of three years), although the growing connections between the digital and physical environment means that many more organisations should have access to this type of data in the near future.

IDENTIFYING THE BEST MEETING ROOM

WeWork is a company that provides shared workspace, community, and services for entrepreneurs, freelancers, startups, and small businesses. As of April 2016, WeWork has over seventy locations in six countries. Each building has a similar mix of room types—offices, meeting rooms, lounges, phone booths, and other common WeWork amenities—which are customized for the local market.

One critical component of every WeWork location is the meeting rooms (Figure 2). In total, WeWork manages over a thousand meeting rooms spread across seventy locations. Functionally, each meeting room is similar, providing a space for teams to gather and discuss work with the aid of a whiteboard and projector. Although the rooms are functionally similar, they are each customized for their location, differing from other rooms in terms of capacity, furniture, visual appearance, and location within the building.

In 2015, the WeWork Research team was asked to identify whether any of the room variations were more successful than others. In other words, do people prefer meeting rooms with windows? Do they favor large rooms? Do they like the room in Boston that has swings instead of chairs?

Method
In order to understand the components of a successful meeting room, we first set out to identify which meeting rooms the

WeWork members most preferred. Given the scale at which WeWork operates, it was not practical to physically visit the hundreds of meeting rooms and conduct interviews with people spread across seventy buildings. To identify the best meeting rooms we instead examined the data that members of WeWork produce when they book a meeting room.

Members of WeWork are able to reserve meeting rooms for periods of time using either the WeWork website or the WeWork app for iOS and Android. Each reservation is stored in a central database, which, as of April 2016, has over a million reservations going back three years. For this study we were not interested in historic trends and therefore only examined reservations from a three month period (13 weeks) between June 6, 2015 and October 4, 2015—the most recent data at the time of the study. In total we analysed 158,000 meetings from 728 rooms in 44 buildings (WeWork has opened a number of locations since this study was conducted). Based on this data we were able to extract two pieces of information that we assumed may be indicators of which rooms people preferred:

1. Room utilization, which is the number of hours a particular meeting room was booked. We assumed the best rooms would be more popular and therefore utilized more often. To measure the utilization we calculated the percentage of time the room was reserved during peak hours (11 am to 4 pm on weekdays). We excluded meetings that occurred off-peak (before 11 am, after 4 pm, on the weekends, and on public holidays) since these tend to be more sporadic and inconsistent.

2. Lead times, which is the time between when a booking was made and when the meeting occurred. If a person books a room well in advance (a long lead time) they can generally pick their favourite room since most of the rooms will be available. But if a person books a room at the last minute, many of the best rooms will be taken and they'll have to make do with what's left. We therefore assumed that the best rooms would have long lead times. We calculated a room's lead time as the median of all lead times for meetings that occurred during peak hours in that room.

In addition to the lead time and room utilization rates, we asked members to rate meeting rooms. Members were recruited through the WeWork website and app. When they opened the website or app after a scheduled meeting, they were presented with popup that asked them "how would you rate your conference room?" with the option to rate the room on a scale of 1–5 (Figure 3). If a member rated the room 3 or less, they were given the option to leave additional comments. In total, we collected 23,140 ratings between June 6, 2015 and October 4, 2015.

Results

We began our analysis by looking at the individual buildings and seeing if there were any general patterns between buildings in terms of utilization, lead times, and ratings.

There is a strong correlation ($r=0.83$) between the median lead time of meeting rooms in a building and the median utilization of rooms in the building. This means that rooms in heavily utilized buildings generally need to be booked further in advance, which is to be expected since heavily utilized rooms are almost constantly booked and require advance booking to find available time.

3 The interface for booking a room (left) and for rating a room (right).

There is, however, a negative correlation between a building's median utilization and a building's median ratings ($r=-0.4$). This is an interesting finding because it demonstrates that ratings, lead times, and utilizations rate are not analogous and not measuring the same thing. One reason for this negative correlation may be that members tend to always give the room a rating of four or five unless there is an immediate problem with the room (such as the whiteboard markers running out). In this case, buildings with heavy utilization may generally have a lower median rating simply because there are more people using the rooms and therefore more opportunities for short-term maintenance issues to arise.

We also examined each of the 728 meeting rooms in relation to one another. Given the differences between buildings, we normalized the lead times, utilization, and ratings for each building, which let us identify whether a room was utilized more or less than the median room in a particular building.

There was a correlation (r=0.47) between a meeting room's lead times and its utilization, but there was no statistically significant relationship between the room's rating and either its utilization or its lead time. In other words, members did not appear to use highly rated rooms more than any other room. Based on these results, we believe that, in this instance, utilization and lead times seem to be better measures of people's preferred meeting rooms.

Had we done this study without access to the booking data, and instead relied solely on survey data, we would have come to a different conclusion about which rooms were most successful. But because we had both sets of data, the booking data and the survey data, we were able to triangulate a better understanding of the spaces. There are still imperfections to this measurement technique since there are clearly other factors driving a person's decision to book a room that are not accounted for in this iteration of the research (such as the way rooms are presented in the booking app). Nevertheless, this data, latent in WeWork's reservation system, proves to be a quick way to assess hundreds of meeting rooms without installing sensors, physically observing the rooms, or conducting interviews. Based on this research, WeWork was able to identify low performing rooms to upgrade, and update the WeWork design standards based on analysis of which room size was most preferred.

USING MACHINE LEARNING TO UNCOVER COMMON BUILDING ISSUES

When a member at WeWork has a problem, they can inform WeWork of the issue by speaking to a WeWork employee either in-person, through email, or via the WeWork app and website. As soon as the issue is raised, WeWork tracks the issue through to its resolution in a database. In the database, each issue is referred to as a 'ticket,' with each ticket containing data about the time the ticket was created, the subject of the ticket (taken directly from the text in the email or app), and other metadata related to the issue. As of April 2016, this database contained just over 180,000 resolved and open tickets.

The tickets in the database vary widely—there are requests for new keys, members notifying WeWork that their company has outgrown its office, people who can't connect to the wifi, and others who would like a different type of milk for their coffee. A subset of the tickets relate to spatial design—people informing us that they are too cold, that their door is broken, or that it is too noisy in their office.

In early 2016, the WeWork Research team was asked to look back through the tickets to identify whether there were any overall trends in WeWork's spatial design. Were certain HVAC systems leading to less complaints? Were particular buildings

more successful than others? Were there trends in maintenance issues?

Method

To find trends in the tickets, we first had to identify the subject of each ticket. We first categorized a thousand tickets by hand to identify the most common topics. We found five recurring themes: HVAC, noise, lighting, maintenance, and tickets not related to spatial design. Given these topics, we set about categorizing the remaining 180,000 tickets into one of these groups.

We initially tried to categorize the tickets using keywords. The algorithm we developed would find all the tickets that mentioned a particular keyword in the text related to the ticket. For instance, a ticket with the subject "I feel too hot," would be assisted the 'HVAC' category because it contained the keyword 'hot,' whereas a ticket containing the word 'loud' would be assigned the category 'noise.'

We tested the keyword method using 2000 randomly selected tickets, comparing the machine-generated classification to the classification assigned manually by the researchers. Although the keyword method was fairly crude, it was surprisingly accurate, particularly for categories that have keywords distinct from the other categories (HVAC, for instance, has words like freezing, warmer, and air conditioning, which tended to only appear in that category).

For HVAC, the algorithm had a precision of 82% and a recall of 92%, meaning that most HVAC tickets were correctly identified by the algorithm with few false positives. For categories that do not have distinct keywords, such as maintenance, which encompasses a range of issues that share few common words, the keyword method fared much worse. For maintenance it has a precision of 46% and a recall of 27%, meaning that only a quarter of the maintenance tickets were correctly identified and only half of the tickets flagged as maintenance actually were about maintenance.

In an attempt to more accurately classify the tickets, we tested the effectiveness of applying naive bayes classification to the tickets. Naive bayes is a classifier based on Bayes' Theorem that calculates the probability that an object belongs to a class given the features of the object and the frequency with which those features occur on other objects of a particular class. In this case, we wanted to classify the ticket based on the words in the ticket. For textual classification, naive bayes is a popular method often used "to define a baseline accuracy" (Kelleher, Namee, and D'Arcy 2015). We used the 'bag of words' method, whereby each word in the subject and description of a ticket became a feature of the ticket (Zhang and Zhou 2010). We trained the classifier on 3000 hand-categorized tickets, which allowed the classifier to learn the probability that tickets classified as being noise-related also contained the word 'sound' somewhere in the bag of words from the subject and description. Another 2,000 tickets were used as a test set, with each ticket classified by both the classifier and a human so we could understand the accuracy of the classifier.

The naive bayes classifier was generally better than the keyword classifier for every ticket category except those tickets related to noise (where it had a recall of 43% and a precision of 31%). The noise category seemed to be difficult for the naive bayes classifier because many of the noise-related tickets were related to other categories. For example, a ticket that states "my air conditioner is making a loud noise," could be categorized as an HVAC issue, but we categorized it as relating to noise since it is primarily a noise complaint. Noise was the only category the naive bayes classifier struggled with, and in other categories, like maintenance, it performed significantly better than the keyword classified, achieving a recall of 96% and a precision of 70%.

Outcomes

The naive bayes classifier was applied to the 180,000 tickets in the database. Although the classifier isn't perfectly accurate, at a macro level the classification is accurate enough to allow WeWork to identify which aspects of the built environment cause the most problems for WeWork members. This has enabled us to prioritise our research so that it will have the most impact on member experience. We have also setup a dashboard that allows WeWork employees to compare how each building is performing, letting them identify macro issues as they emerge in real-time and get fixes out to them quickly.

There are some limitations to this method. Most notably, members tend to only create tickets for things they know can be fixed, such as the temperature, and they might find it hard write a ticket that articulates more subtle and less actionable feelings about their environment. The data is also biased towards negative experiences of the built environment; it tells us what frustrates members, but it doesn't tell us what delights them since people don't create tickets when everything is going well. The data therefore isn't a complete record of how people feel about their environment. That said, the data does provide a foundation for continuously monitoring how people feel about the built environment without needing to constantly send surveys gauging people's preferences.

DISCUSSION: WHAT THE 'DATA EXHAUST' TELLS US ABOUT BUILDINGS

The term 'data exhaust' has appeared recently as a way of describing the data that a person leaves behind in the digital world as they go about their lives. This includes everything from data stored in server logs when a person visits a website to the data recorded for financial purposes about a credit card transaction. These pieces of information often tell detailed stories about the actions of individuals, and many companies and researchers have been using this data to understand people's preferences for everything from search results to advertisements (Neef 2014).

The two case studies in this paper, one analyzing meeting rooms through booking data, and the other evaluating buildings using a ticketing system, begin to articulate how this data exhaust may be used in the evaluation of architecture. The original contribution is not the methods of analysis used, but rather their application to the built environment. They show how data already produced by the people inhabiting buildings, data that is a latent product of the built environment, can be used to analyse people's preferences for particular aspects of the built environment. It shows how traditional means of post-occupancy evaluation, such as surveys, interviews, and site visits, as well as modern methods of electronic sensing, may soon be complemented by this new way of understanding architecture and the people that inhabit it.

In many ways, this type of analysis seems long overdue. Architects have long experimented with computation in the early stages of design, findings ways to create and manufacture buildings using algorithms—some going as far as to claim that computation has given rise to a new style of architecture (Schumacher 2009). But while the process of designing buildings has evolved to embrace computation, the process of evaluating buildings has remained largely unaffected by the advent of computation and the rise of the data exhaust.

In part, research in this area may be limited because of the challenges associated with creating a data set. Most machine learning techniques require fairly large data sets. In this paper, the meeting room analysis looked at data from 158,000 meetings and the ticket analysis looked at 180,000 tickets. Collecting this data required that data collection systems were implemented across multiple buildings for many years prior to the analysis. A lot of architects and researchers may not have access to this scale of data. Even if they do have access to the prerequisite buildings and data, a major limitation of this type of analysis is that it requires a building to be settled, operational, and producing data before it can be evaluated.

4　Seattle - WeWork South Lake Union in Seattle.

If the right data can be gathered, the potential payoff is that aspects of the building can be analyzed without requiring that the building's inhabitants complete a survey or interview. In certain instances it may even provide a more complete picture. For instance, the tickets related to HVAC issues in the second case study give insight into how people felt about their thermal environment over a number of years. While traditional survey methods give insight into how people feel at a particular moment in time, these new methods lend themselves to longitudinal studies that allow researchers to passively collect data about people's perceptions over a period of years. Analysis of this longitudinal data exhaust potentially opens up new avenues of research, and new ways of understanding buildings that take into account the evolution of the project over time.

CONCLUSION

Most buildings are constructed without any form of post occupancy evaluation. In part, this may be due to the typical methods of post occupancy evaluation, which have not benefited from advances in computation to the same extent as other parts of the design process.

In this paper, we have presented two ways that computation and machine learning can be applied to datasets generated by building inhabitants in order to extract information about how people perceive their built environment. Both of these methods have proved successful in practice and have demonstrated tangible impacts on the way similar buildings are designed and

operated. This research shows, in two instances, that data generated by people through the course of their day often carries with it latent information about how they perceive their physical environment, which can be a valuable tool in the long-term analysis of physical environments.

REFERENCES

Bordass, Bill, Adrian Leaman, and Paul Ruyssevelt. 1999. *PROBE STRATEGIC REVIEW 1999 FINAL REPORT 4: Strategic Conclusions.* London: Department of the Environment, Transport and the Regions. http://www. usablebuildings.co.uk/Probe/ProbePDFs/SR4.pdf

Duffy, Frank. 2008. *Work and the City.* London: Black Dog Publishing.

Federal Facilities Council. 2001. *Learning From Our Buildings: A State-of-the-Practice Summary of Post-Occupancy Evaluation.* Washington, D.C.: National Academy Press.

Hiromoto, Julie. 2015. *Architect & Design Sustainable Design Leaders Post Occupancy Evaluation Survey Report.* SOM: New York. http://www.som. com/FILE/22966/post-occupancy-evaluation_survey-report_update_2.pdf

Kelleher, John D., Brian Mac Namee, and Aoife D'Arcy. 2015. *Fundamentals of Machine Learning for Predictive Data Analytics: Algorithms, Worked Examples, and Case Studies.* Cambridge, MA: MIT Press.

McMahan, H. Brendan, Gary Holt, D. Sculley, Michael Young, Dietmar Ebner, Julian Grady, Lan Nie, Todd Phillips, Eugene Davydov, Daniel Golovin, Sharat Chikkerur, Dan Liu, Martin Wattenberg, Arnar Mar

Hrafnkelsson, Tom Boulos, and Jeremy Kubica. 2013. "Ad Click Prediction: A View From the Trenches." In *Proceedings of the 19th ACM SIGKDD International Conference on Knowledge Discovery and Data Mining*, edited by Inderjit S. Dhillon, Yehuda Koren, Rayid Ghani, Ted E. Senator, Paul Bradley, Rajesh Parekh, Jingrui He, Robert L. Grossman, and Ramasamy Uthurusamy. New York: ACM. 1222–1230. http://doi.org/10.1145/2487575.2488200

Neef, Dale. 2014. *Digital Exhaust: What Everyone Should Know About Big Data, Digitization and Digitally Driven Innovation*. Upper Saddle River, NJ.: Pearson FT Press.

Preiser, Wolfgang F. E., and Ulrich Schramm. 2002. "Intelligent Office Building Performance Evaluation." *Facilities* 20 (7): 279–287. doi:10.1108/02632770210435198.

Schumacher, Patrik. 2009. "Parametricism: A New Global Style for Architecture and Urban Design." *Architectural Design* 79 (4): 14–23.

Zhang, Yin, Rong Jin, and Zhi-Hua Zhou. 2010. "Understanding Bag-of-Words Model: A Statistical Framework." *International Journal of Machine Learning and Cybernetics* 1 (1): 43–52. http://doi.org/10.1007/s13042-010-0001-0

IMAGE CREDITS

Figures 1–4: © WeWork

Daniel Davis – Director of Spaces and Cities Research at WeWork. Based out of WeWork's headquarters in New York, Daniel leads a team of researchers investigating how to design workplaces so that people feel happier, more productive, and more connected to their community. He originally trained as an architect in New Zealand and later did a PhD in computational design at RMIT University's Spatial Information Architecture Laboratory in Australia.

The Tectonic of the Hybrid Real

Data Manipulation, Oxymoron Materiality, and Human-Machine
Creative Collaboration

Laura Ferrarello
Royal College of Art

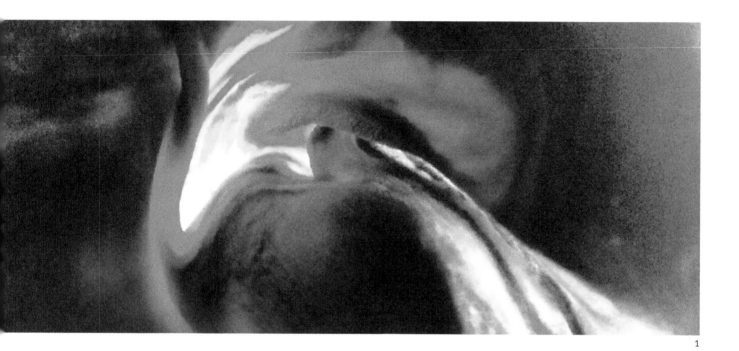

1

ABSTRACT

This paper describes the latest progress of the design platform Digital Impressionism (DI), created
by staff and students in the Information Experience Design programme at the Royal College of
Art in London. DI aims to bridge human creative thinking with machine computation, under the
theoretical method/concept of oxymoron tectonic. Oxymoron tectonic describes the process under
which hybrid materiality, that is the materiality created between the digital and the physical, takes
form in human-machine creative interactions. The methodology intends to employ multimaterial 3D
printers in combination with data manipulation (a process that gives data physical substance), point-
clouds, and the influence of intangible environmental data (like sound and wind) to model physical
forms by interfacing digital and physical making. In DI, modeling is a hybrid set of actions that take
place at the boundary of the physical and digital. Through this interactive platform, design is experienced as a complex, hybrid process, which we call a digital tectonic; forms are constructed via a
creative feedback loop of human engagement with nonhuman agents to form a creative network of
sustainable and interactive design and fabrication. By developing a mutual understanding of design,
machines and humans work together in the process of design and making.

1   Oxymoron materiality. Detail of a
    colour 3D printed and 3D scanned
    apple.

INTRODUCTION

The making of a work of art—poiesis in Greek—is a process in which the embodied relationship between humans and objects is facilitated by machines that mould materials. The act of modeling is a set of creative actions, called design, which humans activate to give shape to physical objects. In such a process, the knowledge of material properties, fabrication techniques, and machine capabilities are pivotal; the poiesis of design is, indeed, an interwoven process, which is informed in real time by the actors that take part in it, and results in the object.

Since design began adopting computational techniques, the relationship among humans, machines, and materials has changed. The physicality of modeling gets simulated in digital space, since to model means to execute commands or manipulate graphic controls created by lines of code. The direct relationship between human creativity and materials is lost; the "informed" act of modeling, given by the material's physical constraints, is replaced by the human's perceived intuition or memory of material physical properties (Titmarsh 2006). The switch from physical contact to mediated manipulation, however, enables designers an extra degree of freedom. As the physical environment is simulated, CAD software does not force the human mind to think through the constraints of physical reality, as normally happens during any fabrication stage. In such a constraint-free environment, digital poiesis grants new forms, which nonetheless would be complex to design by any human brain alone (Hatzelis 2006).

It follows that computers enhance design with a new set of automated skills, like the understanding of large datasets. Computers understand large numbers in fragments of time; they can array them in lines of code. When forms are acknowledged as fields and matrices (Allen 1999), complex and nonlinear geometry enters the practice of design and architecture, which results in shapes "born digital." The granularity of data transforms forms and materials in vectorial flows of substance directed by code; hence a new kind of material sensation that is delivered by the aesthetic quality of the represented object emerges. This code-driven tectonic has integrated new digital material qualities and feedback with design, from the language of machinic procedures. This feedback makes design a procedural set of actions that coordinate process through continuously returned values (Ferrarello, Pecirno, and Spanou 2015).

CONTEXT AND METHODOLOGY

The Materiality of the Hybrid

The emergence of digital tools in design has proceeded in phase with the digitalization of the physical realm. The ubiquitous presence of technology in the form of interfaces has helped to sync the physical with the digital world; sensors transform physical

2

3

2 Digital Materiality. Matter created in ZBrush from a 3D scanned cotton fabric.

3 Oxymoron Perceptions. The shift of materiality of a 3D scanned Bialetti metal coffee machine in the digital real.

4

5

6

information into digital data. The resulting granular understanding of the physical and digital is nonetheless our filter for interpreting our surroundings. If physical objects are the entities we employ to reify the physical world (Žižek 1997), data are the matter by which we reify the digital, and increasingly the physical as well.

At first, the digital world looked at physical metaphors (skeumorphisms) in order to enable human familiarity; for instance the Save icon and the Windows operating system (Gwilt, 2006). However, we are currently witnessing an inversion: we Google physical things, we hashtag topics, we photograph objects as if digital renderings. In other words, the iconography of the digital domain no longer relies on our understanding of the physical world. We live in the world of the hybrid (Ferrarello 2014); images, ubiquitous sensors, and interfaces transmit data to and from the virtual to the extent that our mind dwells in both domains considered as whole.

Digital Material Behavior

The element that most clearly distinguishes the digital from the physical is the human body, due to its capacity to acknowledge surroundings via senses that transform perceptions into tangible experiences (Karana 2015; Malafouris 2013). Although vision plays a primary role, humans use many senses simultaneously, which helps construct our physical reality by assigning unfamiliar intangible entities physical materiality. In such a process, the physical relationship the human body has with its surroundings is pivotal (Ferrarello 2016).

In the digital space of CAD software, two kinds of events happen: on one hand, our traditional understanding of the physical environment is simulated; on the other hand, there is the freedom of being out of the physical environment, where paradoxical association of physical actions can take place—Boolean operations for instance. To model in a gravity- and physics-free environment engages designers with new formal approaches, which can push creativity and imagination beyond physical possibilities.

In the digital environment, physical materials lose their physical properties. Materials become hybrids for the sake of modeling operations (Titmarsh 2006). ZBrush software, for instance, enables "painting" different materials (from wood to gold) onto the same digital mesh, retaining a hyperrealistic visual quality of the chosen material. Indeed, 3D scanned objects keep their material aesthetic quality when imported in the digital space. However, at the granular level, such materiality is lost. Objects are clouds of 3D points connected by meshes. When physical materiality is digitized, it is subject to digital rules, although resembling its physical appearance. In ZBrush, a 3D scanned apple doesn't behave as a physical one; an apple.obj file can be modeled as if clay, while preserving the aesthetic of the physical apple.

But more recently, the freedom of digital modeling has been challenged with digital fabrication machines, like 3D printers, that demand specific rules in order to assign digital objects to physical matter. Such rules direct the modeling process, but do not act as feedback, in which return values become part of the design.

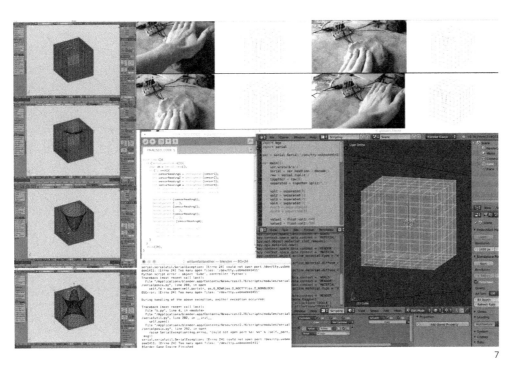

4 UV Flower. Texture of a 3D scanned flowers created with Autodesk 123dCatch software.

5 The Process of the Oxymoron. Physical apple, 3D scanned apple, polymeshed apple, UV texture, digital sculptured apple and apple. obj.

6 Oxymoron Apple. Colour powder 3D printed apple.

7 Interactive Clay. Screenshots displaying the interaction between the cube and human touch transmitted to Blender via an Arduino microcontroller.

7

There is not yet a mutual and interactive communication in terms of design thinking and design process.

The Oxymoron as Methodology
DI follows the concept of the oxymoron as methodology. In literature, the oxymoron is a paradox resulting from the juxtaposition of two elements that appear to be in contradiction. Here, a 3D scanned apple that becomes smooth as clay through the process of digital modeling generates an oxymoron materiality, which is a hybrid composite constructed from the interaction of the digital and physical (Ferrarello 2016). The oxymoron re-establishes communication between humans and materials via the interactive collaboration between human and machines. The oxymoron enables the assemblage of physical paradoxes to generate forms. This oxymoron materiality juxtaposes paradoxical physical properties—read as voxels (3D pixels) in digital computation to create composites. The oxymoron emerges from the understanding machines have of materials, juxtaposed with the human sensory expectation of it. Humans understand an apple because of the physical memory of touching it, while computers read it as a meshed cloud of 3D points with texture mapped onto it.

RESULTS
Under the guide of the oxymoron, DI has been tested through various investigations undertaken by IED students.

Interactive Clay
With William Fairbrother, we investigated the concept of hybrid materiality using a pressure sensor, embedded in a piece of clay, connected to Blender 3D software via an Arduino microcontroller. When the human hand moulds the clay, a digital cube displayed in Blender reacts accordingly. The human input is then stored as Python variables in Blender, with the intention of transmitting it back to the physical clay as vertical force via an actuator. The intention is to explore the computer's understanding of human touch by using physical modeling materials as interface (Ferrarello and Fairbrothe 2015). The problem faced in this experiment is how to translate the physical behavior of clay to the digital one.

Designing with the Instagram API and "Booleaning" Pointclouds
Ker Siang Yeo used the Instagram API to model shapes using RGB colors. Through a hashtag, he collected Instagram posts of Battersea Park in London. He then created a pointcloud from the RGB data extracted from the images through GMic software, which use RGB as xyz coordinates. The pointcloud was then imported into MeshLab to create a mesh. This experiment uses a pointcloud as "sharable" tool for digital modeling. The problem we are facing next is how to control the projection of the RGB data to extract a clean pointcloud.

Ker Siang used a similar methodology to model an apple from image-based models and 3D scanned objects. Using Boolean difference, the pointcloud extracted from the RGB and a 3D scanned apple. This experiment focused on the quality of digital

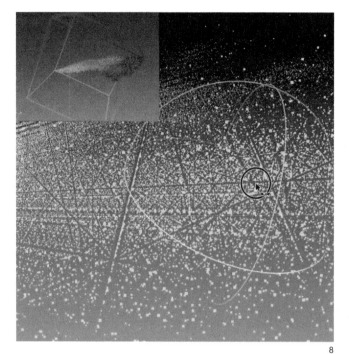

8

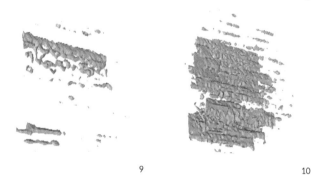

9   10

8   Instagram API. Battersea Park RGB pointcloud.

9   Voxel Modeling.

10 Voxel Modeling.

matter (Ferrarello 2014) as modeling information. This case shows a better way to use the RBG pointcloud for modeling purposes.

Modeling with Voxels

Suramya Kedia experiments with modeling processes that employ voxels as material placeholders. She uses the variation of sound to time to aggreagate forms. Her intention is to understand how properties can pass from material to material via boundaries. By using VSpace software, she creates an abstract environment based on material properties interactions (boundaries), which confront behavior that triggers formal aggregation out of the input information. In her experiment, Kedia used sound and time as starting information. She modulated the concentration of sound in relation to time, which created different kinds of

formal aggregation. By reversing the design approach, which seeks forms before material performance, she uses data flow as a medium of formal material aggregation that shapes a new kind of physical/digital tectonic.

CONCLUSION

Creativity is the human faculty that drives innovation for its capability to adapt, reroute, and challenge surrounding conditions. Digital fabrication machines have helped to bridge digital design with the physical world. Under the guide of the oxymoron, DI intends to constitute an ecosystem in which humans and machines equally participate in a dialectic creative design cycle. By bringing together the procedure of digital design—i.e., variables and nested arrays—and the constraints of physical material properties, this paper described a series of examples that tackle the topic through different aspects.

The challenge this paper proposes goes beyond the mere use of machines. The paper draws a possible route for the profession of the architect/designer; as artificial intelligence improves the capacity to learn, DI proposes to construct a common territory where both can operate by challenging each other's skills. The result is a design language that enables creative dialogue, communication, and interaction between humans and machines. DI pursues a system where collaboration between humans-machines-humans triggers innovation (Geels 2005). Data help to create a loop in this collaboration by linking the physical to the digital environment. The next step is to make complexity—data, pointcloud, voxels, multimaterial 3D printers—a looping information that fosters sustainability in design and manufacturing.

By treating data as matter, DI augments human senses in data modeling techniques; data are no longer stored in databases but treated as the matter of the everyday. DI is a platform that sees design as a body/mind experience of our surroundings, and that leverages human senses and machinic processes to construct forms through an oxymoron tectonic. The process of digital modeling and fabrication becomes a looping sequence of dynamic actions where designers and machines remodel the environment by taking into account the complexity of digital and physical information.

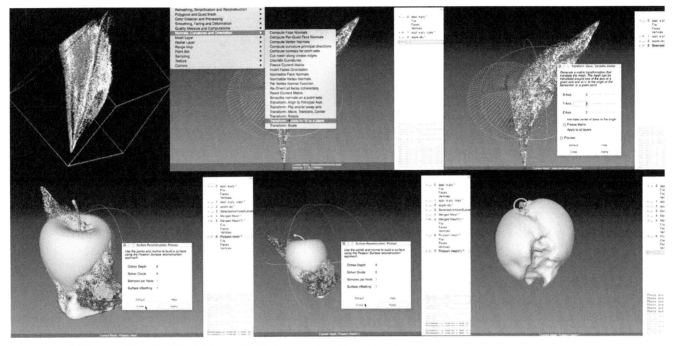

12 Instagram API. The apple.JPG and its RGB pointcloud and the Boolean operation of the pointcloud.

ACKNOWLEDGEMENTS

Thanks to Information Experience Design (IED) and Kevin Walker for giving the project the space to operate. Thanks to William Fairbrother for the help with the Interactive Clay experiment. Thanks to IED students Ker Siang Yao (Designing with Instagram and Booleaning with Pointclouds) and Suramya Kedia (Modeling with Voxels).

REFERENCES

Allen, Stan. 1999. *Points + Lines*. New York: Princeton University Press.

Ferrarello, Laura. 2016. "The Oxymoron of Touch: Tactile Perception of Hybrid Reality Through Material Feedback". In *Digital Bodies: Creativity and Technology in the Arts and Humanities*, edited by Sue Broadhurst and Sara Price. London: Palgrave Macmillian, forthcoming.

Ferrarello, Laura, Michael Pecirno, and Kelly Spanou. 2015. "Digital Impressionism." In *Digital Factory. Proceedings of DADA 2015: International Conference on Digital Architecture*, edited by Weiguo Xu and Weixin Huang. Shanghai: DADA. 361–379.

Ferrarello, Laura, and William Fairbrother. 2016. "Processes of Artefact Creation in the Hybrid-Reality Engaging with Materials though Material Oxymoron". In *Artifacts*. Forthcoming, December 2016.

Geels, Frank. W. 2005. *Technological Transitions and System Innovations: A Co-Evolutionary and Socio-Technical Analysis*. Cheltenham, UK: Edward Elgar.

Gwilt, Ian. 2006. *MadeKnown: Digital Technologies & The Ontology of Making*, edited by Ian Gwilt. Sydney: University of Technology Sydney. 61–79.

Malafouris, Lambros. 2013. *How Things Shape the Mind: A Theory of Material Engagement*. Cambridge MA: MIT Press.

Karana, Elvin. 2015. "Design for Material Experiences." In *Tangible Means: Experiential Knowledge Through Materials*, edited by Anne L. Bang, Jacob Buur, Irene Alma Lønne, and Nithikul Nimkulrat. Kolding, Denmark: EKSIG. 11.

Žižek Slavoj. 1997. *The Plague of Fantasies*. London New York: Verso.

IMAGE CREDITS

Figures 1, 3–4: © Ferrarello, 2016

Figure 2: © Ferrarello, 2013

Figures 5–7: © Ferrarello, 2015

Figures 8, 12: © Yeo, 2016

Figures 9–10: © Kedia, 2016

Laura Ferrarello is an architect, designer and researcher. She currently teaches at the Royal College of Art in London and looks at the interaction between digital/physical reality through design processes capable of enhancing human senses through machine interaction. Laura received a PhD in Architectural Design at IUAV University of Venice (2010).

Beetle Blocks

A New Visual Language for Designers and Makers

Duks Koschitz, PhD
Pratt Institute

Bernat Ramagosa
Arduino, Spain

Eric Rosenbaum, PhD
Massachusetts Institute of Technology

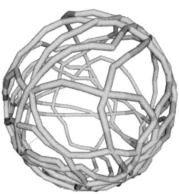

1

ABSTRACT

We are introducing a new teaching tool to show designers, architects, and artists procedural ways of constructing objects and space. Computational algorithms have been used in design for quite some time, but not all tools are very accessible to novice programmers, especially undergraduate students. 'Beetle Blocks' (beetleblocks.com) is a software environment that combines an easy-to-use graphical programming language with a generative model for 3D space, drawing on 'turtle geometry,' a geometry paradigm introduced by Abelson and Disessa, that uses a relative as opposed to an absolute coordinate system. With Beetle Blocks, designers are able to learn computational concepts and use them for their designs with more ease, as individual computational steps are made visually explicit. The beetle, the relative coordinate system, follows instructions as it moves about in 3D space.

Anecdotal evidence from studio teaching in undergraduate programs shows that despite the early introduction of digital media and tools, architecture students still struggle with learning formal languages today. Beetle Blocks can significantly simplify the teaching of complex geometric ideas and we explain how this can be achieved via several examples. The blocks-based programming language can also be used to teach fundamental concepts of manufacturing and digital fabrication and we elucidate in this paper which possibilities are conducive for 2D and 3D designs. This project was previously implemented in other languages such as Flash, Processing and Scratch, but is now developed on top of Berkeley's 'Snap!'

1 Example of:
– a stack of blocks (the program)
– the 3D rendered object
– a 3D print

INTRODUCTION

The audience for Beetle Blocks is comprised of architects, designers, and artists, who have no prior knowledge of programming and wish to learn how to use computational concepts as part of their design process. The courses and workshops that have been taught with Beetle Blocks in preparation for this paper have taken place in undergraduate departments of architecture and design.

Programming is not yet fully integrated in art and architecture programs as a foundational skill and students often find that the learning curve is steep. A graphical language that is easy to learn is a first step in making programming more accessible (Kelleher and Pausch 2005). Beetle Blocks is based on Scratch (Maloney et al. 2010) and now implemented on top of Snap!, both graphical control flow languages. This approach to programming requires the user to formulate every explicit step of an algorithm in a visual way, without the frustrating syntactic pitfalls users experience with typical text-based programming languages. Unlike other graphical programming paradigms, it also positions the fundamental concepts of computation early in the learning process, thus making it more likely that learners will build on these concepts and move on to tasks of greater complexity. The goal is to get to the teaching of computational ideas very quickly in a curriculum and to not be bound by commercial software environments that are more suitable for the production of design projects.

The 3D environment of Beetle Blocks is intuitive because it is modeled after 'turtle geometry' (Papert 1993). Turtle geometry uses a relative rather than an absolute coordinate system, allowing learners to relate their knowledge of body movements to the movements of the computer's representation of the turtle. The combination of a graphical control flow language and turtle geometry is not new, but we believe that the specific deployment of our software, together with fundamental lessons in geometry and digital fabrication, can make a difference in teaching programming early in design education.

Thinking about the construction of space procedurally is a powerful alternative to analog thinking and we believe that both approaches should be taught in architecture programs. Design is often judged on its coherence and how well a formal concept is thought through. Using procedures to construct a spatial constellation requires rigor in the design process and that in turn assists designers with a way to create coherent designs. Another goal consists of teaching how to create machine instructions for digital fabrication, which we address visually via the use of a control flow language.

There is no doubt that relational or parametric modeling is

powerful for architectural design and we believe that a deeper understanding of how to control geometric relationships in a procedural way will become helpful for designers. By learning how to create procedures, we hope that designers will acquire an understanding of how computation can be used for a design project. After learning how to design with Beetle Blocks, students and users in general can transition to commercial software to realize complex projects, where they can apply the acquired knowledge.

THE GEOMETRY PARADIGM

In order to teach computational steps expediently, we decided to link a visual graphical language to a 3D environment via turtle geometry. The local coordinate system can be controlled step-by-step to move about in 3D.

Drawing With a Turtle

In the Logo programming language for children (Papert 1993), the turtle is an object on the screen that is moved around with commands for moves and turns such as 'forward 10' and 'right 90.' Turtle geometry relies on relative movements as opposed to absolute coordinates common on CAD software. Relative movement is intuitive, because we can draw on knowledge of our own bodies' movement in space. This identification between the turtle and one's own body is a form of 'body syntonicity,' which Papert determined to be crucial for learning. It is easy to forget that the idea of an absolute coordinate system (or any invariant) is something that must be constructed or acquired, which we generally do when we are very young or learn CAD software for the first time. We use our intuitive knowledge of a relative coordinate system to construct the idea of an absolute system. Local instruction sets are also concise and easy to read. Relative coordinate systems are therefore a good starting point for learning geometry. In Logo, turtle movement was restricted to a plane (Figure 2) and Papert built a physical corollary in the form of a tethered floor roamer that could drop a pen and draw as it moved (Figure 3).

Abelson and diSessa elaborated on turtle geometry, exploring the possibilities of turtle movement in 3D space in mathematics (Abelson and diSessa 1986). They invented many algorithms that relate continuous geometry to turtle geometry. There have been various implementations of Logo that enable turtle drawing in 3D (Petts 1988), such as Elica (www.elica.net) and Logo3D (logo3d. sourceforge.net), but none of these use a graphical language. StarLogo TNG (education.mit.edu/starlogo-tng) (Colella, Klopfer, and Resnick 2001) has both a graphical language and a 3D environment, but it is designed for creating games and simulations, not for exploring geometry. It allows one to move the turtle up and down, but one cannot rotate it out of the plane for example.

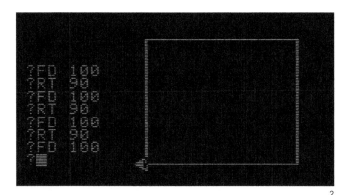

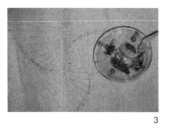
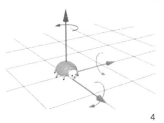

2 Logo instructions for a square (rectangular image due to CRT screen).
3 Floor roamer drawing on paper (Seymour Papert).
4 Positive directions and positive Beetle rotations.

We refer to the turtle as 'beetle' and use conventions of positive relative movements and turns known in CAD software. Red, green, and blue indicate the main directions in x, y, and z (Figure 4).

The Necessity for a Discrete Approach

In mathematics and geometry, shapes are generally described in continuous ways and eighteenth-century mathematics have allowed us to understand the world as a construction of smooth functions. When dealing with computational systems, we are typically bound by the capabilities of machines with discrete registers or switches.

The example of a circle helps to illustrate the difference between continuous and discrete representations. A circle is a simple shape of Euclidean geometry consisting of those points in a plane, which are equidistant (r) from a given point, the center (c). There are several representations that invoke the mental image of a circle, for example a circle drawn on paper, or the mathematical definition for its circumference $2r\pi$. In discrete steps we can describe a circle by moving forward 1 unit and turning one degree to the right 360 times.

All three representations have deep relations to the way we teach design: some artistic, some engineering-oriented, and some maybe more scientific. It is important that designers become acquainted with all of them and can control the relevant methods to express what they want to design. We will focus on the discrete approach, as it allows us to break down geometric constructs into

individual steps and these steps are conducive to writing code. Moving the local coordinate system through the use of many steps can be thought of as writing algorithms for geometry.

DESIGNING A PROGRAMMING LANGUAGE

In this section, we elucidate our decisions as they pertain to the type of formal language we wanted to create. As the goal is to teach computational ideas quickly, we are faced with questions about syntax, visual feedback, and occluding information that might be unnecessary.

Control Flow Versus Data Flow

Control-flow languages have an execution model in which commands are run in a sequence. There is generally a single program counter that points to the next command to be run. The movement of this counter is the flow of control. In data-flow languages, by contrast, all parts of a program may be executing simultaneously. Data flows from one component to another. Control-flow languages are typically represented as lists of instructions, while data-flow languages are typically represented as nodes connected by lines. Data flow is a popular paradigm for graphical languages because its topological structure is naturally represented as a graph.

Beetle Blocks uses the control-flow execution model, because our emphasis is on making programming understandable. The control-flow model allows us to show the execution of a program explicitly, making it easier for users to see what the program is doing at every single step (Figure 5).

Black Boxes

One challenge in the design of graphical languages for learning is exactly where to put the 'black boxes.' By black box we mean a primitive unit that cannot be opened up in order to see what is inside and learn how it works. All programming languages above the level of assembly code are structured hierarchically, with each command encapsulating a set of other commands, until the language bottoms out at its primitives, which are its black boxes.

For example, a programming language might have a command for drawing a circle. This command might encapsulate lower level commands within the very same language for geometric calculations that the user can access. However, these lower level commands might encapsulate primitives, such as commands for rendering pixels on the screen that are not accessible to the user. The primitives, written in a lower level language, are thus black boxes.

The design question of where to put the black boxes is really the question of which concepts users should have access to, and which should remain hidden. A language designed for learning

Beetle Blocks Koschitz, Ramagosa, Rosenbaum

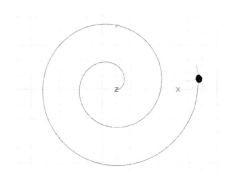

5

should consist of a set of commands that combine accessibility and flexibility, and hide the lower-level commands that provide an irrelevant level of detail or are not sufficiently intuitive (Papert 1993).

In Beetle Blocks, the commands are for manipulating the Beetle's properties (such as its position and rotation), generating a set of basic shapes, and controlling the flow of the program. We chose this set of commands because it makes simple tasks such as drawing a shape very easy and immediate, while also making available a broad landscape of expressive algorithmic possibilities. All the low-level processes that render shapes to the screen, including their shading, textures, and lighting, are hidden from the user.

Expanding the Scratch paradigm

Beetle Blocks makes use of several of the design principles developed for the Scratch programming language (Resnick et al. 2009). These include immediate feedback, commands as visual building blocks, and error prevention. The choice for Snap! is motivated by its expanded functionality and Javascript basis. The language draws heavily from Mitchell Resnick's Scratch project and has been created in collaboration with Brian Harvey, Jens Moenig, and Bernat Ramagosa. A salient feature of Beetle Blocks is its Lambda functionality (Harvey and Mönig 2010), a specific way to express complex computational ideas, which is particularly useful in a pedagogical context.

Clicking on a block has an immediate effect in Beetle Blocks, and most blocks have a visible result. This encourages a tinkering process through which users can learn what blocks do simply by trying them out. Clicking on a stack of blocks causes it to run immediately, which makes it easy to rapidly test different possibilities without waiting for the compilation step required by many programming languages.

The blocks themselves have connectors at the top and bottom, making it obvious that they snap together in a stack. This visual affordance makes it clear that the basic interaction in Beetle

Blocks is stacking together blocks into a program (Figure 6). Beetle Blocks has several properties that prevent frustrating errors. The blocks have certain shapes that only allow them to connect in ways that make syntactic sense (i.e., the stack of blocks will always do something, even if it is not what was intended). There is nothing like a 'compiler error' at all in Beetle Blocks, and no error messages are necessary. Beetle Blocks also uses a 'fail soft' principle. This means that all parameters in blocks have sensible default values, and will do something sensible even if the parameters are left empty. These properties eliminate many of the frustrations that can be the main reason for novices to give up learning to program.

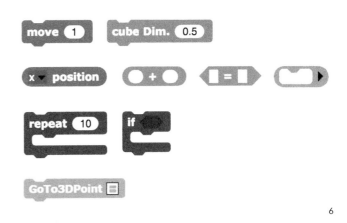

6

5   The spiral is an effect of using a scaling function on moves and turns.

6   Selection of blocks (movement, shape; position, check for equal value, addition, transform procedure to be used as input; repeat block; user defined block).

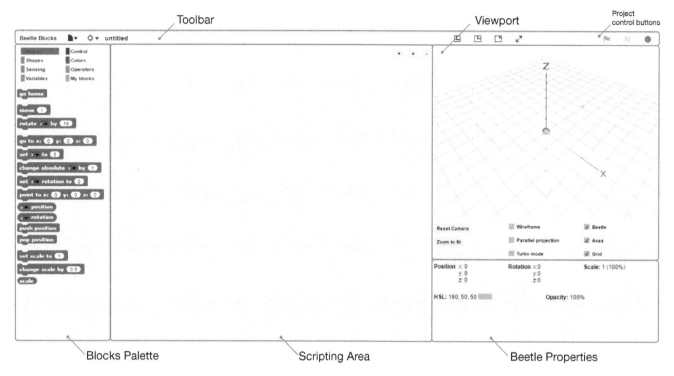

Toolbar — Viewport — Project control buttons

Blocks Palette — Scripting Area — Beetle Properties

7 The new UI.

The User Interface

The new version of Beetle Blocks that is based on Snap! uses a user interface (UI) that will seem familiar to Scratch and Snap! users. The Toolbar at the top has two menus for file management and settings. On the right, we kept the control buttons that allow users to run or stop a program (Figure 7).

The user can drag and drop blocks from the Blocks Palette on the left into the scripting area and can see the 3D model in a Viewport on the top right. The bottom right area allows users to modify properties of the Beetle and the Viewport, and also displays information about the current state of the Beetle.

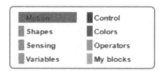

8 Selection of blocks (movement, shape; position, chec

Regarding the color scheme for the blocks, we decided to keep the color coding for movement, variables, and operators. The control blocks are grey and custom procedures are in a lighter blue (Figure 8). Snap! allows us to use many additional features that are available beyond the functionality of Scratch, such as advanced list manipulation and lambda functionality.

TEACHING COMPUTATIONAL IDEAS

We aspire to connect mathematical and computer science ideas to design (Eisenberg 2002) and use fundamental ideas of constructionism (Rusk, Resnick, and Cook 2009). We demonstrate how our language can be used to convey computational ideas to novice programmers by including several examples we have used in workshops and other courses.

Combining Trigonometry and Turtle Moves

When teaching fundamental ideas in computation and geometry, it is productive to show students how turtle geometry can be used to construct mathematically defined shapes via discrete steps. The example of a polygon relies on trigonometric relations of the distance between its center and a vertex [radius], the distance of a move forward [moveDim], and the angle between the center and two adjacent vertices (Figure 9). A user can specify the size of the radius and how many edges the polygon should have. The comments inside the yellow blocks tell users which values can be altered. This discrete circular polygon around a center uses turtle moves and trigonometry to compute the relevant rotations and steps.

Other trigonometric functions can be used in conjunction with beetle moves to describe mathematical objects in 3D. The following example uses the extrude command and is based on incrementing two angles, [theta] and [phi]. The positioning of

Beetle Blocks Koschitz, Ramagosa, Rosenbaum

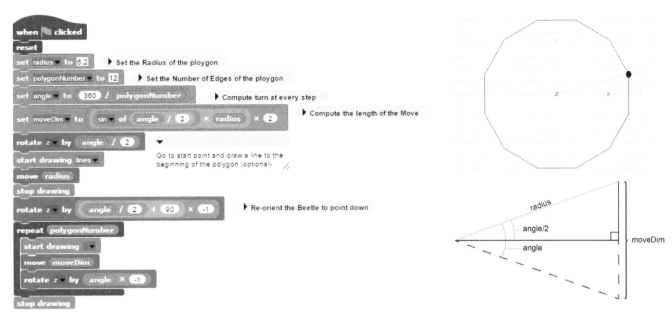

9   Program that draws a polygon (user input: radius, number of edges).

the beetle occurs via the [go to (x) (y) (z)] command, which uses absolute coordinates rather than local moves. Sine and cosine function are used to compute the positions along a torus and the program completes two cycles around a torus (Figure 10). The [go to (x) (y) (z)] allows for global positioning and instructors can teach the relevance of absolute coordinates in conjunction with trigonometric functions.

A Boolean Operation that Relates to Geometry

Boolean operations mean different things in geometry and computer science. This example demonstrates how an 'if-statement' can be used to define an area that should not be filled with

cubes (Figure 11). A user can specify the size of the square and the circular boundary that is to remain empty. A boolean check is used here as a way to control a boundary within which the geometric boolean effect can be visualized.

Nested 'For-Loops'

The following example uses a nested for-loop to move and turn the Beetle such that it walks along an undulating path within a certain range in z. The extruded path creates a loop that is made of smaller loops and in a way visualizes the code structure of the program. All moves are local turtle moves and no computation of absolute values is needed to determine the path (Figure 12).

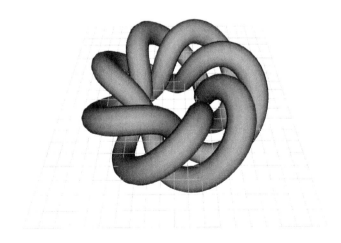

10   Program that draws 2 paths on a torus.

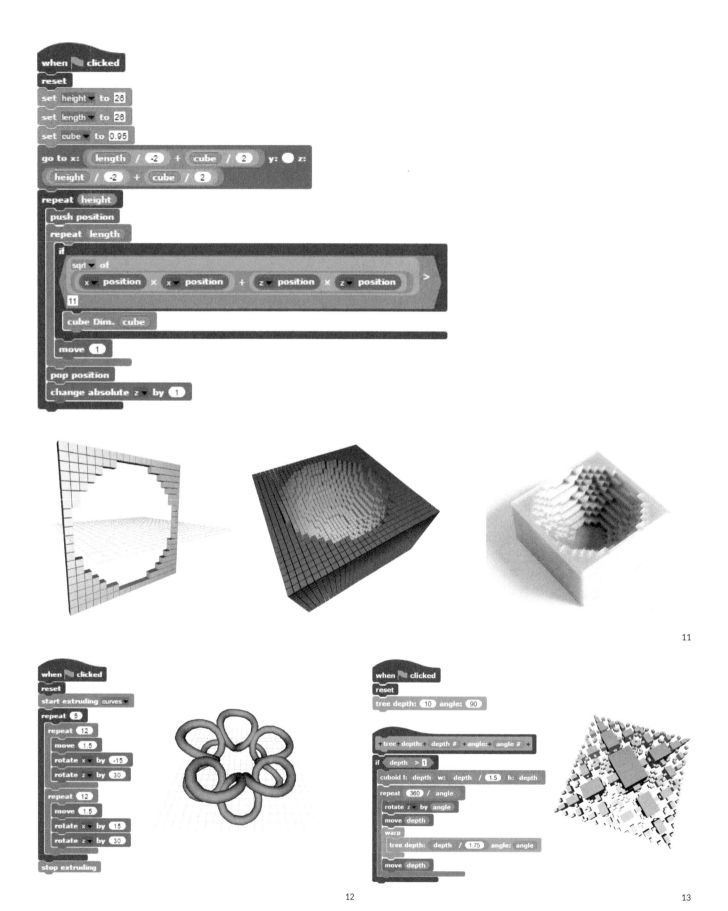

11

12

13

Beetle Blocks Koschitz, Ramagosa, Rosenbaum

Recursion and Fractals

The relation between recursion and fractals is well known and can be used to teach the computational idea of a function-call within a function. The next example uses the numeric values that are used to count down within the recursive algorithm as a variable for the sizes of the used cuboid. The result is a surprisingly short program, here shown with a custom Block that allows users to create their own methods (Figure 13).

Using Lambda (λ) in Design Pedagogy

With the new implementation in Snap!, we can make use of Lambda functionality, which is a computational concept that was introduced in Lisp. The idea is based on message passing and permits programmers to use function calls as input for other functions (Figure 14). This is a powerful idea that can teach how one can write a program that writes a program.

The next example draws a tree-like structure and uses four items in a list [treeParts] that all execute something individually. The custom Block [branch] uses a recursive structure by selecting random items in [treeParts], which consist of 'ringified' calls of [branch]. The random selection within the list has a bias toward the branch function as it is used twice. The two spheres have different sizes.

The counterpart [run()] allows users to execute functions that were ringified, used twice in the example. All moves are controlled via turtle geometry. The variable [branchAngle] only exists within the [branch] procedure. This is a way to teach scoping and availability of variables (Figure 15).

CONNECTING PROGRAMMING TO MAKING

We use the studio model in our courses, which means that students make physical artifacts of their own virtual 3D designs to tinker with (Resnick and Rosenbaum 2013). These prototypes provide the basis for Papert's construction of thought and making, and manipulating materials becomes an intrinsic part of the shared learning environment (Eisenberg et al. 2003). A further ambition lies in teaching foundational manufacturing processes, such as forming, machining, additive manufacturing, and joining. A powerful way of relating manufacturing processes and digital fabrication can be achieved by using Beetle Blocks programs to write machine instructions. We have included a few student examples in this section that show how we link programming to making.

1D to 2D transformation and shaping

Forming is a manufacturing technique that transforms a material without taking any material away or adding to it. We focus on 1D to 2D transformations and use metal wire that can be bent into a 2D shape with a digital wire bender (www.pensalabs.com/

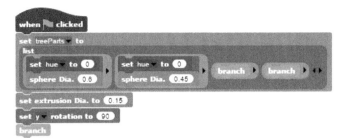

14

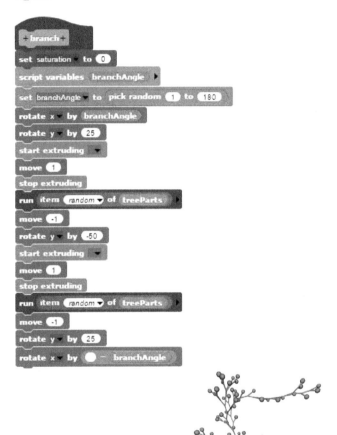

15

11 Program that fills a square with cubes, but avoids a circular zone at its center; 3D version of same algorithm with a spherical boundary and a 3D print.

12 Example of extruded path made by a nested for-loop.

13 Recursive algorithm with a custom Block.

14 Ringify, a block that allows a procedure to be turned into an input.

15 An example of the use of Lambda functionality in Beetle Blocks.

16

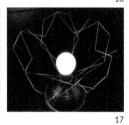

17

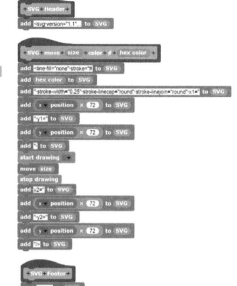

18

diwire-overview/). The machine takes simple instructions that can easily be coded using Beetle Blocks moves to form a closed polygon (Figure 17). The resulting shapes were arranged to form the framework for a lamp shade.

2D instructions and SVG code generating

Regarding 3D manufacturing processes, we want to teach how machines use instruction sets that move a gantry. In order to demystify machine instructions, one can use file types that are human-readable such as the SVG file format. The goal is to demonstrate how one can create a file-writer that writes machine instructions. The example below shows how to record beetle moves and to create a file header and footer. The file is subsequently sent to a vinyl cutter with a 2D gantry and customized heads for various pen types (Figure 18). The student decided to use two different pens to create the drawing below.

3D printing

The last example focuses on additive manufacturing, in this case 3D printing. The pedagogical goal consists of showing an example that essentially works similar to g-code. The design below, by one of our students, consists of cubes that follow two curves that are altered with a sine function (Figure 20).

CONTRIBUTION

Beetle Blocks revisits the combination of two paradigms that are highly accessible for learners: turtle geometry translated into 3D and a graphical programming language adapted from Scratch, now based on Snap! The result is a novel teaching tool that is more accessible than other systems for algorithmic design.

19

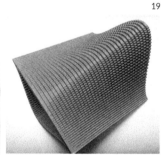

20

16 Wire bender.

17 An example of a design with bent wire (Schuyler Klein, Alyssa Bearoff).

18 Program for a file writer.

19 Vinyl cutter with a custom pen holder, drawing made with two pens (Jackie Hsia).

20 A 3D print of an example in Beetle Blocks (Joseph Kim).

Beetle Blocks Koschitz, Ramagosa, Rosenbaum

Drawing and sketching have been used to teach design for a long time and should not be replaced by other methods, but algorithmic design systems represent a powerful addition to that realm and allow us to address STEM learning in the context of architecture and design schools.

We believe that Beetle Blocks is a teaching tool that makes the power of this form of procedural design available to a broad audience of artists, designers, and architects. Beetle Blocks introduces a procedural way of thinking when constructing designs, which enriches the design vocabulary for architecture and design students.

It is essential to relate programming to making meaningful objects that learners use to construct mental models. We demonstrate in three case studies how we relate Beetle Blocks algorithms to digital fabrication and foundational manufacturing methods. Future steps consist of creating a sharing platform and expanding the geometry kernel.

REFERENCES

Abelson, Harold, and Andrea A. diSessa. 1986. *Turtle Geometry: The Computer as a Medium for Exploring Mathematics*. Cambridge, MA: The MIT Press

Colella, Vanessa, Eric Klopfer, and Mitchel Resnick. 2001. *Adventures in Modeling: Exploring Complex, Dynamic Systems with StarLogo*. New York: Teachers College Press.

Eisenberg, Michael. 2002. "Output Devices, Computation, and the Future of Mathematical Crafts." *International Journal of Computers for Mathematical Learning* 7 (1): 1–44.

Eisenberg, M., A. Eisenberg, S. Hendrix, G. Blauvelt, D. Butter, J. Garcia, R. Lewis, and T. Nielsen. 2003. "As We May Print: New Directions in Output Devices and Computational Crafts For Children." In *Proceedings of the 2003 Conference on Interaction Design and Children*. Preston, England: IDC. 31–39.

Harvey, Brian, and Jens Mönig. 2010. "Bringing 'No Ceiling' to Scratch: Can One Language Serve Kids and Computer Scientists?" In *Proceedings of Constructionism 2010: The 12th EuroLogo Conference*. Paris: EuroLogo. www.cs.berkeley.edu/~bh/BYOB.pdf

Kelleher, Caitlin, and Randy Pausch. 2005. "Lowering the Barriers to Programming: A Taxonomy of Programming Environments and Languages for Novice Programmers." *ACM Computing Surveys* 37 (2): 83–137.

Maloney, John, Mitchel Resnick, Natalie Rusk, Brian Silverman, and Evelyn Eastmond. 2010. "The Scratch Programming Language and Environment." *ACM Transactions on Computing Education* 10 (4): Article no. 16.

Papert, Seymour A. 1993. *Mindstorms: Children, Computers, And Powerful Ideas*. New York: Basic Books.

Petts, Malcolm. 1988. "Life After Turtle Geometry With a 3D Logo Microworld." *Mathematics in School* 17 (5): 2–7.

Resnick, Mitchel, John Maloney, Andrés Monroy-Hernández, Natalie Rusk, Evelyn Eastmond, Karen Brennan, Amon Millner, Eric Rosenbaum, Jay Silver, Brian Silverman, and Yasmin Kafai. 2009. "Scratch: Programming for All." *Communications of the ACM* 52 (11): 60–67.

Resnick, Mitchel, and Eric Rosenbaum. 2013. "Designing for Tinkerability." In *Design, Make, Play: Growing the Next Generation of STEM Innovators*, edited by Margaret Honey and David Kanter. New York: Routledge. 163–181.

Rusk, Natalie, Mitchel Resnick, and Stina Cook. 2009. "Origins and Guiding Principles of the Computer Clubhouse." In *The Computer Clubhouse: Constructionism and Creativity in Youth Communities*, edited by Yasmin B. Kafai, Kylie A. Peppler, and Robbin N. Chapman. New York: Teachers College Press. 17–25.

IMAGE CREDITS

Figure 3: Papert, 1993
Figure 17: Klein and Bearoff, 2015
Figure 19: Hsia, 2015
Figure 20: Kim, 2015

Duks Koschitz is Associate Professor and Director of the Design Lab at Pratt Institute. His wrote dissertation at M.I.T. on Curved-crease Paperfolding and had research positions at M.I.T. and the ETH. He worked at NMDA, Office da, Morphosis, Asymptote, Coop Himmelblau and Ian Ritchie Architects. He graduated from the T.U. Wien in 1998.

Bernat Ramagosa is a software engineer at Arduino SRL. He developed an online programming school and a social knowledge management system at the Citilab (Barcelona). He is the author and lead developer of Snap4Arduino and the lead developer of Beetle Blocks. He holds a Bachelor's and Master's degree from the Open University of Catalonia.

Eric Rosenbaum wrote his doctorate "Explorations in Musical Tinkering" at MIT Media Lab's Lifelong Kindergarten group. He is co-inventor of the MaKey MaKey invention kit. His software projects include Singing Fingers (finger painting with sound), Glowdoodle (painting with light) and MelodyMorph (creating musical instruments and compositions). He holds a Bachelor's (psychology) and a Master's degree from Harvard University.

Ivy

Bringing a Weighted-Mesh Representation to Bear on
Generative Architectural Design Applications

Andrei Nejur
Technical University of
Cluj-Napoca

Kyle Steinfeld
University of California, Berkeley

1

ABSTRACT

Mesh segmentation has become an important and well-researched topic in computational geometry in recent years (Agathos et al. 2008). As a result, a number of new approaches have been developed that have led to innovations in a diverse set of problems in computer graphics (CG) (Shamir 2008). Specifically, a range of effective methods for the division of a mesh have recently been proposed, including by K-means (Shlafman et al. 2002), graph cuts (Golovinskiy and Funkhouser 2008; Katz and Tal 2003), hierarchical clustering (Garland et al. 2001; Gelfand and Guibas 2004; Golovinskiy and Funkhouser 2008), primitive fitting (Athene et al. 2006), random walks (Lai et al.), core extraction (Katz et al.), tubular multi-scale analysis (Mortara et al. 2004), spectral clustering (Liu and Zhang 2004), and critical point analysis (Lin et al. 2007), all of which depend upon a weighted graph representation, typically the dual of the given mesh (Shamir 2008). While these approaches have been proven effective within the narrowly defined domains of application for which they have been developed (Chen 2009), they have not been brought to bear on wider classes of problems in fields outside of CG, specifically on problems relevant to generative architectural design (GAD).

Given the widespread use of meshes and the utility of segmentation in GAD, by surveying the relevant and recently matured approaches to mesh segmentation in CG that share a common representation of the mesh dual, this paper identifies and takes steps to address a heretofore unrealized transfer of technology that would resolve a missed opportunity for both subject areas. Meshes are often employed by architectural designers for purposes that are distinct from and present a unique

1   The Elephetus project by Anders Holden Deleuran (CITA/KADK) and David Reeves (Spatial Slur/ZHA Code) offers an example application of mesh segmentation in generative architectural design. Note that while this project exhibits many of the design qualities intended to be supported by the software proposed in this paper, and its development likely required some of the processes discussed here, the Ivy tool was not employed.

set of requirements in relation to similar applications that have enjoyed more focused study in computer science. This paper presents a survey of similar applications, including thin-sheet fabrication (Mitani and Suzuki 2004), rendering optimization (Garland et al. 2001), 3D mesh compression (Taubin et al. 1998), morphing (Shapira et al. 2008) and mesh simplification (Kalvin and Taylor 1996), and distinguish the requirements of these applications from those presented by GAD, including non-refinement in advance of the constraining of mesh geometry to planar-quad faces, and the ability to address a diversity of mesh features that may or may not be preserved.

Following this survey of existing approaches and unmet needs, the authors assert that if a generalized framework for working with graph representations of meshes is developed, allowing for the interactive adjustment of edge weights, then the recent developments in mesh segmentation may be better brought to bear on GAD problems. This paper presents recent work toward the development of just such a framework, implemented as a plug-in for the visual programming environment Grasshopper.

INTRODUCTION

This paper describes the motivations for the development of a platform for mesh segmentation suited for the requirements of contemporary generative architectural design (GAD). The context for this endeavor is the increased relevance of mesh-based form-finding and simulation techniques in architectural design and the maturation of programing and visual scripting, as the related software tools find more widespread use. In recent years, a number of new techniques for form-finding via mesh-based simulation have taken hold in GAD. These include spring-based physical simulation models, node-based structural simulations, and thermodynamic analysis. Such tools have increased the demand for approaches to mesh creation and manipulation, with meshes beginning to even challenge the relevance the now-dominant non-uniform rational Basis spline (NURBS) surface representation in GAD. It is also notable that the increased relevance of each of these examples has been enabled by the advent of visual programming, and by the Grasshopper programming environment in particular. While meshes have become more widespread for this audience, approaches to mesh segmentation are not well-studied in the context of GAD, nor are they well-supported by existing tools. This research seeks to identify existing relevant techniques in computer graphics (CG), adapt these techniques to the unique needs of the generative architectural design audience, and to produce a framework for their transfer such that they may be effectively applied in this new domain.

The first part of the paper reviews the relevant literature in both CG and GAD. It first presents an abbreviated survey of

technical approaches to mesh segmentation in the context of CG, including a discussion of the applications for which these approaches were developed. As will become apparent, many of the relevant approaches rely on a common representation of the mesh dual, and proceed through the manipulation of a weighted graph. The common use of the weighted mesh dual allows for a number of complementary and overlapping approaches to be implemented via a generalized approach, and forms the basis of the development of the software framework proposed below. The same section details the extent to which mesh segmentation represents an unmet need in architectural design, and would benefit from the algorithms developed for CG applications. Several sympathetic approaches like weaving (Xing et al. 2011a) and mesh stripification (Xing et al. 2011b) are presented here as a survey of architectural projects that have employed mesh segmentation without the benefit of a supportive toolkit. From this survey, the requirements for mesh segmentation that are unique to GAD are derived, and an account of the specific tasks in design that would benefit from mesh segmentation is presented. The second part of the paper presents the methods by which the proposed software framework has been developed. First, the implementation details of the software tool created for mesh segmentation and fabrication are presented. The modular workflow of Ivy for Grasshopper is explained alongside the data structures and algorithms employed. A few typical workflows are detailed in this section, which, given the modular nature of the software and the common representation of the generalized weighted-graph representation, may be combined in a number of ways. Following this, we speculate upon the advantages of this modular technique in connection to the specific needs of GAD practice and research.

Mesh Segmentation in Computer Graphics

The boundary representation of the three-dimensional mesh has practically been a steady companion for the digital embodiment of form since the advent of computer generated imagery. In order to make use of this geometric data type on the computer screen, ways to meaningfully depict an otherwise featureless collection of mesh faces had to be devised. Among those, mesh segmentation stands as one of the most important. Its applications span the entire spectrum of mesh use in CG: mesh interpretation, feature detection, parametrization, multi-resolution modeling, mesh editing, morphing, animation, and compression all rely on some form of mesh segmentation to exist (Shamir 2008). Different applications that carry distinctive requirements have prompted a number of distinct techniques of mesh segmentation that have been well-articulated in previous work (Agathos et al. 2008; Shamir 2008; Chen et al. 2008). The first part of the literature review to follow functions as a meta-survey of these techniques, extracting the common representations and procedures that suggest appropriate transfer to GAD.

Mesh Segmentation in Generative Architectural Design

The architectural use of discrete surface descriptions in the service of form generation precedes the invention of the computer. As far back as the beginning of the last century, Antoni Gaudí used discrete surface representations in the physical computation of form. Later, Frei Otto brought the physical equivalent of a three-dimensional mesh to bear on the elaborate form-finding techniques that structured much of his design research. Early applications of the computer in GAD remained limited to academic research contexts, and regularly employed the mesh representation, as there were few other options for the representation of three-dimensional free-form geometry prior to the development of NURBS. With the advent of commercially-available CAD platforms, GAD transitioned from academic labs to applications in practice. As design practitioners were less likely to develop bespoke geometric routines than CAD researchers, many of the technologies employed were direct transfers from CG. This resulted in the occasional mismatch between the audience for which these technologies were developed and the manner in which they found application in GAD.

For example, a central concern (Pottmann et al. 2015) of GAD is free-form surface rationalization, a broad topic that includes panelization, surface approximation, and constraint-aware design. As such, those mesh operations destined for use in GAD must account for a host of additional properties directly derived from their eventual existence in the physical world. These concerns are qualitatively different than those applications most often targeted by CG, even those applications that involve physical fabrication. Geometric properties such as face size, dimensional proportion, an ability to produce offsets, number of faces, fairness, singularities, and node valence can often be overlooked by applications in CG, but have a tremendous impact in their architectural applications. As comprehensive surveys of mesh use in free-form architectural design have been well-articulated in previous work (Pottmann et al. 2015; Glymph et al. 2004; Liu et al. 2004), in the literature review to follow, we focus on the geometrical and topological traits of meshes that are most relevant to the present research. In summary, this survey finds that mesh segmentation remains relevant in the context of GAD. Whether the design manifests smooth surfaces or discrete collections of polyhedral flat surfaces, the ability to rationalize this geometry—that is to say, to reasonably realize the geometric design within the limits posed by a given a method of fabrication—is paramount. As is presented below, the present research relies heavily on the strong body of research dedicated to the panelization of architectural surfaces, and to the modeling of

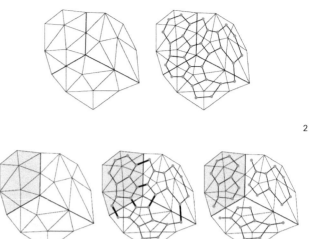

2

3

2   Creating a dual graph on a triangle mesh. Each face becomes a node in the dual graph and each non-naked edge becomes an edge in the same dual graph.

3   A weighted graph diagram and its use in dual graph/mesh segmentation.

4   A simple diagram of the Kruskal's algorithm, the basis of the HMC.

5   Diagram of basic K-Means algorithm steps. 1) Select roots and calculate regions based on weight of graph edges. 2) Relocate roots to central points of regions. 3) Recalculate regions based on new roots.

meshes with fabrication-aware constraints (Glymph et al. 2004). Also brought to bear are the most recent developments related to the rationalization of meshes for fabrication, most notably related to papercraft (Mitani and Suzuki 2004), that suggest application in GAD. Taken together, these two surveys will allow us to discern the need for and provide the basis of a generalized framework for mesh segmentation in GAD.

LITERATURE REVIEW

By surveying the most relevant approaches in CG, this section characterizes the state of the art in mesh segmentation from technical and theoretical points of view. Then, by surveying the existing use of meshes for surface rationalization in GAD, we articulate the requirements of the appropriate transfer of mesh segmentation techniques from applications adjacent to CG to those relevant to GAD. A number of algorithms are presented that were originally developed in CG and that have subsequently been adapted by the current scope of work. The expression of graph theory in each of these cases is highlighted, as the shared reliance on edge-weighted graphs forms the basis of the common framework proposed below.

Survey of Mesh Segmentation Techniques in CG

While specific variants of routines for the segmentation of

meshes abound in computer science research—a testament to the widespread utility of segmentation in general—only a limited number of algorithms have achieved a strong status, variations of which recur often in the most current approaches. In the meta-survey conducted here, 45 papers were examined that either survey segmentation techniques in general in order to characterize the state of the art, or that compare two or more techniques in the context of a specific application. Nearly all those techniques surveyed implement a weighted-graph representation, and an overwhelming number relate strongly to applications in computer graphics, with the techniques demonstrated very often adapted to the specific needs of a relatively narrow set of concerns. As the common foundation of these techniques informs the development of the toolkit proposed below, thereby allowing for the coordinated application of different algorithms in the context of GAD, we highlight here the way in which each technique manifests a weighted graph representation. To illuminate the mesh segmentation algorithms discussed below, we employ a common language and notation for each routine presented. For the sake of clarity, some notions in the referenced papers will be renamed to fit this unifying convention. Among the central concepts are the dual graph and the weighted graph, which we define here. A *dual graph* (Figure 2) is a concept central to graph theory, and is the central operation of mesh segmentation using graph techniques. In the context of each example below, the faces of the mesh form the nodes of the graph, and the bounding edges between faces form the edges of the graph. A *weighted graph* (Figure 3) is one in which nodes and/or edges are assigned numeric values that are interpreted in cost functions. The "weight" or "cost" associated with a graph element is used in order to direct a walk on the graph, and thereby to prioritize certain paths over others. Any number of

processes are used to map costs to elements. Only rarely would such values be set directly by an end-user. More often, specific processes are directed by the algorithm and rely on information found in the geometry of the mesh. In most of the algorithms examined here, the determination of weights is "hardwired" into the algorithm.

Hierarchical Mesh Clustering (Garland et al. 2001)

Hierarchical mesh clustering (HMC) (Figure 4) is among the simplest of the algorithms in this survey, and, like the others presented here, operates on the dual graph of a mesh. HMC applies a greedy clustering routine based simply on edge weights determined by planarity. The algorithm is a straightforward and relatively simple interpretation of a standard Kruskal or Disjoint Set minimum spanning tree algorithm on a graph (Skiena 1998). The most important difference from these standard routines is the iterative recalculation of the cost for the edges at each step. First, a Disjoint Set approach is employed, where the dual graph in the first step is atomized into its connecting nodes. In this step, each cluster has only one node. In subsequent steps, using a greedy approach, one edge at a time is contracted and the clusters separated by the edge are merged into a single cluster. As this proceeds, the algorithm evaluates each edge in a greedy fashion based on the cost of contraction. The cost of each edge is calculated at each step, based on a planarity measure function of the cluster post-merger. In a variant of this approach, a measure of the compactness of the cluster shape may be introduced in the cost calculation, using a value calculated by dividing the squared perimeter of the cluster by its area. This achieves a better approximation of the clustering intent. The time complexity of HMC is usually O(n×log (n)) depending on the test function.

Several spin-offs of this algorithm have been developed, each with some improvement or a change geared towards a specialized use. One of the most remarkable innovations is an HMC based on fitting primitives (Athene et al. 2006). The extended algorithm uses several geometric primitives described through mathematical functions to decide in the cluster joining process. From an architectural fabrication point of view, this is important because the primitives can be used at the end to approximate, and thus help fabricate, the surface through parts easily created in standard processes. Moving beyond the particularities of this routine, any other metric can be used to calculate the cost of contracting an edge. Such modifications have been applied in the service of faster radiosity calculation, collision detection, and surface simplification. A number of similar modifications have been employed as described below in the service of goals specific to GAD.

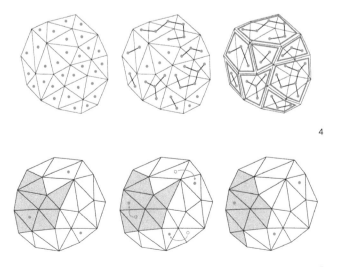

4

5

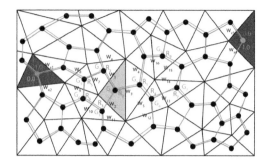

K-Means Clustering
(Shlafman et al. 2002; Funkhouser et al. 2004)

Like HMC, K-means (Figure 5) clustering is an iterative segmentation algorithm valued in CG applications for its relatively simple implementation and robust results. The input for the algorithm is typically an integer number defining the desired number of segments into which the original mesh will be split. Again, the dual graph comes into play at several crucial steps. First, in order to split the mesh, a set of edge weights is computed. Both papers referenced here make combined use of two properties, face-angle and edge-traversing-distance, in calculating the weight value for each edge. Based on these assigned weights, a central best-connected-point is computed using an all-pairs minimal-path algorithm, essentially a Dijkstra's algorithm run for every node in the original dual graph. Once the first root is selected, a second is identified as the farthest node from it. The distance is computed based on the stored values in the cached all-pair minimal path results. The process continues for any roots not yet selected, and the last step is repeated by looking for the most distant node from all the already selected root nodes. Then, after the root selection process is complete, each node is allotted to a mesh segment defined by one root, using the same all-pair weight-based distance calculation. After the initial segmentation is complete for each segment, a new root is computed. The new root is the best connected node in the segment nodes, based on the same all-pairs minimal-path algorithm, but this time, it is calculated only amongst the nodes in the segment. The last two steps are then repeated iteratively until the new root selection stabilizes for each segment on the same node. The time complexity of the algorithm is $O(n^2 \times \log(n))$ where n is the number of faces in the mesh, for the initial all-pair calculation, plus the time required for the computation of the new roots for each segment at every subsequent step.

Random Walks (Lai et al. 2009)

Like K-Means and HMC, the random walk segmentation operates upon a weighted graph, but in contrast with these iterative approaches, the segmentation occurs in a single step. The algorithm requires the identification of a number of starting

faces to be used as roots. Based on these, a weight value is computed for each non-root face that quantifies the likelihood that a random walk starting from the non-root face will eventually end up in a given root face (Figure 6). Each non-root face, then, stores one likelihood value for each root. Once these are computed, each non-root face is assigned to a cluster associated with the root with the highest likelihood. The calculation of each likelihood value proceeds by summing a weighted version of the likelihoods of each neighboring face. As we might expect, the weighting factor is directly related to the corresponding edge-weight, and is calculated in a similar fashion to the edge-costs in other algorithms. In the paper referenced here, the edge-weights are assigned by a combination of dihedral angle, edge-length, and traversal distance. Two variants are presented that respond to the needs of differing applications: one style of calculation is more suited for graphical models, while the other better addresses technical three-dimensional data. The authors also describe a number of methods for seed selection that add a supplemental degree of automation to the algorithm. One approach uses a similar method with the K-means algorithm of Shlafman et al. (2002), where just a number of desired seeds is required, which are then distributed according to the weight of the edges. The other employs a system of particles distributed across the mesh as connected by virtual springs, the energy of which are iteratively minimized and distributed, and seed faces are selected after the system converges. The strength and speed of this approach is rooted in the way values are calculated for all the faces of the mesh in one step. The algorithm solves a set of equations (one calculation for each seed face) through a sparse linear system. Solving the system yields the values for all the faces at once. This gives the algorithm a typical running time of $O(n \times m \times \log(m))$, where n is the number of seeds and m the total number of faces in the mesh. As in the previous two algorithms, this segmentation employs a weighted dual graph at a number of stages and for assorted processes, and differs essentially in the particularities of weight calculation.

Feature Point and Core Extraction (Katz et al. 2005)

A different approach to mesh segmentation, albeit one based on the same ground level concepts from graph theory, is the Feature Point and Core Extraction technique (FP+CE) proposed by Sagi Katz, George Leifman and Ayellet Tal. The algorithm works in multiple steps, all of which make use of the weighted dual graph of the mesh.

Before the actual calculation starts, the model is preprocessed using multidimensional scaling (MDS), a technique that transforms the graph such that the Euclidean distances between vertices approximate their geodesic counterparts. The first formal step is the prominent feature detection. This happens

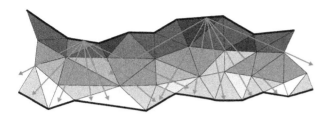

6   Diagram of likelihood calculation for a mesh face that a random walk started in the face will reach a certain root. Here the start face (light gray) and the two end faces (dark gray) are connected by two graphs, red and green, which show part of the chain of potential for a few adjacent faces.

7   The shape diameter function. Rays are shot from every face of the mesh in the opposite direction of the normal face. Some of the rays are discarded based on the angle at which they land on the other side, or the length they travel inside the mesh. The length values of the kept rays are averaged as the shape diameter value.

by checking if the sum (computed using a version of Dijkstra's algorithm) of the geodesic distances of a node to all other nodes in the graph is larger than any of its neighbor's sums. If so, the node is considered a feature node. Second, the mesh "core" is extracted. The core is considered the body of the mesh devoid of features and protruding elements. For this, a spherical mirroring of the mesh is computed and a convex hull is created, such that the nodes close to the hull in the mirrored version of the mesh are considered "core" nodes. The initial collection of these core nodes is extended iteratively until all features have been separated into individual segments. Then, using a standard graph traversing technique (such as depth-first-search), all segments are walked and connected to the core if they don't contain at least one feature node.

Although created specifically for application in graphical model meshes, this method presents a number of elements easily applicable to more general uses, as is discussed in a section to follow.

Shape Diameter Function (Shapira et al. 2008)

The Shape Diameter Function (SDF) represents a unique take on the general approach of segmenting a mesh based on a set of values assigned to each face. The main differences are in the calculation of the values used for the segmentation. The segmentation algorithm itself is a k-way graph cut employed in other research projects related to mesh segmentation (Shlafman et al. 2002). The distinguishing feature is the two-step calculation of the segmentation values.

In the first step, a value is computed for each face of the mesh using a shape diameter function, which serves to describe how

deep the mesh model is at a certain point. For this, a set of rays are constructed in a conical configuration centering on the evaluated point, and oriented in the direction of the inverted face normal. Based on the validity of the landing point on the other side of the model, rays are either stored or discarded. The mean value of any remaining valid rays is computed as the "shape diameter value" (Figure 7) of the face, which may then be clustered in a soft partition using a Gaussian mixture model and an expectation maximization algorithm. The second step uses a k-way graph cut to create hard boundaries and adapt the segmentation to the features of the mesh by factoring a series of geometrical determinants of the mesh (such as dihedral angle and edge length) into the calculated face values. The method described in the paper is again highly dependent on the type of mesh model presented, and can produce meaningful results only on closed and non-noisy meshes. However, within this limited subset of meshes, it is effective at producing meaningful partitions and, in subsequent steps, to extrapolate those partitions into mesh skeletons. In this way, SDF represents a potentially applicable approach to mesh segmentation in GAD, but the meshes in GAD do not necessarily belong to this limited subset, only insofar as a designer is able to understand the limits of its application.

Randomized Cuts (Golovinskiy and Funkhouser 2008)

Although not a novel technique per se, the process detailed in this paper is of interest because it proves the underlying compatibility of mesh segmentation techniques based on dual graph representation. The research employs a number of algorithms from the most commonly used ones (all of them already detailed above in the survey) and randomly switches between techniques in an attempt to find the most consistent cuts in a dual graph. The mesh is cut into functional parts using the same dual graph support and the multiple segmentations are analyzed by a function to determine the most consistent cuts. The process uses hierarchical clustering, k-means clustering, and min-cut clustering (a version of hierarchical clustering) to split the graph multiple times with different random input variables. Each of the cuts is evaluated by the function, and a score is assigned and in the end the best scoring cut is selected. This produces the most consistent possible cut, because no single algorithm and variable set can produce meaningful segmentations every time (Athene et al. 2006).

This statement acknowledges the fact that the wealth of mesh segmentation algorithms and the multitude of research undertaken in this field related to CG is made up mostly of individual specialized research geared toward very specific goals. Even if most of the algorithms employed are general, their implementation is based on the iterative exploration of their outcome. In order to produce meaningful results, the tools' development are

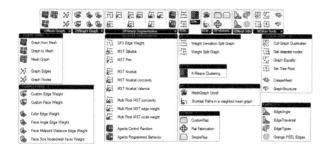

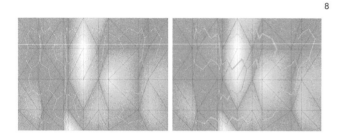

8

9

8   The tool menu in one of the wip versions of Ivy for Grasshopper. 1|Create
    and decompose MeshGraphs tools; 2| Add weight to MeshGraph; 3|Primary
    Segmentation (tree making); 4|Secondary Segmentation; 5|Iterative segmenta-
    tion; 6|Special Segmentation; 7|Fabrication; 8|Mesh Info; 9|Other Tools.

9   Mesh with attached MeshGraph (left) and tree MeshGraph on the same mesh
    calculated with Djikstra's algorithm using dihedral angle as edge weight.

often directed towards a certain behavior. In order to secure
consistent results with every use, many of the values are hardwired in the algorithm, thus forsaking exploration with the tool.

Identification of Need in Generative Architectural Design

Applications in CG regard the mesh quite differently than applications in GAD. In general, CG applications tend to treat a mesh as
a nominally smooth surface that happens to be described using
discrete faces in order to take advantage of the related discrete
computational methods. In contrast, GAD applications more
often rely on the network of connected faces and edges of the
mesh as a simplified representation of architectonic elements,
such as structural framing or facade panels. This basic distinction leads to a number of important requirements for mesh
segmentation that are unique to GAD. A major exception to this
distinction is physical simulation models, such as energy models
or spring-systems, which are treated as discrete descriptions
of nominally smooth surfaces in GAD. Insofar as GAD regards
meshes as simplified representations of architectonic elements,
many of the requirements for subdivision concern fabrication.
Take, for example, the work surrounding the definition of planar
quadrilateral (PQ) meshes (Glymph et al. 2004; Pottmann et
al. 2015; 2008; 2007; Liu et al. 2004), in which a constrained
mesh finds significance in its ability to represent the curtain wall
panels of a glass building. Some of the routines related to PQ

meshes rely on the discovery of developable strips, a procedure
that is described as a subset of mesh segmentation in CG, and
may be handled using a weighted-graph approach. Another
more modest example in scope of execution may be found in
the work of Marc Fornes, whose creative practice specializes in
artistic installations. While there are no publications that detail
the processes employed in the design of these projects, we can
surmise that the subdivision of meshes is central to their realization in that the number of mesh faces (along with the number
of "stripes," which we may presume to be a unit of subdivision)
is listed prominently in the project credits. A related concern
unique to GAD regards the preservation of the specific features
of a mesh, such as folds, creases, and geometric textures. These
concerns are not well addressed in the existing CG literature,
but as is demonstrated below, are possible to support using a
generalized weighted graph approach. Using mesh segmentation
to enable architectural production (both at full-scale and in the
service of architectural scale models) is an important defining
feature of GAD. This particular issue is rarely a goal in CG, and
as a result, there are hardly any tools explicitly developed for it in
the larger field of computer science. Notable exceptions to the
lack of existing segmentation routines that address the unique
needs of GAD include primitive-fitting routines and applications
in papercraft. The primitive fitting technique (Athene et al. 2006)
suggests application to full-scale architectural fabrication, as
does the mesh-segmentation used for part-grafting pieces of
a mesh model using elements selected from a given library of
forms (Funkhouser et al. 2004). Routines in CG designed for the
fabrication of meshes center on small-scale papercraft (Mitani
and Suzuki 2004; Massarwi et al. 2007), which includes small
models made of thin-sheet materials such as paper, cloth (Julius
et al. 2005), metal, or plastic. The aim of this research is to reproduce a natural "look" without the constraint of preserving the
exact input geometry.

Perhaps the most significant mismatch between existing tools
developed for CG and GAD concerns the difference in the
expected cultures of use, and the exploratory nature of the
design process. In GAD applications, software tools are used
primarily as a means for exploration well before any production
takes place. In early stages of the design process, techniques
and approaches are revised often, and multiple algorithms are
often employed in novel combinations to produce results far
beyond what could be anticipated in advance. For this reason,
frameworks that allow access to low-level controls are preferred
by this community over packaged software tools or routines
presented as black-boxes.

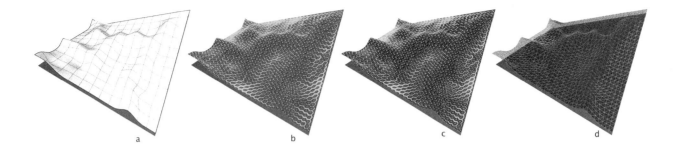

10 One step segmentation of hyperbolic paraboloid with added large-scale surface noise. Step (a) the surface and the guide surface; (b) the mesh version of the surface with a forest of trees resulted from a Kruskal's algorithm calculation. The weight of the MeshGraph is based on the dihedral angle and the proximity to the base (blue-gray) surface. In (c) the same forest of graph trees is separated in single-node trees and multi-node trees. Step (d) shows the mesh segmentation in 4 parts and a separation area of faces in red.

IVY

Presented above are two surveys: the first details the existing approaches to mesh segmentation routines in CG, while the second describes the unique requirements and unmet needs in GAD for such routines. Here, we bring the common foundational representation of the weighted graph dual—identified by the first survey—to bear in the construction of a framework that addresses the needs and requirements identified by the second survey.

We assert that if mesh segmentation routines are to be brought to bear on the unique requirements of GAD, then a comprehensive and modular framework is needed. The wide range of specialized algorithms already used for specific narrow tasks in CG need to be generalized and adapted such that the resulting framework may be applied more situationally. For this, a common representation is required so that the different algorithms may be coordinated and combined by the end-user. Ensuring data transference between algorithms enables custom aggregation of tools that go beyond simple input value changes, and into new and unimagined uses of mesh segmentation. Detailed in this section are the steps taken towards creating a weighted-graph mesh segmentation framework for use in GAD, implemented as an extension for the popular visual scripting platform Grasshopper. We describe here the core data structures, routines, and expected workflows supported by this extension, which is named Ivy, and discuss the extension's unique synthesis of the mesh segmentation routines presented above.

Implemented Data Structures & Routines

A weighted graph dual of a mesh, and its reconfiguration as a minimum-spanning tree via various routines, forms the common basis of many of the routines surveyed from CG. Here we describe the implementation of the required data-structures to represent and manipulate this graph, including the MeshGraph, MNode, and MEdge types, and the related routines that assign weights and perform segmentations in a generalized way that is able to reproduce the routines discussed above. The toolset, as it was implemented in Grasshopper, is organized according to an expected workflow for mesh segmentation. This is reflected by the ordering of the nine groups of components seen in the nearby diagram (Figure 8). Following the tool groups on the Grasshopper ribbon in Ivy equals following the intended general chaining of tools for an expected mesh segmentation.

MeshGraph

The dual graph concept is implemented by Ivy as a data object called MeshGraph. This provides the core functionality of the toolset and allows the interconnection of different tools into a coherent linear flow. A MeshGraph stores a collection of nodes (MNode) and edges (MEdge) that typically correspond to the mesh faces and non-naked edges of a given mesh. It also stores a copy of the given mesh geometry at construction, which allows for operations such as unfolding and fabrication to occur after segmentation.

Edgeweights and Nodeweights

The essence of mesh segmentation is the identification of variable properties of the mesh for the purpose of assigning values to each mesh edge or mesh face. These values, often called weights or costs, serve as a guide for the segmentation routines. In Ivy, there is a dedicated tool group that contains routines for assigning and manipulating weights, including assignment via dihedral angle, distance between face center points, face size, and mesh color. Importantly, there are also tools for assigning arbitrarily calculated weights, which enable the use of any numerically expressible property.

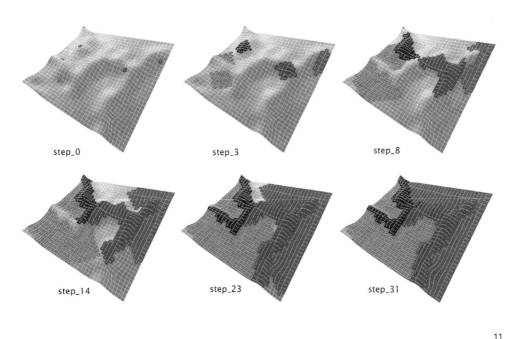

step_0   step_3   step_8

step_14   step_23   step_31

11 One-step segmentation of hyperbolic paraboloid with added surface noise. (a) the surface and the guide surface; (b) the mesh version of the surface with a forest of trees resulted from a Kruskal's algorithm calculation. The weight is based on the dihedral angle and the proximity to the base surface. In (c) the same forest of graph trees is separated in single-node trees and multi-node trees. Step (d) shows the mesh segmentation in 4 parts and a separation area of faces in red.

12 A two-step segmentation routine. The base mesh (a) is used to create a MeshGraph with edges weighted according to the dihedral angle between the faces. Making a tree from the dual MeshGraph (b) results in cup-like surfaces connected with one edge (the least sharp edge from the rims connecting the cups). By removing the edges with the largest weight in the tree (c) the parts are segmented. No faces from the original mesh (d) remain isolated.

11

Segmentation Routines

Routines that segment a mesh given its weighted graph dual represent the largest group of tools in the framework, and also the most important. Virtually all tools that operate and modify a graph in Ivy are considered MeshGraph segmentators. Here we find graph-processing algorithms that make use of already-defined edge or node weights in the service of selecting two MeshGraph nodes, disconnecting them by removing an edge, and then updating the graph structure. At times, this requires operation on a tree, which is expressed as a special case of a graph. This is why the tool sections in Ivy are divided between Primary Segmentation, Secondary Segmentation, Iterative Segmentation, and Special Segmentation.

Primary Segmentation: Tree-Making Routines

The tree-making routines are the standard graph algorithms often employed for this purpose, including Prim's, Kruskal's, Djikstra's (Figure 9), and Depth First Search. The construction of a tree graph is the first step in either splitting a MeshGraph, or preparing it for segmentation in a subsequent step, as discussed in the section below on one-step segmentation. This is the main reason why in the Ivy workflow, converting a dual graph into a tree is considered primary segmentation. Another reason consists in the fact that as a result of the dual graph to tree conversion, a forest of tree graphs can emerge, thus effectively creating a segmentation of the original mesh (Nejur 2016, 4–8). Figure 10 shows an example of Disjoint Set tree making and resulted segmentation and Figure 11 depicts a primary segmentation example based on multiple root minimum span tree algorithms (MRMSTs).

Secondary and Iterative Segmentation

The next section of segmentation routines includes tools that segment readymade tree graphs. Secondary segmentation happens through the selection and removal of tree edges (Ivy Manual, 9). Through a singular edge removal—a simply connected graph—a tree is split into two segments (Figure 12).

Another section is dedicated to self-contained iterative algorithms that operate on the original dual MeshGraph. Here we find the implementation of many of the segmentation routines identified in the literature review above, including K-Means

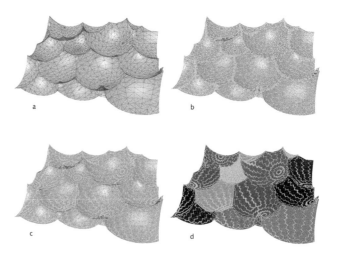

a   b

c   d

12

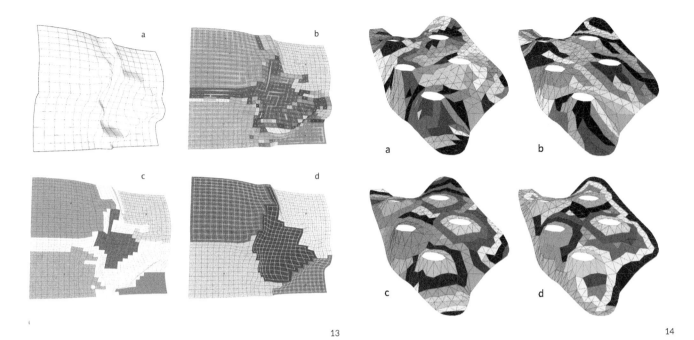

13

14

and HMC, each of which has been adapted to modular visual programming environment of Grasshopper. Figure 13 shows an example of the iterative segmentation workflow.

DISCUSSION

In this paper we demonstrated that generative architectural design, although deeply embedded in the contemporary techniques of digital form-making and representation, still lacks adapted tools for mesh segmentation. It is clear though that mesh segmentation is useful and needed in the context of GAD. However, the specific conditions of generative design claim a different approach than those of CG, for instance where a lot of research has already been conducted.

As a result of the identified need for specific mesh segmentation in GAD, we introduce Ivy, a platform based on the underlying mathematical concepts that power most of the research in the field. Ivy brings together a number of tools ported from CG-based research or developed specifically for the tasks in GAD. The separate tools and technologies are compatible through the graph representation of the mesh, which is already the default data object for most of the research in the field.

The platform implementation for mesh segmentation in Grasshopper brings many benefits to generative design beyond mere availability of tools. The possible aggregation of functionalities facilitated by the visual scripting platform of Grasshopper and the common graph language proves that the sum of tools working together is really much more potent than the simple

13 An enhanced K-Means segmentation setup in Ivy with automatic detection of the number of regions. The base, a doubly curved surface with a large over-lapped noise (a) is overly segmented with a disjoint set (Kruskal's) algorithm with a weight limit. This produces a forest of trees with a diverse number of nodes (b). The weight limit is set so that low weight landscapes can coagulate in larger trees. By extracting those large trees and calculating a weight center for each of them we get a very good approximation of the final placement of the k-means seeds (c). Starting from those seeds, the iterative part of the classic K-means algorithm reaches a stable state in very few steps. This drastically cuts the run time of the algorithm, depending on the size and the features of the mesh, sometimes to less than a third.

14 Different unrolling strategies of a parametrically generated architectural mesh. The meshes are segmented in Ivy with the aim to reduce the final number of pieces needed to unroll the geometry on a flat surface. Shown are four variants resulted from different weight settings for the MeshGraph. From top to bottom, left to right: dihedral angle, edge distance from a median curve, orange peel edge classification starting from the holes, and finally the same orange peel classification started from the edge.

addition of individual algorithmic benefits. For now, implementation of mesh segmentation tools in GAD in general, and Grasshopper specifically, is fairly new, and Ivy was started mostly as a proof of concept. As a result, there are still speed issues and limitations when it comes to meshes with many faces. Optimization of the code to extend usability is one of the goals for the authors. At the time when this paper was written, only a few of the algorithms developed for CG research had been ported to Ivy. The authors hope to extend the range of components that bring useful mesh segmentation to GAD. Along with new segmentation algorithms, the focus of future research and subsequent papers will fall on mesh segmentation usage in GAD. Within this more practical area of investigation, a special place will be dedicated to mesh unrolling. For reasons pertaining to the size and scope of this paper, the tools and workflow descriptions

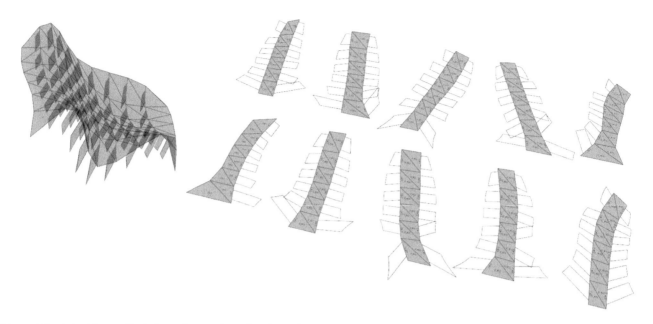

15 A speculative study of the use of Ivy in the service of mesh unrolling. Connecting tabs of arbitrary geometry may be added at each connection edge.

were rather brief. A more in-depth explanation and detailing of the tools and functionalities of Ivy can be found in the Ivy User Manual (Nejur 2016). Because Ivy was designed to be a platform, it enables future research into mesh evaluation, segmentation, and fabrication as part of the Grasshopper ecosystem. This has manifested fabrication techniques and mesh exploration strategies that are included with Ivy, but are only briefly mentioned in this paper. Aside from a few standalone uses mentioned above, a number of workflows, examples, and practical applications remain to be presented. This positions the present body of text as an introduction to the tool and a sample of practical mesh segmentation use inside the GAD paradigm.

ACKNOWLEDGEMENTS

The paper presents research related to a Fulbright grant that supported Andrei Nejur as a visiting scholar hosted by Kyle Steinfeld at UC Berkeley. The development work for the toolset was undertaken using student feedback from the class of Simon Schleicher.

REFERENCES

Agathos, Alexander, Ioannis Pratikakis, Stavros Perantonis, Nikolaos Sapidis, and Philip Azariadis. 2007. "3D Mesh Segmentation Methodologies for CAD Applications." *Computer-Aided Design and Applications* 4 (6): 827–841.

Attene, Marco, Bianca Falcidieno, and Michela Spagnuolo. 2006. "Hierarchical Mesh Segmentation Based on Fitting Primitives." *The Visual Computer* 22 (3): 181–193.

Chen, Xiaobai, Aleksey Golovinskiy, and Thomas Funkhouser. 2009. "A Benchmark for 3D Mesh Segmentation." *ACM Transactions on Graphics* 28 (3): Article 73.

Fornes, Marc. "MARC FORNES & THEVERYMANY." Accessed April 26, 2016. https://theverymany.com/.

Funkhouser, Thomas, Michael Kazhdan, Philip Shilane, Patrick Min, William Kiefer, Ayellet Tal, Szymon Rusinkiewicz, and David Dobkin. 2004. "Modeling by Example." In *Proceedings of the 31st International Conference on Computer Graphics and Interactive Techniques*, edited by Joe Marks. Los Angeles, CA: SIGGRAPH. 652–663.

Garland, Michael, Andrew Willmott, and Paul S. Heckbert. 2001. "Hierarchical Face Clustering on Polygonal Surfaces." In *Proceedings of the 2001 Symposium on Interactive 3D Graphics*. Research Triangle Park, NC: SI3D. 49–58.

Gelfand, Natasha, and Leonidas J. Guibas. "Shape Segmentation Using Local Slippage Analysis." In *Proceedings of the 2004 Eurographics/ACM SIGGRAPH Symposium on Geometry Processing*. Nice, France: SGP. 214–223.

Glymph, James, Dennis Shelden, Cristiano Ceccato, Judith Mussel, and Hans Schober. 2004. "A Parametric Strategy for Free-form Glass Structures Using Quadrilateral Planar Facets." *Automation in Construction* 13 (2): 187–202.

Golovinskiy, Aleksey, and Thomas Funkhouser. 2008. "Randomized Cuts for 3D Mesh Analysis." *ACM Transactions on Graphics* 27 (5): Article 145.

Ivy Nejur, Steinfeld

Julius, Dan, Vladislav Kraevoy, and Alla Sheffer. 2005. "D-Charts: Quasi-Developable Mesh Segmentation." *Computer Graphics Forum* 24 (3): 581–590.

Kalvin, Alan D, and Russell H. Taylor. 1996. "Superfaces: Polygonal Mesh Simplification with Bounded Error." IEEE Computer Graphics and Applications 16 (3): 64–77.

Katz, Sagi, and Ayellet Tal. 2003. "Hierarchical Mesh Decomposition Using Fuzzy Clustering and Cuts." *ACM Transactions on Graphics* 22 (3): 954–961.

Katz, Sagi, George Leifman, and Ayellet Tal. 2005. "Mesh Segmentation Using Feature Point and Core Extraction." *The Visual Computer* 21 (8): 649–658.

Lai, Yu-Kun, Shi-Min Hu, Ralph R. Martin, and Paul L. Rosin. 2009. "Rapid and Effective Segmentation of 3D Models Using Random Walks." *Computer Aided Geometric Design* 26 (6): 665–679.

Lin, Hsueh-Yi Sean, Hong-Yuan Mark Liao, and Ja-Chen Lin. 2007. "Visual Salience-Guided Mesh Decomposition." *IEEE Transactions on Multimedia 9* (1): 46–57.

Liu, Rong, and Hao Zhang. 2004. "Segmentation of 3D Meshes through Spectral Clustering." In *Proceedings of the 12th Pacific Conference on Computer Graphics and Applications*, Seoul: PG. 298–305.

Mangan, Alan P., and Ross T. Whitaker. 1999. "Partitioning 3D Surface Meshes Using Watershed Segmentation." *IEEE Transactions on Visualization and Computer Graphics* 5 (4): 308–321.

Massarwi, Fady, Craig Gotsman, and Gershon Elber. 2007. "Papercraft Models Using Generalized Cylinders." In *Proceedings of the 15th Pacific Conference on Computer Graphics and Applications*. Maui, Hawaii: PG. 148–157.

Mitani, Jun, and Hiromasa Suzuki. 2004. "Making Papercraft Toys from Meshes Using Strip-based Approximate Unfolding." *ACM Transactions on Graphics* 23 (3): 259–263.

Mortara, Michela, Giuseppe Patané, Michela Spagnuolo, Bianca Falcidieno, and Jarek Rossignac. 2004. "Plumber: A Method For a Multi-Scale Decomposition of 3D Shapes Into Tubular Primitives and Bodies." In *Proceedings of the Ninth ACM Symposium on Solid Modeling and Applications*, Genova, Italy: SM. 339–344.

Nejur, Andrei. 2016. "Ivy for Grasshopper Manual: Version 0.802." *Digital Design Research Repository Andrei Nejur*. Accessed July 01, 2016. http://research.n2arh.ro/ivy/manual.

Pottmann, Helmut, Michael Eigensatz, Amir Vaxman, and Johannes Wallner. 2015. "Architectural Geometry." *Computers & Graphics* 47: 145–164.

Pottmann, Helmut, Alexander Schiftner, Pengbo Bo, Heinz Schmiedhofer, Wenping Wang, Niccolo Baldassini, and Johannes Wallner. 2008.

"Freeform Surfaces from Single Curved Panels." *ACM Transactions on Graphics* 27 (3): Article 76.

Pottmann, Helmut, Yang Liu, Johannes Wallner, Alexander Bobenko, and Wenping Wang. 2007. "Geometry of Multi-layer Freeform Structures for Architecture." *ACM Transactions on Graphics* 26 (3): Article 65.

Shamir, Ariel. 2008. "A Survey on Mesh Segmentation Techniques." *Computer Graphics Forum* 27 (6): 1539–1556.

Shapira, Lior, Ariel Shamir, and Daniel Cohen-Or. 2008. "Consistent Mesh Partitioning and Skeletonisation Using the Shape Diameter Function." *The Visual Computer* 24 (4): 249–259.

Shlafman, Shymon, Ayellet Tal, and Sagi Katz. 2002. "Metamorphosis of Polyhedral Surfaces Using Decomposition." *Computer Graphics Forum* 21 (3): 219–228.

Skiena, Steven S. 1998. *The Algorithm Design Manual*. Santa Clara, CA: TELOS—the Electronic Library of Science.

Stork, David G., Richard O. Duda, and Elad Yom-Tov. 2004. *Computer Manual in MATLAB to Accompany Pattern Classification*. Hoboken, NJ: Wiley-Interscience.

Taubin, Gabriel, and Jarek Rossignac. 1998. "Geometric Compression through Topological Surgery." *ACM Transactions on Graphics* 17 (2): 84–115.

Xing, Qing, Gabriel Esquivel, Ergun Akleman, Jianer Chen, and Jonathan Gross. 2011. "Band Decomposition of 2-Manifold Meshes for Physical Construction of Large Structures." In *Posters of the 38th International Conference and Exhibition on Computer Graphics and Interactive Techniques*. Vancouver, BC: SIGGRAPH.

Xing, Qing, Gabriel Esquivel, Ryan Collier, Michael Tomaso, and Ergun Akleman. 2011. "Spulenkorb: Utilize Weaving Methods in Architectural Design." In *Proceedings of the 14th Annual Bridges Conference: Mathematics, Music, Art, Architecture, Culture.*, edited by Reza Sarhangi and Carlo H. Séquin. Coimbra, Portugal: Bridges. 163–170.

Yamauchi, Hitoshi, Stefan Gumhold, Rhaleb Zayer, and Hans-Peter Seidel. 2005. "Mesh Segmentation Driven by Gaussian Curvature." *The Visual Computer* 21 (8–10): 659–668.

IMAGE CREDITS

Figure 1: Deleuran and Reeves, 2015
Figures 2–15: © Nejur, 2016

Andrei Nejur is an Assistant Lecturer in the Architecture Department at the Technical University of Cluj-Napoca

Kyle Steinfeld is an Assistant Professor specializing in digital design technologies in the Department of Architecture at the University of California, Berkeley.

Generative Robotics

Sean Ahlquist
University of Michigan, Taubman College of Architecture and Urban Planning

Generative design can be defined, in part, as a process of exploration spurred by the integration of codified *actions* as opposed to explicit constraints. D'Arcy Thompson provides the foundation for tools for generative design in his description of *homologies*.[1] These are understood as fields of possible outcomes emanating from geometric relationships that encapsulate the organization of physical forces into form. The evolution of this approach in computational design has focused on ever more accurate simulation—rather than representation—of form as the consequence of imposing forces, material properties, and rules of assembly.

It is this last item, assembly, that the work in this section addresses. The projects showcase a clear entrenchment of generative design in their approach towards robotic fabrication. While often confined to the exploration of virtual forms, the work in this section applies the concepts of generative design to the activities of robots themselves. The robot shifts from being the executor of the final static design iteration, emerging from a computational exploration, to one that is fully submerged within the cyclical processes of generative design. The making of material and the assembly of components become active processes of integrating agents and continually adapting to emerging aggregate behaviors. Therefore, design no longer involves the production of a static series of instructions or conceptual design intentions; rather, it entails the determination of a dynamic scientific model, which is based on the relationships of changing systems, that best negotiates design criteria, material action, and machinic operations. Here, the machine's capacity to process complex interrelations can be employed to embed material data in the form-generation process.

Of the work in this section, Brugnaro's construction strategies, driven by material behavior, exhibit a generative design space that is actively redefined as the assembly, context, and aggregate behavior emerges. Braumann, Moorman, and Vasey each present work with a similar approach, while adding feedback through human interaction as a transformational design agent of fabrication. Where Brugnaro and Vasey provide a material-specific implementation of this approach towards generative robotics, Braumann and Moorman each discuss the accessible nature of the programming environment, empowering a broader range of human-robot design explorations.

Advancing the level of intelligence in a robot-centric generative design framework also requires more highly tuned controls over material explorations that are increasingly multi-hierarchical and non-standard. Devadass's timber structure, in adapting raw timber into large-scale structural components, questions the prototypical steps for processing material. Where robotic fabrication moves from defining constraints to establishing a field of possible operations, the concept of material processing can be completely re-thought. Devadass does not simply assume a role within multiple steps of material production from sourcing to construction-grade material. Rather, sourcing, engineering, and robotic manufacturing are set at critical axes of a multi-dimensional design space. Schwinn's work in robotic sewing leverages this approach by allowing the vast hierarchies that form material-driven architectures to be a part of the generative design process. This is expressed in constructions tailored to the micro scale of material make-up, the meso scale of interacting material assemblies, and the macro scale of system behavior.

1. D'Arcy Thompson, *On Growth and Form*, abridged edition, (Cambridge: Cambridge University Press, 1961).

Generative Robotics

Robotic Softness

An Adaptive Robotic Fabrication Process for Woven Structures

Giulio Brugnaro
ICD, University of Stuttgart
The Bartlett, UCL

Ehsan Baharlou

Lauren Vasey

Achim Menges
ICD, University of Stuttgart

1

ABSTRACT

This paper investigates the potential of behavioral construction strategies for architectural production through the design and robotic fabrication of three-dimensional woven structures inspired by the behavioral fabrication logic used by the weaverbird during the construction of its nest. Initial research development led to the design of an adaptive robotic fabrication framework composed of an online agent-based system, a custom weaving end-effector and a coordinated sensing strategy utilizing 3D scanning.

The outcome of the behavioral weaving process could not be predetermined a priori in a digital model, but rather emerged out of the negotiation among design intentions, fabrication constraints, performance criteria, material behaviors and specific site conditions. The key components of the system and their role in the fabrication process are presented both theoretically and technically, while the project serves as a case study of a robotic production method envisioned as a soft system: a flexible and adaptable framework in which the moment of design unfolds simultaneously with fabrication, informed by a constant flow of sensory information.

1   The robotic soft system unfolds
    simultaneously in the physical and
    digital realms through a real-time
    updating interface.

2   Southern Masked Weaver (*Ploceus
    Velatus*) (Photo by Chris Eason,
    2008. Distributed under the
    Creative Commons Attribution 2.0
    Generic license).

154

INTRODUCTION

Current methods for architectural production are based on notational and geometrical representation, where only what can be drawn and measured can be built (Carpo 2011). Moreover, these methods are usually organized in a linear progression from design intention to materialization, which impedes any feedback among the different stages of the realization of a project. This hinders the design and realization of systems which require a more explorative approach based on a constant exchange of information between the stages of design and fabrication (Sharif and Gentry 2015). In order to move beyond a linear design and realization process, this paper proposes the use of robotic fabrication processes for architectural production not as mere standard fabrication environment, but as a soft system, according to the definition given by S. Kwinter:

> "A system is 'soft' when is flexible, adaptable, and evolving, when it's complex and maintained by a dense network of active information or feedback loops, or, put in a more general way, when a system is able to sustain a certain quotient of sensitive, quasi-random flow" (Kwinter 1992).

Within this understanding, the resulting object is not a separate entity from the process that produces it: the whole workflow is envisioned as an integrated system that adapts and evolves according to design intentions, fabrication constraints, performance criteria, material behaviors and specific site conditions.

In this regard, the habitats built by animals are a great source of inspiration for fabrication processes. Rather than following a predefined blueprint, their strategies are based on a set of building behaviors triggered by external and internal stimuli, medium-low level assumptions and a constant sensing of the environment (Gould 2007), which altogether create an evolving feedback-based system for construction. In order to investigate these concepts in an actual fabrication case study, the research focused on the design and robotic production of three-dimensional woven structures inspired by the behavioral fabrication logic used by the weaverbirds (Ploceidae) (Figure 2) during the construction of their nests (Collias and Collias 1962).

CONTEXT

Behavior-based robotic strategies, within which this project finds precedence, affirm the importance of observation and sensing procedures in contrast to symbolic representation (Brooks 1990). Pure reactive systems are exclusively based on sensory information with no internal representation of the world, yet they are able to generate even complex behavioral responses responding to a set of stimuli or conditions (Matarić 2007). Furthermore, the importance of biological role models for this field of research was

2

identified by R. C. Arkin, "the study of behavior-based robotics should begin with an overview of biological behavior" (1998).

In recent years, a series of pioneering projects introduced robotic fabrication procedures based on sensor-actuator feedback loops into the field of robotic architectural production. In 2012, K. Dorfler and R. Rust, with the project "Interlacing," presented a digital interface equipped with a network of sensors that allowed on-line control of the robot (Dorfler and Rust 2012). In 2014, R. L. Johns presented the project "Augmented Materiality," in which real-time updating fabrication tools dealt with unpredictability of the material system (melted wax) which had not previously been possible to simulate accurately (Johns 2014).

More recently, A. Menges introduced the concept of cyber-physical making in architecture, defined as construction processes which intensively link the realm of physical production with the virtual domain of computation. Within this context, behavior-based construction is based on "real-time physical sensing and computational analysis, material monitoring, machine learning and continual (re)construction" (Menges 2015), instead of relying on explicit instructions and predictive modelling. One of the first practical applications of these concept in the architecture field is represented by the "ICD/ITKE Research Pavilion 2014–2015," where an adaptive robotic fabrication process was devised to construct a composite compression shell with a pneumatic formwork monitored in real-time during the construction. (Vasey et al. 2015).

Within the project "TailorCrete," the University of Southern

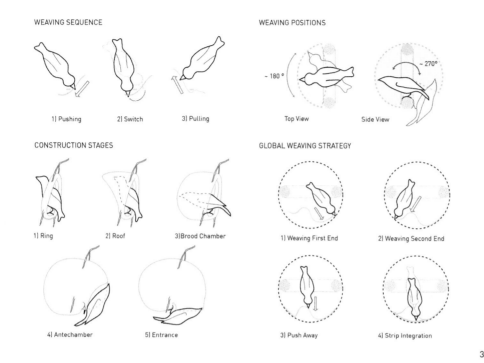

WEAVING SEQUENCE

1) Pushing 2) Switch 3) Pulling

WEAVING POSITIONS

~ 180°   Top View   ~ 270°   Side View

CONSTRUCTION STAGES

1) Ring 2) Roof 3) Brood Chamber

4) Antechamber 5) Entrance

GLOBAL WEAVING STRATEGY

1) Weaving First End 2) Weaving Second End

3) Push Away 4) Strip Integration

3   The principles abstracted from the the analysis of the weaverbird's construction behavior were used to inform the robotic fabrication process. The main reference for the diagrams are the studies by Nicholas and Elsie Collias (1962).

4   An adaptive robotic fabrication process for woven structures inspired by the behavioral fabrication logic used by the weaverbird during the construction of its nest.

5   Rattan was chosen as suitable weaving material for its flexibility, bending behavior and variable length.

6   Rattan's flexibility was determined through material tests which involved fixing the two ends of a stick at different distances in order to compute the range of possible bend radii. In this image, for a stick diameter of 8 mm, the bend radius goes from 215 to 20 cm.

3

Denmark presented an offline robotic fabrication process to model unique concrete reinforcement structures made out of bent metal bars, creating double-curved three-dimensional meshes (Cortsen et al. 2014). However, in this case, the bars were not woven together in an active-bending structure, but were plastically deformed and kept in place by metal wire knots. Although the structure and the robotic actuation were predefined before the fabrication process, the project results are relevant because of the integrated use of sensor feedback: laser positioning was used to enable adjustments to tolerances between horizontal and vertical rebars.

BIOLOGICAL ROLE MODEL

The analysis of the biological role model determined a catalogue of relevant abstracted principles that informed the robotic soft system both locally and globally (Figure 3). On the local level, the sequence of how the bird holds the grass strip and then weaves it into the woven mesh inspired the design of the robotic end-effector and the local weaving logic of actions. Globally, the different weaving positions and the sequence of construction stages, together with responsiveness to the environment and boundary conditions, influenced the overall global robotic behavior. Studying the construction principles utilized by the bird to construct the nest and their variation throughout the process (e.g. the varying density of the woven mesh to filter light and protect the interior) played a key role in informing the robotic actuation with performance criteria that are constantly evaluated and gradually determine the unfolding of the weaving procedure.

SYSTEM DEVELOPMENT

Following the analysis of the biological role model, the key features of the weaverbird's building behavior were abstracted and applied within a robotic fabrication environment for the production of three-dimensional woven structures (Figure 4). The properties and shape of these architectures keep evolving as the process gradually unfolds; one stick is added after another, thus creating a stable active-bending morphology after each step. The studies of the weaverbird's construction process revealed the importance of finding a suitable material for weaving, especially considering flexibility, bending behavior and variable length. Rattan, widely used for furniture and basket production, proved to be a good choice in regard to these parameters. The material derives from a family of palms named Calamoideae, mostly harvested in South-East Asia, with long and slender stems that grow like a vine over other local vegetation. Compared to bamboo, rattan has a full circular section and it is much more flexible (Figure 5). Its bendability depends significantly on its diameter (Figure 6), which ranges from a few millimeters up to approximately several centimeters (Johnson and Sunderland 2004). The diameter used for this project ranges from 5 to 12 mm and was determined empirically according to the size of the robotic fabrication prototypes and tools.

The mechanism implemented in the weaving robotic end-effector is derived from the weaverbird's behavior, though the dual actuation deviates from the weaverbird's single articulated beak. Threading in and out is achieved by pushing the fiber or stick into the woven system, releasing it and catching it in on the

4

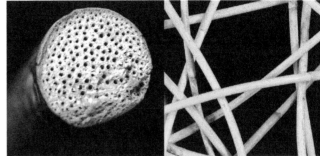

5

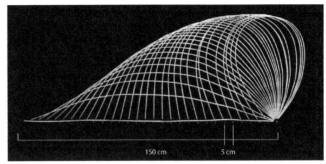

150 cm   5 cm

6

other side. and then pulling it away toward the intended weaving direction. During the weaving process, the rattan stick is bent and passed through the woven mesh, deforming and shaping the ongoing flexible structure. At the end of each weaving sequence, the stick assumes a position impossible to simulate precisely beforehand because of the complex interaction of forces.

To deal with the system tolerances, the main strategy has been to establish a tight loop of sensor-actuator feedback, where information extracted directly from the real woven mesh was processed and utilized within the soft system to drive the robotic actuation. The designers maintain some control over the system and can observe and inform the evolution of the woven structure through a set of computational tools which compile sensor data and visualize in real-time the robotic movements and effector actuation. In this way, the project created a direct link between the digital and physical, and established a feedback system that informed the computational system with sensor data from the real world.

The research developed through a series of small-scale experiments and prototypes, and culminated in a final prototype as a partition of a larger woven structure. The main construction concept for this prototype, derived significantly from the weaverbird's construction logic, was to start from an initial stable core of boundary conditions generated by the robot, then extend beyond them by using a loose hierarchical structure that evolved as each member would rely reciprocally on its adjacent neighbors to stand.

BEHAVIORAL ROBOTIC FABRICATION

While many standard fabrication processes are organized as a linear progression from design intention to materialization, behavior-based fabrication strategies depend on constantly updating sensor-actuator feedback loops.

At the core of the soft system is an Agent-Based System (ABS), where robots become "autonomous-decision making entities called agent" able to assess the environment where they operate, in this case the woven structure as reconstructed through sensor data, and make decisions according to a set of predefined rules (Bonabeau 2002).

For this research, the architecture of the soft system is composed of two integrated loops operating at two different scales, one acting locally (1) and the other globally (2), constantly exchanging information (Figure 7). The local loop (1) deals with the direct manipulation of the material with the weaving robotic end-effector and its coordination with local robot movement. The sensor data is gathered at this level with a depth camera (Microsoft Kinect 360) and is only related to a small portion (about 50 x 50 cm) of the woven mesh ahead of the effector's weaving direction. The global loop (2) utilizes scanning to process the density of the current overall woven structure, iteratively comparing it to the initial intended densities. which act as loose guidelines for the development of the overall system. Thus in each iteration, the global ABS utilizes a behavior-based process to decide where to begin the weaving sequence for the next piece or rattan.

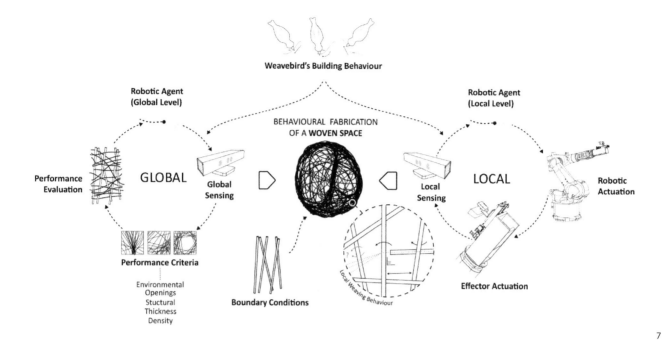

Because behavioral fabrication processes significantly rely on real-time data streams between all the components of the loop, it was very important to implement network communication that would allow computation and motion planning to happen smoothly and efficiently. Building upon online control methods which had been developed previously at the Institute for Computational Design, a system of communication was utilized including three main components: the "Client," where the Agent-Based System (ABS) is running in real-time within an interface designed in Rhinoceros3D and Grasshopper/RhinoPython, the "Server," responsible for maintaining the consistent pace of information exchanged, and finally, the "Robotic Controller," which controls the actuation of a standard 6-axis industrial robot, a KUKA KR 125/2.

The overall coordination of the system was made possible using the KUKA RSI module (Robotic Sensor Interface), which allows the exchange of XML packages over Ethernet. This allowed the client and design environment to receive information regarding the actual position and orientation of the robotic Tool Center Point (TCP) and send back the next target frame calculated by the ABS.

Local Weaving Agent-Based System
The ABS is an online computational design tool responsible for local relative motion planning for the robot as well as synchronized effector actuation. At any moment, the ABS calculates a vector direction towards the next ideal weaving position, and solves for the vector which would allow the effector to stay perpendicular to the pieces of rattan currently being woven. The

target orientation of the effector is in a plane perpendicular to the overall weaving direction and the local tangent of the stick where the weaving procedure is acting (Figure 9).

In coordination with this movement, the ABS simultaneously outputs synchronous serial commands for the effector actuation. These outputs are specifically timed according to the local positions of the woven pieces where the weaving procedure will be executed. Because the precise positions of all elements are not known, and the rattan sticks shift during the process of weaving, these outputs must be generated in real-time while the robot is operating based on the sensor data. The weaving procedure is achieved when a series of actuations is executed in order around pre-existing rattan elements (Figure 8). Particular attention was given, on one hand, to ensuring that sequential curvilinear movements would not collide with the ongoing woven structure, and on the other, to the effector constraints and tolerances: ideally the two sticks being woven were locally close to perpendicular and the distance between them was greater than three times the tolerance of the local scanning process.

Weaving Robotic End-Effector
The weaving end-effector is the device responsible for manipulating the material during the robotic actuation. As already mentioned, though actuated gripping is common place, there were no industrial precedents for weaving in three dimensions. It was thus necessary to design and fabricate a custom robotic end-effector to enable an automated and mechanical weaving procedure (Figure 10).

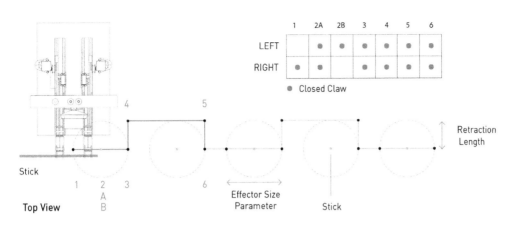

	1	2A	2B	3	4	5	6
LEFT		●	●	●	●	●	●
RIGHT	●	●		●	●	●	●

● Closed Claw

Retraction Length

Stick

Top View

1   2   3   6   Effector Size
     A   Parameter   Stick
     B

7 The soft system is composed of two coordinated feedback loops, one acting on a local level while the other on a global one, unfolding together in real-time.

8 The weaving sequence, composed by a series of key points in sequential order, adapts each time according to the specific stick's configuration captured by sensor data.

8

One of the main required features to operate within a behavioral fabrication process is the inherent adjustability of the tools, which can be attuned as the system unfolds and do not require a high level of precision. In order to achieve this inherent adjustability, the device is equipped with two coordinated robotic claws that are significantly inspired by the weaverbird's strategy to hold the grass strips with its beak during the construction. Because of their ability to incrementally close, they could grab different diameter sizes of rattan sticks (Figure 11). To further enhance the adjustability of the tool, the claws are mounted on two linear rails that actuate at different times to grab and release the piece currently being woven on the other side of a preexisting rattan member.

Effector control was achieved through serial commands from the client to an Arduino board, controlling four servo motors (two for the linear rails, two for opening and closing the claws). Together with the mechanical weaving procedure, the devices responsible for the sensing strategy were integrated on the effector itself, following at any time the robotic trajectory.

Performance Criteria
The performance input criteria investigated in the project is related to the modulation of material density within the woven system. This quality is translated into useful information through a gradient density mapping of the current and target density map. This acts as a loose guideline of intended density for the ABS, which uses agent vision and a weighting system to negotiate design intentionality with the fabrication constraints and material parameters previously described. Though the final prototype only included a limited number of rattan elements, if the system was developed further, this strategy could allow the system to achieve specific light gradients, to reinforce local areas, and to control the level of openness and visual permeability.

A scalar field generated as a color gradient mapping (in this case in grayscale) offers an intuitive way to visually understand and control density. By utilizing a scalar field mapped over the entire system, it is possible to instruct the ABS (global) with a simple behavior, for instance to avoid lighter areas and prefer darker ones, in order to vary the amount of material distributed in the structure (Figure 13).

While this performance mapping is initially defined at the beginning of the design process, the system development is continuously informed by this mapping. An "In-progress Density Map" reflecting the current status of construction is generated with two different methods: 1) Recording the previous robotic movements and relating them to the ongoing woven structure. 2) Directly from sensor data which calculates the physical material density in each discretized area of the woven system. Furthermore, the designer could also intervene and decide to manipulate the system development by altering this intended density map during production.

Global Weaving Agent-Based System
While the local agent's environment acts only locally, a global agent acts on a digital reconstruction of the whole structure to determine where to place the next piece of rattan. This virtual agent navigates the boundary conditions to find a position in which the intended final density is highest, but the actual constructed density is low, by using the values of the in-progress density map. This virtual representation is updated at the end of every weaving sequence. After locating a start position, the agent then determines its direction as a weighted average of scalar values within its field of vision, determining where the rattan stick will be woven next (Figure 12).

Sensing Strategy
To obtain the necessary information regarding the physical

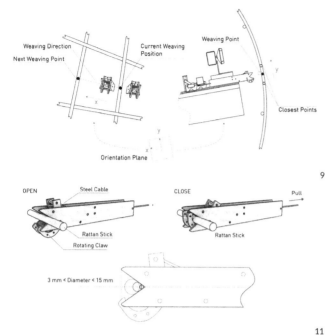

9

11

10

evolving structure, a depth camera with a lower accuracy but large field of vision (Microsoft Kinect 360) was preferred over more accurate sensors which return only a single numeric distance or point, such as a simple proximity sensor or laser range finder. The Kinect, mounted on the weaving robotic effector, returns into the design environment a raw point cloud representing the scanned geometry located in three-dimensional space relative to the robot base. Because this importation is computationally heavy, in the local behavioral loop, the Kinect scans only a specific area (about 50 x 50 cm) in front of the effector's moving direction. As in the reactive control paradigm, which enables responsivity without internal representation of the world (Mataric 2007), this data is only temporary: it is used to compute the next weaving sequence in real-time and is not stored in the system (Figure 14).

For the global sensing strategy, the Kinect was attached on the top plate of the weaving effector and oriented along a plane orthogonal to the TCP (Figure 15). Every session started with the calibration of the sensor with a known geometry and a first explorative scan to approximately locate the woven structure in space. Because the field of vision of the Kinect is small, and because of the difficulty completely synchronizing the position of the robot with the incoming data, it was not possible to scan the whole structure from either a single position or in a single continuous movement. For this reason, it was necessary to develop a procedure that creates a series of scanning positions in front of the structure and, in coordination with the robotic movements, triggers the Kinect when it reaches one of these locations. Because the orientation plane of the sensor and its relation with

the TCP is known, it has been possible to correctly re-orient each scan in the digital space even if the position of the Kinect was constantly changing (Figure 16). Subsequently, the point clouds are combined together, duplicated points removed and the point cloud reduced to a minimal collection of points describing the linear axis of the sticks.

RESULTS AND REFLECTIONS

The research succeeded in developing a behavioral fabrication strategy for the production of three-dimensional woven structures based on a series of principles abstracted from the analysis of the weaverbird's building behavior. The framework of the soft system, even if specifically tailored for the weaving procedure, is potentially transferable to other fabrication tasks with different materials and robotic production workflows.

In order to automate a process heretofore only present in vernacular handmade processes or biological construction methods, a custom robotic weaving effector was constructed and utilized to successfully enable robotic weaving. Understandably, the process had some difficulties: while at the beginning of the fabrication the woven structure is really flexible and unstable, after several iterations it becomes progressively rigid. The friction generated to weave a stick across the ongoing mesh also increases. Both conditions could determine significant deformations in the structure while the weaving procedure is happening, therefore it could be challenging to complete the sequence of the effector's actuations without any error. Further improvements to the mechanical parts and reduction of the effector's size would improve the usability of the tool to enable even more complex geometrical

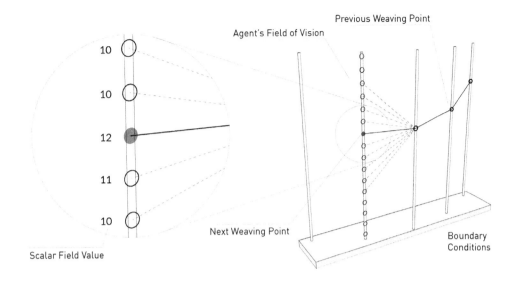

Agent's Field of Vision

Previous Weaving Point

10

10

12

11

10

Scalar Field Value

Next Weaving Point

Boundary
Conditions

9   Behavioral re-orientation of the
    weaving end-effector according
    to the local configuration of the
    woven structure.

10 The design of the end-effector
    enabled the development of
    the robotic weaving procedure,
    constantly adjusting to operate in
    different local configurations of the
    woven mesh.

11 The shape of the gripping claws,
    closing incrementally, allows to hold
    sticks of different sizes.

12 The robotic weaving agent travels
    across the woven mesh adjusting
    its direction according to the
    user-defined performance field,
    determining this way how the
    weaving sequence will unfold.

12

configurations and constructed morphologies. Significantly, the process implied the development of a bridge between the physical and digital realm, maintained by a series of integrated sensor-actuator feedback loops acting at different scales, where the moment of design blurs with fabrication, unfolding together in real-time (Figure 17). The agent based system enabled the system to use behavioral rule based logics to develop, integrating initial design intentions with material behavior and fabrication limitations. The sensing strategy represented one of the key components of the system and enabled the behavioral feedback loop, however, it was quite challenging to extract extremely accurate information from the Kinect 360 sensor and process it in real-time to the ABS. From time to time, the low level of precision and errors in the calibration undermined the quality of the behavioral responsiveness and, in this regard, a depth camera with higher resolution and precision would be beneficial to the consistency of the overall process.

From an architectural design perspective, the fabrication process outlined a series of potential applications and relevant features for robotic fabrication processes. First, the construction process could be initiated from only a few boundary conditions and generate a complex doubly curved interwoven system without the need of an expensive formwork. Furthermore, the system could be locally tailored to achieve specific architectural features, for instance, openings or reinforcement ribs, in relation to structural and environmental considerations, while the overall system remains flexible and structurally redundant.

CONCLUSIONS

The main driver of this research was the investigation of behavioral fabrication processes and their potential application in the field of architecture, opposing standard linear construction methods and geometrically static notational systems of design representation and construction. This expands possible methods and production systems for architectural production thanks to a novel use of already existing computational tools and robotic technologies, including machine vision and online control.

Fusing together design and production, the system got closer to natural construction processes which evolve based on behavioral adaptation to external and internal stimuli (Vasey et al. 2015). The project more broadly suggests an alternative approach for production where the design process is not centered on realizing a predefined solution, but instead embraces explorative and experimental processes. In this regard, one of the future research's directions is to envision possible ways of utilizing these methods in larger scale projects to reconsider the role of robots in construction processes.

ACKNOWLEDGEMENTS

The work here presented was developed in a year-long individual Master Thesis by Giulio Brugnaro and supervised by Prof. Achim Menges, Ehsan Baharlou, Lauren Vasey and Tobias Schwinn within "ITECH: Integrative Technologies and Architectural Design Research" Master Program at the University of Stuttgart, Germany.

REFERENCES

Arkin, Ronald C. 1998. *Behavior-Based Robotics*. Cambridge, Mass: MIT Press.

Bonabeau, E. 2002. "Agent-Based Modeling: Methods And Techniques For Simulating Human Systems". *Proceedings Of The National Academy Of Sciences* 99 (Supplement 3): 7280–7287.

Brooks, Rodney A. 1990. "Elephants Don't Play Chess." *Robotics and Autonomous Systems* 6 (1): 3–15: 31.

Carpo, Mario. 2011. *The Alphabet And The Algorithm*. Cambridge, Mass.: MIT Press.

Collias, Nicholas E., and Elsie C. Collias. 1962. "An Experimental Study Of The Mechanisms Of Nest Building In A Weaverbird". *The Auk* 79 (4): 568–595.

Cortsen, J., J.A. Rytz, L.-P. Ellekilde, D. Sølvason, and H.G. Petersen. 2014. "Automated Fabrication Of Double Curved Reinforcement Structures For Unique Concrete Buildings". *Robotics And Autonomous Systems* 62 (10): 1387–1397.

Gould, James L., and Carol Grant Gould. 2007. *Animal Architects*. New York: Basic Books.

Johnson, Dennis Victor and Terry C. H Sunderland. 2004. *Rattan Glossary*. Rome: Food and Agriculture Organization of the United Nations.

Kwinter, Sanford. 1992. *Soft Systems*. New York: Princeton Architectural Press: 211.

Matarić, Maja J. 2007. *The Robotics Primer*. Cambridge, Mass.: MIT Press.

Menges, Achim. 2015. "The New Cyber-Physical Making In Architecture: Computational Construction". *Architectural Design* 85 (5): 28–33.

Sharif, Shani and T. Russell Gentry. 2015. "Design Cognition Shift from Craftsman to Digital Maker". In *Emerging Experience in Past, Present and Future of Digital Architecture, Proceedings of the 20th International Conference of the Association for Computer-Aided Architectural Design Research in Asia (CAADRIA 2015)*. Daegu. 683–692.

Vasey, L., E. Baharlou, M. Dörstelmann, V. Koslowski, M. Prado, G. Schieber, A. Menges, and J. Knippers. 2015. "Behavioral Design and Adaptive Robotic Fabrication of a Fiber Composite Compression Shell with Pneumatic Formwork". In *Computational Ecologies: Design in the Anthropocene, Proceedings of the 35th Annual Conference of the Association for Computer Aided Design in Architecture (ACADIA)* edited by L. Combs and C. Perry. University of Cincinnati, Cincinnati OH. 297–309.

IMAGE CREDITS

Figures 1: Giulio Brugnaro, Institute for Computational Design, University of Stuttgart, 2014–2015.
Figure 2: Chris Eason, 2008 (Distributed under the Creative Commons Attribution 2.0 Generic license).
Figures 3–17: Giulio Brugnaro, Institute for Computational Design, University of Stuttgart, 2014–2015.

_____

Giulio Brugnaro is currently Research Assistant at The Bartlett School of Architecture in London as part of the InnoChain Research Network. His research focuses on developing adaptive robotic fabrication processes and sensing methods that could allow designers to engage with the explorative dimension of making and the potential of materials agency. Previously, he received a B.Arch in "Architectural Sciences" (cum laude) at IUAV University of Venice and a M.Sc. in "Integrative Technologies and Architectural Design Research" at the University of Stuttgart.

_____

Ehsan Baharlou is a doctoral candidate at the Institute for Computational Design (ICD) at University of Stuttgart. He holds a Master of Science in Architecture with distinction from the Islamic Azad University of Tehran. Along with pursuing his doctoral research, he has taught seminars at the ICD since 2010.

_____

Lauren Vasey is a Research Associate at the Institute for Computational Design (ICD) at University of Stuttgart. She received a Bachelor of Science (cum laude) in Engineering from Tufts University, and a Masters of Architecture with distinction from the University of Michigan. She has taught several workshops and courses in computational design and robotic fabrication and has worked previously at the University of Michigan Taubman College FAB Lab as well at the Swiss Federal Institute of Technology (ETH-Zurich), Chair for Architecture and Digital Fabrication.

_____

Professor Achim Menges, born 1975, is a registered architect and professor at the University of Stuttgart, where he is the founding director of the Institute for Computational Design since 2008. He is also Visiting Professor in Architecture at Harvard University's Graduate School of Design since 2009.

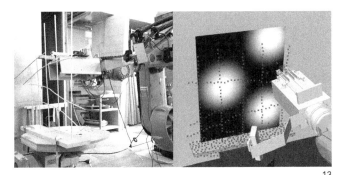

13

15

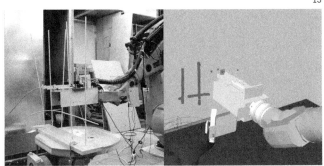

14

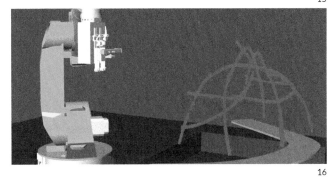

16

13 The robotic weaving process is guided by a user-defined scalar map, which represents intended material density.

14 The depth camera scans only a small portion of the mesh ahead of the effector's weaving direction, reducing the computational intensity and allowing real-time responsiveness.

15 The robotic agent scans the ongoing woven structures before each weaving sequence to obtain a description of the current configuration.

16 The global pointcloud is updated with sensor data of each partial scan and subsequently processed to inform the ABS.

17 The robotic agent performs the weaving sequence based on previously acquired sensor information which locates the position of each woven stick with an accuracy of 1 cm.

17

Towards New Robotic Design Tools

Using Collaborative Robots within the Creative Industry

Johannes Braumann
Robots in Architecture
UfG Linz

Sven Stumm
IP RWTH Aachen

Sigrid Brell-Cokcan
Robots in Architecture
IP RWTH Aachen

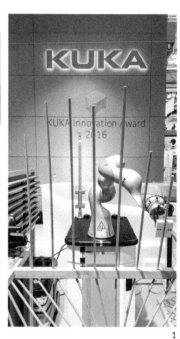

1

ABSTRACT

This research documents our initial experiences of using a new type of collaborative, industrial robot in the area of architecture, design, and construction. The KUKA LBR-iiwa differs from common robotic configurations in that it uses seven axes with integrated force-torque sensors and can be programmed in the Java programming language. Its force-sensitivity makes it safe to interact with, but also enables entirely new applications that use hand-guiding and utilize the force-sensors to compensate for high tolerances on building sites, similar to how we manually approach assembly tasks.

Especially for the creative industry, the Java programming opens up completely new applications that would have previously required complex bus systems or industrial data interfaces. We will present a series of realized projects that showcase some of the potential of this new type of collaborative, safe robot, and discuss the advantages and limitations of the robotic system.

1 Sensitive robotic assembly informed through haptic programming. Showcased at the Hannover Fair 2016.

INTRODUCTION

Ten years ago, the use of industrial robots within architectural research was considered to be high-end, often depending on robotic engineers that worked with designers to realize projects. Today, these robots have turned into well-researched tools that can be found at many larger universities, and are also starting to leave the field of research towards full-scale applications, as demonstrated in the work of Gramazio & Kohler, Achim Menges, and others (Figure 2).

Companies such as Branch Technology are now basing their innovative applications on reliable robotic platforms, thus allowing them to focus on their construction-specific tasks, rather than having to develop an entire robotic system.

Whereas robots are often used to perform tasks similar to existing CNC machines, our main research interest lies in the core of construction strategies and how we can advance robotic software technology to assist in assembling tasks (refer to the project section and Stumm et al. 2016).

Currently, significant development in the area of software can be observed in our field, enabling tablet-like interaction with the control panel and new networked devices to act as a robot controller through software options like KUKA mxAutomation (Munz et al. 2016, Braumann and Brell-Cokan 2015). Within the context of architecture and design, the community of creative robot users itself has become the main enabling factor to greatly lower the entry bar for new users of robotics by exchanging ideas on the same level and providing accessible but powerful tools such as KUKA|prc, HAL, and others.

Recently, we can observe changes happening within the robot industry as well, as companies are moving their focus away from traditional robot installations within safety fences towards safe human-robot collaboration, while also investigating new applications beyond the automotive sector.

INTELLIGENT WORK ASSISTANT

For this purpose, a number of robotics companies have developed a new generation of lightweight robotic arms such as KUKA's LBR iiwa (intelligent industrial work assistant), with similar "safe" machines being available from ABB (YuMi), UR, and a range of smaller robotic startups, building upon research such as by the DLR (Albu-Schäffer et al, 2007). The iiwa (Figure 3) differs from the common robotic-arm template. Immediately visible is its design, which is optimized to reduce the potential of harming the user by minimizing the amount of sharp edges and reducing crushing hazards, thus making it possible to safely interact with the robot in a haptic way.

2a

2b

2c

3a

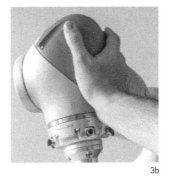

3b

2 Architectural robotic projects: ICD/ ITKE Research Pavilion 2015–16 (a), Branch Technologies (b), Echord ETH Zurich (c).

3 Save interaction with a collaborative robot: Compliant mode (a), collision detection (b).

Furthermore, the LBR iiwa employs seven axes, enabling the robot to move the kinematic chain behind its tool, without moving the tool itself. The additional redundant axis greatly increases the robot's kinematic flexibility, offering mathematically infinite possibilities to move to a defined point in space, which can be used to avoid obstacles, and to offset the comparably limited axis range. The force-torque sensors within every axis constitute another step-change in robotic technologies, enabling the robot to feel and react to contact and pressure. Previously,

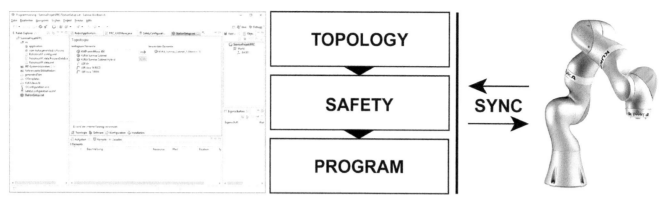

4 SunriseWorkbench: Defining robot topology, safety and program within the IDE and synchronizing it directly to the LBR iiwa robot.

force-torque sensors for industrial robots were mostly mounted directly between the tool and the flange, or the drive currents for individual axes were used to provide only a rather coarse feedback. Adding these sensors to every axis does not just enable force-sensitive applications, but makes the robot potentially safe to work with, as it can now feel collisions at every joint, rather than just at the tool tip.

The sensors are not limited to a Boolean yes/no collision, but provide fine-grained data that can be used for many different applications—e.g., those that require information about contact state or process force. Additionally, the robot can be moved around manually in compliant mode, which does not simply release the brakes of the motors, but compensates gravity and measures the force applied at each joint to support any user-given movement. Even the weight of workpieces can be compensated in a similar fashion, so objects mounted onto the robot can be moved as if they weighed nearly nothing.

Within the scope of architecture and design, we see a special potential for interactive design parametrization and force-guided assembly, which is further discussed in the project section.

Robot as an App

While similar applications have been possible before, the iiwa's new programming provides accessible libraries that allows the creative user to utilize these features at a high level without needing external data processing or specialized software.

Unlike previous LBR iterations such as the LBR4, which relied on the regular KUKA Robot Language (KRL), the iiwa introduced SunriseOS, which uses Java as the robot's programming language. This allows us to use state-of-the-art programming strategies, such as object-oriented programming, and most importantly, to implement external libraries that range from geometric functions to image processing and cloud networking. As an integrated development environment (IDE), KUKA provides the Sunrise

Workbench, a modified version of the open-source software Eclipse. In addition to making complex tasks easier to program, it also greatly opens up the scope of robot programming. Rather than being limited to certain interfaces and tech packages, we can define custom ways of communication and interaction, e.g., using high-performance technologies developed by the game industry for Android, which also uses Java for its apps.

An exemplary programming workflow via Sunrise Workbench (Figure 4) starts with the user creating a new Java project and writing the logic using the provided KUKA-specific libraries for movements, force conditions, etc. Tools and coordinate systems are also created within the Sunrise Workbench and can be intelligently nested, so that, for example, multiple tool-tip coordinate systems are assigned to a single tool. Finally, safety-specific properties are set in the Safety Configuration and the IO configuration is loaded from WorkVisual, as with regular KRC4 robots. The project can now be synced with the robot, replacing the previous program. On the robot, the previously set-up tools and coordinate system now show up and can be calibrated, thus matching the digital environment with the actual physical space. The next synchronization between robot and IDE then pushes these changes back into the Workbench. A simulation in advance is not possible, only syntax errors are checked and highlighted.

Initial Experiences

As the iiwa robot has only recently been introduced and is priced comparatively high, we could not build upon the knowledge of many other, creative robot users. For example, applications in the creative industry are robotic scale model testing at the BRG/ETH Zurich (2016) and camera motion control for movies at CMOCOS (Shepherd and Buchstab 2014). The findings below represent our initial conclusions and subjective evaluation after slightly less than a year of using iiwa robots.

- **Acceptance:** While negative connotations for robotic arms are common, the design of the iiwa seemed to greatly

Towards New Robotic Design Tools Braumann, Brell-Cokcan, Stumm increase public acceptance as it distanced itself from regular industrial robots, while also not too directly emulating human arms. People were quick to interact with the iiwa without hesitations or safety concerns, while students were curious to work with a new machine.

- **Mechanics:** Immediately noticeable is the iiwa's low weight of less than 30 kg, which makes it significantly easier to move and set up than, for example, 50 kg Agilus robots. At the same time, the larger iiwa can manipulate up to 14 kg, while the Agilus series currently tops out at 10 kg, resulting in a very favorable weight to payload ratio. The relatively low axis-speed of 57–144°/sec can be limiting for some applications, but most collaborative setups require significantly lower speeds for safety reasons. The iiwa's main drawback compared to other robots is the very limited range of its axes (e.g. +-140 at A1 and +-152.5 at A7), which is most likely the result of the effort to reduce crushing hazards, while also leading internal wiring and tubing up to the flange. Even with presumably easy movements, it has been very common for us to hit the axis limits of the machine, usually requiring fine-tuning of the posture as well as the redundant, seventh axis.

- **General Programming:** The idea of adapting an established IDE for robot programming definitely shows merit, as does the choice of Java as a well-documented language that many students have been exposed to as early as in secondary school. The KUKA libraries are well integrated and come with a certain amount of documentation, making it possible to quickly create an initial program. A significant challenge arises from the decision to only allow a single active project on the robot (though it can contain multiple sub-programs). This leads to particular challenges when multiple users are collaborating on a project, as the current system lacks any kind of version control, which gives users the choice between synchronizing the project to their local machines (thus overwriting the local project) or moving data the other way around. Therefore, any versioning has to happen before the project is uploaded to the robot.

- **Motion Programming:** While Java offers many advantages for the efficient programming of tasks, thus allowing us to work with parallel threads and asynchronous tasks, the physical robot itself can only do a single operation at once. This leads to problems with blended movements, as the robot has to know the next position in order to create a smooth trajectory without stopping at every programmed position. The user can

either put multiple commands into a MotionBatch, or execute motiontasks asynchronously, both of which caused problems with very large, complex toolpaths. Troubleshooting is further complicated by the fact that the only feedback provided on the robot control panel are console messages as well as exceptions. While the older KRC4 controller presents the user a pointer at the currently active line, SunriseOS only allows the user to step through a program if a debugging process is started through the Sunrise Workbench from a remote PC. Any kind of "block selection" has to be programmed in Java, e.g., through a custom debugging interface.

- **Expandability:** Through additional libraries, it was quickly possible to integrate additional features such as a custom GUI based on the Java.Swing toolkit, as well as geometric operations that allow a similar interaction with geometry as in CAD software. See the project section for more details.

- **Safety:** For applications that involve direct interaction between the user and robot, such as collaborative assembly, but also for the initial programming of robotic tasks, especially in education, the integrated safety features of the iiwa offer significant advantages over conventional robots. With the relevant safety package, it is possible to set a global collision stop that will be applied irrespective of the particular program that is being used, and to protect that setting with a password. While these features do not make any application automatically safe—e.g., when sharp tools are involved—many applications do not require any further, external safety equipment.

NEW PROGRAMMING INTERFACES FOR COLLABORATIVE ROBOTS

In previous research we have developed KUKA|prc (parametric robot control – Braumann and Brell-Cokcan 2011), allowing creative users to quickly and intuitively program and simulate robotic arms within a visual programming environment. For many users, this environment proved essential towards rapidly prototyping new fabrication strategies, but also for quickly engaging students. In our own teaching we have experienced it to be highly beneficial to provide students with an accessible interface that allows them to explore the capabilities and constraints of robotic arms in both a virtual environment, immediately followed by actual, physical experiments. Depending on their skills and interests, students can then either work very "deep" within the programming of the robot, or at a higher level through the visual programming environment—and of course in any combination of these two.

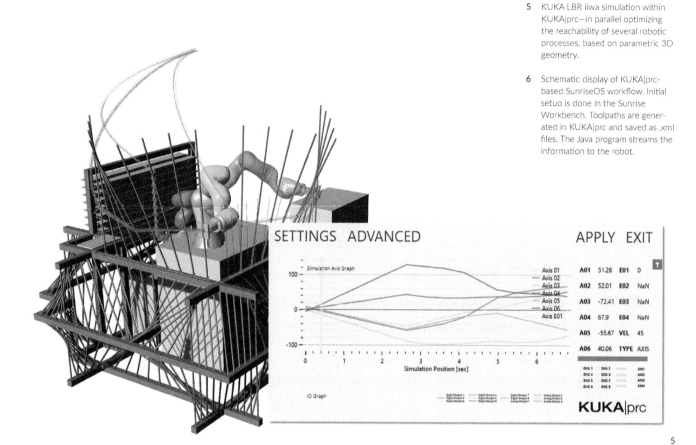

5   KUKA LBR iiwa simulation within KUKA|prc—in parallel optimizing the reachability of several robotic processes, based on parametric 3D geometry.

6   Schematic display of KUKA|prc-based SunriseOS workflow. Initial setuo is done in the Sunrise Workbench. Toolpaths are generated in KUKA|prc and saved as .xml files. The Java program streams the information to the robot.

5

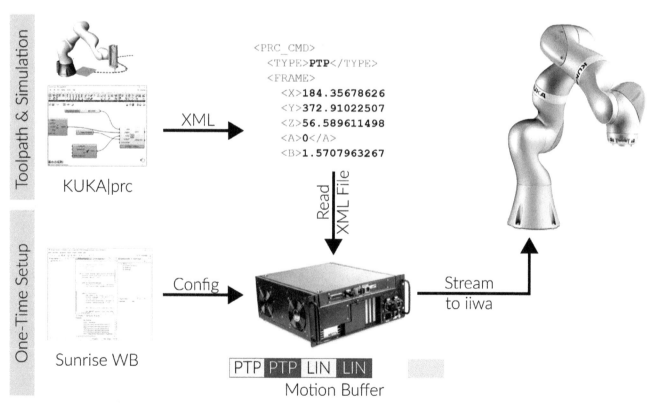

Toolpath & Simulation

KUKA|prc

XML

```
<PRC_CMD>
  <TYPE>PTP</TYPE>
  <FRAME>
    <X>184.35678626
    <Y>372.91022507
    <Z>56.589611498
    <A>0</A>
    <B>1.5707963267
```

Read XML File

One-Time Setup

Sunrise WB

Config

Stream to iiwa

PTP PTP LIN LIN

Motion Buffer

6

We were unable to find any suitable offline programming environment capable of simulating and controlling an iiwa arm;the KUKA SimPro software was only capable of simulating the robot, but unable to output code, and a direct link for ROS would require very deep changes to the controller, as well as software that is not common outside of the core-robotics sector. Thus, we decided to integrate the iiwa robot into the KUKA|prc environment (Figure 5).

The first challenge was the definition of a new kinematic model that supports a single kinematic chain of seven rotary axes, as opposed, for example, to a six-axis robot with an external turntable. This step was taken in preparation of the iiwa's arrival, building upon our experience with the iiwa's predecessor. As literature did not provide any fully formulated solution, we experimented with using an evolutionary solver towards creating series of axis movements with the same toolframe but different position of the redundant axis. Using this empirical data, we were able to formulate and test geometric hypotheses within Grasshopper that finally resulted in a fast and reliable inverse kinematic model of the iiwa robot.

Next, we had to consider on how to transfer data from Grasshopper to the visual programming environment. We first evaluated "hacking" the synchronization process of the Sunrise Workbench to directly sync projects to the robot, but ultimately saw too many practical as well as safety issues. Therefore, we chose to use a "Firmata" by creating a standardized, base program that is capable of communicating with outside sources through defined means. A similar approach has been chosen by Andy Payne and Jason Kelly Johnson for their Firefly (2016) environment, where a firmata is uplloaded onto an Arduino controller which can then communicate with Grasshopper.

For the transfer of data between the offline programming system and the iiwa robot, we decided to utilize a human-readable format to allow quick and easy changes to the file, finally choosing XML due to the availability of powerful libraries both on the .NET side as well as for Java.

As discussed above, the blending of complex toolpaths is comparatively complicated for the iiwa, as it has to put the movement commands either into a MotionBatch, or process them asynchronously, which can lead the program to terminate early, despite several asynchronous movements commands still awaiting processing. Similar to our work with mxAutomation, which deals with several similar limitations, we therefore implemented a custom buffering system that would not process all commands in advance, but selectively group them into MotionBatches when needed and watch over the execution of asynchronous movements (Figure 6).

Rather than uploading a new project for each element, users are now simply confronted with a filebrowser that allows them to select XML files from all sources accessible to the Windows site of the controller, from network shares to USB drives. This allows us to very quickly deploy new projects to the iiwa robot, as well as effectively share a single robot between multiple groups of students.

INITIAL PROJECTS

In the past years, we have used the KUKA LBR iiwa collaborative robot in a series of research and teaching activities. The following projects were chosen to showcase some of the iiwa/Sunrise specific features that set the robotic system apart from other industrial robots.

Robotic Calligraphy was developed as a showcase installation for KUKA Robotics and the Ars Electronica Center (AEC), a digital media museum in Europe. It builds upon the concept of generating different greyscale values by rotating an asymmetric calligraphy pen that produces pure black when its wide side is normal to the toolpath, and a lighter shade the more it deviates from the normal. In previous projects and workshops (Brell-Cokcan and Braumann 2014), we have implemented the strategy within the Grasshopper environment and were therefore able to immediately test it with the iiwa by changing the code generation from KRL to Sunrise/XML. Building upon this proof of concept, it was our goal to create a completely integrated program that runs entirely on the Sunrise controller, with its own graphical user interface and without requiring an external PC (Figure 7).

The first step was to create a simple geometric library—based on the default javax.vecmath library—that allows us to work with geometric objects such as planes in a similar way as within the RhinoCommon framework. It was then very easy to layout the toolpaths and apply the rotation based on the brightness of a raster image. The comparison with the CAD output enabled an immediate error checking of the results.

Once the pathplanning was finished, we looked into the capturing of photos through a camera. Instead of depending on an industrial camera, we were able to simply attach a Logitech C920 consumer webcam to the Sunrise controller and install the accompanying driver software. We then captured an image through OpenCV, used Haar feature-based cascade classifiers (Viola and Jones 2001) to recognize faces, and cropped the image to ideally fit the face within the aspect ratio of the paper. Finally, we implemented a graphical user interface based on javax.swing that allowed the user to fine-tune the brightness and contrast of the captured image.

Both calligraphy pen as well as the webcam were then mounted

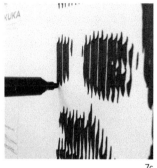

7a 7c

7   Image capturing with a regular webcam (a), image processing through a custom made GUI running on the smartPAD for easy interaction (b), calligraphy process (c).

8   LBR iiwa "cell" at the Ars Electronica Museum, without requiring additional safety (a). Calligraphy result (b).

8a

7b

8b

on the iiwa through a 3D-printed endeffector. The process was laid out in a way that someone could manually position the iiwa through hand-guiding, have their "selfie" taken by the robot, which the operator could then fine-tune on the control panel before starting the drawing process.

The developed system was active for more than a month at the Ars Electronica Center, operated exclusively by non-expert staff without any previous robot experience (Figure 8).

DIANA: In previous research, we have experimented with the fabrication of ruled surfaces through wooden rods, with the goal of using a small-scale robot to create a large-scale installation. Until recently, a KUKA Agilus robot only performed the cutting and multi-axis drilling of the support structure, while the rods would have to be cut and placed manually (Figure 9). The reason for that can be found in the fact that wooden rods are a natural material with very high tolerances. These tolerances can be due to improper storage and humidity, or simply because of the used

wood type, and may amount to more than 10 mm on a 1000 mm rod. DIANA, the dynamic interactive assistant for novel applications, is a robot installation developed by teams from RWTH Aachen University (Chair for Individualized Production in Architecture and the Cybernetic-Cluster IMA/ZLW & IfU) and Robots in Architecture as part of the KUKA Innovation Award for the Hannover Fair 2016 that showcases the challenges of using robots in the construction industry, demonstrating concepts such as mass customization and strategies towards dealing with environments with high tolerances.

For this project, we digitally designed and built a rod-structure out of 45 x 45 mm wooden slabs and 12mm diameter birch rods. The robot setup provided by KUKA consisted of a iiwa 14 R820 robot mounted onto a flexFELLOW platform—a robot base that integrates the controller and can be quickly relocated. An important design tool from the very beginning was our iiwa simulation through KUKA|prc that allowed us to optimize the geometry in regards to reachability and collisions. While we

Towards New Robotic Design Tools Braumann, Brell-Cokcan, Stumm

9c

9b

9 Robotically fabricating the supporting structure with a regular, non-complaint robot (a). Previous, manual assembly of rods (b). New, sensitive assembly informed through haptic programming presented at the Hannover Fair 2016 (c).

were able to incorporate the position and orientation of each mounting point into the Java program, these values assume an ideal, digital environment and take neither production tolerances, nor an imprecise placement of the robot itself into account.

First of all, the user takes the iiwa robot and manually guides it from one side of the base structure to the other side, a process we refer to as "haptic programming" (Stumm et al. 2016). By capturing the movement, a curve is parametrized that informs the fabrication process, making every generated structure unique. Based on this data, the robot generates a list of tasks for that fabrication process: First of all, the robot takes a rod out of the supply station, which groups rods within a range of 50 mm. In order to calculate the exact length of the rod, the robot moves each tip of the rod onto the flexFELLOW platform until a contact is established. This contact is established solely through the force-torque sensors, without requiring additional sensory equipment. The robot records the distance between the gripper fingers and the measuring position, thus being able to calculate the length in each direction. Using a common circular saw, the rod is then cut according to the previously generated curve. For the actual assembly, the robot moves the rod into the approximate position of each mounting position as known from the design. Due to the high material tolerances, it then searches for the exact position via the force sensors. Once it feels a significant drop in the force values, it concludes that the rod has slipped into mounting position and continues the assembly process by retrieving the next rod (Figure 10).

In the past, such systems have mostly relied on external measuring equipment that calculates the offset between the ideal and the actual toolframe. However, such systems are complicated to use and not ideally suited to the rough, changing environments of a construction site. The force-sensitivity of the iiwa provides a well-integrated way towards implementing such functionality into an assembly process, without requiring external equipment or special software.

CONCLUSION AND OUTLOOK

The KUKA LBR iiwa represents a significant step change not only for the robotics industry, but also for the construction industry, as it enables completely new ways of interacting with robots;both in regards to physical and digital interaction. Even more than previous machines, the robot becomes a universal platform that can be adapted through software for specific tasks and implemented into existing systems, from the cloud to Building Information Modeling.

Using the prepared "firmata" allows users to quickly prototype processes and check reachability through the KUKA|prc and the visual programming environment, where toolpath layouts can be more easily constrained to geometries. Within Sunrise Workbench, the user can continue working with the automatically generated XML file, adding, for example,additional safety and interaction features around it.

At the moment, the iiwa's working range limits larger-scale applications within a construction context. However, the transfer of

the robotic system onto a mobile platform (KMR iiwa, Figure 11), will allow us to navigate autonomously within a workspace that is limited only by the capacity of its batteries.

ACKNOWLEDGEMENTS

This research was performed as part of the "Robotic Woodcraft" research project (FWF-PEEK AR238) and the KUKA Innovation Award. The DIANA team consisted of Sven Stumm, Martin von Hilchen, Elisa Lublasser and Prof. Sigrid Brell-Cokcan (IP-RWTH Aachen), Philipp Ennen and Prof. Sabina Jeschke (IMA/ZLW & IfU RWTH Aachen) and Johannes Braumann (Robots in Architecture).

REFERENCES

Albu-Schäffer Albin, Sami Haddadin, Christian Ott, Andreas Stemmer, Thomas Wimböck, and Gerd Hirzinger. 2007. "The DLR Lightweight Robot: Design and Control Concepts for Robots in Human Environments." *Industrial Robot: An International Journal* 34 (5): 376–385.

Braumann, Johannes and Sigrid Brell-Cokcan. 2011. "Parametric Robot Control: Integrated CAD/CAM for Architectural Design." In *Proceedings of the 31st Annual Conference of the Association for Computer Aided Design in Architecture (ACADIA).* Banff, Alberta, Canada: ACADIA. 242–251.

Braumann, Johannes, and Sigrid Brell-Cokcan. 2015. "Adaptive Robot Control - New Parametric Workflows Directly from Design to KUKA Robots." In *Real Time: Proceedings of the 33rd eCAADe Conference,* vol. 2, edited by B. Martens, G. Wurzer, T. Grasl, W. E. Lorenz, and R. Schaffranek. Vienna, Austria: eCAADe. 243–250.

Brell-Cokcan, Sigrid and Johannes Braumann. 2014. "Robotic Production Immanent Design: Creative Toolpath Design in Micro and Macro Scale." In *ACADIA 14: Design Agency—Proceedings of the 34th Annual Conference of the Association for Computer Aided Design in Architecture,* edited by David Gerber, Alvin Huang, and Jose Sanchez. Los Angeles: ACADIA. 579–588.

ETH Zurich. "Robotic Scale-Model Testing". Accessed May 5th 2016. http://www.block.arch.ethz.ch/brg/research/robotic-scale-model-testing

Munz, Heinrich, Johannes Braumann, and Sigrid Brell-Cokcan. 2016. "Direct Robot Control with mxAutomation." In *Rob | Arch 2016: Robotic Fabrication in Architecture, Art and Design 2016,* edited by Dagmar Reinhardt, Rob Saunders, and Jane Burry. Vienna: Springer. 440–447.

Payne, Andy and Jason Kelly Johnson. "Firefly." Accessed May 5th 2016. http://www.fireflyexperiments.com

Shepherd, Stuart and Alois Buchstab. 2014. "KUKA Robots On-Site." In *Rob | Arch 2014: Robotic Fabrication in Architecture, Art and Design,* edited by Wes McGee and Monica Ponce de Leon. Vienna: Springer. 373–380.

Stumm, Sven, Johannes Braumann, Martin von Hilchen, and Sigrid Brell-Cokcan. 2016. "On-Site Robotic Construction Assistance for Assembly Using A-Priori Knowledge and Human-Robot Collaboration." In *Proceedings of the International Conference on Robotics in Alpe-Adria-Danube Region.* Belgrade, Serbia: RAAD.

10 DIANA: Capturing geometry through manual guiding (1), measuring exact length (2), cutting (3), force-sensitive assembly (4).

11 KMR iiwa platform at the Chair for Individualized Production in Architecture at RWTH Aachen.

11

Viola, Paul, and Michael Jones. 2001. "Rapid Object Detection using a Boosted Cascade of Simple Features." In *Proceedings of the 2001 IEEE Computer Society Conference on Computer Vision and Pattern Recognition*, vol. 1. Kauai, HI: CVPR. 551–560.

IMAGE CREDITS

Figure 2a: © University of Stuttgart

Figure 2b: © Branch Technologies

Figure 2c: © ETH Zurich

All other figures: © Robots in Architecture and RWTH Aachen University

Sigrid Brell-Cokcan and Johannes Braumann founded the Association for Robots in Architecture in 2010 with the goal of making industrial robots accessible to the creative industry. Towards that goal, the Association is developing innovative software tools such as KUKA|prc (parametric robot control) and initialized the Rob|Arch conference series on robotic fabrication in architecture, art, and design which–following Vienna in 2012, Ann Arbor in 2014, and Sydney in 2016 – will be held 2018 in Zurich. Robots in Architecture is a KUKA System Partner and has been validated as a research institutions by national and international research agencies such as the European Union's FP7 program. Recently, Sigrid founded the new chair for Individualized Production in Architecture at RWTH Aachen University. Johannes is heading the robotics lab at UfG Linz and leading the development of KUKA|prc. Their work has been widely published in peer reviewed scientific journals, international proceedings, and books, as well as being featured in formats such as Wired, Gizmodo, FAZ, and RBR.

Sven Stumm is a computer scientist at the Chair of Individualized Production in Architecture (IP) at RWTH Aachen University with a focus on intelligent systems and backgrounds in humanoid, mobile and stationary industrial robotics. His core competences are in robot controller development, sensor data processing, probabilistic modelling as well as software development for programming and simulation of robotic applications. At IP he researches assembly and production processes within construction, and novel interactive interfaces for robot programming.

RoboSense

Context-Dependent Robotic Design Protocols and Tools

Andrew Moorman

Jingyang Liu
Dept. of Architecture, Sabin
Design Lab, Cornell University
AAP

Jenny E. Sabin
Dept. of Architecture, Sabin
Design Lab, Cornell University
AAP / Jenny Sabin Studio

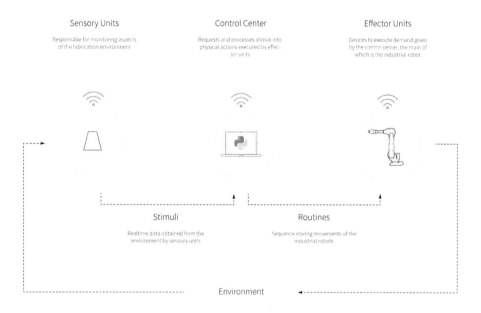

1

ABSTRACT

While nonlinear concepts are widely applied in analysis and generative design in architecture, they have not yet convincingly translated into the material realm of fabrication and construction. As the gap between digital design model, shop drawing, and fabricated result continues to diminish, we seek to learn from fabrication models and natural systems that do not separate code, geometry, pattern, material compliance, communication, and form, but rather operate within dynamic loops of feedback, reciprocity, and generative fabrication. Three distinct, but connected problems: 1) Robotic ink drawing; 2) Robotic wine pouring and object detection; and 3) Dynamically Adjusted Extrusion; were addressed to develop a toolkit including software, custom digital design tools, and hardware for robotic fabrication and user interaction in cyber-physical contexts. Our primary aim is to simplify and consolidate the multiple platforms necessary to construct feedback networks for robotic fabrication into a central and intuitive programming environment for both the advanced to novice user. Our experimentation in prototyping feedback networks for use with robotics in design practice suggests that the application of this knowledge often follows a remarkably consistent profile. By exploiting these redundancies, we developed a support toolkit of data structures and routines that provide simple integrated software for the user-friendly programming of commonly used roles and functionalities in dynamic robotic fabrication, thus promoting a methodology of feedback-oriented design processes.

1   Pipeline for coordination of information exchange between sensory units, control center and effector units.

INTRODUCTION

Recent advances in computation, visualization, material intelligence, and fabrication technologies have begun to fundamentally alter our theoretical understanding of general design principles as well as our practical approach towards architecture and research. This renewed interest in complexity has offered alternative methods for investigating the interrelationships of parts to their wholes, and emergent self-organized material systems at multiple scales and applications. The advantages of researching and deploying such methodologies in the field of architecture are huge, as they impact aspects of sustainable design, optimization, construction, and novel design expression and material systems.

Existing digital fabrication techniques such as CNC (computer numerically controlled) milling and cutting are useful tools, but they are also severely limited by their 2D and reductive constraints. 3D CNC tools offer a much more adequate and versatile approach to issues of shape, material, and geometry, but these tools are particularly underexplored when it comes to fabrication at the architectural scale. Two of the most promising technologies are 3D printing and rapid assembly via robotics for manufacturing of individual and continuous component parts or fibrous assemblages, skin systems, and nonstandard architectural elements. Together, these technologies are geared towards becoming indispensable tools for nonlinear manufacturing as well as complex form-making, but they raise the question: How might advancements in robotic fabrication and design in alternate industries and disciplines impact the architectural design process? Needed now are rigorous multi-directional and multi-disciplinary investigations that can help shape the future trajectories of these material innovations and technologies for architecture.

This paper focuses on one area of ongoing design research in the Sabin Design Lab at Cornell AAP: the development of novel custom-design tools and interfaces featuring environmental feedback for user interaction with a 6-axis industrial robot. Our goal is two-fold: 1) The design and implementation of a user friendly pipeline and interface for the seamless programming of an ABB IRB 4600 45 kg, 2.05 m reach industrial robot to operate in concert with cyber-physical devices; and 2) To develop a flexible tool kit that will foster a design process and methodology enmeshed in feedback, and will engage the human hand and digital handcraft in the co-production of robotically steered materiality and novel tectonic systems.

BACKGROUND

The Sabin Design Lab at Cornell AAP is an experimental design research lab that investigates the intersections of architecture and science, and applies insights and theories from biology, physics, engineering, computer science, and mathematics to the design of material structures. We ask, "How might architecture respond to issues of ecology and sustainability whereby buildings behave more like organisms in their built environments? What role do humans play in response to changing conditions within the built environment?" The Sabin Design Lab is interested in probing the human body for design models that give rise to new ways of thinking about issues of adaptation, change, and performance in architecture. Our research projects are diverse and operate across multiple scales from the nano to the macro. The Sabin Design Lab specializes in computational design, data visualization, and digital fabrication.

There are now certainly many established and internationally recognized research and design units engaged in these topics and questions, including the Mediated Matter Group at MIT led by Neri Oxman; the Self-Assembly Lab at MIT led by Skylar Tibbits; and CITA at the Royal Danish Academy of Fine Arts led by Mette Ramsgaard Thomsen. Several cutting-edge research units are advancing this type of work primarily as it touches upon issues of construction, fabrication, and large-scale architectural applications, including pavilions, walls, and envelopes. The work of Achim Menges and his students at the Institute for Computational Design (ICD) at University of Stuttgart operates at a large scale through the explicit exploration of natural systems for novel structures in the context of computational matter. Recently, the ICD and the Institute of Building Structures and Structural Design (ITKE) at the University of Stuttgart have constructed a biomimetic fiber-reinforced polymer (FRP) research pavilion, whose building method accommodates changing stiffness in a pneumatic formwork for robotic fiber placement through an integrated sensor for toolpath-adjustment based on contact force with the formwork. Designed, fabricated, and constructed over one-and-a-half years by students and researchers within a multi-disciplinary team of biologists, paleontologists, architects, and engineers, this project investigates natural fiber composite shells alongside the development of cutting-edge robotic fabrication methods for FRP structures (Menges and Knippers 2015; Vasey et al. 2015).

Similarly, Gramazio and Kohler of ETH Zurich focus on additive digital fabrication techniques used for building non-standardized architectural components including bricks, mesh structures and smart dynamic casting with concrete and other plastic materials. One such investigation into Remote Material Deposition (RMD) studied adaptation to changing construction conditions for assembling structures with thrown clay, using digital point cloud data to adjust tool-paths based on material divergences due to the clay's malleability and the unpredictability in landing conditions (Dörfler et al. 2014).

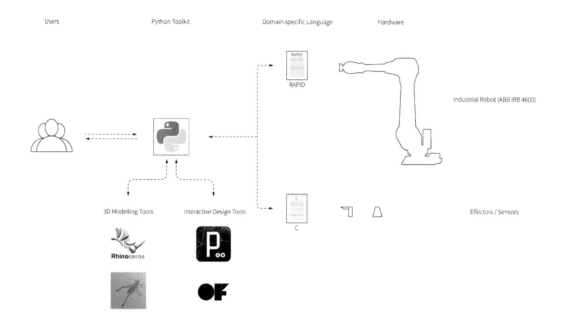

Users   Python Toolkit   Domain-specific Language   Hardware

RAPID

Industrial Robot (ABB IRB 4600)

3D Modelling Tools   Interactive Design Tools   C   Effectors / Sensors

2

2   Python toolkit allowing users to auto generate rapid code for programming contextually informed robotic tool paths through user-friendly interfaces.

While both of these exemplary research units engage similar trajectories in non-standard dynamic robotic manufacturing, we are also invested in design processes that employ the human hand, personalization, and even empathy. Closer to this is the work of Guy Hoffman, who is at the forefront of human-robot interaction that engages joint activities between humans and robots and human-robot collaboration (Hoffman 2004).

In each case, a simultaneous transition in design methodology occurs by shifting the production machine from the executor of an explicit and comprehensive set of commands to an actor in a dynamic and reciprocal relationship with its fabrication environment. This is evident in recent architectural research that incorporates moments of feedback in online robotic fabrication processes. However, little elucidation is provided on the complexity inherent in programming these dynamic feedback networks, and few dedicated explorations exist in making their configuration accessible to a non-technically-experienced audience. To facilitate a wider adoption of this approach to design and fabrication, we believe it also necessary to unlock the tools of this methodology to all designers, not just the digitally inclined.

This domain of research is not without contemporaries: similarly oriented robotics software frameworks, like the popular Robotic Operating System (ROS) Industrial (Quigley et al. 2009) and Robo.op (Bard et al. 2014), provide helpful resources for live-programming industrial robotics in tandem with computer vision techniques, potentially for architectural use, but fall short

either in their accessibility to the general design community or the extensiveness of their scope of application. In contrast, select geometric modelling software add-ons compliant with ABB industrial robots—namely HAL Robotics, Onix (Elashry et al. 2014), and Lobster for Rhino 3D's Grasshopper plugin—leverage existing graphical algorithm editors to make code generation and IR simulation compatible with geometric information. Although these programs offer a more user-friendly and intuitive pipline for robotic programming, they do not integrate workflows for communication with cyber-physical devices. These software add-ons are not able to communicate directly and simultaneously with multiple forms of hardware, like industrial robots, sensors, and effectors, since they operate on different domain-specific languages. The gap between software and hardware becomes a hurdle for building a dynamic feedback loop in robotic fabrication. As a result, the rapid development by designers of online robotic fabrication processes integrated with feedback networks remains an intimidating, if not unfeasible, venture.

METHODS
Contemporary feedback networks in robotic fabrication often emphasize interactivity between cyber-physical elements and flexibility against indeterminate processes, promising a much higher level of integration between the physical processes of making and the virtual domain of information (Menges 2015). However, the emergence of cyber-physical systems in robotic fabrication imposes an intense challenge to the inexperienced or uninitiated. User interfaces for real-time control—like the Quipt

gesture interface and Live (Batliner 2016) control platform—aid designers in creating fruitful human-robot interaction settings. But even once acclimated to or abstracted from the burdensome programming of industrial robotics, incorporating information feedback may require additional interconnected sensors, actuators, and processing software, potentially distributed across multiple platforms, each programmed in a different, domain-specific language.

Our objective was to simplify and consolidate the multiple platforms necessary to construct feedback networks for robotic fabrication into a central programming environment in order to minimize prerequisite knowledge while maintaining the full body of tasks otherwise handled by a more visibly complicated system. Our experiments prototyping feedback networks for use with robotics in design practice suggested that the application of this knowledge often follows a remarkably consistent profile. By exploiting the redundancies exposed by this profile, we developed a support toolkit of data structures and routines, providing simple, integrated software for the user-friendly programming of commonly-used roles and functionalities in dynamic robotic fabrication, promoting a methodology of feedback-oriented design processes.

Similar to the homeostatic processes for regulatory loops in biological systems, our prototype interfaces make explicit use of three interdependent software components, emphasizing the coordination of information exchange between sensory units—responsible for monitoring aspects of the fabrication environment—and a control center, which requests and processes these "stimuli" into physical actions executed by effector units, whose primary member is the industrial robot. Our intervention was to explicitly encapsulate these roles into designer-friendly customizable data structures housed in a single Python programming toolkit. Meanwhile, all inter-component communication and the domain-specific programming of hardware and our ABB IRB 4600 robotic arm are invisibly handled in the background within the toolkit, accessed intuitively from a collection of custom libraries (Figures 1 and 2).

By leveraging these common feedback roles into a designer-friendly programming interface, our ambition is that designers may more fully engage the orchestration, ordering, and design of dynamic information exchange among their environment, themselves, and their fabrication systems. The structures defined, protocols (a means for unrelated objects or systems to communicate with each other) created, and routines written were intended to be intuitive and to encourage the implementation of robotically achieved, materially driven fabrication, but generalizable enough to not restrict the users' scope of development.

To realize this indeterminate exchange of sensory updates and instructions—necessary for dynamic feedback—our pipeline makes extensive use of the client-server communication model, implemented across all integrated cyber-physical components. In this model, clients and servers exchange messages in a request–response messaging pattern: to communicate, a relay of client requests and server responses are exchanged in a common language dictated by a predefined protocol, so that both parties know how to interpret their received messages. Accordingly, our toolkit employs a set of custom request-response messaging patterns and communication protocols appropriate for the interpretation and handling of the sensor-effector and robotic programming that would otherwise be conducted by users from scratch. In our pipeline, this occurs passively as a consequence of the user's programming.

RESULTS

Prototype Feedback Scenario 1: Robotic Ink Drawing

In our first feedback-driven design scenario—a test in robotic ink drawing—particular attention was given to the dependency of the operation and programming of Effectors on finessing a stylized physical output through ink deposition (Figure 3). The grey scale information of various reference images was used to affect the location of target points for directing our industrial robotic arm's motion: each pixel's grey scale value was extracted and then sorted to find the image's darkest points, generating initial centroids to attract 8,700 target points randomly populated within its bounds. These target points were clustered according to their closest centroid. For each cluster, a new centroid was calculated based on the mean of the points within it. By looping this process, target points were continually reorganized into clusters based on the grey scale value of the reference image, until none of the cluster assignments changed. The final set of points was then drawn by the industrial robotic arm (Figure 4).

In this drawing model, the style is considerably affected by the industrial robot's speed, as well as the rate of ink flow from an effector—a customized drawing instrument—which consists of a brush and an automated ink feeder assembly. A valve attached to tubing that supplies ink to the brush controlled the flow rate of the ink. By manipulating the flow rate and the speed of the industrial robot's motions, multiple dot styles were generated: with low industrial robot (IR) speed and high flow rate, ink was deposited in excess to the target point, creating an emissive effect, while a combination of high robotic arm speed and low flow rate contributed to dots of irregular shape. Alternatively, a dripping impression could be obtained by high IR speed and high flow rate, while low industrial robot speed and flow rate generated precise round dots. Here in particular, our developing toolkit facilitated the programming of specific robotic motions to

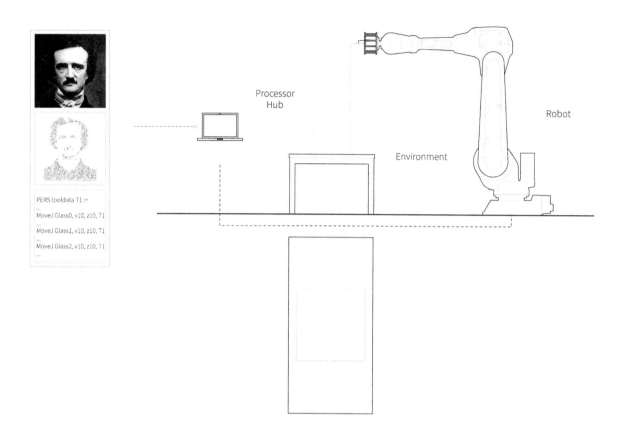

PERS tooldata T1 :=
...
MoveJ Glass0, v10, z10, T1
...
MoveJ Glass1, v10, z10, T1
...
MoveJ Glass2, v10, z10, T1
...

Processor Hub

Robot

Environment

3

4

3   Robotic Ink Drawing Loop.

4   Completed ink drawing of a portrait of Edgar Allan Poe, signed by Sulla, our ABB IRB 4600 industrial robotic arm.

operate in tandem with the above-mentioned drawing parameters produced by the effector, thus synchronizing drawing media, style of physical production, and image reference data into a single system. Moreover, programmed together in one consolidated file, we were able to engender more coordinated control of the industrial robot and the end effector in the precise management of ink deposition quality. This means of creating various drawing styles by manipulating IR speeds would later influence the generation and design of heterogeneity in material deposition (Scenario 3). Nevertheless, as an experiment in online robotic programming and effector coordination, our uni-directional flow of information had not yet demonstrated an integration of feedback within the design process.

Prototype Feedback Scenario 2: Robotic Wine Serving

In the subsequent model, our client-server structure was more fully utilized to build a reciprocal connection between cyber and physical spaces. Here, an indeterminate environmental factor—the location and presence of an arbitrary number of wine glasses—was incorporated into the execution process to guide the robot's actions in serving wine (Figures 5 and 6).

Our sensor, a ceiling-mounted RGB-D camera, was used to capture RGB color information paired with per-pixel depth information through requests to the object for updates. Using the resulting RGB color values, the boundaries of wine glasses were detected by comparing color properties with surrounding regions and grouping clusters of like-pixels into circular "blobs," allowing their centroids to be extracted as target points for the industrial robot. The corresponding depth values captured for these centroids allowed us to grab coordinates in three-dimensional space for the glasses, which were transformed according to the coordinate system native to the industrial robot. Finally, a parameterized toolpath for the IR was generated by connecting these target points in a minimal route. This begins with the robot's rest position while "on break" and then is altered live at the incident of a guest's adjustment of the position of a wine glass. The entire process was enclosed within a continuous loop instigated by the absence or presence of wine glasses on a serving table and on call throughout a welcoming reception for the college (Figures 7 and 8).

In this project, we underscore the interdependence of sensory updates with online robotic programming to accommodate an unpredictable environment. Execution loops for wine pouring were parameterized around an indeterminate, perhaps empty or even unstably positioned, list of glass locations in 3D space. Finesse and refinement of the extrusion loop was important in order to avoid spilling or harming guests. Yet, in a problem space of such volatility, the live adjustment of complex behavioral processes could be programmed concisely and entirely in the Python programming language as an intuitive description of data transformation, commands, and execution loops. We believe this flow of information from an erratic human environment, through a digital environment for processing, and then reactively fed back to the physical realm is emblematic of a new paradigm for fabrication.

Prototyping Feedback Scenario 3: Dynamically Adjusted Extrusion

An ongoing third case study synthesizes the approaches explored in the robotic wine serving and ink drawing investigations with a stronger relationship to the feedback-based manufacturing of non-standard physical elements. The ink drawing behavior previously described is translated to a robot-mounted extrusion system, whose heat and deposition speed are dynamically adjusted by the user while material is deposited. The interaction protocol for adjustment relies on a block-based interface which situates the user and interaction system behind a glass wall, proximal but safely distant from the robotic arm's operation (Figure 9). Deposition properties, as well as stopping commands, are mapped to colored sides of the block, read by a camera, while a continuous extrusion profile is parameterized by the position of the block relative to an infrared distance sensor (Figure 10).

Practice-oriented, our developing work sees the production of custom-modified components as a real-time engagement between fabrication tool and user, pairing an indeterminate updating of commands with the potential embedding of qualitative assessment into the process of design making.

DISCUSSION

With the rise in emphasis of pseudo-cognitive, reactive machines, a convergence between design and making is surely forthcoming—cyber-physical systems, even purely linguistically, predicate a marriage of digital and physical realms. This holds great promise to not just make possible, but make seamless the co-evolution of material and digital complexity in the built environment. Positioning and maintaining the design fields at the forefront of this trend in innovation must involve an update to the methodologies of our tools, which describe and reveal the intertwined processes for design and making, mostly through code. Critical to staying engaged in this shift is the design and generation of tools that incorporate complex behavioral programming in a user friendly and approachable tooling environment. Our intent was to develop a kit of programs and software interfaces to facilitate this shift, both in our lab and across disciplines whose research might benefit from its adoption.

Consequently, the described development of this toolkit was undertaken as a series of recognizably playful scenarios, both to eliminate the non-essential hurdles often encountered in more practice-oriented investigations, but also because their principles could be extrapolated into architecturally applied digital fabrication techniques. For example, we may re-assess the live adjustment of ink deposition—dripping or brushing properties, flow rate, deposition speed—as a direct antecedent for dynamic fine-tuning in additive manufacturing processes. The real-time modification of extrusion properties—viscosity, extrusion profile, or deposition speed—is a close reality, either as a corrective measure for otherwise "bad prints" or perhaps on the basis of qualitative assessments, decided by the designer in media res. Similarly, the interaction between humans and robots engaging in cooperative activity, as in our wine serving scenario, anticipates those feedback-based fabrication environments where robotic adaptation to indeterminacy is preeminent. This is especially true in person-shared fabrication settings where human engagement implies both a concern for safety and a promise to include receptivity in digital handcraft.

Importantly, any rigorous application, including both those undertaken and those suggested, usually requires several iterations of prototyping with equivalent rigor: ours involved the continual formulation, construction, assessment, and re-formulation of behavioral processes and information exchange. Programming

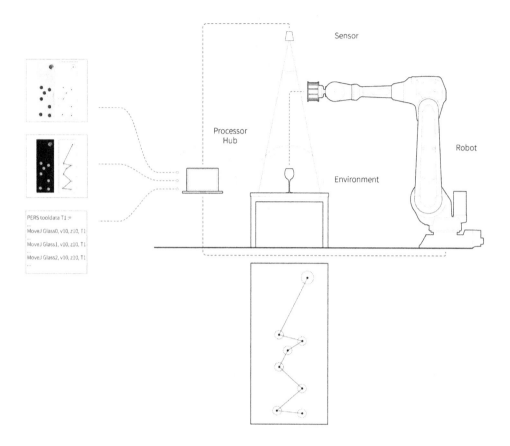

5

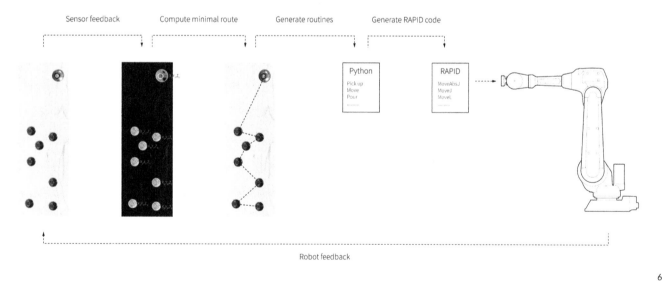

6

5   Robotic Wine Serving Loop.

6   Robotic Wine Serving Workflow Diagram.

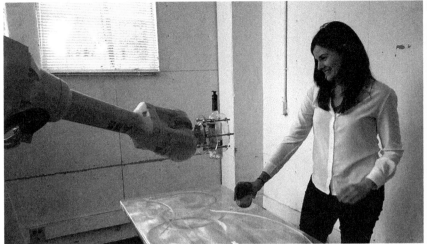

7   Dynamic wine pouring with object detection, shortest path calculation between objects and immediate tool path generation for pouring. This test highlights two levels of feedback: 1. Through a user moving their glass in space and 2. Between multiple glasses in shifting locations on a table. The tool path adjusts accordingly in real time for both cases.

8   Dynamic tracking of glass object in an unpredictable environment for accurate pouring.

and testing models for environmental feedback constituted the design process. Access to our software toolkit with a predilection for feedback-driven design allowed the process to proceed with efficient use of time and energy.

As acknowledged, the ambition of our toolkit is primarily software-oriented. In opening industrial robotics to the design fields, our foremost aim is to alleviate the difficulty in programming feedback into cyber-physical systems, and thereby also improve the convenience of building their hardware accompaniments. Nevertheless, to accommodate lapses in electronics-building experience, we recommend using existing modular, ready-to-use tool sets for the Arduino platform, which simplify and condense the learning process significantly. One popular tool set, Grove, takes a building block approach to assembling electronics, providing plug-and-play modules for rapid physical assembly, which intuitively integrates with the programming of Sensor and Effector objects through our toolkit.

CONCLUSION

This work aims to develop a toolkit including software, custom digital design tools, and hardware for robotic fabrication and intuitive user interaction. While nonlinear concepts are widely applied in analysis and generative design in architecture, they have not yet convincingly translated into the material realm of fabrication and construction. In this paper, we demonstrate three case study projects that probe cyber-physical interactions in the context of robotic fabrication. This allows for the generation of dynamic and immediate tool path responses in shifting environments. These projects present useful scenarios that have immediate translation into a materially directed generative design process in the context of robotic fabrication and assembly. The main thrust of this work concerns the evolution of material and digital complexity in the built environment through prototypical design experiments that rigorously abstract, extend, and translate dynamic behaviors and models with the end goal of generating adaptive architecture and material assemblies that operate at the

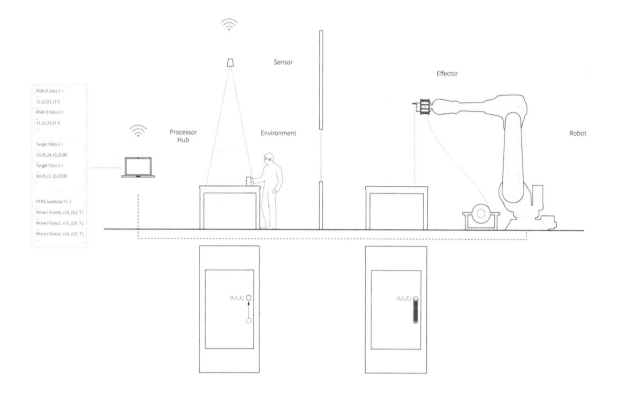

9

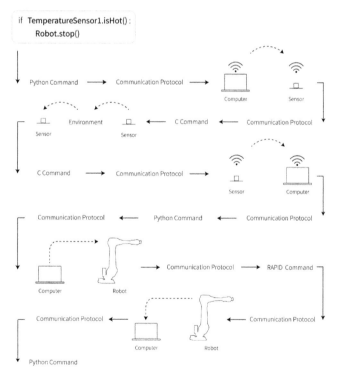

human scale. Our design process moves fluidly between analog and advanced digital procedures, often inserting the human hand or digital handcraft in the meaningful and rigorous negotiation of scale and complex behavior. As the gap between digital design model, shop drawing, and fabricated result continues to diminish, we seek to learn from fabrication models and natural systems that do not separate code, geometry, pattern, material compliance, communication, and form, but rather operate within dynamic loops of feedback, reciprocity, and generative fabrication. This paper presents one approach to the negotiation of this gap through methods, tools, hardware, and software that are now being implemented in 6-axis robotic additive manufacturing as well as dynamic object detection, placement, and assembly.

ACKNOWLEDGEMENTS

This material is partially supported by the National Science Foundation under NSF EFRI ODISSEI -1331583. Any opinions, findings, conclusions or recommendations expressed in this material are those of the author(s) and do not necessarily reflect the views of the National Science Foundation. This work is also supported by the Department of Architecture, Cornell College of Architecture, Art and Planning, Cornell University.

9   Dynamically Adjusted Extrusion Loop.

10   Diagram Of the Complete Execution Flow.

REFERENCES

Bard, Joshua, Madeline Gannon, Zachary Jacobson-Weaver, Michael Jeffers, Brian Smith, and Mauricio Contreras. 2014. "Seeing Is Doing: Synthetic Tools for Robotically Augmented Fabrication in High-Skill Domains." in *ACADIA 2014: Design Agency—Proceedings of the 34th Annual Conference of the Association for Computer Aided Design in Architecture*, edited by David Gerber, Alvin Huang, and Jose Sanchez. Los Angeles: ACADIA. 409–416.

Batliner, Curime. 2016. "Robot UI: User Interfaces for Live Robotic Control." In *Robotic Fabrication in Architecture, Art and Design 2016*, edited by Dagmar Reinhardt, Rob Saunders, and Jane Burry. Cham, Switzerland: Springer. 376–387. doi: 10.1007/978-3-319-26378-6_30.

Dörfler, Kathrin, Sebastian Ernst, Luka Piskorec, Jan Willmann, Volker Helm, Fabio Gramazio, and Matthias Kohler. 2014. "Remote Material Deposition: Exploration of Reciprocal Digital and Material Computational Capacities." In *What's the Matter: Materiality and Materialism at the Age of Computation*, edited by Maria Voyatzaki. Barcelona: ENHSA-EAAE. 361–377.

Elashry, Khaled, and Ruairi Glynn. 2014. "An Approach to Automated Construction Using Adaptive Programing." In *Robotic Fabrication in Architecture, Art and Design 2014*, edited by Wes McGee and Monica Ponce de Leon. Cham, Switzerland: Springer. 51–66. doi:10.1007/978-3-319-04663-1_4.

Hoffman, Guy, and Cynthia Breazeal. 2004. "Collaboration in Human-Robot Teams." In *AIAA 1st Intelligent Systems Technical Conference*. Chicago, IL: AIAA. doi:10.2514/6.2004-6434.

Menges, Achim. 2015. "Fusing the Computational and the Physical: Towards a Novel Material Culture." *Architectural Design* 85 (5): 8–15.

Menges. Achim, and Jan Knippers. 2015. "Fibrous Tectonics." *Architectural Design* 85 (5): 40–47. doi: 10.1002/ad.1952.

Quigley, Morgan, Ken Conley, Brian P. Gerkey, Josh Faust, Tully Foote, Jeremy Leibs, Rob Wheeler, and Andrew Y. Ng. 2009. "ROS: An Open-Source Robot Operating System." In *ICRA Workshop on Open Source Software* 3 (3.2): 5.

Vasey. Lauren, Ehsan Baharlou. Moritz Dörstelmann, Valentin Koslowski, Marshall Prado, Gundula Schieber, Achim Menges, and Jan Knippers. 2015. "Behavioral Design and Adaptive Robotic Fabrication of a Fiber Composite Compression Shell with Pneumatic Formwork." In *Computational Ecologies: Design in the Anthropocene—Proceedings of the 35th Annual Conference of the Association for Computer Aided Design in Architecture*, edited by Lon Combs and Chris Perry. Cincinnati: ACADIA. 297–309.

IMAGE CREDITS

All Figures: Moorman, Liu, Sabin, 2016

Andrew Moorman is an undergraduate student in Cornell University's B.Arch program, where he serves as a research associate in the Sabin Design Lab at Cornell AAP and teaching assistant within the Department of Computer Science. His current research trajectories involve generative design strategies and tools for robotic fabrication.

Jingyang Liu is a research associate and teaching associate at Cornell AAP. Liu graduated in 2015 with a post-professional Master degree of Architecture from Cornell AAP. He is now working on robotic fabrication and computational design in the Sabin Design Lab at Cornell AAP.

Jenny E. Sabin is an architectural designer whose work is at the forefront of a new direction for 21st century architectural practice—one that investigates the intersections of architecture and science, and applies insights and theories from biology and mathematics to the design of material structures. Sabin is the Arthur L. and Isabel B. Wiesenberger Assistant Professor in the area of Design and Emerging Technologies and the Director of Graduate Studies in the Department of Architecture at Cornell University. She is principal of Jenny Sabin Studio, an experimental architectural design studio based in Ithaca and Director of the Sabin Design Lab at Cornell AAP, a trans-disciplinary design research lab.

Collaborative Construction

Human and Robot Collaboration
Enabling the Fabrication and Assembly
of a Filament-Wound Structure

Lauren Vasey, Long Nguyen
ICD, University of Stuttgart

Tovi Grossman
Autodesk UI Research

Heather Kerrick
Autodesk Applied Research

Danil Nagy
The Living, Autodesk Research

**Evan Atherton, David
Thomasson, Nick Cote**
Autodesk Applied Research

Tobias Schwinn
ICD, University of Stuttgart

David Benjamin
The Living, Autodesk Research

Maurice Conti
Autodesk Applied Research

George Fitzmaurice
Autodesk UI Research

Achim Menges
ICD, University of Stuttgart

1

ABSTRACT

In this paper, we describe an interdisciplinary project and live-exhibit that investigated whether untrained humans and robots could work together collaboratively towards the common goal of building a large-scale structure composed out of robotically fabricated modules using a filament winding process. We describe the fabrication system and exhibition setup, including a custom end effector and tension control mechanism, as well as a collaborative fabrication process in which instructions delivered via wearable devices enable the trade-off of production and assembly tasks between human and robot. We describe the necessary robotic developments that facilitated a live fabrication process, including a generic robot inverse kinematic solver engine for non-spherical wrist robots, and wireless network communication connecting hardware and software. In addition, we discuss computational strategies for the fiber syntax generation and robotic motion planning which mitigated constraints such as reachability, axis limitations, and collisions, and ensured predictable and therefore safe motion in a live exhibition setting.

We discuss the larger implications of this project as a case study for handling deviations due to non-standardized materials or human error, as well as a means to reconsider the fundamental separation of human and robotic tasks in a production workflow. Most significantly, the project exemplifies a hybrid domain of human and robot collaboration in which coordination and communication between robots, people, and devices can enhance the integration of robotic processes and computational control into the characteristic processes of construction.

1   Human and robot collaborative
    building process.

2   Final assembled structure
    composed of over 200 unique
    modules.

3   Tensegrity Module.

4   Pin-hooking detail.

INTRODUCTION

Though robotic fabrication has challenged standardized means of production for the architecture and manufacturing industries, the specialized knowledge and skill set that robots require, and the organization and development of customizable robotic processes, are significant logistical challenges which increase costs, compartmentalize production and assembly tasks, and favor linear, file-to-factory production chains.

Though active research is being done to introduce robots and other fabrication equipment directly onto the construction site (Helm et al. 2012, 2014), the use of robotic fabrication in large-scale projects often necessitates a workflow in which components are processed offsite in a factory or lab by specialists to be later transported, organized, and then assembled onsite by hand. This type of workflow provides little recourse if unexpected tolerances or deviations are encountered during the assembly and construction process.

Augmenting physical construction practices with digital mechanisms and just-in-time production can offer several benefits to existing protocols. By involving humans directly within the robotic fabrication process, and providing them with instructions through wearable interfaces, the separation of tasks within the production pipeline can be specialized according to ability: a robot's precision can be augmented by the fine motor control and cognitive ability of the human, and the monitoring of the process and feedback enabled through user interfaces allows the seamless trade-off of tasks between human and machine.

In this paper, we present a live exhibit and fabrication project that investigated whether humans and robots could work collaboratively and safely towards the common goal of fabricating and assembling a pre-designed large-scale structure. A prototypical robotic fabrication process was created that allowed the production of unique tensegrity modules which aggregated in a system. Tracking and monitoring of the current build status of the assembly enabled just-in-time instructions to be distributed to the participants via a smartwatch interface. The ultimate application of such an investigation is to question whether global system monitoring instructions delivered through devices could enable the organization and coordination of workers, robotic processes, material, and components in a deployment scenario such as a construction site.

CONTEXT

Human and Robot Collaboration

In applications for the architecture, engineering, and construction (AEC) industries, the shift from robots engineered for task specificity towards robots capable of generic tasks has enabled the

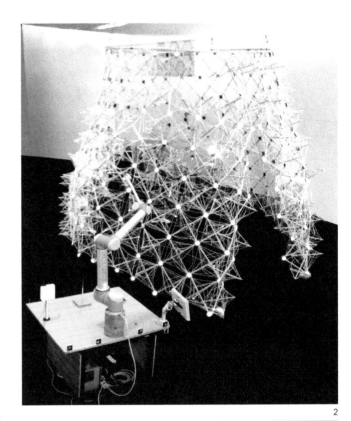

2

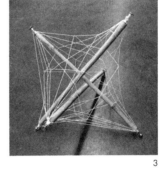

3

4

development of customizable fabrication processes and robotic control protocols (Menges and Schwinn 2012). The ability to further augment these robotic fabrication processes through connected devices and sensor feedback enables increased integration and cross-linking between physical and digital domains (Menges 2015).

While robots in industry were originally purposed for the execution of repetitive tasks, they are becoming increasingly involved in less structured and more complex tasks, including interacting directly with humans, an interdisciplinary field of research broadly considered human and robot interaction (HRI) (Goodrich and Schultz 2007). Production processes in the architecture and construction industries, which involve complex tasks in unstructured environments, are thus a highly relevant application scenario for the investigation of human and robot collaboration.

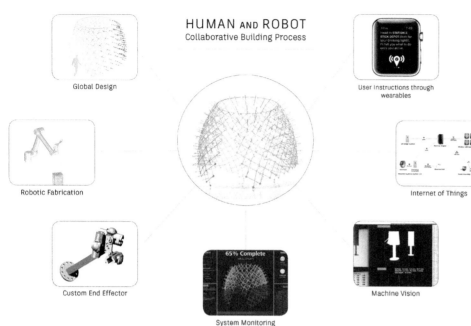

HUMAN AND ROBOT
Collaborative Building Process

Global Design

Robotic Fabrication

Custom End Effector

System Monitoring

65% Complete

User Instructions through wearables

Internet of Things

Machine Vision

5   System monitoring, connected devices, custom tooling, robotic fabrication, and computational design enabled the development and realization of a human and robot collaborative building process.

5

User Interfaces for Fabrication

Within robotic fabrication research, user interfaces have enabled the possibility of connecting robotic actions directly with user input: Dörfler and Rust developed a set of tools and a flexible open interface to enable on-line control of a KUKA robot (Dörfler and Rust 2012). This investigation and others have primarily leveraged user interfaces as a medium by which the designer might influence the outcome of the process according to targeted feedback (Johns et al. 2014). In contrast, the investigation provided here utilizes wearable interfaces as a means to communicate task instructions and protocols to an audience completely unfamiliar with the physical system and task. This investigation also builds upon investigations in industry. For example, sHop Architects utilized RFID tagging and a custom iOS application to organize fabricated building components for the Barclays center (Grogan 2014).

Filament Winding

Coreless filament winding is a fabrication process whereby an industrial robot incrementally lays fibers on a minimal or temporary formwork. This process is highly relevant as a customizable method which allows the formationof filament-based materials, such as glass or carbon fiber. Coreless filament winding significantly differs from other fiber application processes in the aerospace and automotive industries, where a mold often serves as the basis on which fibers are applied. These alternate production methods are very limited, in that they can only be utilized to produce serially identical units. Robotic fabrication expands the limitations of these methods (Prado et al. 2014) and enables the deposition of material precisely where it is needed, thus facilitating the fabrication of highly differentiated structures.

FABRICATION SYSTEM

To investigate whether humans and robots could collaboratively fabricate and assemble a structure, a prototypical fabrication system utilizing a process of robotic filament winding was developed based on the following criteria:

- Complexity of tasks: The fabrication system could not be overly simple, neither so difficult that someone unfamiliar with the task could complete it.
- Minimization of participation time: The total time required for a single participant in an exhibit was limited to 10 to 15 minutes.
- Utilization of machine vision: Adaptive regeneration of robotic control code would enable the use of non-standard materials and compensate for human error.
- Task allocation: A system in which the human user does the tasks that require fine motor control, and the robot those that require high precision, in which tradeoffs are facilitated through instructions delivered via wearable interfaces.
- User safety: Typical robotic safety procedures require that robotic control code is simulated before execution, or run in a safety mode at low speed. The exhibition format necessitated extra precautions to guarantee the reliability of the control code and safety of participants.

Collaborative Construction Vasey, et al.

Module Development

A single tensegrity module was developed whereby three compression elements are held apart in a precise position during fabrication in order to align with each other in a global system. Bamboo was utilized for the compression members due to its energy efficiency and relative stiffness to weight, though it is non-standard material and thus ill-suited to traditional automation techniques. A custom pin detail inserted into the end of the bamboo is utilized to catch and hold the string during winding.

Global Design

The main driver of a global structure's design was to describe a continuous doubly curved surface using modules of non-uniform shape composed of uniform length members. Assuming the modules are placed in the correct positions, the unique geometry of each module ensured that the structure would take the desired final form without any human oversight.

FABRICATION SETUP

A collaborative Universal Robot (UR) 10 robot was utilized because of its precision, its adaptive force control, and its ability to have instructions directly streamed and executed over socket communication (Universal Robots 2009). A CNC-milled custom end effector was developed that could precisely control the unique rotation between two of the three bamboo sticks through the rotation of two kinematically linked gears (Figure 4). When the filament is secured on a pin, and the robot incrementally re-orients the end effector, fiber is pulled from the fiber source and wound onto the bamboo frame. A lightbox and USB-connected webcam enables the scanning and digitization of each bamboo tip (Figure 7).

In robotic filament winding processes, one of the main challenges is to control the system tension, which naturally fluctuates throughout the process. During winding, a simple mechanism utilizes a hanging weight and a spiral compression spring, which provides frictional resistance between the weight and the fiber source to maintain an approximately constant tension on the string source, thus acting when the robot is moving towards or away from the fiber source (Figure 7). This type of system is derived from similar dancing bar tension control systems in industrial extrusion and rolling processes, but has the advantage that no signal processing or actuated braking is necessary (Becker 2000).

For the exhibition, four identical robotic stations are arranged to operate simultaneously. Secondary stations which contain the LED-embedded connection details and the bamboo pieces are located around the perimeter of the exhibit. For user safety, it was necessary that the exhibit itself have a single entrance so that only users registered in the system could enter the space.

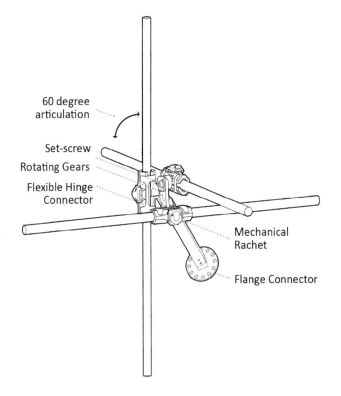

60 degree articulation

Set-screw
Rotating Gears
Flexible Hinge Connector

Mechanical Rachet

Flange Connector

6

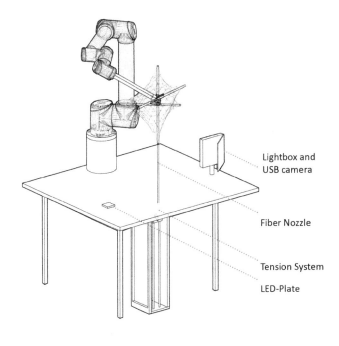

Lightbox and USB camera

Fiber Nozzle

Tension System

LED-Plate

7

6 CNC-milled customizable end effector.

7 Single robot station setup with simple mechanical tension control.

8   The Smartwatch delivered just-in-
time instructions to participants.

9   The CNC-milled customizable end
effector without a part loaded.

10   The effector could be fastened
around an irregular piece of
bamboo through the tightening of
a ratchet.

8

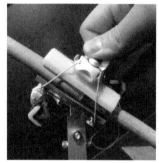

9   10

COLLABORATIVE FABRICATION
Collaborative Fabrication and Assembly Process

Upon entrance to the exhibition, users would be outfitted with an Apple Watch, an Apple iPhone, and safety glasses. The user would then be guided by just-in-time instructions through the exhibition, beginning by obtaining materials. The custom iOS application and the specifics of the user interface are discussed in a separate paper, "Crowd-sourced Fabrication" (2016).

To fabricate a specific module at a robot station, the geometric properties of this unit are accessed from a database and then applied to an instantiation of the module in a CAD design environment. Correspondingly, the robot aligns itself perpendicularly to a reference surface for each bamboo part, so that the correct position and orientation is achieved, which allows the user to load the mechanical end effector by fastening and tightening a mechanical ratchet, a maneuver which utilizes the human's dexterity and the robot's precision. Confirmation that each part has been inserted correctly in the effector is confirmed from the

user's watch before the robot proceeds. The robot then executes a custom scanning routine to re-digitize each tip correctly in space, re-generating the control code based on these deviations. Before uploading this code, the smartwatch interface prompts the user to confirm they are behind a safety line, and the winding routine executes at full speed.

During winding, a preliminary scaffolding layer is wound first to connect each end of bamboo to each of its neighbors without crossing or doubling the previously laid fibers. Final nebulous layers are wrapped, allowing the system to achieve an equilibrium condition, after which the tensegrity module can be removed from the effector by hand without significant deformation. Instructions on the Apple Watch, and LED lights embedded within the connective modules between the parts themselves, then indicate to the user where to fasten the module on the existing structure by tightening two zip ties on the connection detail.

System Communication

The robotic winding process was embedded in global control framework, which coordinated active users, robotic stations, and the current build status. The "Foreman Engine," the central brain in this framework, monitored and managed the overall progress of the construction and assembly, and systematically assigned the next part to the next available robot station (Lafreniere 2016).

The smartwatch devices worn by participants of the exhibit provided step-by-step instructions, and each user's current

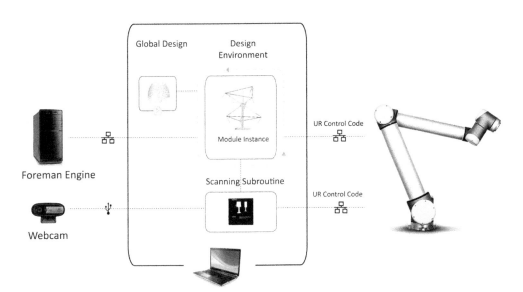

Robot Assembly Station

11

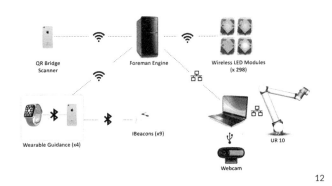

12

humans make mistakes or cannot be precise. For example, by not tightening the ratchet of the end effector holding the part in its entirety, the pieces of bamboo could rotate slightly due to the high applied torque.

To mitigate these sources of error, a scanning process was implemented to individually measure each bamboo tip, recalculate the position of the tip relative to the robot flange, update the object-oriented structure of the module in the design environment, and regenerate all of the robotic control code. A simple image analysis was utilized: for each of the 6 points of interest, the robot moves to a position in alignment with a plane in space. Based on simple image processing of two images, one taken through reflection, and the known mathematical location of the robotic position in space, the measured deviations of this pin are directly applied as a transformation of the pin in its local coordinate system (Figure 13).

Robotic Motion Planning for Filament Winding

In order to develop consistent and predictable robotic motion, it was necessary to develop algorithms that produce robotic winding patterns, methods that simulate these robotic paths, and methods that converted these paths into unambiguous robotic control code. Though software for the development and output of robotic control code exists by third party developers (Braumann and Brell-Cokcan 2014; Schwartz 2012; Elashry 2014), the kinematic complexity of the maneuvers necessary in a winding process, and the rather restrictive nature of the axis limits of Universal Robots, necessitated a stand-alone library for a computer-aided design environment (CAD) that would allow definitive

status within the app was reported to the foreman engine. A series of iBeacons were additionally distributed throughout the exhibit to track the physical location of the workers, so that instructions could be presented automatically when a user arrived at a specific location (ibid.).

The robot server was configured as an in iteratively updating CAD environment on an individual PC (Figures 11 and 12). An update to a file structure on the control PC of each robot station communicated to the design environment to update the data structure of the current part, thereby triggering an update to the control code. Additional changes signified what control code to be immediately uploaded to the robot: either loadStick (A, B, C), scanParts(), or executeWinding().

Machine Vision

There were two main sources for error within this workflow: one being the deviations from an inherently variable material, and the other being the inherent source of error which arises when

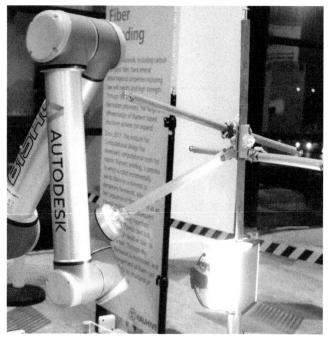

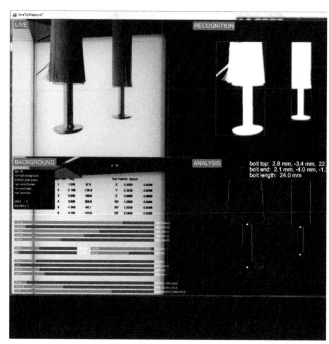

13

14

control over motion type (linearly interpolated or axis specific), interpolation parameters, and inverse kinematic configuration data.

There were three main challenges in the robotic control of the process: (i) maximize reachability of all potential points with tolerances by correctly positioning the fiber source relative to robot and module relative to effector flange (ii) to derive feasible fiber patterns which would not run out of axis due to the discontinuous axis 4 and axis 6 of UR10 robots, and (iii) to guarantee that the robot control code would be executed reliably for each unique unit geometry despite deviations.

To achieve these goals, a geometric inverse kinematic solver for non-spherical wrist robots was implemented to calculate correct joint values for any given target position in each of 8 possible kinematic configurations.

Motion Planning

While typical motion planning positions a single tool coordinate frame in a world coordinate system, in this fabrication process, the position of interest on the end effector of the robot is constantly changing. During a hooking movement, the frame of interest acts at the pin of the end of the bamboo, and during the winding, a plane offset from a position on the module being wrapped. To solve for the planes for the machine code for a specific module, the planes which exist within the flange coordinate system of the end effector are aligned onto the base coordinate system of the fiber source, and through a backwards transformation, the position of a tool coordinate frame is solved

in the world coordinate system (Figure 18).

Each target plane has an additional degree of freedom because it can be rotated about its Z-axis, affectively rotating the end effector around the Z-axis of the fiber source. To limit unnecessary robotic movement, the X-axis of the plane is solved as the plane with the greatest component in the direction of the default flange extension (Z-axis, flange coordinate system), so that the orientation of the robot flange projected in the world XY plane can be precisely controlled.

Reachability

To establish a task-specific reach envelope which would represent the three-dimensional solution set of reachable positions, a two-dimensional set of planes are generated in the flange coordinate system in which the Z-axis points towards the approximate center of the module. This set of planes is tested for reachability in each configuration. The generalized three-dimensional solution set of reachable positions can be achieved by rotating this set of planes around the flange Z-axis, which is the equivalent of rotating the sixth-axis of the robot (Figure 16).
This procedure:

- Enables the precise definition of the ideal fiber nozzle relative to robot base.
- Enables precise position of the length of effector extension.
- Defines the height of the fiber nozzle above the table to avoid collision.
- Determines which kinematic configurations of the eight

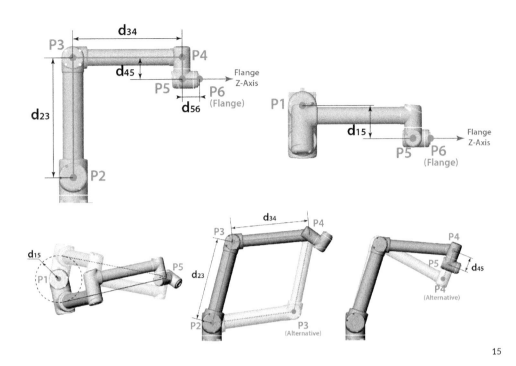

13 Image processing is used for part digitization; For each pin, the robot aligns the local pin frame to a known base frame. The deviation of the pin from the expected position can be applied as a transformation in the local coordinate system of the pin frame.

14 Two bitmaps, one captured through reflection, are utilized to calculate the deviations of the pin.

15 A geometric inverse solution for non-spherical wrist robots is used to solve for the axis positions of any given target position. Three of the axis positions can be mirrored, resulting in 8 unique kinematic configurations for any given position. Depicted are the two possible solutions for joint1, the two possible solutions for joint2, and the two possible solutions for joint3.

15

possible have the most significant envelope and should be utilized.

- Illustrates that the most problematic area on the module to reach are positions close to the Z-axis of the flange where the solution for the X-axis is singular.

Path Planning for Axis Limitations

UR10s pose a kinematic challenge in that their fourth and sixth axes are discontinuous. In many industrial robots, including most KUKA and ABB robots as well as UR5s, the fourth and sixth axes are configured with infinite motion. This capability facilitates any motion or task that would cause a single axis to always increase or decrease, for example, tightening a screw. Though this limitation is specific to this setup, it is worth considering as a general case: any fabrication process with externally connected wires or power sources can similarly not allow an infinite twisting of either the fourth or the sixth axis.

Topological Model for Path Planning

To come up with a feasible fiber sequence that would not run out of axis, an object-oriented topological computational model was produced that included pins, the ends of each bamboo tip, and lines, the set of all lines which connect two adjacent pins. This model embeds the following relationships:

- From a single pin, it is possible to move to 4 lines (the four not connected to the pin)
- From a line, it is possible to move to 2 pins (inverse of rule one) or 4 lines, but only moves in the positive direction of the

previously laid fiber paths are valid; otherwise unwrapping would occur.

Through these embedded relationships, all possible paths between any two pins can be generated combinatorically as a sequence of right (0) and left turns (1) between end pins (-, 00, 01, 11, 000, 111, 101,110, 001, 010, 101....). This method allows all paths to be evaluated, sorted, and compared for their relative effect on the axes during path planning (Figure 19).

Post-Processing Algorithms for Evaluating Sequences

The inverse kinematic solver returned the correct joint position for each axis as a value between -180 and 180, but this value added to any multiplier of 360 is also valid. A set of post-processing algorithms was implemented to evaluate any sequence of robotic positions for the relative impact on axis values to determine if the sequence would stay within axis limits. This algorithm implements the assumption that the UR robot will always move to the axis position closest to the previous position in the same configuration during linear movement. For any sequence of moves, the summation of the relative axis change of all sequences when added with the joint values at the start of the motion must stay within -360 and 360 degrees for each axis.

Fiber Syntax Generation and Unambiguous Control Code

The robotic control code for the filament winding can be divided into two portions: the scaffolding, which connected all pins in one continuous line, and subsequent fiber winding. The scaffolding was in principle more challenging, as previous positions could not be revisited in the case that an axis was close to its

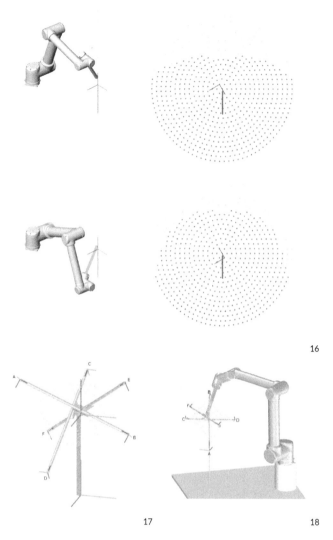

16

17   18

limits. Axis-specific joint movement for each pin was utilized in the control code, so the change in configuration was defined unambiguously as the robot moved from one pin to the next. If a purely Cartesian-based positon strategy for motion control had been utilized, the robot would have run out of axis in three to four moves.

For the final wrapped layers, a sequence was produced utilizing topological path planning which touched all edges at least once and minimized winding time. All robotic positions were defined unambiguously in pseudocode with fully descriptive geometric and kinematic information, re-constructed from the module geometry in the design environment, and then translated into control code linking linearly interpolated positions. The same syntax was used for each module to minimize troubleshooting during the exhibition.

RESULTS AND LIMITATIONS
Though exhibit visitors were shadowed by the exhibition

designers to ensure quality control, this project demonstrated that untrained workers—unfamiliar with both the necessary tasks required of them and the physical system itself—could successfully collaborate with robots to build a large structure when provided with just-in-time instructions delivered through a wearable device. Over 200 visitors came to the site to help build the 12 foot pavilion composed out of 224 unique tensegrity modules. The final structure was on display as a live collaborative building process and exhibit for over three days.

Despite the project's successes, several aspects could have been improved. When an unanticipated error occurred, for example, the tension was too high for the safety settings of the UR robot, it triggered an error, and the entire part had to be discarded. Thus, one next step to improve the process would be to identify common errors, to determine how they are easily detected from sensor feedback or input from the user interface, and to tailor the information sent to the user based on this analysis.

One of the most significant limitations of the project was the use of a robot that had limited axis ranges. To enhance the interactivity of the process, and to make use of the robot's inherent ability to achieve customizability, the design of the fiber layout could have become interactive or differentiated, in which case performance criteria, fabrication constraints, and user choice could have been integrated into a computational design tool and interface.

Scanning and digitizing the parts successfully enabled the utilization of non-standard materials. However, one missed opportunity was to store the deviations of the as-built geometry, and regenerate subsequent control code, allowing tolerances to be compensated for computationally as the construction and assembly process progressed (Vasey, Maxwell, and Pigram 2014).

DISCUSSION
As a demonstrator, the project illustrates that interconnected devices, sensor feedback, and responses enabled through user interfaces enable new possibilities for reconsidering protocols for humans and robots to work together, particularly as applied in construction or fabrication processes which involve the coordination and organization of many parts, processes, and people. More significantly, the project demonstrates the new possibility of a hybrid domain of robot and human collaboration in construction, in which coordination and communication between hardware and software facilitates new possibilities. These possibilities, including just-in-time production, tolerance compensation, and task monitoring and allocation can serve the purpose of more fully integrating robotic processes and computational control into the characteristic processes of construction during all design and production stages.

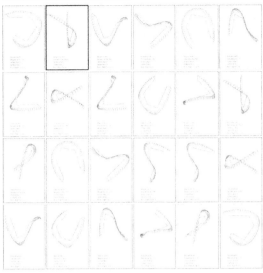

Fiber Syntax Library: Boolean Right Turns: 10

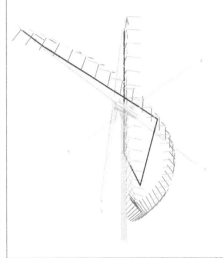

Single Fiber Sequence

16 Reachable positions relative to the robot flange in two kinematic configurations. Every module and all toolpaths in the global design had to fit within this envelope.

17 Frames existing in the coordinate system of the robotic flange.

18 Mapping of the frames onto the base frame of the fiber source and solution of the tool frame location in the space of the world.

19 Topological motion planning and move evaluation: All paths can be generated through embedded topological relationships as a sequence of right and left turns. Any sequence of moves can be evaluated for its relative effect on axis values and sorted during path planning.

19

ACKNOWLEDGEMENTS

This project was an interdisciplinary collaboration between several research groups:

Autodesk Applied Research Directed by Maurice Conti. Concept Development: Heather Kerrick, David Thomasson, Evan Atherton, Nicholas Cote, Lucas Prokopiak, Arthur Harsuvanakit.

Autodesk Research, User Interface Research & Research Transfer Groups. Interaction Development: Tovi Grossman, George Fitzmaurice, Justin Matejka, Fraser Anderson, Ben Lafreniere, Steven Li, Nicholas Beirne, Madeline Gannon, Thomas White, Andy Nogueira.

The Living – An Autodesk Studio, Directed by David Benjamin. Design System Development: Danil Nagy, James Stoddart, Ray Wang, Dale Zhao.

Institute for Computational Design, University of Stuttgart, Directed by Prof. Achim Menges. Material System & Robotic Fabrication Development: Lauren Vasey, Long Nguyen, Thu Phuoc Nguyen, Tobias Schwinn.

LED System Development: Marcelo Coelho Studio

Funding: Autoddesk Inc.

Robots: UR Robots

REFERENCES

Becker, Bruce. 2000. "Selecting the Right Tensioning System." *Machine Design*, February 1. http://machinedesign.com/technologies/selecting-right-tensioning-system

Braumann, Johannes, and Sigrid Brell-Cokcan. 2012. "Real-Time Robot Simulation and Control for Architectural Design." In *Digital Physicality:* *Proceedings of the 30th eCAADe Conference*, vol. 2, edited by Henri Achten, Jiri Pavlicek, Jaroslav Hulin, and Dana Matejdan. Prague: eCAADe. 479–486.

Dörfler, Kathrin, Florian Rist, and Romana Rust. 2012. "Interlacing: An Experimental Approach to Integrating Digital and Physical Design Methods." In *Rob | Arch 2012: Robotic Fabrication in Architecture, Art and Design*, edited by Sigrid Brell-Çokcan and Johannes Braumann. Vienna: Springer. 82–91.

Elashry, Khaled, and Ruairi Glynn. 2014. "An Approach to Automated Construction Using Adaptive Programing." In *Robotic Fabrication in Architecture, Art and Design*, edited by Wes McGee and Monica Ponce de Leon. Cham, Switzerland: Springer. 51–66.

Goodrich, Michael A., and Alan C. Schultz. 2007. "Human-Robot Interaction: A Survey." *Foundations and Trends® in Human-Computer Interaction* 1 (3): 203–275.

Grogan, Abi. 2014. "The Architecture Industry Takes a Cue From Manufacturing." *Engineering and Technology Magazine* 9 (2). Last modified 10 February 2014, http://eandt.theiet.org/magazine/2014/02/big-league-dreams.cfm

Helm, Volker, Selen Ercan, Fabio Gramazio, and Matthias Kohler. 2012. "In-Situ Robotic Construction: Extending the Digital Fabrication Chain in Architecture." In *ACADIA 12: Synthetic Digital Ecologies—Proceedings of the 32nd Annual Conference of the Association for Computer Aided Design in Architecture*, edited by Jason Kelly Johnson, Mark Cabrinha, and Kyle Steinfeld. San Francisco: ACADIA. 169–176.

Helm, Volker, Jan Willmann, Fabio Gramazio, and Matthias Kohler. 2014. "In-situ Robotic Fabrication: Advanced Digital Manufacturing Beyond the

20 Final structure composed out of 224 unique modules fabricated and assembled by human and by robot.

20

Laboratory." In *Gearing Up and Accelerating Cross-Fertilization Between Academic and Industrial Robotics Research in Europe.*, edited by Florian Röhrbein, Germano Veiga, and Ciro Natale. Vol. 94 of *Springer Tracts in Advanced Robotics*. Cham, Switzerland: Springer. 63–83.

Johns, Ryan Luke, Axel Kilian, and Nicholas Foley. 2014. "Design Approaches Through Augmented Materiality and Embodied Computation." In *Rob | Arch 2014: Robotic Fabrication in Architecture, Art and Design*, edited by Wes McGee and Monica Ponce de Leon. Cham, Switzerland: Springer. 319–332.

Lafreniere, Benjamin, Tovi Grossman, Fraser Anderson, Justin Matejka, Heather Kerrick, Danil Nagy,Lauren Vasey, Evan Atherton, Nicholas Beirne, Marcelo Coelho, Nicholas Cote, Steven Li, Andy Nogueira, Long Nguyen, Tobias Schwinn, James Stoddart, David Thomasson, Ray Wang, Thomas White, David Benjamin, Maurice Conti, Achim Menges, and George Fitzmaurice. 2016. "Crowdsourced Fabrication." In *Proceedings of the 29th Annual ACM Symposium on User Interface Software & Technology*. Tokyo: UIST. Forthcoming conference.

Menges, Achim. 2012. "Morphospaces of Robotic Fabrication: From Theoretical Morphology to Design Computation and Digital Fabrication in Architecture." In *Rob | Arch 2012: Robotic Fabrication in Architecture, Art and Design*, edited by Sigrid Brell-Çokcan and Johannes Braumann. Vienna: Springer. 28–47.

Menges, Achim. 2015. "The New Cyber-Physical Making in Architecture: Computational Construction." *Architectural Design* 85 (5): 28–33.

Menges, Achim, and Tobias Schwinn. 2012. "Manufacturing Reciprocities." Architectural Design 82 (2): 118–125.

Prado, Marshall, Moritz Dörstelmann, and Tobias Schwinn. 2014. "Coreless Filament Winding." In *Rob | Arch 2014: Robotic Fabrication in Architecture, Art and Design 2014*, edited by Wes McGee and Monica Ponce de Leon. Cham, Switzerland: Springer. 275–289.

Schwartz, Thibault. 2012. "HAL: Extension of a Visual Programming Language to Support Teaching and Research on Robotics Applied to Construction." In *Rob | Arch 2012: Robotic Fabrication in Architecture, Art and Design*, edited by Sigrid Brell- Cokcan and Johannes Braumann. Cham, Switzerland: Springer. 92–101.

Vasey, Lauren, Iain Maxwell, and Dave Pigram. 2014. "Adaptive Part Variation A Near Real-Time Approach to Construction Tolerances." In *Rob | Arch 2014: Robotic Fabrication in Architecture, Art and Design 2014*, edited by Wes McGee and Monica Ponce de Leon. Cham, Switzerland: Springer. 67–81.

Universal Robots. 2009. *User Manual: UR10/CB3 Version 3.0 (rev. 15965)*. Odense, Denmark: Universal Robots A/S. http://www.universal-robots.com/media/8764/ur10_user_manual_gb.pdf

IMAGE CREDITS

All figures: ICD University of Stuttgart/ Autodesk Research/ Autodesk Applied Research

Lauren Vasey is a Research Associate at the Institute for Computational Design (ICD) at University of Stuttgart. She received a Bachelor of Science (cum laude) in Engineering from Tufts University, and a Master's of Architecture with distinction from the University of Michigan. She has taught several workshops and courses in computational design and robotic fabrication and has worked previously at the University of Michigan Taubman College FAB Lab as well at the Swiss Federal Institute of Technology (ETH-Zurich), Chair for Architecture and Digital Fabrication.

Long Nguyen, born in 1988, holds a Bachelor's degree in Computer Science from Cambridge University and a Master's degree from University College London with focus on 3D Computer Graphics and Virtual Environments. With a great passion for teaching programming, mathematics, computational design, and thinking, Long joined the Institute for Computational Design as a Research Associate in October 2014 where he conducts research in evolutionary design, robotic fabrication, and geometric modelling.

Tovi Grossman is a Distinguished Research Scientist at Autodesk Research, located in downtown Toronto. His research is in HCI, focusing on input and interaction with new technologies. Tovi received a Ph.D. in Human-Computer Interaction from the Department of Computer Science at the University of Toronto.

Heather Kerrick is a Senior Research Engineer at the Autodesk Applied Research Lab where she explores human and machine interaction. She received her Bachelor's in Mechanical Engineering at the University of Maryland and her Master's in Product Design from Stanford University.

Danil Nagy is a Lead Designer and Senior Research Scientist with The Living group within Autodesk Research in Brooklyn, New York. He received his Master's in Architecture and Master's of Science in Urban Planning from Columbia University. His work and research focuses on computational design, generative geometry, advanced fabrication, machine learning, and data visualization. Danil was project manager of the Hy-Fi installation at the MoMA PS1 courtyard in Queens, New York, and is the design lead on the long-term collaboration between Autodesk and Airbus, including the Bionic Partition project.

Evan Atherton. Originally from Southern California, Evan moved to San Francisco to study Mechanical Engineering at UC Berkeley, where received a B.S. in 2011 and M.S. in 2012. He is currently a Senior Research Engineer at Autodesk's Applied Research Lab exploring the future of robotics, digital fabrication, and storytelling.

David Thomasson is a Principal Research Engineer, leading the robotics initiative at the Autodesk Applied Research Lab. One aspect of his research is investigating real-time feedback and control for industrial robot arms. David studied electrical engineering at the University of Sydney and The University of Queensland.

Nick Cote received his Master of Architecture from the Rhode Island School of Design (RISD), 2015, where he also studied Printmaking and Illustration. His training in robotics comes from coursework, assistantships, and research positions concerned with robotics in art and design. He has been a Design Robotics Researcher at Autodesk since June, 2015.

Tobias Schwinn is research associate and doctoral candidate at the Institute for Computational Design (ICD) at the University of Stuttgart, Germany. In his research, he focuses on the integration of robotic fabrication and computational design processes. Prior to joining the ICD in 2011, he worked as a Senior Designer for Skidmore, Owings and Merrill in New York and London.

David Benjamin is Founding Principal of The Living, an Autodesk Studio. He and the studio have won awards for their innovative approach to design from the Architectural League, the American Institute of Architects, Architizer, the Museum of Modern Art, New York Foundation for the Arts, and Fast Company, and Holcim. David currently teaches at Columbia Graduate School of Architecture, Planning and Preservation. Before receiving a Master of Architecture from Columbia, he received a Bachelor of Arts from Harvard.

Maurice Conti is an innovator, futurist and creative visionary. His ideas, projects and creative leadership have been leveraged by Fortune 100 companies, government agencies, boutique manufacturing studios and renowned artists. He is director of strategic innovation at Autodesk. There he is defining and leveraging technological, economic and cultural trends, especially in the context of exponential change. His mission is to help technology bring more creativity into the world.

George Fitzmaurice, Ph.D. is a Director of Research and runs the User Interface Research Group for Autodesk. He has been with Autodesk (including Alias) for over 15 years conducting research in 2D and 3D UIs including: input devices, large displays, two-handed interaction, multi-touch, pen-based UIs, TrackingMenus, spatially-aware displays, 3D navigation and tangible UIs. Fitzmaurice received a B.Sc. in Mathematics with Computer Science at MIT, an M.Sc. in Computer Science at Brown University and a Ph.D. in Computer Science at the University of Toronto.

Professor Achim Menges, born 1975, is a registered architect and professor at the University of Stuttgart, where he is the founding director of the Institute for Computational Design since 2008. He is also Visiting Professor in Architecture at Harvard University's Graduate School of Design since 2009.

Robotic Fabrication of Structural Performance-based Timber Grid-shell in Large-Scale Building Scenario

Philip F. Yuan

Hua Chai

Chao Yan (Corresponding Author)
Tongji University

Jin jiang Zhou
SUZHOU CROWNHOMES CO. LTD

1

ABSTRACT

This paper investigates the potential of a digital geometry system to integrate structural performance-based design and robotic fabrication in the scenario of building a large-scale non-uniform
timber shell. It argues that a synthesis of multi-objective optimization, design and construction
phases is required in the realization of timber shell construction in architecture practice in order to
fulfill the demands of building regulation. Confronting the structural challenge of the non-uniform
shell, a digital geometry system correlates all the three phases by translating geometrical information between them. First, a series of structural simulations and experimentations with different
objectives are executed to inform the particular shape and tectonic details of each shell component
based on its local condition in the geometrical system. Then, controlled by the geometrical system,
a hybrid process of different digital fabrication technologies, including a customized robotic timber
mill, is established to enable the manufacture of the heterogeneous shell components. Ultimately,
the Timber Structure Enterprise Pavilion as the demonstration and evaluation of this method is
fabricated and assembled on site through a notational system to indicate the applicability of this
research in practical scenarios.

1   Timber Structure Enterprise Pavilion
    in Horticultural Expo, construction
    process.

INTRODUCTION

Since the first industrially patented use of glue-laminated timber in the early 20th century of Germany (Muller 2000), modern timber production technology has completely transformed the situation of wood material from its ancient scenario. With technological development, a variety of defects of raw wood material are solved such as knots, heterogeneous density, limited length, corrosiveness, etc., and timber has become a building material with properties of large-scale adaptability, high structural performance and long durability, which are all highly required in building construction. Moreover, with the great awareness of environmental issues in our society, wood as a renewable resource with negative carbon footprint and low embodied energy (Kolb 2008), has been broadly reconsidered to establish a low carbon emission pre-fab construction process in architecture practice. And due to its natural color, light reflection and material texture, timber has been applied into various building types to create glamorous interior experiences. With the development of structural engineering in the 20th century, timber has become a material frequently utilized for grid-shell structures to realize enormous large-span buildings. In the early stage of the development, in examples such as Weald and Downland Open Air Museum designed by Buro Happold and the Multihalle (multi-purpose hall) in Mannheim by Frei Otto, timber material was usually fabricated into small-dimension components to form a double layer or even multi-layer shell system. By utilizing the flexibility of timber, the components can be easily assembled into free-form structure. Nevertheless the timber shell composed by small-dimension components will not only require more time in fabrication and assemblage, but also conflict with current fire regulations in timber building.

Actually, the timber shell structure is a complicated system in architecture design and construction. Unlike concrete or steel material, which can be easily shaped by a customized mold, the fabrication of wood material usually involves a complicated drilling and cutting process and a relatively more sophisticated joint system, which can lead to a variety of difficulties in its realization. In timber shell structures, the aim for high structural performance of the shell geometry at the macro scale and the simplification of the design and fabrication of its internal joint system at the micro scale are usually two mutually contradictory pursuits. In the geometrical optimization process of pursuing a uniform stress distribution along the shell, the geometrical result usually tends to be a free-form surface with heterogeneous curvature conditions, which results in a great variety among the shell components. And this heterogeneity within both shell components and their joints has no doubt deeply challenged traditional methods of timber construction. Even more, in an architectural practice scenario, unlike relatively small-scale installations, the design and construction of timber shell structures is highly restricted by building regulations. The shell structure not only has to distribute the gravity into all the components evenly, but also has to have enough structural stability to confront wind and seismic load. This multi-objective design requirement brings different types of parameters into the structural evaluation of timber shell systems. The shape, dimension and orientation of each timber component will be determined by both the local curvature of the shell geometry and the particular requirements of structural performance on stability. As a consequence, more variation will be brought into the structural components, which are difficult to simplify by a geometrical rationalization process and can only be fabricated and assembled through a highly customized construction method.

Since the digital turn in the field of architecture, the innovative development of digital fabrication technology has brought great advancement to traditional ways of building construction. Compared with the standardized mechanical approach, digital construction is characterized by customization and personalization. In the digital paradigm, code becomes a medium translating geometrical information directly between performance simulation, virtual shell modeling and material fabrication, which breaks the limitation of mechanical construction and offers a new possibility for the heterogeneous fabrication of timber shell structures. With this development, the construction of shell structure in contemporary timber buildings, like Centre Pompidou-Metz and Nine Bridges Golf Club designed by Shigeru Ban and supported by design-to-production, can be controlled with high precision and efficiency. However, the deployment of digital technology involved in these cases only occurred in the construction phase and rarely supported the design process by an interactive network.

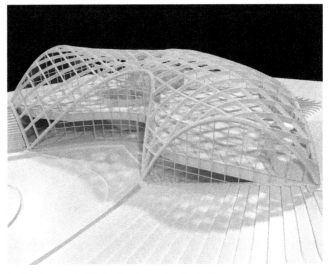

2   Physical Model of Timber Structure Enterprise Pavilion.

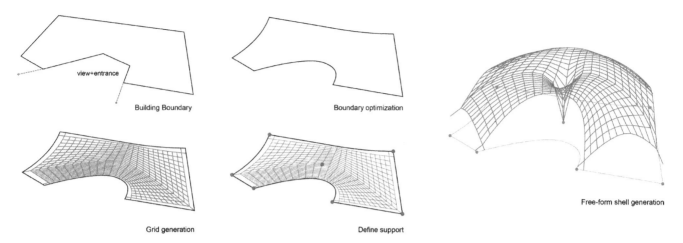

3 Form Finding Process of Free-form Structure.

Based on this situation, the research in this paper addressed the design and construction of the Timber Structure Enterprise Pavilion in the Horticultural Expo (Figure 1,2). Confronting the difficulties of non-uniform timber shell structures in architecture practice discussed above, the aims of this research are as follows:

- to establish a design-fabrication platform for large-scale building scenarios, in which all the architectural regulations and requirements, like stability in multiple loading situations, the fire-proofing of structural components (limiting the minimal size of each component), waterproofing, etc., can be balanced and fulfilled in a single system.
- to correlate the structural performance-based design with the construction process in timber shell building, in which the structural simulation and experimentation are utilized not as a mere evaluation tool, but as a design driver to generate and optimize both shell geometry and the joint system.
- to create a parameter-based feedback loop to integrate the simulation, design, digital modeling and physical fabrication of shell structures, in which all the phases can inform each other.

In order to accomplish the research aims, the approach was subdivided into three parts. The first part was a series of structural simulations, experiments and optimizations, studied through a gravity-based shell form-finding method, to improve the structural stability in confronting horizontal loads. In the process, a particular grid-beam system was set up for the shell structure first, and then experiments on a series of joint prototypes were performed to test their structural performance and to use the results to determine their distribution on the shell. Second, according to the local geometrical condition of the shell surface and the dimension requirement for structural strength, prototypes of steel joint and timber beam, which were chosen in the first phase, were then transformed into different variations

through a geometrical information system to construct a virtual model of the shell structure in a digital environment. Third, the geometrical system in the digital model was translated into different types of information through a coding process to instruct both CNC timber lamination and robotic fabrication. Further, as the result of this research, all the shell components were assembled on site based on a notational system to accomplish the construction of the pavilion. As such, the structural performance-based design and fabrication provided a solution for the construction of a timber shell pavilion, and reciprocally the realization of the pavilion became a proof of concept to demonstrate the performance of the working system developed in the research.

BACKGROUND

Two scenarios define the research scope in this paper: non-uniform timber shell structure design and large-scale architecture practice. In the former scenario, digital technology was requested to establish a platform for the design and fabrication of the heterogeneous components. In the latter context, the digital platform then has to be turned into a geometrical information system to coordinate all the aspects of realizing a large-span timber building. Facing the two scenarios together, the first challenge is that the structure's performance has to fulfill the multiple requirements of structural regulation. A timber shell usually must be simulated and evaluated on different objectives like strength and stability, and in different scopes like overall shape and particular joints. So the performance-based design process is no longer linear, and the parameters extracted from the result of different simulations have to be balanced and coordinated to generate the geometry system of the timber shell. The second challenge is that the structure usually cannot be constructed by a homogeneous fabrication and assemblage logic due to the large building scale and complicated regulation demands. The large span of the

Timber Structure Enterprise Pavilion Yuan, Chai, Yan, Zhou structure and fire-proofing regulations of wood material limit the minimal size the wooden component, so the large scale timber shell is usually designed by using a grid-beam system, which might contain different scale components according to stability requirement. Consequentially, the different scale components in the timber shell require a hybrid of various manufacturing methods, requiring the digital geometrical system to be turned into an adaptive medium, through which the virtual model can be translated into different languages for instructing different types of machines. In general, according to these challenges, what this paper explores is not merely a particular method or workflow of digital design and fabrication, but a hybrid and adaptive system that could combine different methods and workflows into an integrated entity.

METHODS

Research on the Joint System of Structural Performance-Based Timber Grid-Shell

Due to the large amount of interconnections in a grid-shell, the joint system usually plays the most important role in determining its structural performance. In this phase, starting with a free form shell geometry generated by a gravity-based form finding process, the mechanical properties of different joint systems were examined through digital simulations and physical experiments to inform the distribution of different joint prototypes into a hybrid grid-shell system.

Structural Performance-Based Form-Finding and Optimization

The compression-based shell form-finding could be traced back to the pre-digital age in architecture. For example, in the works of Antoni Gaudi, Heinz Isler and Frei Otto, the forming process of shell geometry could be conducted through a series of physical experiments. In contemporary design, lots of digital simulation tools for compression form finding are available, such as "Rhinovault" designed by Philippe Block, to be able to conduct the process precisely in the digital environment. In this research, the initial geometry of the timber structure was generated through a gravity-based form finding process in Rhinovault1 (Figure 3). Then, according to the fabrication methods for timber material and the fireproof regulation, the initial geometry was translated into a particular grid-beam system.

Based on the site boundary, the two axes in the grid shell were designed to intersect with a particular angle, rather than being perpendicular to each other. In this situation, as the stability of the whole shell was relatively low, the structure was optimized into a primary-secondary beam system to increase the internal rigid connectivity. In the beam system, the cross-section of both primary beam and secondary beam were defined as rectangular,

which is inherent in glue-laminated timber production technology. Generally, the primary beams were comprised of long and continuous curved timber, and each adjacent pair was connected by a row of short straight beams (Figure 4). In order to balance the simplification of the assembly process and the smoothness of shell geometry, the primary beams were defined as always being perpendicular to the ground while the orientation of secondary beams varied according to the local surface norm along the shell.

The form-finding process in Rhinovault mainly calculates the static equilibrium of stress distributed along the geometry based on gravity, regardless of the performance of structural stability including buckling, bending and shear (Adriaenssens, et al. 2014).

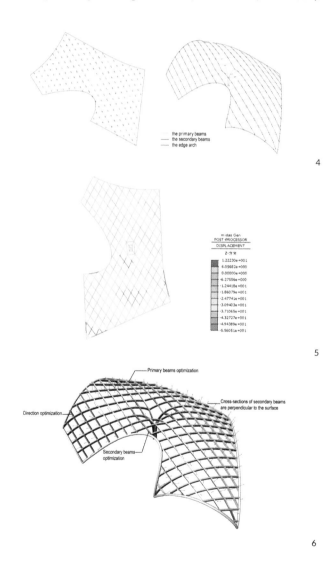

4

5

6

4   Primary-Secondary Beam System within Oblique Grid.

5   Structural Displacement Simulation with Multi-Directional Payload.

6   Optimization of Beam Dimensions.

However, in architecture practice, the structural stability in confronting horizontal loads, like wind and seismic, is also a main aspect in evaluating shell structure performance. So a series of multi-objective structural simulations with material properties were conducted in the subsequent optimization process (Figure 5), and the parameters extracted from their result were then correlated together to adjust the dimension of each beam to reinforce its local structural stability (Figure 6). After the optimization process, the dimension of the beam sections were set in a continuous variations from 650 mm x 250 mm to 350 mm x 250 mm.

Experimentation on Joint Prototypes

According to structural regulation, the joint between two timber components can only be considered as a semi-rigid or articulated connection. For grid-shell structures, the level of joint stiffness has a direct impact on the overall structural stability. Therefore, a series of digital simulations and physical experimentations were conducted to demonstrate the requirement in order to meet the level of overall stability. The result of these experiments became the main parameter to inform joint prototype design and distribution.

In the design of the timber joints, the simulations of structural stress and deformation were conducted to compare the shell structure with the ideal condition of full rigidity and the current condition of semi-rigidity (Figure 7). Taking the full rigidity condition as a contrast sample, the stress simulation result showed that the difference between the two systems is in an allowable range, which indicated that the stress was already evenly distributed along the beams in the current condition. However, structural deformations in the two systems were only close in several areas, where the local curvature of shell geometry was relatively large (Figure 8). So in order to fulfill the overall stability requirement of the semi-rigid system, an appropriate joint stiffness had to be achieved to control deformation within the allowable design range (Figure 9). As a result, to meet the building regulations, the average joint stiffness had to be more than 500 kNm/rad, and if considering the situation of a hybrid of semi-rigid joints and articulated joints, the minimal stiffness would have to be over 850 kNm/rad.

According to the stiffness requirement, three typical joints were designed for further physical experimentation: bolt plate joint, overlapping plate joint and planting bar joint. In the stiffness tests of full scale joints carried out in the structures laboratory, five measuring devices were placed on different parts of each joint prototype to record the amount of deformation under a payload ranging from 5 kN to 50 kN, and then the joint stiffness was calculated based on the test result (Figure 10, 11). As the calculation shows, the joint type with the best stiffness performance was the planting bar joint (Table 1). Through a feedback loop between comparative experimentation and joint prototype adjustment, a series of optimizations were made to improve

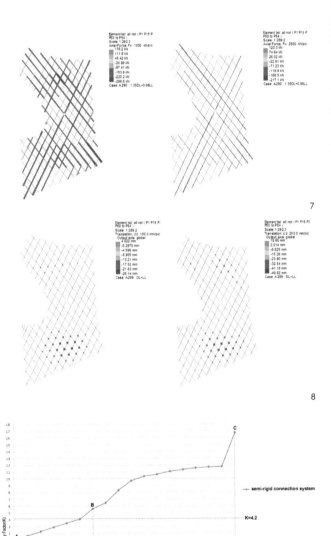

7

8

9

7   Simulation of the Axial Stress of Semi-rigid Joint System (left) and Rigid Joint System (right).

8   Simulation of the Structural Deformation of Semi-rigid Joint System (left) and Rigid Joint System (right).

9   The Relationship between Joint Stiffness and Structure Stability (A: Hinge Joint, B: Semi-rigid Joint, C: Rigid Joint).

	displacement meter 1(mm)	displacement meter 2(mm)	displacement meter 3(mm)	displacement meter 4(mm)	displacement meter 5(mm)	Stiffness (kNm/rad)
5KN	115	118	108	99	71	1727.62
10KN	182	190	174	165	115	2195.64
15KN	276	286	263	247	175	2237.27
20KN	411	420	385	365	257	2163.66
25KN	542	551	506	488	337	2209.34
30KN	673	682	626	602	419	2157.77
35KN	855	865	791	765	529	1996.92
40KN	1175	1183	1083	1046	728	1711.67

Table 1

Timber Structure Enterprise Pavilion Yuan, Chai, Yan, Zhou

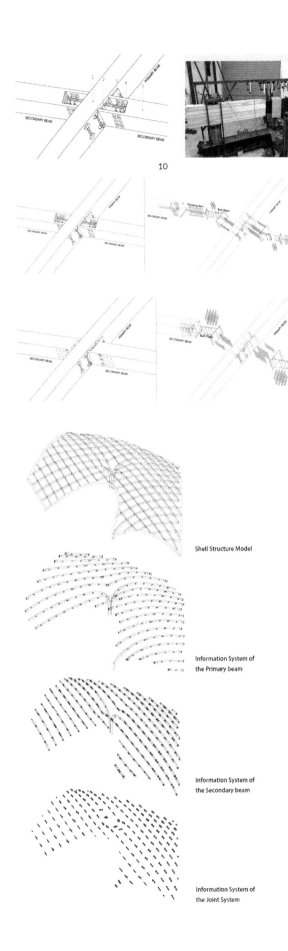

10

11

12

13

Shell Structure Model

Information System of
the Primary beam

Information System of
the Secondary beam

Information System of
the Joint System

14

the stiffness of this joint type further, including the planting bar length, etc. Finally, by combining the planting bar joint with the overlapping plate joint, the connection stiffness was enhanced to 1500 kNm/rad, which was considerably larger than the design value 850 kNm/rad, to ensure a certain degree of safety redundancy for the structure.

In the optimized joint, two T-shaped plates were connected together with high strength bolts. One of the T-shaped plates was fixed to the primary beam by high strength bolts, while another T-shaped plate was fixed to the secondary beam by planting bars (Figure 12). Due to the heavy steel components in the new joint, although the stiffness was highly enhanced, the large weight of the joint would introduce more load to the shell itself. In the structural experimentation, the bolt plate joint performed with much lower stiffness. However, because it connects two beams together only by bolts and clamping members (Figure 13), its weight would be much lighter too. So, in order to balance the structural load and stability, these two types of joint were distributed into the structural system according to the result of structural analysis; in the areas requiring only low joint stiffness, the bolt plate could effectively meet the rigidity requirement as well as reduce the structure weight.

In addition to joint stiffness experiments, several other experiments regarding different aspects of structural performance were also conducted during the research process, such as material properties testing, moisture control testing, coating durability testing, etc. The results from all the simulations and experiments were correlated together to inform the establishment of the geometrical system of the grid-shell structure.

Structural Geometry System

In the research, a digital information model was constructed to represent the structural simulation results for design visualization and optimization. Further, the geometrical system, which included the shape, dimension and orientation of steel joint and timber components, was translated into different types of information for different fabrication machines in the construction

10 The Position of Measuring Devices.

11 Physical Experiment on the
Stiffness of Typical Beam Joint.

12 Combination of Bolt Plate and
Planting Bars.

13 Bolt Plate Joint.

14 Structural Geometry System.

15

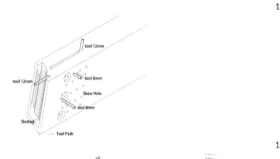

16

17

18

phase. The whole process from form finding and optimization to material fabrication could be integrated and controlled through a digital information feedback loop.

The structural geometry system in this research includes three layers of information (Figure 14). The first is the shell surface, which provides the local geometry condition for orienting structure components. The second layer is the structure model, which stores the geometrical information of the beam and joint components. The final layer is the fabrication processing information, which is the convergence of structural geometry and fabrication methods and includes a series of tool-path generating codes.

In the geometrical system, the numeric information of each structural component was converted into a parametric relationship. For example, the dimension and orientation of each joint was parameterized into the geometrical relation between the initial shell surface and timber beams by using the Grasshopper program. The main bolt plate was always parallel to the side surface of the primary beam; the axis of the planting-bar was always perpendicular to the cross-section of secondary beam, etc. With this parametric system, the initial beam model as input parameters can be processed through code to generate joint variations automatically. And as so, the numeric information of the structural geometry can be easily translated between the three layers.

Digital Fabrication

With the embedded code in the structural geometry system, three types of geometrical information were generated as output to control different digital fabrication machines for three different structural components: primary beam, secondary beam and steel joint. According to the limitations of different fabrication tools, the primary beams and the steel joints were prefabricated by CNC machine in a factory while the cutting and drilling of the secondary beams were implemented by robot.

CNC Prefabrication of Curved Beam and Steel Joint

Limited by the dimension of glue-laminating devices and digital fabrication tools, the fabrication process of the primary beams employed CNC machines to control the shaping and drilling process. At the beginning, geometric information, like the outline of the curved beams and the position of the drilling holes, was first transmitted to cutting machine through G-code to cut templates for the glue-laminated process. The templates were not only used to inform the curvature of timber beams, but also to position the bolt-holes for steel joints. Since the glue-laminating devices can move flexibly to shape the beams with different curvature conditions (Figure 15), a template-based fabrication method with fully adaptivity was established in the

15   Glue Laminating Process of Curved Beam.

16   Translation from Geometrical Model to Robotic Tool-path.

17   Simulation of Robotic Fabrication Process.

18   Robotic Fabrication Process.

factory. In general, although large-scale curved beams have to be fabricated by the glue-laminating devices, the digital information in the geometry system could be effectively transmitted into the process through templates to conduct the indirect translation of geometrical information.

The mass production of the steel joints was accomplished in a factory as well. Due to the variety and complexity of these components, the working planes in the digital environment were first redefined to adapt to the unfolding process for each joint. Then the geometric information was outputted in the form of 2D cutting paths to a CNC machine to fabricate all the plates of the joints. Further, as the plates were melted together according to the angle information, large quantities of joint variations could be produced in a relatively short time.

Customized Robotic Fabrication

In order to resolve the difficulties in fabricating the large amount of secondary beam variations, six-axis robotic arms were employed to connect the digital model and the beam fabricating process. The robot was equipped with a milling spindle with speed of 18000 rpm and an ER32 type gripper, which was able to process all the tectonic details on a beam by switching between different cutters (diameters vary from 3.5 mm to 20

mm). Then the Grasshopper plug-in KUKA|prc 2 was utilized to translate the geometric information to robotic motion as machine code to control different tool paths (Figure 16). First, a rough milling tool-path was set to cut out the general shape of each timber beam. Then a series of additional tool-paths were generated to control the fabrication of the tectonic details, such as bolt-holes, slotted openings, etc. The whole process of fabricating beam variations can be done without any sophisticated geometric drawings. The customized robotic fabrication process effectively ensured the large quantities of the shell components, improving the fabrication accuracy, and reducing fabrication time and costs (Figure 17, 18). Even more, the fabrication method also influenced the joint design due to the physical limitations. For example, since the inner corner of the slot openings was filleted in the physical environment according to the diameter of the milling cutters, the joint plate had to be designed smaller than the original geometry.

On Site Assembly

As all the components of the timber grid-shell could be prefabricated in the robot lab and factory, the on-site construction process only involved a series of location and assemblage tasks. At the beginning of construction, all the beams and steel joints were accurately numbered and labeled according to the axis

19

20

21

19 Assembly Process.

20 Structural Components Assembly.

21 Interior View of Timber Shell System.

notation system in the digital model. Then a full-space scaffolding system with adjustable devices was introduced to enable the three-dimensional location of each beam in real space (Figure 19). In order to reduce the accumulated error, a real-time tracking mechanism was employed into the assembly process. When each beam was located in space and supported by the scaffolding system, a total station was utilized to track and measure the survey points on the beam. The data recorded by the total station was then returned to the digital space and compared with the corresponding data in the geometry system to check the displacement of the actual beam location. Further, the initial dimension generated by the geometry system for the scaffolding would be updated to adapt to the construction errors. In general, the intelligent, data-oriented assembly technology ensured the accomplishment of the project with a high-degree of precision in a fairly short time period (Figure 20, 21).

RESULTS AND REFLECTION

By utilizing the adaptive geometry system, the design and construction of the Timber Structure Enterprise Pavilion in Horticultural Expo was accomplished with a high degree of precision. The building is around 2000 square meters and its maximal span is about 40 meters. The timber shell system of the building contains 27 long curved beams, 184 short beams and 368 steel joints, in which no two components are completely the same due to their different local geometry condition and structural performance (Figure 22). The whole process of structure fabrication and assemblage controlled by the geometrical system only required four months, which is considerably short for a building in this scale. As a result, the accomplishment of the building construction can be considered as proof to demonstrate the adaptivity, capability and efficiency of the system developed in this research to resolve the complicated problems of timber shell structures.

CONCLUSION

The research in this paper reflects the challenges of contemporary digital design and fabrication in the scenario of large-scale architecture practice. According to these challenges, a structural performance-based geometrical system was developed to integrate the multi-objective structural evaluations, the parameter based design and the hybrid of multiple fabrication methods. As the demonstration of the system, the process of design and construction of the Timber Structure Enterprise Pavilion shows that architecture should be interpreted as a balance of contradictory factors in terms of not only form and function but also structure and tectonics. This research proposes that a hybrid of multiple strategies within an adaptive system to solve one problem, rather than a single homogeneous approach to solve multiple problems, would be able to provide an alternative thinking

in the contemporary exploration of digital design and fabrication, and might eventually bring advanced digital technologies from the academic context to the practical field of architecture.

ACKNOWLEDGEMENTS

The work presented in this paper was funded by "The 13th Five-Year" National Development Plan "Green Building and Building Industrialization" Special Program, National Natural Science Foundation of China, Sino-German Science Foundation, and Shanghai Digital Construction Engineering Research Center–Tongji Architectural Design (Group) Co.,Ltd. Key Project Research Foundation. And it is part of the collaborative project between College of Architecture and Urban Planning Tongji University, Archi-Union Architects, Suzhou Crownhomes CO.,LTD and Fab-Union. The authors would like to express their gratitude to all contributors in this project.

NOTES

1. The Rhinoceros® Plug-In RhinoVAULT which is developed by Block Research Group, emerged from research on structural form finding using the Thrust Network Analysis (TNA) approach to intuitively create and explore compression-only structures.
2. KUKA|prc is a parametric robot control tool developed by Association for Robots in Architecture which enables one to program industrial robots directly out of the parametric modeling environment, including a full kinematic simulation of the robot. The generated files can be executed by the KUKA robot, without requiring any additional software.

REFERENCES

Adriaenssens, Sigrid, Philippe Block, Diederik Veenendaal, & Chris Williams. 2014. *Shell structures for architecture: form finding and optimization.* London: Routledge.

Block, Philippe. 2009. "Thrust Network Analysis: Exploring three-dimensional equilibrium." PhD diss., Massachusetts Institute of Technology. Dept. of Architecture.

Kolb, Josef. 2008. *Systems in Timber Engineering: Load-Bearing Structures and Component Layers.* Basel: Birkhäuser.

Lachauer, Lorenz, Matthias Rippmann, and Philippe Block. 2010. "Form Finding to Fabrication: A digital design process for masonry vaults." *Proceedings of the International Association for Shell and Spatial Structures (IASS) Symposium.* Shanghai: IASS.

Menges, Achim. 2011. "Integrative Design Computation: Integrating material behaviour and robotic manufacturing processes in computational design for performative wood constructions." *Proceedings of the 31st Annual Conference of the Association for Computer Aided Design in Architecture (ACADIA).* edited by J.S. Johnson, B. Kolarevic, V. Parlac, and J. Taron. Calgary: ACADIA. 72–81

22 Interior View of Timber Shell System.

Muller, Christian. 2000. *Laminated Timber Construction*. Basel: Birkhäuser.

Rippmann, Matthias and Philippe Block. 2013. "Funicular Shell Design Exploration." *Proceedings of the 33rd Annual Conference of the Association for Computer Aided Design in Architecture (ACADIA)*. edited by P. Beesley, O. Khan and M. Stacey. Cambridge: ACADIA. 337–346.

Rippmann, Matthias, Lorenze Lachauer, and Philippe Block. 2012. "Interactive Vault Design." *International Journal of Space Structures*. 27(4): 219–230.

Veenendaal, Diederik and Philippe Block. 2012. "An overview and comparison of structural form finding methods for general networks." *International Journal of Solids and Structures*. 49(26): 3741–53.

Willmann, Jan, Michael Knauss, Tobias Bonwetsch. et al. 2015. "Robotic timber construction-Expanding additive fabrication to new dimensions." *Automation in Construction*. 61:16–23.

IMAGE CREDITS

Figures 1, 2, 16, 18, 19, 20, 21, 22: Philip F. Yuan, 2015–2016
Figures 5, 7, 8, 9, 10, 11, 15: Jinjiang Zhou, 2015, © SUZHOU CROWNHOMES CO. LTD

Philip F. Yuan is a Professor in Architecture at Tongji University in Shanghai, and the director of the Digital Design Research Center (DDRC) at the College of Architecture and Urban Planning (CAUP), Tongji University. He is also the founding director of Archi-Union Architects. As one of the founders of the Digital Architectural Design Association (DADA) of the Architectural Society of China (ASC), his research and practice focuses on digital design and fabrication methodology with the combination of Chinese traditional material and craftsmanship.

Hua Chai holds a Bachelor degree of architecture from Tongji University, Shanghai. Currently he is a Postgraduate student at Tongji University. His research focuses on the digital design and robotic fabrication of wood tectonics based on structural performance.

Chao Yan holds a master degree in architecture from Southern California Institute of Architecture (SCI-Arc). Currently he is PhD candidate in Tongji University, Shanghai. He also teaches as a visiting lecturer at China Academy of Art, Hangzhou. His research focuses on digital design theory and the relation between digital technology and architectural perception.

Jinjiang Zhou is the director of the Technical department of Suzhou Crownhomes. He is also the founding director of Suzhou Chuang zhi Engineering Consultant. His research and practice focuses on lightweight structure, form-finding and timber structure.

Robotic Fabrication of Non-Standard Material

Pradeep Devadass
The Architectural Association

Farid Dailami
Bristol Robotics Laboratory

Zachary Mollica

Martin Self
The Architectural Association

1

ABSTRACT

This paper illustrates a fabrication methodology through which the inherent form of large non-linear timber components was exploited in the Wood Chip Barn project by the students of Design + Make at the Architectural Association's Hooke Park campus. Twenty distinct Y-shaped forks are employed with minimal machining in the construction of a structural truss for the building. Through this workflow, low-value branched sections of trees are transformed into complex and valuable building components using non-standard technologies. Computational techniques, including parametric algorithms and robotic fabrication methods, were used for execution of the project. The paper addresses the various challenges encountered while processing irregular material, as well as limitations of the robotic tools. Custom algorithms, codes, and post-processors were developed and integrated with existing software packages to compensate for drawbacks of industrial and parametric platforms. The project demonstrates and proves a new methodology for working with complex, large geometries which still results in a low cost, time- and quality-efficient process.

1   Wood Chip Barn's Truss.

INTRODUCTION
Timber forks as a material

Commonly used for structural building components, wood is typically treated as a rectilinear material—its natural forms reduced to square sawn timber and various sheet materials. While producing complex forms, recent production of doubly curved glulaminated timber components furthers this treatment. Without consideration of their internal fibres, logs are sawn to regular dimensions by manufacturers for ease of processing; in the process, wood's internal fibers are cut, thus sacrificing strength (Carpo 2011, 105). A wasteful redundancy becomes apparent, in which material is processed two or more times to achieve characteristics that may already be present in the original material. Analogous to the intuitive assembly of specifically curved timbers in traditional ship building, alternative conceptions of design and fabrication processes have recently been proposed (Monier et al. 2013; Stanton 2010) that address this redundancy. The intention of using timber in its natural form is to harness the strength of its continuous grain—chains of cells that are optimally aligned to transmit force (Desch and Dinwoodie 1996). New non-standard technologies address the conventional reasons for avoiding timber in its natural shape and form in the construction industry; complexity in handling, processing data, and difficulties encountered during the fabrication process. The project began with the a need to understand a 'good fork.' Slater and Ennos showed that tree forks become stronger with increase in inclination away from the vertical axis, which leads to the formation of more elliptical branches in cross-section (Buckley, Slater, and Ennos 2015). This principle promoted exploitation of this embedded strength in its inherent form. With this as a guide, a number of potential structural systems were developed based on the fork's Y-shaped geometry. In the built work, 20 forks are used to achieve a Vierendeel truss-like structure spanning 25 meters without the need for customized steel sections.

Within this paper, four key explorations are elaborated: the development of a precise geometric referencing system to ensure consistent placement of each component, independent of its irregular surface features; simplifying the information acquired from photogrammetric 3D scan models; development of geometries which are independent of inaccuracies caused by scanning techniques; and analysis of existing software solutions and development of milling strategies and robotic toolpath generation.

GEOMETRIC REFERENCE SYSTEM

Working with a complex and irregular material, an important consideration throughout the project was ensuring the precise handling of the forks. The whole fabrication workflow was

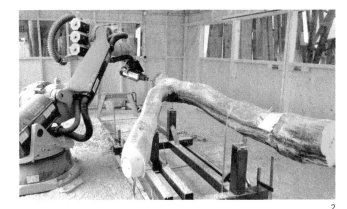

2

3

4

2, 3, and 4 Robotic Milling.

organised around a consistent referencing system to ensure a precise translation of data between digital and physical environments. This referencing system consisted of three points that were physically drilled in to each fork component—defining a local origin point, orientation axis, and plane analogous to a local construction plane in 3D-modelling software. These holes were picked up in the 3D-scanning process so that they could be incorporated in the digital modelling processes and ultimately transferred back to the physical realm by being used as the supporting points when the fork component was mounted on a 3-point jig which is pre-calibrated by the robot inside the robot cell.

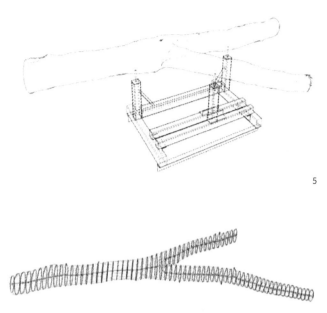

5

7

6

5   Diagram of a fork with 3 points/pins, trolley, rail.

6   Extraction of centre-line.

7   Fabricated geometries: Truncated cone shaped tenon and mortise.

SIMPLIFIED INFORMATION

The photogrammetric method, one of most cost-effective solutions to extract 3D models from 2D imagery, was used for the scanning of the forks. The main constraint of working with this method, which used Autodesk's 123D Catch software, was the inaccuracy in the acquired measurements (Erickson, Bauer, and Hayes 2013). From the data collected by scanning 24 forks, an average error from +/-3 mm to +/-15 mm was found. Another drawback in using the photogrammetric method was that the forks scanned could be extracted only in the form of polygonal meshes, which increased the processing and computational time, especially when each fork geometry is represented by a mesh with an average of 60,000 planar faces. Hence the information was required to be reduced to minimum without compromising on the obtained accuracy.

Traditional timber-framing methods include the projection of straight centrelines and axes onto irregular pieces of wood. Measurements are then taken outwards from this arbitrarily introduced centre geometry to ensure that variations in the tree's form have no bearing on the overall organization. For this project, rather than straight centerlines, centre curves are defined using a polygon-based method. Transverse sections are cut through each fork at regular intervals to obtain the outer profile of their geometry and then local best-fit diameters and centroids are calculated for each profiles' section, which are then interpolated to generate the medial curves. Points and vectors could then be extracted from the centreline throughout the entire geometry of the fork, which was essential for the development of geometries to be fabricated.

DEVELOPMENT OF FABRICATION GEOMETRIES

Due to the inaccuracies of the 3D-scanned fork mesh, all geometries to be fabricated were developed with respect to the three-point reference system and centrelines. Therefore, on a global scale, the geometric connections of one fork will match with the geometric connections of another, though the fabricated geometries of a fork might not have been at a precise location with reference to its own scanned mesh volumes. Also, two more important precautions were taken while developing the geometries. First, an average thickness of 15 mm was added to the external surfaces of all geometries to avoid damaging the milling cutter when the scanned geometry was smaller than the actual physical size of the fork. While this process saved the cutter from damage, it caused unnecessary air milling where the physical material is less than the size of the digital geometry. Secondly, for situations when the scanned geometry was larger than the actual size of the fork, all finishing or inner surface geometries were deeper offset by 15 mm, which resulted in subtracting more material than required.

USER-INTERFACE

Recent developments in architectural programming interfaces for computational design and fabrication, especially in Rhinoceros3D and Grasshopper software, have allowed the user to graphically code and observe real-time results as the code develops. No ready-made plugin was available off the shelf for providing milling strategies for a given geometry. This parametric interface was an undeniable necessity where repetitive fabrication methods were required to be applied on 20 forks which were considerably

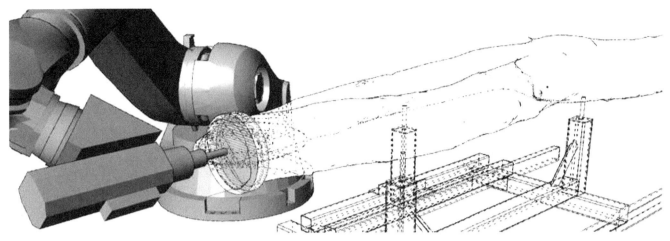

8   Robotic Simulation using in-house algorithms, Kuka-PRC in Grasshopper3d and Rhinoceros3d platform.

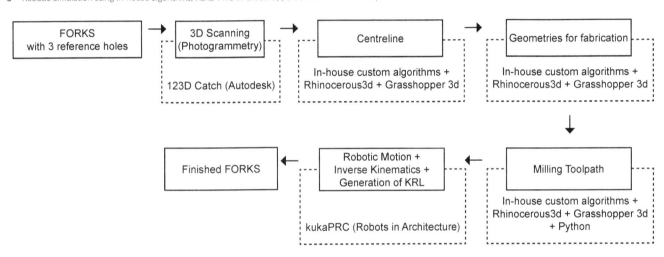

9   Work-flow of the fabrication process.

different in shape, size, and form. The kukaPRC plugin provided an inverse kinematic (IK) solution within the Rhinoceros-Grasshopper interface, which allowed checking and prediction of the problems encountered during the robotic simulation. The challenge using this IK solver was that the predicted robotic simulation errors could not be automatically rectified. Industrial software packages—like Mastercam/Robotmater—were taken into consideration, and which offered highly competent milling strategies and had the capability to tackle the problems stated, but lacked flexibility and were limited in the transferring, handling, development, and manipulation of irregular geometry. Hence, the main aim was to develop a single platform for all specified tasks.

GENERATE MILLING STRATEGIES

The most efficient approach for generating a toolpath that defines the precise location of the tool is through Cartesian Coordinate Programming using an offline method (Braumann and Cokcan 2011). Having a model of each scanned fork, with the known location of its geometries to be fabricated, makes this method more suitable than an online programming method. The kukaPRC plugin developed by Robots in Architecture, Vienna offers three command types for Cartesian Coordinate Programming; LIN (Linear Movement), PTP (Point to Point Movement), CIR (Circular Movement). The basic input required for these commands is a plane comprised of XYZABC information where XYZ defines the location and ABC defines the orientation of the end-effector. This information can be easily extracted from planes in Grasshopper where all geometries and data need to be referenced to the robot base. This laid the foundation for the development of in-house toolpath generation for milling strategies: a 3-axis roughing toolpath to remove maximum material at high speed; multi/5-axis finishing toolpath to provide an accurate finish for single or double curvature surface; swarf-milling for fabricating sides of a geometry; and a drilling toolpath for creating holes.

10 Wood Chip Barn's truss.

Basic inputs that include a closed polyhedron—with a minimum of three sides, including its base surface—are required, while parameters like tool length, retraction height, direction of toolpath, tool direction, tool offset, and stepover can be controlled easily. There were three major advantages of this process: 1) The milling toolpath could be quickly modified as the entire algorithm was parametrically defined. 2) Being an in-house developed tool, it gave complete control to the fabricator. Since timber is not a uniform material, quick edits of the toolpath could be performed. For example, milling along the grain direction was found to give a better finish and could by altering the direction of toolpath and stepover. 3) Compatibility issues were prevented as all tasks could be performed in this single interface.

AN ALTERNATIVE APPROACH

Although the process gave rise to a single interface for most tasks, an automated IK solver was still a necessity. With the drawbacks explained previously, Robotmaster still provides a very powerful optimization interface that not only detects the robotic simulation errors, but also provides an intelligent solution to the problem. Hence, we at Hooke Park wanted to harvest this potential by combining the advanced techniques of Robotmaster for optimising robotic simulations and the power of custom-coding in Grasshopper. A process was developed

in multiple stages: First, on the Grasshopper platform, the custom toolpaths that also included handling and manipulation of data were executed. Then, a post processor was written on the same platform to generate XYZIJK of the toolpath in Automatically Programmed Tool (APT) code format. The format was represented in XYZIJK, where XYZ defines the Cartesian coordinates of Tool Centre Point (TCP) and IJK defines the vector for tool orientation. Thereafter, the code is imported into the Robotmaster using the Robotmaster Import Utility Tool. Although this provides a two-stage process using two different interfaces, this offers a robust and automated output.

ANALYSIS & CONCLUSION

Robotic fabrication complete, the elements of the truss were pre-assembled in Hooke Park's Big Shed before being erected on site. While a number of small inaccuracies were noted during the assembly process, given the large irregular nature of these components, none were significant enough to have an impact on the global truss geometry. Errors of between 2 mm to 4 mm were primarily attributed to the manual positioning of the trolley along its rail and robotic tool errors. On the later forks to be machined, these errors were significantly reduced and found to be a maximum of 2 mm. With its scaffolding removed, the truss was left self-supporting—its rigid, naturally formed forked

11 Wood Chip Barn Building.

components working together to allow a non-triangulated truss to stand stably. Spanning 25 m from front to back and 10 m side to side, the arching truss rises to 8.5 m at its zenith. The project establishes a workflow for exploiting low-value branched sections of tree into complex and valuable building components without a significant increase in production time when compared to traditional methods. As, we believe, the first full-scale materialization of the concepts described, the Wood Chip Barn project sets out an innovative workflow within which each individual process might be further developed and refined.

REFERENCES

Braumann, Johannes and Sigrid Brell-Cokcan. 2011. "Parametric Robot Control: Integrated CAD/CAM for Architectural Design." In *Proceedings of the 31st Annual Conference of the Association for Computer Aided Design in Architecture (ACADIA)*. Banff, Alberta: ACADIA. 241–252.

Buckley, Gareth, Duncan Slater, and Roland Ennos. 2015. "Angle of Inclination Affects the Morphology and Strength of Bifurcations in Hazel (*Corylus avellana* L.)." *Arboricultural Journal* 37 (2): 99–112.

Carpo, Mario. 2011. *The Alphabet and the Algorithm*. Cambridge: MIT Press.

Desch, H. E. and J. M. Dinwoodie. 1996. *Timber: Structure, Properties, Conversion and Use.*, 7th ed. London: Macmillan.

Erickson, Mark S., Jeremy J. Bauer, and Wilson C. Hayes. 2013. "The Accuracy of Photo-Based Three-Dimensional Scanning for Collision Reconstruction Using 123D Catch." SAE Technical Paper 2013-01-0784. doi:10.4271/2013-01-0784.

Monier, Vincent, Jean-Claude Bignon, and Gilles Duchanois. 2013. "Use of Irregular Wood Components to Design Non-Standard Structures." *Advanced Materials Research* Vol. 671–674: 2337–2343. DOI: 10.4028/www.scientific.net/AMR.671–674.2337

Slater, Duncan and Roland Ennos. 2015. "Interlocking Wood Grain Patterns Provide Improved Wood Strength Properties in Forks of Hazel (*Corylus avellana* L.)." *Arboricultural Journal* 37 (1): 21–32.

Stanton, Christian. 2010. "Digitally Mediated Use of Localized Material in Architecture." In *Proceedings of the XIV Congress of the Iberoamerican Society of Digital Graphics*. Bogotá, Colombia: SIGraDI. 228–231.

IMAGE CREDITS

Figure 1: Mollica, 2015

Figures 2–6, 8–9: Devadass, 2015

Figure 7: Vegesana, 2015

Figure 10: Bennett, 2015

Figure 11 &12: Bennett, 2016

Pradeep Devadass is an Architect & Robotic Fabrication Specialist, currently working at The Architectural Association's Hooke Park Campus, collaborating with Bristol Robotics Laboratory to develop innovative robotic fabrication methods for manufacturing complex geometries from organic timber elements. He holds a Master of Architecture degree in Emergent Technologies & Design. He is the co-founder of rat[LAB]-Research in Architecture & Technology, an independent research & development organisation based in India. He has worked with various prominent organisations in UK, USA, China, and India. His research work also includes adaptive technologies, advanced fabrication techniques, and algorithmic design methods which have been published and exhibited internationally.

Farid Dailami is Associate Professor for Knowledge Exchange in Manufacturing at Bristol Robotics Laboratory, UWE. Over the past twenty-five years, Dailami has been engaged on a number of projects related to Robotics, Automation, Manufacturing, and Mechatronics. He is also the currently European Co-Ordinator of three Robotics Innovation Facilities (RIFs) located in the UK, France, and Italy, part of the ECHORD++ programme. He has been engaged in developing robotics solutions for automated loading of turbine blades in machining fixtures and a number of other projects concerned with application of automation and robotics in manufacturing.

Zachary Mollica is a Canadian architect and maker whose work explores the integration of innovative digital methods alongside traditional craft knowledge. Zac completed his undergraduate studies at Dalhousie School of Architecture, and has since worked for a number of architecture and design practices in Amsterdam, Lunenburg, Toronto and Vancouver. Completing the Architectural Association's Design + Make programme in Hooke Park over the past two years, Zac led the development of the Tree Fork Truss within the Wood Chip Barn student project.

Martin Self is Director of Hooke Park and Programme Co-Director of the MArch Design + Make programme based at Hooke Park, the Architectural Association's woodland campus. Holding degrees in aerospace engineering and architecture theory, he worked as a consultant engineer at Ove Arup & Partners between 1996 and 2007, where he was a founding member of its Advanced Geometry Group. Projects at AGU included collaborations with architects including Alvaro Siza, OMA, UNStudio, and Shigeru Ban, and artists Anish Kapoor and Chris Ofili. He has taught students in realising design-build projects at the Architectural Association since 2005.

12 Wood Chip Barn Building.

Use of a Low-Cost Humanoid for Tiling as a Study in On-Site Fabrication

Techniques and Methods

Mathew Schwartz
Advanced Institutes of
Convergence Technology,
Seoul National University

1

ABSTRACT

Since the time architecture and construction began embracing robotics, the pre-fab movement has grown rapidly. As the possibilities for new design and fabrication emerge from creativity and need, the application and use of new robotic technologies becomes vital. This movement has been largely focused on the deployment of industrial-type robots used in the (automobile) manufacturing industry for decades, as well as trying to apply these technologies into off-site building construction. Beyond the pre-fab (off-site) conditions, on-site fabrication offers a valuable next step to implement new construction methods and reduce human work-related injuries. The main challenge in introducing on-site robotic fabrication/construction is the difficulty in calibrating robot navigation (localization) in an unstructured and constantly changing environment. Additionally, advances in robotic technology, similar to the revolution of at-home 3D printing, shift the ownership of modes of production from large industrial entities to individuals, allowing for greater levels of design and construction customization. This paper demonstrates a low-cost humanoid robot as highly customizable technology for floor tiling. A novel end-effector design to pick up tiles was developed, along with a localization system that can be applied to a wide variety of robots.

1 The NAO robot with motion capture markers attached while inside the capture volume of the motion capture studio.

INTRODUCTION

Construction Injury

While robotics plays an important role in increasing efficiency and quality in manufacturing, it is also vital in reducing workers' exposure to dangerous and hazardous manufacturing and construction processes. Although the displacement of jobs continues to be a highly controversial topic, the reduction of construction injuries is of great benefit, as manufacturing and construction industries represent a large share of both fatal and non-fatal work accidents (OECD 2007). While many factors, such as regulations, have reduced workplace deaths, the early 1990s had seen a risk of death in construction three times higher than all other workers combined (Jeong 1998). Based on current conversations within manufacturing about compliant robotics being necessary for human-robot coexistence, workplace injuries in construction have been in large part accidents caused by slipping or falling objects, rather than direct contact with machinery (Jeong 1998). In 2002, falling totaled 56% of injuries, while being hit by falling objects was 17%, suggesting that such injuries continue to be a common on-site issue (Haslam et al. 2005). Furthermore, injuries are not limited to an immediate loss of life or health, but include long-term effects from sicknesses and diseases from exposure to lead, asbestos, or even radiation. As seen from the March 11, 2011 earthquake disaster in Japan, human exposure during reconstruction and debris removal is extremely dangerous and results in a wide range of health conditions. For these and other reasons, it would be highly beneficial for society to develop strategies for on-site robotic construction, even without considering economic and design benefits.

Teleoperation

In this regard, the robot can be controlled with either one-to-one mapping or from a remote site with some degree of autonomy. A popular application of teleoperation is in the medical field of surgery. The surgeon is able to control the robot through an interface, such as a haptic feedback pen, allowing for more accurate and fine-tuned control during the procedure (Marohn and Hanly 2004). While this is close to a one-to-one mapping, other aspects of medical robotics, such as wrist injuries during ultrasounds, may benefit from a more intelligent teleoperated robotic assistant (Conti, Park, and Khatib 2014). In the case of surgery and ultrasound imaging, the robotic system is in a fixed location. In unstructured environments, such as buildings, a different type of robot would be required.

Immediately after the Japan disaster, teleoperated systems were originally deployed using unmanned construction technologies (OSUMI 2014). These remote-control technologies may include greater autonomy, to the degree that a single human operator could concurrently control multiple machines (Bock, Linner, and Ikeda 2012).

During 2014 and 2015, researchers had prepared for the DARPA Robotics Challenge (DRC) (Johnson et al. 2015), which aimed to advance the state-of-the-art for robotics research. Based on the disaster in Fukushima, the need for robots to navigate unstructured environments that were designed for people became apparent. Often, the general public views humanoid robots as futuristic servants dealing with house chores and taking coats from guests. Even so, current household robots for chores such as vacuum cleaning have been reduced to simple cylinders on wheels.

However, as the built environment is designed for people, the simplest way for robots to utilize the same space is to share similar features and general physical qualities of humans. This was seen as a critical part of the DRC, since robots were required to climb stairs or even a ladder, and be able to operate (shut off) a valve. A secondary aspect to the DRC was the intermittent communication between the operator and robot. The operator was able to receive a limited amount of data from the robot, requiring the implementation of three general approaches: long periods of waiting (in which a time rule was implemented), a fully autonomous robot able to complete the tasks, or a combination of both. In all cases, the human was removed from danger, and in each case, various levels of research were needed. While this research is still ongoing, the goal is to develop diversified robots to navigate an unspecified space with various obstacles, rather than develop special machinery for each task in which it may rarely need to be teleoperated (Hasunuma et al. 2002).

On Site

One reason pre-fab has more successfully implemented robotics is the presence of structured environments. When robots move on site to an unstructured environment, similar tasks become more difficult (Feng et al. 2015). While the use of a robot on site was acknowledged in the research of Ariza and Gazit (2015), the robot relied on a structured environment through predefined rotations.

The discussion on using robots for on-site construction is not new. In 1985, while recognizing that commercial robots for construction in any sense were non-existent, the feasibility of an on-site robot was doubted due to the need for "programming and installation for each particular case" (Warszawski and Sangrey 1985). Understandably, the idea of automation in terms of robotic autonomy was not yet sufficiently developed to the point that it would seem valuable.

Additionally, manufacturing of large complex systems can be viewed in a similar way as on-site construction, as in the case of airplanes. Recognizing the need for versatile lightweight robots to navigate the structure, AirBus Group has begun research into implementing humanoids for their airplane manufacturing (Stasse et al. 2014). In this type of situation, a wheeled robot may not be the most practical choice, but rather one with some type of legs. Construction sites in general are dangerous areas, regulated by governmental bodies and laws (Weil 1992). The utilization of robots on site, albeit through autonomy or teleoperation, can be lifesaving. Further, the use of humanoid robots allows for a versatile robot to use a variety of tools to accomplish a variety of tasks within a space developed for humans. This use of humanoids has implications for where the utilization of robotic manufacturing begins and ends, as these robots are not limited to manufacturing, and may be permanent or semi-permanent agents in the space.

Another clear distinction with commonly used robotic techniques in architecture and the proposed on-site methods is within the generation of toolpaths. Unlike the precisely pre-planned motions used in CNC or robotic manufacturing, an on-site robot is given a task, not a specific motion (such as joint angles over time). This task is processed by the robot, and what would be considered a toolpath is generated in real-time. Real-time generation is needed for the robot to navigate unstructured environments, where modifications to the motion control are based on this new or changing environment, in order to accomplish the given task. As an example, the arm motion of picking or placing a tile is dependent on the robot's orientation to that tile. The ability to calculate the robot's arm motion based on any orientation, rather than forcing the robot's overall orientation into an exact match each time, allows the robot to adapt within the space. This workflow is further detailed in the methodology section.

At Home
Just as the cost of 3D printed manufacturing dropped dramatically as it moved from industry to retail market, so has the cost of robotics. Robots such as the humanoid NAO (Gouaillier et al. 2009) have begun entering a somewhat affordable price point while maintaining high levels of functionality, as demonstrated in the RoboCup soccer league (RoboCup Technical Committee 2015). As with the use of humanoids for disaster situations and specialized manufacturing, the benefit of a humanoid robot at home, beyond the social aspect, is the ability for navigation and manipulation in a built-for-people environment. Beyond the difficulties of a wheeled robot traversing ground with raised sections, the bipedal form reduces the footprint required of the robot, allowing for a smaller turning radius and ability to fit into more difficult spaces. Furthermore, the current state of low-cost humanoids, and robots in general, is comparable to the availability of 3D printing in the late 1990s.

Tiling Robots
There has been a sporadic history in the use of robots for tiling or mosaic operations. When the panel is already available, albeit a single ceramic tile or a composed mosaic, researchers have explored the process of laying the tile on site. In the work of Ahamed Khan et al. (2011), the floor tiling system is semi-autonomous. In this case, the robotic system is a method of reducing strain on the worker as well as speeding up the process. In contrast, other researchers theorize an autonomous machine to lay tiles by comparing specific aspects of a robot's capability with small-scale experiments with studies of human labor cost and fatigue (Navon 2000; Apostolopoulos, Schempf, and West 1996). These robots are large machines intended for large-scale installations and have not demonstrated the capability for actual development and implementation.

Current State
This research demonstrates the current efficacy and capacity of a low-cost humanoid meant for consumer experimentation. The basis for replacing humans on the construction site is due to the dangers of the construction industry, and the benefits of robots being able to perform construction tasks directly on site affords value beyond safety. In keeping with the target point of the robot used, a novel end effector for the tile pickup was designed for temporary use and implementation with 3D printing and low cost manufacturing. Although the system used here is based on motion capture, the methods in which the robot's kinematic frame is found within the space can be transferred to other localization systems as well as other robots.

While the robots in the DRC used a variety of onboard vision sensors, this work relies solely on an external localization system and kinematic calculations. For single-use robots, an external localization system may be ideal, and a network of construction robots or a permanent live-in robot would greatly benefit from a data center to coordinate and control robot positions and tasks. As an initial step, this research focuses on the ability to rely purely on kinematics based on an external localization system, making this a visually unassisted on-site robot.

LOCALIZATION METHODOLOGY
System Overview
This research uses the V4 NAO robot from Aldebaran, controlled using the C++ API and the inverse kinematics solution as detailed in the work of Kofinas, Orfanoudakis, and Lagoudakis (2013). The method for controlling the humanoid walk is similar to the methods afforded through the API with a few variables modified to compensate for the tile weight.

The motion capture system is from Vicon, with (12) T-160

Markers on Rigid Body

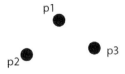

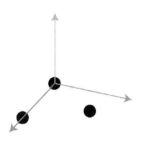

2 Diagram representing three markers or tracking points on the robot in any configuration creating a frame.

cameras. Real-time streaming is accomplished with the Vicon RealTime SDK through Nexus (Vicon 2014). The motion-capture system is capable of sub millimeter accuracy, offering a viable testing platform for localization.

Localization

Two methods are used for localization of the robot. While in perfect conditions they would produce the same results, slight errors in the encoders and manufacturing of the robot result in a difference of a few degrees between the calculated frames. Both methods provide advantages and disadvantages. In the first method, the head joint is calibrated within the space. This can be done with the robot placed directly in the space without regard for orientation as the markers are placed on that joint. In the second method, a predefined orientation within the space is designated as the initial robot end effector configuration.

With an external localization system, it is necessary to calibrate the robot position of a known joint in order to correctly locate the end effector in the space. In the case of a humanoid, the head joint is suitable as it can have markers placed on it without occlusion during motion. If an industrial arm is used, the base frame would also be a viable candidate.

In order to begin calibration, three markers must be placed on the robot. In a different localization system, these markers would be replaced with beacons, active markers, or tracking markers for camera-based systems. From these three points (P1, P2, P3), an initial frame can be created, as seen in Equation 1 and graphically in Figure 2.

$$\hat{y} = \|P_1 - P_3\|$$
$$\hat{x} = \|P_1 - P_2\|$$
$$\hat{z} = \hat{x} \times \hat{y}$$
$$\hat{x} = \hat{y} \times \hat{z}$$

$$^w_h T = \begin{bmatrix} \hat{x}_x & \hat{y}_x & \hat{z}_x & P_{1.x} \\ \hat{x}_y & \hat{y}_y & \hat{z}_y & P_{1.y} \\ \hat{x}_z & \hat{y}_z & \hat{z}_z & P_{1.z} \\ 0 & 0 & 0 & 1 \end{bmatrix}$$

(1)

Once this initial frame is created, the frame of the robot in world space must be created in order to save the offset between the two frames, and continuously use the initial frame as a reference. To do this, the head joints are rotated one after the other. During each rotation on a single axis, any of the three original markers are stored at three points in time. With these three points, the axis of the motor can be found by circumscribing a circle. The center point of the circle, which will be the center of the motor, can be found using Equation 2.

$$\alpha = \frac{\|P_2 - P_3\|^2 (P_1 - P_2) \cdot (P_1 - P_3)}{2\|(P_1 - P_2) \times (P_2 - P_3)\|^2}$$
$$\beta = \frac{\|P_1 - P_3\|^2 (P_2 - P_1) \cdot (P_2 - P_3)}{2\|(P_1 - P_2) \times (P_2 - P_3)\|^2}$$
$$\gamma = \frac{\|P_1 - P_2\|^2 (P_3 - P_1) \cdot (P_3 - P_2)}{2\|(P_1 - P_2) \times (P_2 - P_3)\|^2}$$
$$P_c = \alpha P_1 + \beta P_2 + \gamma P_3$$

(2)

The motor axis can be found using the normal of the circumscribed circle using Equation 3.

$$\hat{n} = \frac{\vec{P_1} - \vec{P_2}}{\|\vec{P_1} - \vec{P_2}\|} \times \frac{\vec{P_2} - \vec{P_3}}{\|\vec{P_2} - \vec{P_3}\|}$$

(3)

At this point, one of the head motor axes is known. The secondary axis can be found the same way, represented in Figure 3.

With both motor axes found, two of the three required vectors exist. However, due to manufacturing tolerances and the slight but existent noise in the motion capture system, these axes are not perfectly perpendicular. To estimate the origin of the neck frame, the closest point to both vectors are found using Equation 4.

Rotate Motor	Circumscribed Circle	Center	Normal

 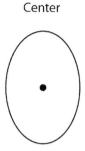 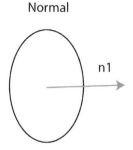

3 Using one of the marker references in three locations after motor rotation, a circle center and normal can be extracted, corresponding to the motor position and axis.

$$A = (\hat{n}_2 + P_{2c}) - (\hat{n}_1 + P_{1c})$$

$$B = P_{2c} - (\hat{n}_2 + P_{2c})$$

$$C = (\hat{n}_1 + P_{1c}) - P_1c$$

$$ma = \frac{(A \cdot C)(C \cdot B) - (A \cdot B)(C \cdot C)}{(B \cdot B)(C \cdot C) - (C \cdot B)(C \cdot B)}$$

$$mb = \frac{ma(C \cdot B) + (A \cdot C)}{C \cdot C}$$

$$P_a = P_1 + (ma * B)$$

$$P_b = P_3 + (mb * C)$$

$$P_{mid} = \frac{P_a + P_b}{2} \qquad (4)$$

To calculate the final orientation, the vectors found previously are used to generate the third vector, and the cross product is used again to maintain the primary axis as perpendicular (Equation 5).

$$\hat{y} = \hat{n}_2 \times \hat{n}_1$$

$$\hat{z} = \hat{n}_1 \times \hat{y}$$

$$\hat{x} = \hat{y} \times \hat{z} \qquad (5)$$

To create the frame, the midpoint found in Equation 4 is used in the translation (last) column of a 4x4 matrix, and the orientation vectors found in Equation 5 are used for the rotation. Equation 6 shows the final homogeneous transformation matrix from the world(w) to the calculated(c) frame.

$$_c^w T = \begin{bmatrix} \hat{x}_x & \hat{y}_x & \hat{z}_x & P_{mid.x} \\ \hat{x}_y & \hat{y}_y & \hat{z}_y & P_{mid.y} \\ \hat{x}_z & \hat{y}_z & \hat{z}_z & P_{mid.z} \\ 0 & 0 & 0 & 1 \end{bmatrix} \qquad (6)$$

This calculated frame exists in world space but must be relative to the original marker frame. For this, a transformation matrix from the head markers to the calculated frame is stored by Equation 7.

$$_c^h T = {_h^w}T^{-1} {_c^w}T \qquad (7)$$

When the robot moves throughout the space, the marker positions are queried, the marker frame is created, and the stored transformation matrix is applied, giving the neck frame in world space. As the Denavit-Hartenberg (DH) parameters from this stored frame are known, the end effector location in world space can be calculated. In the next section, a faster method for calculating the robot frame is described if the end effector is able to be placed in a known location within the localization system.

While the use of the base frame as a calibration joint allows for the operator to place the robot within the space without regard for orientation or position, it does take more time and calculations than beginning at a known configuration. This second calibration method requires the robot end effector to be placed in a known configuration and calibrates using the DH parameters and encoders.

With the end effector in a known location, the frame of the end effector {p} is known in world space {w}, and the transformation to the base frame {h} can be calculated (Equation 8).

$$_0^h T = {_p^w}T {_2^3}T {_1^2}T {_0^1}T$$

Where,

$$_p^w T = {_3^w}T \qquad (8)$$

As with the head rotation method, this stored transformation is applied during the tiling operation.

TILING METHODOLOGY

End Effector

While many methods for vacuum systems exist, the mobile nature of a tiling robot requires a portable and light-weight solution. Therefore, an externally powered vacuum system with tubing was not appropriate, and a custom end effector for the robot was developed. As seen in Figure 4, the components are

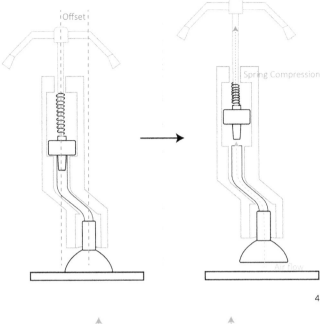

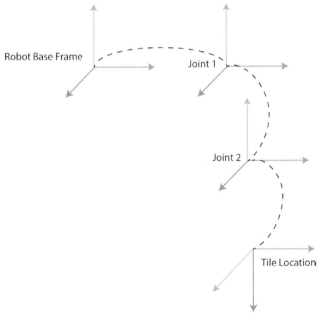

4   Diagram of passive suction end effector for robot.

5   Frame configuration for pick/drop position of robot arm. The tile location is coincident with the target position of the end effector. The base frame is the calibrated frame discussed in the localization section, Joint 1 is the shoulder, and Joint 2 is the elbow frame.

a 3D printed housing (which can be seen in the Results section) with puller, silicon tube, spring, stopper, and suction cup.

In the inactive state, the vacuum can always be created, as a normal suction cup works as the spring pushes against the stopper and into the tubing. When the robot fingers

are activated, the stopper is pulled away from the tubing, compressing the spring, and allowing air-flow, releasing the vacuum. The second aspect to the design is the use of the flexible tubing. As can be seen in Figure 4, the pulling lever is offset from the location of the suction cup. The fingers of the robot are not in the center; as such, using a straight line for the end effector would create another offset in the kinematic definition. To simplify calculations, the flexible design is used to align the suction cup with the center of the wrist and elbow joints. With this, the DH parameters can remain the same, with the final frame distance being extended by the length of the end effector. This system can be applied to a variety of robots in order to simplify the kinematics. Additionally, the passive vacuum allows for a reliable and low energy solution, a necessity in mobile robotics where battery life is vital and closely related to the voltage demand of the motors.

The spring used in the system must be light enough for the robot to compress, while being strong enough to overcome friction of the silicon tube in order to complete the seal. In this configuration, a spring was used which takes approximately 9 Newtons to compress 6 mm, the distance needed to release the vacuum. With a clean press of the end effector, a solid vacuum is formed, lasting at least one hour during testing. However, an off angle press of the tile drastically reduces the vacuum formed and can result in a drop within a few minutes. Hence, it is vital for the end effector to be sufficiently pressed against the tile in this passive system.

Tile Planning

Once the robot is calibrated and reaches the tile location, the end effector must press against the tile to pick it up. The tile frame is in world space and is the desired location of the end effector. For simplicity, the tile frame is then used as the end effector frame for the inverse kinematics solution. As shown in the next section, State Machine, the frame is calculated before the robot picks up or drops the tile. If no solution is found, the robot re-aligns to the tile. Once crouched in place for picking/dropping the tile, the frame from the tile is copied vertically by a threshold under 30 mm. Multiple positions are queried by rotating the frame in order to find an inverse kinematics solution such that the path planning is approached vertically.

State Machine

The state machine diagram shows the logic behind the robot tiling. As seen in Figure 6, calibration occurs once at the beginning. The methods for picking or dropping a tile are the same. Throughout the system, the robot queries the current location of the markers in order to calculate the robot frame in the world, using the offset described in the Methodology of this paper.

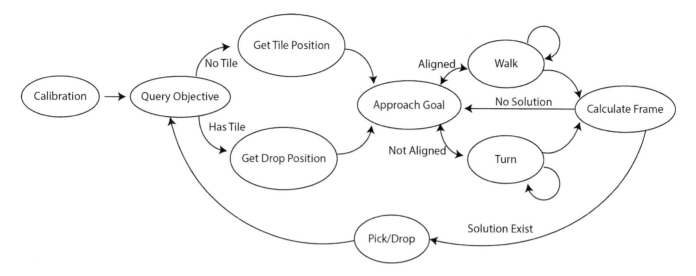

6 State machine diagram of the robot during the entire experiment. This state machine is independent of robot type and can be implemented with an industrial arm as well.

A main difference between much of the work done in pre-fab architectural/design research and this paper is the implementation of a state machine as the controlling factor. In many cases, a CAD file or G-code is sent to the robot control system to be interpreted as coordinates to move to. In this system, the robot receives only a file listing the colored tiles to be used, and autonomously navigates to the goal positions. All steps are handled within the state machine, including communication delays and errors.

RESULTS

Although the main goal of implementing robots on the construction site is for safety and reduction of human injuries, both accuracy and length of time for achieving the goals must be considered. At the same time, implementing this system in a humanoid robot for at-home assistance may present different challenges, such as a case where a person is unaware how to tile, or there is no time constraints on their project.

As this project is not focused on the development of a commercial robot, but rather research and experimentation for future methods, accuracy and time-scale are not solved, but instead are calculated to demonstrate the efficacy of the system, and what would be required to solve them.

Experiment

The experimental results can be seen through multiple images during a tiling action. The robot was set in the initial configuration on a wooden platform. The robot was commanded to autonomously navigate to the tile pickup location, as seen in Figure 7. Black tape was used to repeatedly place the tile in the same position within the motion capture space during

experiments. After pickup of the first tile (Red), the robot was to move 1.5 meters to the back left and drop the tile.

After picking the tile, the robot reconfigures the arm position based on the state, ie. turning or walking. As the robot approaches the goal position, the tile frame is calculated as the end effector frame in the inverse kinematics equation. If a solution is found, the robot will crouch down in order to pick/drop the tile as seen in Figure 8.

When dropping the final tile, the robot accurately places it. As seen in Figure 8, the fingers grasp the lever, which releases the vacuum on the tile. However, retraction of the arm due to motor limitations is not directly vertical, shifting the orientation of the tile by a few degrees.

Finally, the robot returns to the tile pickup area to retrieve the next tile (Green), as seen in Figure 9. Due to battery limitations and accuracy of the end effector, the demonstration experiment concluded when the robot returned to the tile pickup location.

Accuracy

While the localization system without the use of onboard vision is shown to be effective, the accuracy of picking a tile prevents a full mosaic from being created. The encoders on the robot are within a reasonable tolerance, while the motors are either lacking torque, or lack the appropriate low level drivers for incremental motions. When commanding a joint a specific rotation value, the joint does not always achieve the desired location, and while the encoder value acknowledges this, compensating for this error is extremely difficult. A technique for overcompensating in the opposite direction of the motors failed attempt was

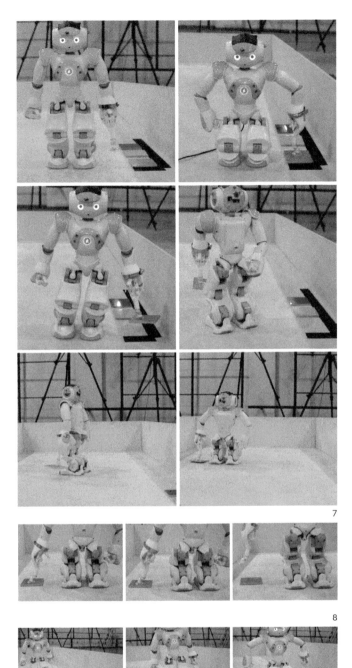

implemented, but did not reduce the error to a useful degree. For this reason, multiple attempts to pick up and drop a tile are done, increasing the time needed for a tile cycle. While the robot would be adequate as a placement system, a higher quality motor is required for completely accurate tiling.

However, applications of a limited accuracy robot may still be realistic. As recycled or damaged tiles can be used in mosaic patterns, a combination of computer vision and the robotic tiling system can produce results similar to that of a non-specific mosaic tile placement. Furthermore, input of the damaged tile's shape and color can be processed by the robot to generate numerous designs that fit within the accuracy limits of the robot.

Times

The time to move from the calibration location to a tile, then a drop location and back to the next tile is broken down in Figure 10. By comparing this to the state diagram, it is clear that most of the time spent is approaching the goal position. While the robot itself can walk at a moderate rate, as seen in both Tile Approach sections, the inclusion of a tile changes the mass location. To compensate for this, the tile is held in the front of the robot during a forward walk, and low to the side for rotation and minor adjustments.

The second most time consuming aspect occurred during the Drop Attempt. As the motors lack the torque to command small movements, the final end effector position is not exact. Using the encoders, the achieved position can be calculated, and if it is outside of the valid zone, the arm retracts and attempts again, causing multiple Drop Attempts. As a conservative estimate using the failed attempts in the timeline, it takes 494 seconds to pickup, drop, and return to the pickup location. If this was sustained, it would take 13 hours to tile 100 tiles, a 1x1 meter square of 10 cm tiles—clearly not a reasonable length of time for a construction site, but still achievable at a house in order to make a mosaic floor pattern autonomously overnight. Furthermore, these times can be drastically reduced by a slight increase of torque in the motors, reducing the required Drop Attempt times. Second, further research into the mass compensation while carrying a tile can reduce the approach and adjustment times. Assuming a 50% reduction in approach time, and a single Drop Attempt, the single cycle would take 339 seconds, and 100 tiles 9.4 hours.

CONCLUSION

While this paper demonstrates the ability for a low cost humanoid robot to place tiles in the correct locations, it is neither efficient nor accurate enough to be considered a viable option at this point. However, by either increasing the joint accuracy or

7   Snapshots of the robot in the motion capture space performing the experiment. The black tape on the head secures the three markers described in the methodology. The end effector is 3D printed and secured around the hand with a metal clasp.

8   Robot crouches down and places tile in the desired location. The first frame shows the robot fingers relaxed and spread out, while the second frame shows the fingers clenched, lifting the end effector lever, and the final image showing the tile dropped.

9   The robot returns to the tile pickup location after dropping the Red tile at the goal position.

10 Breakdown of the timings of the experimental results.

including an onboard vision feedback system, this work shows that a solution is within the near future. Additionally, the method for calibrating the robot's position using an external localization system can be used in the future. The end effector design is a low cost and low power consuming option for use with any type of robot. Furthermore, integration as a human-robot collaboration may increase efficiency for aiding in drop placement while providing a safe location for the laborer.

As buildings have autonomy built into them, the use of external localization systems become common place. Either through built in LIDAR or UWB for use in wayfinding, the cost of on-site robotics decreases dramatically. When the robots need limited or no vision capabilities due to control from a data center, the cost per robot decreases while the cost per building increases, at the same time affording exponentially more possibilities during the life cycle of the building.

ACKNOWLEDGEMENTS

This research was funded in part by grant #AICT-2014-0011. The author would like to thank the many interns of the Digital Human Research Center over the last two years that have contributed their knowledge and helped in the code development.

REFERENCES

Ahamed Khan, M. K. A., K. I. Saharuddin, I. Elamvazuthi, and P. Vasant. 2011. "A Semi-Automated Floor Tiling Robotic System." In *2011 IEEE Conference on Sustainable Utilization and Development in Engineering and Technology*. Semenyih, Malaysia: STUDENT. 156–159.

Apostolopoulos, Dimitrios, Hagen Schempf, and Jay West. 1996. "Mobile Robot for Automatic Installation of Floor Tiles." In *Proceedings of 1996 IEEE International Conference on Robotics and Automation*, vol. 4. Minneapolis, MN: ICRA. 3652–3657.

Ariza, Inés, and Merav Gazit. 2015. "On-Site Robotic Assembly of Double-Curved Self-Supporting Structures." In *Proceedings of the 19th Conference of the Iberoamerican Society of Digital Graphic Design*, edited by Alice Theresinha Cybis Pereira and Regiane Trevisan Pupo. Florianópolis, Brazil: SIGRADI. 746–753.

Bock, T., T. Linner, and W. Ikeda. 2012. "Exoskeleton and Humanoid Robotic Technology in Construction and Built Environment." *The Future of Humanoid Robots: Research and Applications*, edited by Riadh Zaier. InTech Open Access Publisher. 111–146.

Conti, François, Jaeheung Park, and Oussama Khatib. 2014. "Interface Design and Control Strategies for a Robot Assisted Ultrasonic Examination System." In *Experimental Robotics: The 12th International Symposium on Experimental Robotics*, edited by Oussama Khatib, Vijay Kumar, and Gaurav Sukhatme. Berlin: Springer. 97–113.

Feng, Chen, Yong Xiao, Aaron Willette, Wes McGee, and Vineet R. Kamat. 2015. "Vision Guided Autonomous Robotic Assembly and as-Built Scanning on Unstructured Construction Sites." *Automation in Construction* 59: 128–138.

Gouaillier, David, Vincent Hugel, Pierre Blazevic, Chris Kilner, Jérome Monceaux, Pascal Lafourcade, Brice Marnier, Julien Serre, and Bruno Maisonnier. 2009. "Mechatronic Design of NAO Humanoid." In *Proceedings of 2009 IEEE International Conference on Robotics and Automation*. Kobe, Japan: ICRA. 769–774.

Haslam, Roger A., Sophie A. Hide, Alistair G. F. Gibb, Diane E. Gyi, Trevor Pavitt, Sarah Atkinson, and A. R. Duff. 2005. "Contributing Factors in Construction Accidents." *Applied Ergonomics* 36 (4): 401–415.

Hasunuma, Hitoshi, Masami Kobayashi, Hisashi Moriyama, Toshiyuki Itoko, Yoshitaka Yanagihara, Takao Ueno, Kazuhisa Ohya, and Kazuhito Yokoi. 2002. "A Tele-Operated Humanoid Robot Drives a Lift Truck." In In *Proceedings of 2002 IEEE International Conference on Robotics and Automation*, vol. 3. Washington, DC: ICRA. 2246–2252.

Jeong, Byung Yong. 1998. "Occupational Deaths and Injuries in the Construction Industry." *Applied Ergonomics* 29 (5): 355–360.

Johnson, Matthew, Brandon Shrewsbury, Sylvain Bertrand, Tingfan Wu, Daniel Duran, Marshall Floyd, Peter Abeles, Douglas Stephen, Nathan Mertins, Alex Lesman, John Carff, William Rifenburgh, Pushyami Kaveti, Wessel Straatman, Jesper Smith, Maarten Griffioen, Brooke Layton, Tomas de Boer, Twan Koolen, Peter Neuhaus, and Jerry Pratt. "Team IHMC's Lessons Learned from the DARPA Robotics Challenge Trials." *Journal of Field Robotics* 32 (2): 192–208.

Kofinas, Nikos, Emmanouil Orfanoudakis, and Michail G. Lagoudakis. 2013. "Complete Analytical Inverse Kinematics for NAO." In *Proceedings of the 13th International Conference on Autonomous Robot Systems*. Lisbon, Portugal: Robotica. 1–6.

Marohn, Col. Michael R., and Capt. Eric J. Hanly. 2004. "Twenty-First Century Surgery Using Twenty-First Century Technology: Surgical Robotics." *Current Surgery* 61 (5): 466–473.

Navon, R. 2000. "Process and Quality Control with a Video Camera, for a Floor-Tilling Robot." *Automation in Construction* 10 (1): 113–125.

OECD. 2007. "Work Accidents." In *Society at a Glance 2006: OECD Social Indicators 2006 Edition.* Organisation for Economic Co-operation and Development.

Osumi, Hisashi. 2014. "Application of Robot Technologies to the Disaster Sites." In *Report of JSME Research Committee on the Great East Japan Earthquake Disaster.* Tokyo: The Japan Society of Mechanical Engineers. 58–74.

RoboCup Technical Committee. 2015. "Robocup Standard Platform League (NAO) Rule Book." Last modified July 1, 2015. http://www.informatik.uni-bremen.de/spl/pub/Website/Downloads/Rules2015.pdf

Stasse, Olivier, F. Morsillo, M. Geisert, Nicolas Mansard, Maximilien Naveau, and Christian Vassallo. 2014. "Airbus/future of Aircraft Factory HRP-2 as Universal Worker Proof of Concept." In *2014 IEEE-RAS 14th International Conference on Humanoid Robots.* Madrid, Spain: RAS. 1014–1015.

Vicon. 2014. "Nexus 1.8.5." https://www.vicon.com/downloads/core-software/nexus/nexus-185-installer.

Warszawski, Abraham, and Dwight A. Sangrey. 1985. "Robotics in Building Construction." *Journal of Construction Engineering and Management* 111 (3): 260–280.

Weil, David. 1992. "Building Safety: The Role of Construction Unions in the Enforcement of OSHA." *Journal of Labor Research* 13 (1): 121–132.

IMAGE CREDITS
All Photography: Schwartz, 2016

Mathew Schwartz has a BFA and a MSc. in Architecture with a focus on digital technology from the University of Michigan. He is currently at the Advanced Institutes of Convergence Technology in South Korea, where he is a research scientist who focuses on human factors. His work, which bridges science and engineering with art and design, makes use of cutting-edge robotics and motion capture technology to mimic human characteristics which he then incorporates into commercial applications, architecture, and models used in scientific research.

Robotic Sewing

A Textile Approach Towards the Computational Design and
Fabrication of Lightweight Timber Shells

Tobias Schwinn
Oliver David Krieg
Achim Menges
ICD University of Stuttgart

1

ABSTRACT

Unlike any other building material, timber has seen numerous innovations in design, manufacturing, and assembly processes in recent years. Currently available technology not only allows architects to freely shape building elements but also to define their micro- or macroscopic material make-up and therefore the material itself. At the same time, timber shells have become a focus of research in wood architecture by rethinking both construction typologies and material application. Their main advantage, however, also poses a challenge to its construction: As the shell is both the load-bearing structure as well as enclosure, its segmentation and the individual segment's connections become increasingly important. Their complex and often differentiated geometries do not allow for standardized timber joints, and with decreasing material thickness, conventional connection techniques become less feasible.

The research presented in this paper investigates textile strategies for the fabrication of ultra-lightweight timber shells in architecture. Specifically, a robotic sewing method is developed in conjunction with a computational design method for the development of a new construction system that was evaluated through a large-scale prototype building.

1   Interior view of the ICD/ITKE Research Pavilion 2015-16. The finished large-scale prototype building serves as an architectural demonstrator for research into robotic sewing.

INTRODUCTION

The development of new building products such as Laminated Veneer Lumber (LVL) and Cross-Laminated Timber (CLT) not only enabled the use of planar elements but also drove timber construction towards a homogenization of wood properties for the sake of calculability. The unique and specific characteristics of wood, such as its anisotropy and elasticity—which both depend on grain direction—are leveled out through the cross-lamination of individual anisotropic layers. Whereas conventional timber construction relies on thick cross sections and metal fasteners for connections, the aim of this research is to construct very thin shells made from plywood by taking advantage of the specific material characteristics in the design and fabrication process and by using alternative connection strategies suited for very thin sheet material. Consequently, form-finding principles have to be established that incorporate materiality and fabrication possibilities.

New developments in digital fabrication, particularly in robotics, allow for the implementation of design principles such as heterogeneity, hierarchy, or anisotropy in architecture, which are characteristic principles found in nature (La Magna et al. 2013). Enabled by computational design strategies, the premise of these high-level principles is that, similar to biology, more geometric variability can lead to higher performance of the structure, given that its elements are individually adapted to specific requirements and less energy and material is required in their construction. Consequently, the research presented in this paper is to a large extent based on the abstraction and transfer of morphological principles from biology to the design and fabrication of thin shells in architecture.

The availability, analysis, and abstraction of relevant biological principles and the aspects related to structural analysis and simulation, both of which are discussed in Bechert et al. (2016), lie outside the scope of this paper. Rather, this paper focuses on the technical transfer of these principles and on its methods of implementation in the design and fabrication of thin shells; specifically, on the joint design and custom robotic fabrication approaches. Its main hypothesis is that the design and construction of very thin, geometrically differentiated timber shells requires alternative connection and fabrication methods and that, consequently, they can be built entirely without the need of traditional timber fasteners (Fig. 1).

CONTEXT AND RELATED WORK
Thin Timber Shells

Continuous thin concrete shells have been a popular typology in architecture and engineering from the 1930s to the 1980s. However, they seem to have lost their appeal in recent decades due to the comparatively high effort necessary for their in-situ construction, which involves the manufacturing of complex formwork and extensive manual labor, even with today's high degree of automation. An alternative is the construction of thin timber shells made from segments, as demonstrated in previous work in the Landesgartenschau Exhibition Hall (Krieg et al. 2015) and the ICD/ITKE Research Pavilion 2011 (Menges and Schwinn 2012). However, the dimensional limitations of plywood, as well as the subsequent need for transportation, limits the size of the segments from which the shell is to be constructed. In addition, as the previous case studies have demonstrated, the planarity of the available timber sheet material leads to the requirement of elaborate planarization strategies in order to approximate complex double-curved surfaces while respecting fabrication constraints (Schwinn et al. 2014; Schwinn et al. 2013).

Another major challenge in the development of segmented shells is the design of the joints, as the stiffness continuity of the shell is interrupted (Li and Knippers 2015). Therefore, a particular focus of any design approach for segmented shells has to be the joint design. This has led to a considerable research effort directed towards the design and fabrication of so-called integrated joints (Krieg et al. 2015; Robeller and Weinand 2015). Contrary to conventional timber construction, the complexity of the joint is integrated into the element through elaborate digital fabrication techniques. However, this poses a problem for thin shells with a highly reduced material thickness. Not only can the geometric complexity of integral joints not be embedded within an element's edge below a certain material thickness, but building codes also regulate the required distance between metal fasteners like screws and the material surface. Both aspects limit the further reduction of shell thickness. Furthermore, although integrated connectors such as finger and dovetail joints are form- and force-fit connections, they are usually the result of subtractive fabrication processes such as CNC milling. In these processes, the continuity of the cellulose fibers in the wood is disrupted, potentially weakening the material, particularly where the main forces are transferred between building elements. These aspects have led to the search for alternative joining techniques for thin timber and to the investigation of a possible transfer of textile processing techniques into timber construction.

Textile Techniques

From a microscopic point of view, wood, like most biological constructions, can be considered a natural fiber composite exhibiting similar properties to man-made composites such as fiber-reinforced polymers (FRP). In both cases, the fiber direction in the lay-up determines the anisotropic material behavior. In the case of technical textiles, this allows users to design the material properties to meet structural or other requirements in a

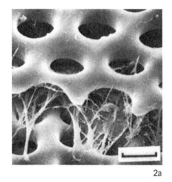

2a

2b

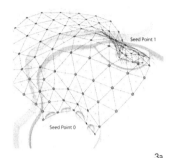
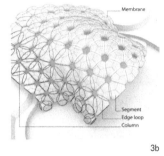

3a

3b

4

bottom-up manner. This is why FRPs play such a crucial role in automotive and aerospace industries, where a high stiffness-to-self-weight ratio is required. In the case of wood, however, the anisotropic material behavior is usually considered undesirable in the context of the building industry, leading to the development of plywood or fiberboards with mostly homogeneous characteristics.

The objective of this research is consequently to re-consider and re-interpret the material properties of wood in light of its textile nature. The field of technical textiles and in particular the field of fiber-reinforced polymers provides established methods for manipulating anisotropic sheet material with the potential of being transferred into an application in timber architecture. Sewing is identified as a particularly promising approach, where initially planar sheets are connected along their edges using thread, resulting in complex three-dimensional forms based on two-dimensional cut patterns. The goal is therefore to apply techniques such as patterning, sewing, lacing, and lamination to the fabrication of lightweight building components made from timber, and to exploit the material's elasticity during fabrication, resulting in self-supporting and stable structures.

The parallels between timber and textiles become more evident at material thicknesses much lower than those usually used in timber construction. Given the thickness of rotary sliced veneer of about 1 millimeter, the material can be manipulated like a textile. Furthermore, sewing has similar advantages in timber construction as it has in textiles, as many small connections are usually preferable to few large connectors (Herzog et al. 2003), which lead to force concentrations and possibly material failure. While textile approaches have previously been proposed by Weinand and Hudert (2010), as well as Fleischmann et al. (2012), they have mainly been limited to weaving and intersecting plywood strips.

Despite the possible benefits and advantages of sewing timber, examples seem extremely rare. The Couture armchair by Färg and Blanche (2015) is made with a stationary industrial grade sewing machine used to connect layers of plywood of approximately 4 mm each. The fabrication is restricted to two-dimensional sewing in the plane on a large table and the work piece is guided manually. However, it proves the viability to sew plywood using machines from the leather and upholstery industry and indicates the high quality of sewing that can be achieved. An industrial example of robotic sewing with applications mostly in the automotive industry is the 3D-Robot Sewing unit by KSL for the "application of decorative seams to automotive interiors" (Keilmann 2014). In general, however, sewing has mostly eluded automation, most likely due to the complications of simulating the complex behavior of textiles. It remains a process where human sensory capacity and manual dexterity cannot be easily replaced by mere robotic precision.

Yet, robotic sewing does provide the opportunity to sew three-dimensionally, unconstrained by any work plane, and with high precision. It opens up areas of application thatin turn would lie outside the range of human capacity. Using an industrial robot for

Robotic Sewing Schwinn, Krieg, Menges

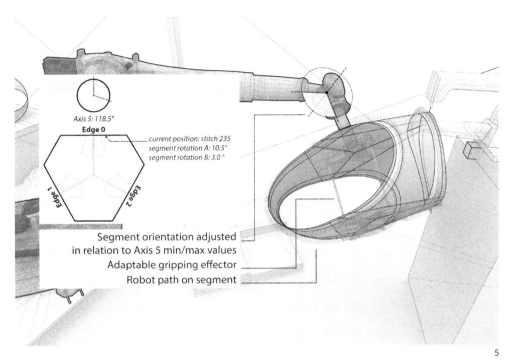

Axis 5: 118.5°
Edge 0

current position: stitch 235
segment rotation A: 10.5°
segment rotation B: 3.0 °

Edge 1

Edge 2

Segment orientation adjusted
in relation to Axis 5 min/max values
Adaptable gripping effector
Robot path on segment

2   A: Microscopic view of the collag-
     enous fibers in the connections
     between plates (Telford 1985).
     B: Fibrous connection between
     segments in the built demonstrator.

3   A: The circles represent the particle
     system that originates from two
     seed points. B: The resulting
     mesh provides the basis for the
     subsequently added geometric
     parameters that define each
     segment shape and type.

4   Front view of the built
     demonstrator with the visible
     double-layered construction.

5   The robot guides the segment
     through a stationary sewing
     machine. Its orientation can be
     adjusted according to certain
     out-of-reach or collision criteria,
     which are mainly in relation to axis
     5 and can therefore be solved by
     rotating the segment out around
     the needle's center point.

5

handling timber veneer through a stationary sewing machine was therefore investigated as a fabrication approach in this research project.

Biomimetics

In previous research, biology has been an extensive concept generator for the design of segmented shells (Grun et al. 2016). Biomimetics serves as a design strategy to analyze and transfer the mechanical behavior, constructional morphology, and functional properties of load-bearing systems from nature to technology. Different biological role models have been repeatedly used as concept generators for the design of thin shells. This includes the aforementioned timber shells, the ICD/ITKE Research Pavilion 2011 and the Landesgartenschau Exhibition Hall, but also the ICD/ITKE Research Pavilions 2012 (Reichert et al. 2014) and 2013–14 (Dörstelmann et al. 2014). In the latter two examples, the fibrous morphology of natural fiber composite shells of lobsters and flying beetles, which consist of chitin fibers in a protein matrix, was translated into extremely lightweight shell structures.

Research Objectives

Based on the observations regarding the architectural and structural potential of segmented thin shell structures, the anatomy of wood, the textile approaches in lightweight constructions using FRP, as well as the opportunities provided by a biomimetic design approach, the objectives of the research within the scope of this paper are as follows:

1. Transferring textile processing methods from the field of technical textiles to digital fabrication and construction in architecture;

2. Utilizing the specific material characteristics of wood as drivers for the development of a construction system including multi-material joints;

3. Investigating alternative approaches to planarization for segmenting shell surfaces taking the bending elasticity of wood into account;

4. Synthesizing the design principles, material characteristics, and textile methods in an integrated computational design and robotic fabrication approach; and finally

5. Testing the above hypotheses through the construction of a full-scale architectural prototype.

METHODS
Biomimetic Design

Biological examples of segmented shells as they can be found in the regular and irregular sea urchins in the class of *Echinoidea* have been investigated and analyzed with a focus on the structural morphology of the shells on different levels of hierarchy, ranging from the microscopic level to the whole organism. In the context of this research, the most relevant novel principles that have been identified and translated into design principles for segmented shells include (1) the fibrous connections between the plates in the test (Fig. 2); (2) internal supports, which connect the upper and lower shell layers of irregular sea urchins (Fig. 4); (3) shell openings, which can be found in irregular sea urchins in the form of lunules; and (4) the growth principle of plate accretion and addition was transferred for the design process of the shell.

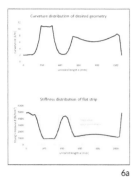
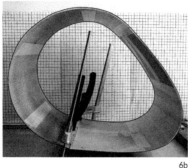

6a   6b

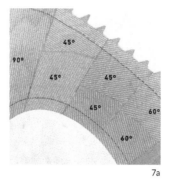
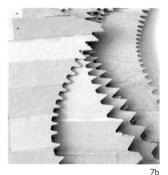

7a   7b

The fibrous collagenous material acts as an additional connecting element, which can be observed in the joints between the plates of some sea urchins. It is hypothesized that it plays an important role in maintaining the shell integrity during growth as well as for dynamic forces throughout the sea urchin's life (Chakra and Stone 2011). This principle of a multi-material connection has been translated on the level of the connection between elements as well as on the level of connection of layers of veneer during lamination using robot sewing.

Computational Design and Simulation

An integrative computational design approach is essential in order to incorporate different requirements of fabrication, material characteristics, biological principles, architectural design intentions, structural analysis, and ultimately also constraints of time and budget. There are different computational strategies to control and evaluate interdependent parameters within a form-finding process. In this research project, the aforementioned procedural biomimetic principle of plate accretion and addition was used as a basis for the development of a design tool.

In a first step, the design of the spatial configuration of the shell structure—that is, its orientation on site, its span, and its support points—is defined parametrically as a design domain in the form of a non-uniform rational b-spline (NURBS) surface. In the subsequent step, the design domain is populated by a particle system originating at the supports, which represents the segment center points and implements the growing radii of the segments over time. The result is an evenly spaced arrangement of input points with tangent circles indicating the increasing radius of separation between segments away from the seeding origins. This process terminates as soon as the design domain is filled with input points and the kinetic energy of the system is minimized (Fig. 3A). The third step constitutes the generation of a topological relation between previously generated input points on the design domain using a Delaunay triangulation of the UV parameter space. The edges in the resulting mesh form the basis for the generation of the segment loops in the shell (Fig. 3B).

An initial challenge for the design of the loops' edge curves was the approximation of elastic curves of different radii that result from varying the material stiffness of the bent plywood. However, as the required stiffness can be calculated from the desired curvature radius, discretized stiffness values could be deduced from the digital model, resulting in a database of element-specific lamination instructions (Fig. 6). The resulting tri-looped segments each consist of three developable surface patches of varying curvature.

The digital model at this point not only captures the architectural design intent, and the material specifics of bending custom-laminated plywood, but also the constraints of fabrication resulting from the specific robotic fabrication setup. These are continuously monitored by simulating the robotic fabrication process within the digital design environment, which allows for checking for out of reach positions and collisions (Fig. 5). In order to arrive at a configuration that integrates these constraints, design iterations of the parametric model allowed us to locally adjust curvature of the design domain, the spacing and location of the input points for the segment generation, and control the curvature of the segments through the shape of their edge curves. Additionally, each design iteration has been analyzed structurally. The structural analysis has been informed by material tests described in Bechert et al. (2016). This informed model then forms the basis for the generation of the fabrication data: the lamination instructions, the generation of the finger joints along the edge loops where the segments are connected to each other, the NC-code necessary for their 3-axis milling, and finally the generation of the robot code based on the sewing lines (Fig. 5).

Textile Methods: Material Design

Similar to textile fabrication methods, the grain direction and lay-up of the laminate ultimately defines the material's characteristics. In order to approximate the required curvature along one of the elastically bent strips, their bending stiffness has to be programmed through an individualized lamination process (Tamke et al. 2012). The relation between varying grain directions and

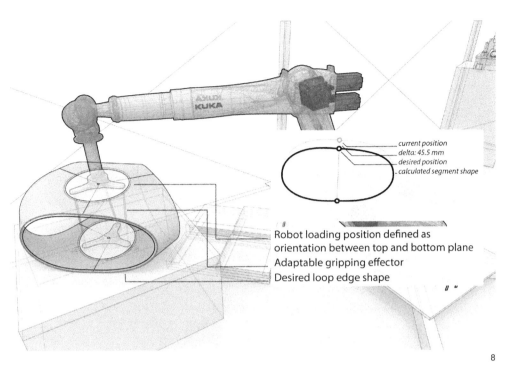

current position
delta: 45.5 mm
desired position
calculated segment shape

Robot loading position defined as
orientation between top and bottom plane
Adaptable gripping effector
Desired loop edge shape

6   A: Digital prototyping through
    data-driven simulation. B: Physical
    prototyping allowed the deduction
    of a database of discretized
    stiffness values depending on the
    material lay-up.

7   A: An elastically bent strip shows
    areas with differently arranged
    veneer. B: A stack of laminated and
    CNC cut strips shows their diversity
    in veneer lay-up.

8   The robot is employed as a
    positioning aid in a first step in
    order to fix the three strips in their
    correct position. The upper part of
    the effector indicates the position
    in which the strips have to be
    attached.

8

layers of veneer, and the resulting bending stiffness was evaluated through physical bending tests with beech plywood (Fig. 6). These experiments also established the lower bound of curvature that the laminate would tolerate before failing. Summarizing, in areas of higher curvature, the material stiffness has to be low, resulting in low material thickness and the grain direction of the laminate running mostly perpendicular to the bending direction. Conversely, in areas of low curvature, a high stiffness is required, resulting in a material thickness of about twice as high and the predominant grain direction in the laminate running parallel to the bending direction.

Each strip is divided into 100-mm-wide areas of discretized stiffness and carries information on the number of veneer layers and their grain direction, depending on the required radius. The strips are then unrolled and nested on the bounding box of available beech veneer and plywood stock material. From here on, the production information is connected to a sheet of stock material that usually includes two to four individual strips. In order to guarantee a precise and efficient lamination process, the instructions are transferred to the stock material using a projector. Since the resulting laminates have locally differentiated material thickness, lamination is achieved through vacuum pressure. The individual strips are then cut out from the laminated sheet using a 3-Axis CNC cutting process. The finger joints along the looped edges are shaped to accommodate the varying angles between segments. As these range from coplanar to perpendicular, the finger joints vary in width and length (Fig. 7).

Textile Methods: Robotic Sewing

The robotic prefabrication process is developed not only to connect three elastically bent strips into a segment, but also to assist in their assembly beforehand. In this first step, the robot is used as a positioning aid in order to assemble the three strips in their correct elastically bent shapes. As there is otherwise no information on the relative position of the upper and lower base plane of the segment where all three strips meet, the robot positions its variable effector accordingly (Fig. 8). All three strips are then glued and connected to the effector, which in turn is fixed in its predefined position in order to stabilize the segment for sewing.

An industrial grade sewing machine is used as an external tool through which the robot then guides the segment (Fig. 9). It is integrated into the robot control and activated through a command in the robot code. When the sewing machine controller receives a command, it initiates a stitch and sends back a signal once the stitch is complete. This method ensures that there is no lateral movement of the segment while the needle is penetrating the material. Instead, the segment is only moved in between stitches. Even with the communication between robot control and sewing machine, one stitch only takes about 500 milliseconds. Sewing lines are predefined in the digital model based on the size of the segment and the robot effector position in order to ensure a distributed connection, and as a consequence, evenly distributed pressure between the strips. In order to accommodate minor differences between the simulated segment model and the actually bent segment, a path correction

9   Robotic sewing process. The robot guides the segment through an industrial sewing machine. Both machines communicate through a common control system. The stitches are placed 10 millimeters apart, and are also used to attach the white membrane strips onto the segment.

routine has been incorporated into the sewing process that allows the online control and adjustment of the robot position during fabrication.

As the robotic sewing ensures that the three strips are tightly connected while the glue is curing, even after the segment has been taken off the robot effector, it is also used to attach an additional membrane element along the segment edges, which allows the connection of neighboring segments through lacing. As a second hierarchy of textile connection, this conventional and well-proven membrane technique ensures easy accessibility during assembly.

On-Site Assembly

After varnishing the segments in order to protect the wood against moisture and to prevent fungi growth, the 151 pre-fabricated segments where assembled on site in 12 days. While the finger joints could be used as a positioning aid, the lacing technique allowed the gradual tightening of the connection between two segments during the assembly process, and thereby adjusted already connected segments when necessary. The overall structure not only showed a high flexibility but also enough rigidity to be built from the ground up without large-scale support, even during critical assembly phases (Fig. 10).

RESULTS

Each of the 151 segments is between 0.5 and 1.5 meters in diameter and has a material thickness between 3 and 6 mm. The extremely low thickness of the material required the

development of the described textile connection approaches. Consequently, the entire shell has been constructed without the need for traditional timber fasteners. The structure ultimately weighs 780 kg while covering an area of 85 square meters and spanning 9.2 meters. With an average material thickness / span ratio of 1/1000, the building has a structural weight of 7.85 kg/m² shell.

The large-scale building prototype is not only a demonstrator for the developed construction system, but is also designed to incorporate surrounding site conditions. The pavilion opens up towards the campus area and provides a view towards the university buildings and the neighboring park. It provides shade and forms a point of attraction for an otherwise rarely used part of the campus (Fig. 11).

Similar to the biological role model, the structure not only functions as a pure shell, but allows for structural situations with higher bending moments through the introduction of columns (internal supports) as an integral part of the construction system. All forces are ultimately translated locally into membrane forces in the plywood segments, but globally, the system allows a transition from a shell system to a column and slab system. This offers entirely new architectural possibilities by expanding the catalogue of structural types and possibly constitutes one of the main contributions to the field of segmented shells. The synthesis of the many different and sometimes-conflicting design requirements allows for the exploration of a specific domain of the solution space that would not be accessible otherwise.

10 Assembly process of the demonstrator. The lacing connection's rigidity allowed the cantilevering of segments during the construction.

10

DISCUSSION AND OUTLOOK

Multi-material fibrous connections have proven to be a valuable technique for thin, segmented timer shells. Being exposed to rain and UV-light, delamination has been expected to be a major concern. However, the sewn connections within each segment have proven to prevent delamination even after the employed glue was heavily exposed to the weather. Instead, the pretension in the elements derived from elastic bending was lost faster than expected due to creeping of the wood. This was presumably due to the unusually wet spring and the repeated wetting and drying of the wood, resulting in the sagging of the structure in the mostly horizontal slab area of the shell. Although the construction system has not been developed as a weatherproof shell, the integration of membranes already hints at the possibility for integrating larger membrane elements, such as a secondary layer functioning as a constructive wood protection, which is expected to mitigate the creeping effect.

The described robotic fabrication process is based on a large-range industrial robot arm with the material attached to the robot flange and a stationary sewing machine. An alternative robotic sewing setup has been tested as part of the Robots in Architecture workshop, "Robotic sewing of timber veneer laminates," where two cooperating robots handle the work piece and guide it through an industrial sewing machine. Another alternative setup where the sewing machine is carried by the robot as an effector is currently under investigation. In each case, the robot's reach and the orientation of the sewing machine define the resulting design solution space. As the fabrication process gets more complex, adaptive robot control is being investigated through the integration of a sensor-based control loop. In order to achieve highly adaptive sewing techniques for thin and flexible materials, sensor input is essential to react to, or encourage and build upon the inherent material behavior during the fabrication process.

ACKNOWLEDGEMENTS

The work presented in this paper was partially funded by the German Research Foundation (DFG) as part of the Collaborative Research Centre TRR 141 "Biological Design and Integrative Structures." The project was also supported by GettyLab.

The authors would like to thank their fellow investigators from the Institute of Building Structures and Structural Design (ITKE)—Prof. Jan Knippers, Simon Bechert, and Daniel Sonntag—and the work group Paleontology of Invertebrates of the department of Geosciences at the University of Tübingen (IPPK)—Prof. Oliver Betz, Prof. Nebelsick, and Tobias Grun—as well as their colleagues Long Nguyen, Michael Preisack, and Lauren Vasey for additional support during design development.

The presented research was conducted at the intersection between research and teaching together with students of the ITECH MSc programme. The authors would like to express their gratitude towards the students Martin Alvarez, Jan Brütting, Sean Campbell, Mariia Chumak, Hojoong Chung, Joshua Few, Eliane Herter, Rebecca Jaroszewski, Ting-Chun Kao, Dongil Kim, Kuan-Ting Lai, Seojoo Lee, Riccardo Manitta, Erik Martinez, Artyom Maxim, Masih Imani Nia, Andres Obregon, Luigi Olivieri, Thu Nguyen Phuoc, Giuseppe Pultrone, Jasmin Sadegh, Jenny Shen,

11 Top view of the finished demonstrator. B: Interior view at night.

11

Michael Sveiven, Julian Wengzinek, and Alexander Wolkow, who strongly contributed to the development of this work.

REFERENCES

Bechert, Simon, Jan Knippers, Oliver David Krieg, Achim Menges, Tobias Schwinn, and Daniel Sonntag. 2016. "Textile Fabrication Techniques for Timber Shells: Elastic bending of custom-laminated veneer for segmented shell construction systems." In *Advances in Architectural Geometry 2016*. Zurich: AAG. Forthcoming publication.

Chakra, Maria Abou, and Jonathon Richard Stone. 2011. "Holotestoid: A Computational Model for Testing Hypotheses about Echinoid Skeleton Form and Growth." *Journal of Theoretical Biology* 285 (1): 113–25. doi:10.1016/j.jtbi.2011.06.019.

Menges, Achim, and Moritz Dörstelmann. 2014. "Leichtbau BW Installation." *Institute for Computational Design (ICD)*. http://icd.uni-stuttgart.de/?p=10992. Accessed May 4, 2016.

Dörstelmann, Moritz, Stefana Parascho, Marshall Prado, Achim Menges, and Jan Knippers. 2014. "Integrative Computational Design Methodologies for Modular Architectural Fiber Composite Morphologies." In *ACADIA 14: Design Agency–Proceedings of the 34th Annual Conference of the Association for Computer Aided Design in Architecture*, edited by David Gerber, Alvin Huang, and Jose Sanchez. Los Angeles: ACADIA. 219–228.

Färg, Fredrik, and Emma Marga Blanche. 2015. "COUTURE Armchair." Färg & Blanche. http://fargblanche.com/COUTURE-armchair-produced-by-B-D-BARCELONA-DESIGN. Accessed May 3, 2016.

Fleischmann, Moritz, Jan Knippers, Julian Lienhard, Achim Menges, and Simon Schleicher. 2012. "Material Behaviour: Embedding Physical Properties in Computational Design Processes." *Architectural Design* 82 (2): 44–51.

Grun, Tobias B., Layla Koohi, Tobias Schwinn, Daniel Sonntag, Malte von Scheven, Manfred Bischoff, Jan Knippers, Achim Menges, and James H. Nebelsick. 2016. "The Skeleton of the Sand Dollar as a Biological Role Model for Segmented Shells in Building Construction: A Research Review." In *Biological Design and Integrative Structures*. Berlin: Springer. Forthcoming publication.

Herzog, Thomas, Julius Natterer, Roland Schweitzer. 2003. *Holzbau Atlas*. Basel: Birkhäuser.

Keilmann, Robert, and Yongwu Chen. 2015. "Basic 3D-Robot Sewing Unit KL 500." KSL. http://www.ksl-lorsch.de/en/products/automotive/composites/basic-3d-robot-sewing-unit-kl-500/. Accessed May 4, 2016.

Krieg, Oliver David, Tobias Schwinn, Achim Menges, Jian-Min Li, Jan Knippers, Annette Schmitt, and Volker Schwieger. 2015. "Biomimetic Lightweight Timber Plate Shells: Computational Integration of Robotic Fabrication, Architectural Geometry and Structural Design." In *Advances in Architectural Geometry 2014*, edited by Philippe Block, Jan Knippers, Niloy J. Mitra, and Wenping Wang. Cham: Springer International Publishing. 109–125. doi:10.1007/978-3-319-11418-7_8.

La Magna, Riccardo, Markus Gabler, Steffen Reichert, Tobias Schwinn, Frédéric Waimer, Achim Menges, and Jan Knippers. 2013. "From Nature to Fabrication: Biomimetic Design Principles for the Production of Complex Spatial Structures." *International Journal of Space Structures* 28 (1): 27–39. doi:10.1260/0266-3511.28.1.27.

Li, Jian-Min, and Jan Knippers. 2015. "Segmental Timber Plate Shell for the Landesgartenschau Exhibition Hall in Schwäbisch Gmünd—the Application of Finger Joints in Plate Structures." *International Journal of Space Structures* 30 (2): 123–139.

Menges, Achim, and Tobias Schwinn. 2012. "Manufacturing Reciprocities." *Architectural Design* 82 (2): 118–125. doi:10.1002/ad.1388.

Reichert, Steffen, Tobias Schwinn, Riccardo La Magna, Frédéric Waimer, Jan Knippers, and Achim Menges. 2014. "Fibrous Structures: An Integrative Approach to Design Computation, Simulation and Fabrication for Lightweight, Glass and Carbon Fibre Composite Structures in Architecture Based on Biomimetic Design Principles." *Computer-Aided Design* 52: 27–39. doi:10.1016/j.cad.2014.02.005.

Robeller, Christopher and Yves Weinand. 2015. "Interlocking Folded Plate—Integral Mechanical Attachment for Structural Wood Panels." *International Journal of Space Structures* 30 (2): 111–122.

Schwinn, Tobias, Oliver David Krieg, and Achim Menges. 2013. "Robotically Fabricated Wood Plate Morphologies." In *Rob | Arch 2012: Robotic Fabrication in Architecture, Art and Design*, edited by Sigrid Brell-Çokcan and Johannes Braumann. Vienna: Springer. 48–61. doi:10.1007/978-3-7091-1465-0_4.

Schwinn, Tobias, Oliver David Krieg, and Achim Menges. 2014. "Behavioral Strategies: Synthesizing Design Computation and Robotic Fabrication of Lightweight Timber Plate Structures." In *ACADIA 14: Design Agency—Proceedings of the 34th Annual Conference of the Association for Computer Aided Design in Architecture*, edited by David Gerber, Alvin Huang, and Jose Sanchez. Los Angeles: ACADIA. 177–188.

Smith, Andrew B. 2005. "Growth and Form in Echinoids: The Evolutionary Interplay of Plate Accretion and Plate Addition." In *Evolving Form and Function: Fossils and Development*, edited by Derek E. G. Briggs. New Haven, CT: Yale Peabody Museum. 181–195.

Tamke, Martin, Paul Nicholas, Mette Ramsgard Thomsen, Hauke Jungjohann, and Ivan Markov. 2012. "Graded Territories: Towards the Design, Specification and Simulation of Materially Graded Bending Active Structures." In *ACADIA 12: Synthetic Digital Ecologies—Proceedings of the 32nd Annual Conference of the Association for Computer Aided Design in Architecture*, edited by Jason Kelly Johnson, Mark Cabrinha, and Kyle Steinfeld. San Francisco: ACADIA. 79–86.

Telford, Malcolm. 1985. "Domes, Arches and Urchins: The Skeletal Architecture of Echinoids (Echinodermata)." *Zoomorphology* 105 (2): 114–124.

Vincent, Julian. 2009. "Biomimetic Patterns in Architectural Design." *Architectural Design* 79 (6): 74–81. doi:10.1002/ad.982.

Weinand, Yves, and Markus Hudert. 2010. "Timberfabric: Applying Textile Principles on a Building Scale." *Architectural Design* 80 (4): 102–107. doi:10.1002/ad.1113.

IMAGE CREDITS

Figure 2a: © Telford, 1985
All other figures: © ICD/ITKE University of Stuttgart

Tobias Schwinn is research associate and doctoral candidate at the Institute for Computational Design (ICD) at the University of Stuttgart, Germany. In his research, he focuses on the integration of robotic fabrication and computational design processes. Prior to joining the ICD in January 2011, he worked as a Senior Designer for Skidmore, Owings and Merrill in New York and London, applying computational design techniques to parametric form-finding, rationalization, complex geometry, automation, and environmental design. Tobias studied architecture at the Bauhaus-University in Weimar, Germany and at the University of Pennsylvania in Philadelphia as part of the US-EU Joint Consortium for Higher Education. He received his diploma-engineering degree in architecture in 2005.

Oliver David Krieg is a research associate and doctoral candidate at the Institute for Computational Design at the University of Stuttgart. With the completion of his Diploma degree in 2012, he also received the faculty's Diploma Prize. Prior to that, he worked as a Graduate Assistant at the Institute's robotic prototype laboratory "RoboLab" since the beginning of 2010. With a profound interest in computational design processes and digital fabrication in architecture, he participated in several award-winning and internationally published research projects. In the context of computational design, his research aims to investigate the architectural potentials of robotic fabrication in wood construction.

Achim Menges is a registered architect and professor at the University of Stuttgart, where he is the founding director of the Institute for Computational Design. Currently he is also Visiting Professor in Architecture at Harvard University's Graduate School of Design and Visiting Professor of the Emergent Technologies and Design Graduate Program at the Architectural Association, London. Achim Menges' research and practice focuses on the development of integrative design processes at the intersection of morphogenetic design computation, biomimetic engineering, and digital fabrication. His projects and design research have received numerous international awards, and have been published and exhibited worldwide.

Programmable Matter

Sean Ahlquist
University of Michigan, Taubman College of Architecture and Urban Planning

The design of programmable matter seeks to achieve performative articulation through the vicissitudes of increasingly minute and multi-hierarchical scales of material formation. Often, this design pursuit triggers a discussion and examination of natural systems, implicit throughout the work in this chapter and explicit in the work of Körner's "Bioinspired Kinetic Curved-Line Folding" and Huang's "Durotaxis Chair." To address the conceptual nature of the research documented here, though, it is necessary to define the relationship between natural and artificial (architectural) systems. This is best captured by George Jeronimidis, an expert in biomimetics and composite structures, who provides a general understanding of biological formation that can be appropriated to programmable matter: "morphogenesis of biological organisms—the animation of geometry and material that produce form."[1]

Yet, to discuss this comparison also demands the discussion of how concepts of morphogenesis are applied to the formation of architectural systems. Jeronimidis points out the need for developing a *technological transformation*. Instead of enacting a replication or representation of biological formation, often referred to as biomimicry, computational design becomes a manifold process of investigating the specifics of a natural system, transforming those logics into codified methods, and enacting such methods within other constraining and contextual systems.

Referred to more appropriately as *biomimetics*, it is the integration of the specific contextual parameters for material processing and performative parameters that defines the distinction between computationally generative processes and morphogenesis. Even though both natural and artificial systems are often referred to as bottom-up processes, certain material parameters and methods of formation will impose top-down constraints. Biological structures are formed, extraordinarily, from only four basic polymer fibers,[2] while in manufactured material systems, a constrained *morphospace* based on material and manufacturing parameters always exists.[3] The work in this section exemplifies various approaches for how to define and implement this distinction between natural and artificial systems, and the means by which emergent and top-down conditions are negotiated.

Pineda and Ramirez-Figueroa each address the transformation of biological systems to design logics. Both Pineda's crystallographic material symmetry and Ramirez-Figueroa's exploration of bacterial spores unfurl natural ordering systems as design generators. Schleicher and Nicholas both discuss the opportunities afforded by higher-order design systems. Schleicher hybridizes form finding and form conversion to explore bending-active behavior as a meso-scale tiling condition. The interdependencies of multi-scale material operations are explored through Nicholas's incremental sheet forming.

The research by Sharmin, Wang, and Ramsgaard Thomsen confront material design, shifting from the manipulation of homogenous materials to the explicit formation of material differentiation through textile design. Yu, Huang, and Retsin do so through means of 3D printing, while Wit explores the negotiation of bespoke shape generation and fabrication through robotic winding of carbon fiber. Comparatively, these various avenues of research test the extensiveness by which differentiated performance is designed and materialized in morphologically based systems that are completely seamless in nature.

1. George Jeronimidis, "Biodynamics," *Architectural Design* 74, no. 3 (2004): 90–95.
2. Jeronmidis cites the four polymer fibers as cellulose in plants, collagen in animals, chitin in insects and crustaceans, and silks in spiders' webs.
3. A *morphospace* is a multi-axial graph which houses the descriptions of forms both real and possible. Cited from George R. McGhee Jr., *Theoretical Morphology: The Concept and its Applications* (New York: Columbia University Press, 1999).

Programmable Matter

The Grammar of Crystallographic Expression

Sergio Pineda
Mallika Arora
P. Andrew Williams
Benson M. Kariuki
Kenneth D. M. Harris
Cardiff University

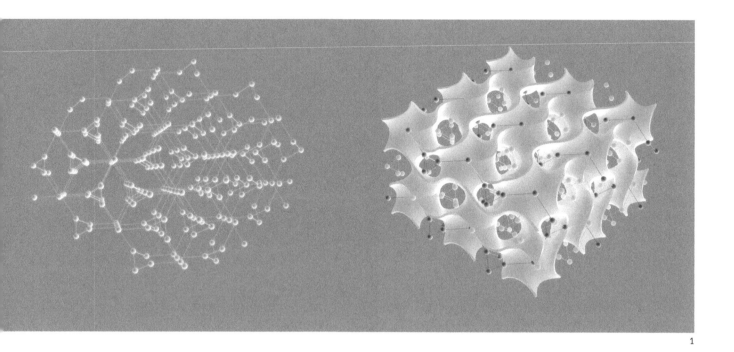

1

ABSTRACT

This paper stems from a research collaboration which brings together two disciplines at different ends of the scale spectrum: crystallography and architecture. The science of crystallography demonstrates that the properties of crystalline materials are a function of atomic/molecular interactions and arrangements at the atomic level—i.e., functions of the form and structure of the material. Some of these nano-geometries are frameworks with special characteristics, such as uni-directional porosity, multi-directional porosity, and varied combinations of flexibility and strength. This paper posits that the symmetry operations implicit in these materials can be regarded as a spatial grammar in the design of objects, spaces, and environments. The aim is to allow designers and architects to access the wealth of structural information that is now accumulated in crystallographic databases as well as the spatial symmetry logics utilized in crystallography to describe molecular arrangements. To enable this process, a bespoke software application has been developed as a tool-path to allow for interoperability between crystallographic datasets and CAD-based modelling systems. The application embeds the descriptive logic and generative principles of crystallographic symmetry. Using this software, the project, *inter alia*, produces results related to a class of geometrical surfaces called *Triply Periodic Minimal* (TPM) surfaces. In addition to digital iterations, a physical prototype of one such surface called the *gyroid* was constructed to test potential applications in design. The paper describes the development of these results and the conclusions derived from the first stage of user testing.

1   The interwoven *srs* nets and the *gyroid* surface dividing these nets.

CRYSTALLOGRAPHICALLY INSPIRED ARCHITECTURE: A NEW PATHWAY FROM NANOGEOMETRY TO DESIGN

This paper has emerged from an ongoing collaboration between architects and crystallographers as part of a 30-month research project funded by the Leverhulme Trust in the UK. We formulated the project with the idea that creative practice, spatial practice, and design can be enhanced through their affiliation with fundamental science.

A handful of architects have published work related to symmetry concepts that resemble or are similar to crystallographic space groups. For instance, architects such as James Strutt and Gulzar Haider investigated complex symmetries and space packing concepts in the 1970s and 1980s. We value this previous work, but we think that our current collaboration goes much further. To our understanding, our project is the first long-term collaboration between crystallographers and architects that proposes instrumental and design outputs.

One goal of the project is for designers and architects to learn from the structure of matter at the atomic and molecular level, recognizing that structural features and symmetry operations found at nanoscales offer possible design diagrams with multi-scale potential. Thus, the project exposes designers and architects to the spatial symmetries utilized in science to understand the structure of matter at molecular scales. Until now, the full extent of the grammar of crystallographic expression has been beyond the disciplinary horizon of the design community worldwide. In advancing our interdisciplinary project, we believe that a full understanding of this grammar—along with tools to operate within the principles of spatial symmetry—will be highly advantageous to designers.

In order to develop this new kind of tool for designers, the project employs the principles of Rapid Application Development (RAD). RAD is an approach based on creating a functioning (bespoke) script as early as possible, and refining the result through feedback and iteration. The feedback comes from the eventual users of the system, and is used to refine the result in successive stages. At the present stage of the project, we have developed version 1.0 of the software application. In this paper, we report on the development of this RAD methodology and how it was used in the production of a spatial/structural prototype. In April of 2016, we tested the application with our first user-group at the Smartgeometry conference in Gothenburg. The user-group comprised architects and architectural students.

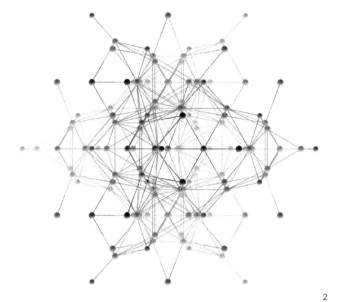

2

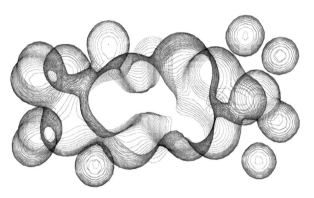

3

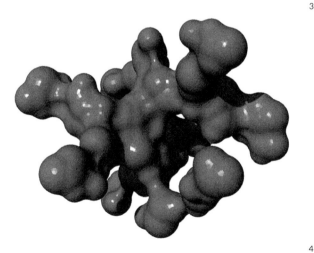

4

2   Expressive possibilities within the bespoke software application, based on a topology of points.

3   Expressive possibilities within the bespoke software application, based on a topology of lines.

4   Expressive possibilities within the bespoke software application, based on a topology of surfaces.

5   The *gyroid* and its unit cell translated into a lattice.

ENCODING THE CRYSTALLOGRAPHIC SPACE GROUPS

In the 20th century, the science of crystallography developed powerful procedures to analyze experimental data (specifically, X-ray diffraction data) in order to elucidate the structural properties of crystalline materials, which are based on two fundamental mathematical principles: periodicity and symmetry.

A crystalline substance exhibits three-dimensional periodicity. In other words, crystals have a high degree of repeat organisation in their structure, such that atoms, molecules, or ions typically repeat in regular arrays to form three-dimensional lattices with long-range order. The smallest repeating unit, which gives the complete periodic structure when translated in three dimensions, is called the unit cell. Further, the unit cell is an imaginary volumetric entity in Cartesian space, which defines within itself the positions of constituent atoms. These atomic positions are often related by symmetry, with symmetry operations that include reflection, rotation, inversion, and combinations of translation/reflection and translation/rotation. There is a finite number of combinations of these symmetry operations which can produce three-dimensionally periodic structures. Each such combination of symmetry operations is called a space group. Mathematics dictates that there are only 230 unique, three-dimensional space groups.

Within our project, we developed a design tool that allows designers to operate directly with any of these 230 space groups. The software application employs the principles of periodicity and symmetry to devise a generative process that simulates the assembly of molecules in crystal structures. It begins by selecting and importing an asymmetric unit from a crystallographic database, then runs through a series of operations to produce a periodic crystal structure. Thus, the interface allows for interoperability between existing crystallographic datasets and vectorial modelling systems in such a way that designers are able to not only simulate but also manipulate the crystalline assembly processes. *Rhino*, as one of the most widely used NURBS based CAD platforms, serves as the host environment and *Grasshopper* serves as its associated visual programming software. Within *Grasshopper*, the bespoke plugin application is programmed with *IronPython*.

Once the fundamental data has been simulated in *Rhino*, the application allows for designers and architects to give new readings to these geometries by generating a range of expressive possibilities. In crystallography, various aspects of a crystal structure are understood through visualization techniques, which give rise to notable expressive potential. The application embeds the descriptive logic of crystal structures to create a language of form based on points, lines, and surfaces (Figures 2–4).

The Grammar of Crystallographic Expression Pineda, Arora, Williams, Kariuki, Harris

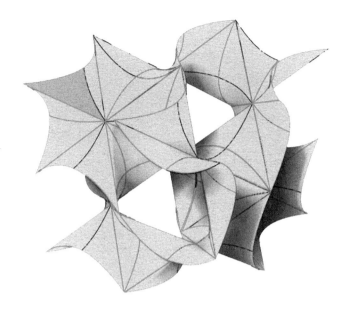

6 Construction of the *gyroid* based on the symmetry operations of space group 230.

TRIPLY PERIODIC MINIMAL SURFACES

The idea of crystallographic space groups can be used as a tool not just to generate the "node and network" topology of crystal structures, but also to generate a class of surfaces called *triply periodic minimal* surfaces (TPM). In these surfaces, which exhibit periodicity in three dimensions, the asymmetric patch is a minimal surface stretched between a framework. The resulting surface is continuous and without breaks. One possible TPM surface is known as the *gyroid* (or the G-surface), which can be constructed by applying the symmetry operations of space group *Ia-3d* (space group number 230) onto a specific triangular surface patch (Figures 5 and 6).

TPM surfaces are of interest in science from several perspectives. On the one hand, TPM surfaces help visualize the more abstract properties of crystal structures. The *gyroid* reveals intriguing topological qualities by partitioning space into regular non-intersecting labyrinthine channels, with interesting optical abilities. The axes of these channels correspond to the vertices of two interwoven *srs* nets. In crystallography, a net or periodic graph is a mathematical abstraction which describes the topological framework of a family of actual molecular networks. One such class of nets is called the *srs* net, in which the vertices are three-coordinated figures. The *srs* net elegantly inter-threads with its enantiomorph in such a way that the structure is chiral— whereas one net is left handed, the other net is right handed (Hyde, O'Keeffe, and Proserpio 2008). The *gyroid* divides this pair of nets, mapping an equipotential surface between the two nets (Figure 1).

Furthermore, the extensive occurrence of TPM surfaces in nature indicates potential applications for their material properties. The *gyroid*, for instance, is now known to underlie the structural characteristics of many different types of material, from lipids in cells to synthetic polymeric molecular melts.

The construction of these surfaces has been known to crystallographers and mathematicians for many years, but the intricacy of their construction has not been widely discussed among designers. Furthermore, in the past, designers have not had access to tools that allow them to explore and utilize the symmetry operations of the crystallographic space groups. Our project aims to help designers and architects expand their understanding of space, structure, and form through these crystallographic processes.

The software application that we have developed was shared with a group of designers during the Smartgeometry conference in Gothenburg in April 2016, within a cluster entitled *Nano-Gyroids*. The workshop focused on learning to use the new software application, understanding how to utilise crystallographic space groups, and constructing *gyroid* and *gyroid*-like geometries. The workflow involved deriving new asymmetric unit patches, often with an affinity to existing molecular structures, and exploring the idea of symmetry inherent in the crystallographic space groups to build TPM constructs or alternative formations derived from the application of crystallographic symmetries. By varying the parameters of the seed patch and generating periodic structures using different space groups, participants produced a broad spectrum of digital iterations (Figure 7).

PROTOTYPE

After developing an understanding of their use in crystallography, participants of the cluster investigated ways of creating aggregates and morphologies inspired by *gyroid* formations and helped to develop a methodology to construct physical prototypes of the same geometry through alternatives to 3D printing. The resulting prototype—a lightweight structure assembled with prefabricated stainless steel members and fabric components—occupied 9 m³ of space (Figures 8 and 9).

The design of the lightweight structural system was partly inspired by the work of artist Alison Grace Martin, who has developed lightweight structures to build objects of mathematical interest. Before the workshop, she shared her structural ideas with the workshop leaders, inspiring us to develop a structural interpretation of *gyroid* and *gyroid*-like formations. This concept was further advanced with workshop participants, who developed morphologies for the final installation and helped to construct it.

7   Wireframe views of constructs developed by workshop participants utilizing different space group symmetries.

The Grammar of Crystallographic Expression Pineda, Arora, Williams, Kariuki, Harris

8　Initial prototype for the construction of a light-weight *gyroid* structure.

CONCLUSION

The tools developed in our project so far have undergone a first round of testing by potential users. The users came from a broad spectrum of backgrounds in practice and academia. The results of the workshop suggest merit in our software, particularly as it systematizes the understanding of form and structure in its fundamental relation to material properties, and further provides a powerful interface for designers. In other words, the bespoke software is highly effective in allowing designers to operate with the grammar of crystallographic expression within a very short time frame.

In future, we believe that there are several significant areas in which the work described here could be developed and implemented. This new design work could operate at the crystallographic scale, at the architectural scale, or both. If regarded as a "blueprint" for designing objects, spaces, and environments, crystallographic geometries are highly significant, with applications such as:

- *Optical modulation:* crystalline structures frequently display properties such as uni-directional porosity, so that the material may appear opaque from most directions of view, but transparent when viewed in a specific direction. Components displaying these optical properties would be highly significant in modulating the levels of privacy of spaces, and their visual relationships to their surroundings.

- *Micro-climatic modulation:* the mono- and multi-directional porosity of crystalline structures sets them out as models with ideal geometries to regulate the enclosure and exposure of spaces with regard to an environment. Light, wind, water, and other environmental phenomena may be regulated through components with these geometrical characteristics.

- *Structural implementations:* among crystalline materials are structures with characteristics associated with strength or flexibility or both.

- *Sensing and actuation:* the combination of material assemblies that respond differently to changes in temperature allows for controlled use of the energy absorbed when they are exposed to sources of energy such as solar radiation. In such instances, sensing and actuation may be carried out directly through the interaction of these assemblies.

ACKNOWLEDGEMENTS

1.　We are thankful to the Leverhulme Trust for their generous support in making this research project possible.

2.　We are thankful to Alison Grace Martin for sharing with us her knowledge in constructing iterative geometries. Our process in constructing a prototype of the *gyroid* was inspired by her work in constructing objects of mathematical interest.

3.　We thank Smartgeometry 2016 and the participants in the

9 Prototype of a *gyroid* morphology as part of the *Nano-Gyroids* cluster at Smartgeometry 2016.

9

Nano-Gyroids cluster who were integral in developing part of the work presented here: Pinar Aksoy, Marie Bartz, Giselle Bouron, Niklas Nordstrom, Marta Pakowska, Mauricio Rodriguez, Sebastian Andersson, and Liv Andersson.

REFERENCES

Gandy, Paul J. F., and Jacek Klinowski. 2000. "Exact Computation of the Triply Periodic G ('Gyroid') Surface." *Chemical Physics Letters* 321 (5–6): 363–371.

Hyde, Stephen T., Michael O'Keeffe, and Davide M. Proserpio. 2008. "A Short History of an Elusive Yet Ubiquitous Structure in Chemistry, Materials, and Mathematics." *Angewandte Chemie International Edition* 47 (42): 7996–8000.

IMAGE CREDITS

Figure 1: Mallika Arora and Sergio Pineda, 2015
Figure 2: Sergio Pineda, 2013
Figures 3, 5: Mallika Arora, 2015
Figure 4: P. Andrew Williams and Mallika Arora, 2015
Figure 6: Mallika Arora, 2016
Figure 7: Clockwise from top left: Sebastian Andersson, Niklas Nordstrom, Marta Pakowska, Marie Bartz, Sebastian Andersson, Sebastian Andersson, 2016
Figure 8: Sergio Pineda, 2016
Figure 9: Daniel Davis, 2016

Sergio Pineda is an architect with experience in design research at the convergence of computation, fabrication, material science and emergent tectonics. After obtaining his Diploma from the Architectural Association in 2004, he worked for five years with practices in London (Foster + Partners, Adjaye Associates and Leit-werk) on a variety of award winning commissions in Barcelona, Denver, New York and Milan. He is currently based at Cardiff University, with a focus on collaboration between creative practice and fundamental science. In the past, he has collaborated with choreographers, sociologists, and hydrologists on initiatives such as the Architectural Association Visiting School in Medellin entitled Hydromeme.

Mallika Arora is an architect and computational designer with professional experience in Europe and Asia. She graduated with an MSc in digitally driven architecture from the TU Delft, and a B.Arch. from New Delhi. Mallika is interested in exploring the potential of computational design through a performative, methodological, and semiotic perspective. She has previously been associated with organisations such as OMA Rotterdam, INTACH Delhi and the TU Delft. Her current research includes a multidisciplinary project funded by the Leverhulme Trust, which brings together architects and crystallographers at Cardiff University to investigate what designers can learn from matter at the nanoscale.

P. Andrew Williams is a postdoctoral research chemist at Cardiff University. He completed a Masters in Chemistry at the University of Oxford prior to undertaking a PhD in solid-state chemistry with Professor Kenneth Harris at Cardiff University. His research has been based on the study of crystallization processes, the ability of molecules to form multiple different packing arrangements and the determination of crystal structures. Currently, he is involved in a collaborative project between crystallographers and architects.

Benson M. Kariuki has vast research experience in Solid State Chemistry with particular interest is the rationalization of relationship between structure and properties. He graduated from the University of Cambridge (PhD, 1990) and obtained postdoctoral and research fellowships at the Universities of Liverpool, Birmingham and University College London and a lectureship at Cardiff University .

Kenneth D. M. Harris graduated from the University of Cambridge (PhD; 1989), before holding academic positions in Physical Chemistry at St Andrews, UCL and Birmingham. He was appointed Distinguished Research Professor in Chemistry at Cardiff University in 2003. He has advanced new directions of research to understand structural properties of crystalline materials, and has developed novel experimental techniques to investigate these properties. His research has been recognized by several awards and prizes, and election to three national/international academies (Royal Society of Edinburgh, Learned Society of Wales, Academia Europaea). He has held appointments as Visiting Professor in Japan, France, USA, Spain and Taiwan.

Bacterial Hygromorphs

Experiments into the Integration of Soft
Technologies into Building Skins

Carolina Ramirez-Figueroa
School of Architecture,
Newcastle University

Luis Hernan
School of Architecture,
Newcastle University

Aurelie Guyet
Medical School,
Newcastle University

Martyn Dade-Robertson
School of Architecture,
Newcastle University

1

1   Spore coated polyimide strips inside
    the dry chamber.

ABSTRACT

The last few years have seen an increase in the interest to bring living systems into the process of
design. Work with living systems, nonetheless, presents several challenges. Aspects such as access
to specialists' labs, samples of living systems, and knowledge to conduct experiments in controlled
settings become barriers which prevent designers from developing a direct, material engagement
with the material. In this paper, we propose a design methodology which combines development of
experiments in laboratory settings with the use of what we call *material proxies*, which refer to materials that operate in analogue to some of the behaviors observed in the target organism. We will
propose that combining material proxies with basic scientific experimentation constitutes a form of
direct material engagement, which encourages richer exploration of the design domain.

We will develop this argument by reporting on our experience in designing and delivering the
primer component of a themed design studio, structured around bacterial spores as hygroscopic
components of building facades. The six-week design project asked students to consider the
behavior of bacterial spores, and to imagine a number of systems in which they could be employed
as actuators of a membrane system that responded to fluctuations in humidity. The module is interesting in that it negotiates some of the challenges often faced by designers who want to develop a
material engagement with living systems, and to produce informed speculations about their potential in architectural design.

INTRODUCTION

In this paper we preset some initial student-led experiments into the use of a new class of hydromorphic material. Traditionally, research on hygromorphic material has focused on wood, programming hydromorphic behaviors through the patterning and combination of laminates and composites. However, recent advances in materials science has led to the development of hygromorphic materials based on bacterial spores attached to passive (hygromorphically unresponsive) layers (Chen et al. 2014; Chen et al. 2015).

These new materials work in a similar way to wood-based hygromorphs, but have distinctive characteristics. Wood laminates, for example, can be made robust and are relatively easily integrated with more traditional forms of construction and manufacture. They can, however, be slow to respond and less efficient (weight for weight) compared to bacterial-based hygromorphs. They also lack sensitivity to small changes in, for example, ambient humidity. Bacterial-spore-based hygromorphs (as they are described throughout this paper) are, in contrast, highly responsive and highly sensitive to even small changes in ambient humidity. They are, however, difficult to scale, with current demonstrators based on a weak passive layer of 8 micrometer polyimide material. They also offer the potential for enhanced programmability in terms of the patterning of the bacteria active layer, in addition to the molecular characteristics of the bacteria. The effects of the hydromorphic response may also be amplified by combining it with other materials, mechanisms, and technologies.

Whilst bacterial spores are a dormant form of the bacteria, they develop from live cells and contain the capacity for life—spores retain the genetic material to germinate into vegetative cells and multiply as environmental conditions becomes favorable. In this context, design exploration with living systems inevitably presents us with challenges. Aspects such as access to specialist labs and samples of bacteria, in addition to the specialized experimental knowledge needed to manipulate them, become barriers that prevent designers from developing a direct, material engagement with such systems. In this paper, we develop a design project that combines the work in laboratory settings with more traditional craft-based material engagements mediated by what we describe as proxies. We show that while there are challenges in engaging with living or near-living technologies, bacteria-based hygromorphs offer insight into a compelling and accessible architectural technology.

BACKGROUND

Hygroscopic materials can be used to produce low-complexity actuators which are operated by changes in moisture content in

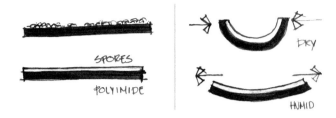

2

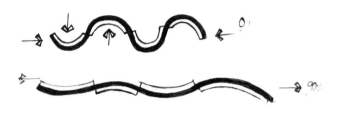

3

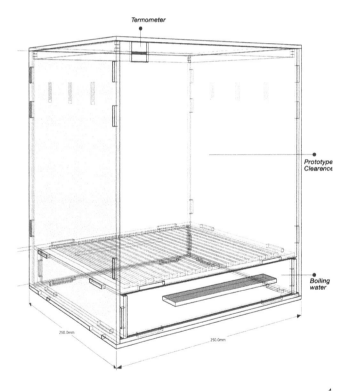

4

2   2A and 2B. Diagram showing programmability of bacterial spores by means of inoculation.

3   Concertina-like behavior achieved by alternative placing of spores.

4   Axonometric diagram of humidity test chamber.

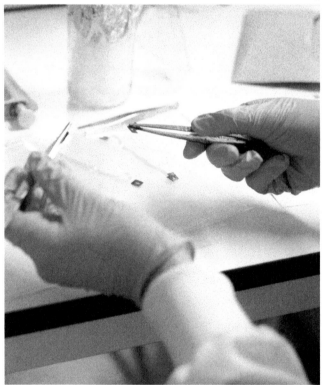

5   Kapton strips prepared with initial coating of Poly-L-Lysine.

6   Process of preparing elastomer strips.

the air—converting, for example, the energy held within evaporating water into mechanical work. These have been of particular interest in architecture, as they are integrated in the research and development of intelligent skins and dynamic envelopes, which aim to produce sophisticated assemblies that respond dynamically to their environment (Velikov and Thun 2013; Wigginton and Harris 2002) In this context, hygroscopic materials respond to open-air environmental conditions without complex mechanical sensing and actuation that require high levels of maintenance and energy input (Hensel and Menges 2006; Hensel, Menges, and Weinstock 2006).

Hygromorphic materials are also interesting for the way in which they constitute a hybrid of computation and actuation. They can be said to compute, as their configuration (output) changes in response to an external input. Traditionally, the capacity for a material to exhibit hygroscopic properties is associated with natural materials such as wood, whose internal cellular structure promotes water exchange, producing physical changes in the material (Holstov et al. 2015). These systems can be programmed through a range of methods, including the lamination of materials in composites of active (hygromorphic) and passive (unresponsive) layers (Holstov, Bridgens, and Farmer 2015; Menges and Reichert 2015; Correa et al. 2015).

We have identified a new type of hygromorphic material which uses bacterial spores to create an active layer coated onto a passive layer. Some bacteria species have a complex life cycle that allows vegetative cells to undergo morphological changes to form spores that display interesting hygromorphic proprieties. A spore is a dormant and resistant stage that allows a bacterium to preserve its biological integrity and genetic material to survive adverse environmental conditions (temperature, starvation, chemicals changes). Spores can expand or shrink in response to relative humidity. Some *Bacillus* species can lose 12% of their diameter in dry environment but recover it with rehydration. This is connected to the mechanics between the different parts of bacterial spores. For instance, *Bacillus subtilis* spores have at their center a dehydrated core containing genetic material. This core is enveloped by different layers—moving outwards from the core, these are: the inner germ wall, made of cross-linked peptidoglycan; the outer cortex; a lipid membrane; coat layers; and a final crust layer (Baughn and Rhee 2014). The function, composition, and organization of these layers are different, but ultimately allow the spore to expand and contract as it absorbs water or dries out. Bacterial hygromorphs described in this report exploit specific proprieties of Bacillus subtilis cotE gerE, a mutated strain that lacks some outer coat proteins, which amplifies the expansion of the spores.

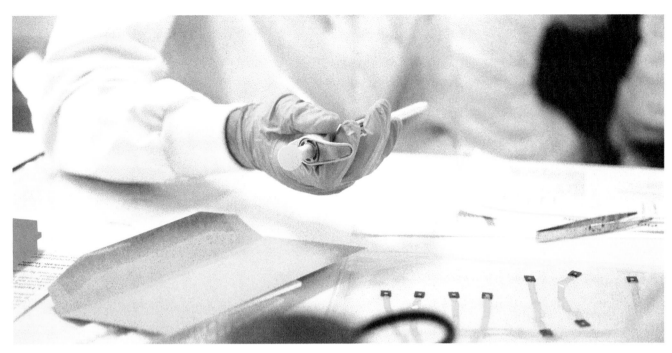

7   Introductory workshop to pipetting.

Recent research has looked to utilize these properties in the generation of evaporation-driven engines and electrical generators. These systems involve mechanical contraptions that integrate spore-coated elastomers which expand and contract in response to fluctuations in humidity (Chen et al. 2015). Applications for bacterial spore actuators have been explored in design contexts, including intelligent fabrics and human-computer interfaces (Yao et al. 2015; Heibeck et al. 2015).

Spore-coated elastomers constitute a potentially powerful technology to produce highly programmable architectural hygromorphic materials. As with wood based hygromorphs, they may represent a low-energy, low-maintenance dynamic building component—with potential application in, for example, advanced building skins. It is estimated that a spore-coated elastomer can operate up to 1,000 cycles before replacement and would not require special conditions for operation (Chen et al. 2015). There are, however, several challenges in designing systems using this emerging technology. Bacterial-based hygromorphs require their mechanical assemblies to be carefully optimized to the scale and power of the delicate coated elastomers. Access is also required to specialist laboratories, and knowledge is needed to grow, isolate, and prepare the spores and the coating solution. To address these challenges, we ran a design studio which included lab-based experimentation combined with *material proxies*, which involved more traditional workshop-based prototyping and that allowed us to negotiate limited access to laboratories and living materials.

METHODS

Set as part of our Stage 3 Program (3rd year undergraduate) we developed a five-week design project based on a brief to develop an actuated system which could open and close an aperture depending on the levels of humidity. This design component was conceived as part of an intelligent building skin.

Laboratory Strategies

An intensive two-day laboratory session was conducted to give the students direct experience working with the material, and to begin the process of prototyping to inform later designs. The central task of the session was the inoculation of bacterial spores onto the elastomer surface. Students initially familiarized themselves with the preparation of the spore solution, consisting of glue and a spore suspension. This was used to coat 15-centimeter-wide strips of eight-micrometer Kapton, a translucent polyimide film manufactured by DuPont. The film is prepared with an initial coat of Poly-L-Lysine, a liquid form of a synthetic polymer which increases adherence of spores to the elastomer surface. Once dry, the spore solution is deposited on the film using pipettes that allow a precise control of the deposition of the spore solution. The patterning of the spore solution allows us to program the hygromorphs with specific behaviors. Figures 2A and 2B show an elastomer strip which has been programmed to react by contracting and expanding as a cantilever—folding as humidity drops and the material dries. This is achieved by coating an elastomer strip on one of its faces. When humidity raises, it triggers a reaction in the spores, which expand to allow larger

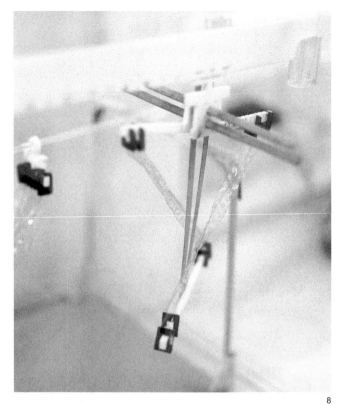

volumes of water within the walls of their membranes. This causes the active layer (a solution of glue and spores) to expand, pushing the elastomer outward. The opposite occurs as the material dries out. Variations of the inoculation pattern allow us to program the hygromorph into different behaviors. Inoculating spores in an alternating pattern on either sides of the polyimide material, for instance, causes the strip to expand and contract in a concertina-like shape (as shown in Figure 3). By coating an elastomer on alternating sides, the pull and push effect of the spores described above interact, resulting in a bending of the material.

Much of the workshop involved familiarizing students with the basic methods to fabricate bacteria-based hygromorphs, which involved producing the coating spore solution and understanding the methods to program behavior through patterns of inoculation (figures 5 to 7). In addition, they also devised their own experiments and novel actuators to test how they related to their initial design ideas. Experiments were performed in a pair of humid and dry chambers, which were custom built for the project (shown in figure 4). The tray of the high-humidity chamber was filled with hot water, which increased the air moisture inside the chamber to levels of 95% relative humidity (dry bulb temperature 25°). To create dry conditions, the low-humidity chamber container was filled with silica gel crystals, bringing relative humidity to 30% (dry bulb temperature 24°). A lid on top of the chamber provided access to the main volume, where prototypes were hung to test. The chambers allowed us to place prototypes inside in order to record their reaction to extreme levels of humidity (figures 8 and 9).

Proxy Strategies

Getting to grips with soft technology is made more difficult by the constraints on its use and the time available for experimentation. While a strip of elastomer inoculated by bacterial spores is estimated to be capable of pulling 50 times its own weight (Chen et al. 2015), at this scale the loads relevant to the actuation of, for example, a window, would require hundreds—perhaps thousands—of strips and other issues begin to come into play. Tiny amounts of friction between components and the stiffness of different materials start to play a part when added up over many parallel components. To better understand this, we needed to incorporate the bacteria in the early stages of the design process. This is challenging, however, as direct experience of the material depends on experimentation that, given potential biological risk, can only be performed in the controlled conditions of a microbiology laboratory.

An alternative to direct use of bacteria hygromorphs is *material proxies*, defined as technologies that mimic some aspect of the material system we are interested in investigating. In this context, we used shape-memory alloys (SMAs), which can be

8   Testing of transversal movement prototype in dry chamber.

9   Prototypes inside test chamber.

Bacterial Hygromorphs Ramirez-Figueroa, Hernan, Guyet, Dade-Robertson

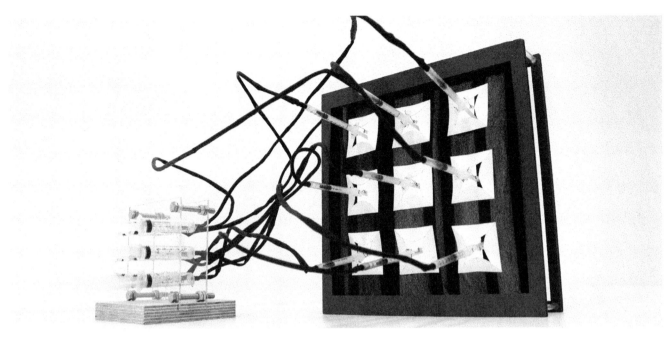

10 Prototype A, concept model showing connection between panel and pneumatic system.

programmed to contract when voltage is applied to them, in the same way that hygromorphs respond when being wet or dry. For example, Nitinol, a nickel and titanium alloy, remains in a martensite phase at normal temperature. In a martensite state, atoms are arranged in a grid, in which one axis is slightly longer than the others. On a macro-scale, this means that the metal is malleable, adopting any form that is forced onto it. When the metal is brought to its transition temperature, the metal shifts to its austenite phase, in which atoms arrange in the most compact and regular pattern possible. On a macro-scale, this snaps the metal back to its original or so-called parent position (Kumar and Lagoudas 2008).

SMAs provided a simple way to integrate actuating components, close to the range of operation of bacterial hygromorphs, into the physical prototypes produced in the design strategies phase. Students were provided with 0.005" diameter Muscle Wire, an SMA produced by Dynalloy Inc. The datasheets suggest a pull force of 0.49 lb for this gauge of wire (Dynalloy n.d.), which allowed us to establish a rough equivalence of one strand of SMA for 37 strips coated with bacterial solution.

RESULTS

Bacterial-spore-based hygromorphs appear to offer potential in architectural design, especially in their capacity to bridge seemingly unrelated scales—the response of a single cell, measuring less than 2 micrometers, could be made to power an actuation system relevant to the design of building components. Our initial experiments have shown some of this potential, but also revealed

some of the challenges in using this new type of system. In all, ten students participated in the studio and, in presenting their designs here, we reflect on some of those challenges and opportunities.

One central challenge in designing with bacterial-based hygromorphs is in bridging their scale of operation to that of the built environment. This requires adapting our current models of actuation, which have often been informed by industrial technologies, to the soft power afforded by these new materials. In approaching this, we initially directed students to 507 *Mechanical Movements* (Brown 1903), a nineteenth-century handbook that compiles mechanisms with different strategies to transform and make use of different sources of power and motion. The mechanisms are relevant in the context of this project, in that they are optimized to take advantage of relatively high-power inputs from steam engines to hydropower. Using them as point of departure involves actualizing their configuration to respond to and harvest smaller sources of power by orchestrating accurate movements. Bacteria-based hygromorphs, in contrast to industrial power sources, generate power at small scales, which must either be amplified or arranged in parallel—as hundreds of actuators in complex combinations—in order to become relevant to the scale of the built environment. In architecture, we simply don't yet have a repertoire of actuated mechanisms that can make use of such sensitive but low-powered components.

Figures 10 to 16 show some of the prototype systems produced by the students. Projects can be broadly classified based on the way they integrate power generation and actuation. A first

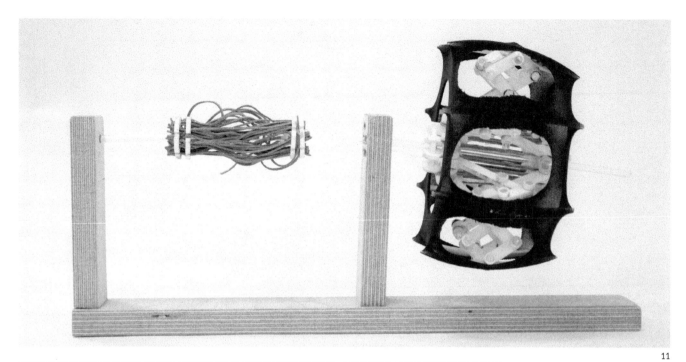

12

11 Prototype B, concept model, the bank of bacteria spore hygromorphs is depicted as flexible bands to the left.

12 Prototype E, concept model showing the combination of several panels, hygromorphs depicted as white strips.

13 Prototype C, concept model of the assembled prototype. Curved elements wrapping the main body represent the bacterial spores as connected to actuate the ball-bearing mechanism.

14 Prototype D, assembled prototype showing the rotating mechanism connected to diagonally arranged banks of spore-coated stripes.

15 Prototype E, front view of early prototype integrating bacteria-spore hygromorphs, depicted in semi-transparent strips, and their connection to opening panels.

16 Prototype F, exploded view of shutter design showing banks of bacteria spores hygromorphs arranged opposite to each other, combining their action to produce organic like motions.

13

14

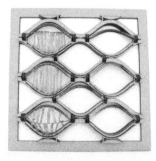

15

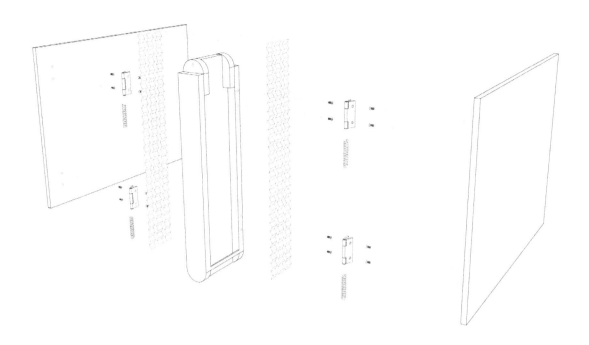

category separate power generation—in this context bacterial hygromorphs— and actuation systems, incorporating different strategies to optimise their interface. A second category interweave actuation and power generation in a single system.

A good example of the first category is prototype A, designed by George Entwistle and shown in figure 10. It comprises a pneumatic system which uses a bank of spore-coated strips to push and pull the linear motion of pistons in a batch of syringe pumps. Fluid inside the pumps acts as interface to transmit power generation of strips to the actuation system, comprised of panels of flexible membranes that expand and contract. A similar approach is followed in prototype B, shown in figure 11 and designed by Michael Bautista-Trimming. The design allows a flexible membrane to expand and contract following the deployment of the mechanism. Actuation is generated by a linear motion, which is interfaced by a piston connected to a battery of bacterial hygromorphs, arranged radially connecting two sliding rings.

Prototypes C and D follow a similar strategy of separating power generation from actuation systems. The interface between both systems, however, is in direct control of the actuator. Prototype C, shown in figure 13 and designed by Adam Kamal Najia, consists of a rotating ball-bearing mechanism that forces a flexible membrane to twist, closing the aperture. Spore-coated strips are placed to the sides of the mechanism. Prototype D, by Adnan Qatan and shown in figure 14, also makes use of a rotating mechanism that operates an array of blades. Spore coated strips are arranged diagonally, pulling a central ring that allows the mechanism to transmit and transform linear into rotational motion.

A second category of projects merge power generation and actuation together, weaving their components and generating geometric strategies that create tight couplings between mechanical parts and spore-coated elastomers. In prototype E, shown in figures 12 and 15, Aldrich Choy creates a system which weaves spore-coated elastomers within the actuating mechanism itself. Groups of strips are connected to a rotating, rigid element. Conditions of closure and aperture are generated by this element and, in part, by the dense arrangement of bacterial hygromorphs, which provide the power to actuate the system and produce a dense, tissue-like mass that opens and closes as they expand and contract. The strips are connected to a number of fixed points which are separated along the panel's depth, which amplifies the condition of closure. Prototype F, designed by Julian Besems and shown in figures 16 to 17, generates a tight coupling between power generation and actuation, blurring the boundary between both functions. As seen in figure 17, the system incorporates banks of strips arranged in parallel that are embedded to the sides of a chain of hanging panels. Banks of strips are arranged in opposite sides, each facing the inside and outside of the panel. This allows a nuanced actuation, which responds to shifting moisture conditions by allowing banks of strips to counteract each other with jittery, small motions that resemble organic movements.

Other explorations also hinted at different ways in which spore-coated components can, potentially, actuate without association to mechanical assemblies. This is the case of explorations shown in figure 18, designed by Iona Haig, which experiment with the properties of flexible materials, such as fabric, and in the way that creases and folds can take advantage of the hygromorphic

properties of spores to create movements that, aggregated within a dense mass, might produce conditions of closure and aperture.

CONCLUSION

What is notable about the designs presented in this paper is that whilst ingenious and elegant, they have developed inevitably in isolation from the hydromorphic materials. Although the lab experiments informed the overall form factor of the actuators, the students consistently underestimated the number of strips which would be required to power their system. The use of the proxy, whilst giving a sense of the power scale, didn't account for the volume of strips required for even modest mechanical power. The use of proxies in developing prototype models needs to be refined further to better reflect the scale of power generation. Also, a further avenue of exploration would not only work to develop more efficient mechanical assemblies that minimize points of friction, but also to develop systems which merge tasks of power generation and actuation into the same component. This is an ambitious task, as it requires new interface and motion logics that depart fundamentally from those produced for industrial sources of energy and motion. A few elements of the exploration presented in this paper hint towards this direction. Notwithstanding, we believe the methodology followed and the resulting prototypes provide useful approaches to the design of living and semi-living materials, which hold the potential to constitute highly programmable, soft computational architectural components.

ACKNOWLEDGMENTS

We would like to thank our Stage 3 students whose projects are included in this paper: George Entwistle, Julian Besems, Aldrich Choy, Michael Baustista-Trimming, Adam Kamal Najia, Adnan Qatan, Bradley J.W. Davidson, Simon Quinton, Iona Haig and Kimberly Baker.

REFERENCES

Baughn, Anthony D. and Kyu Y. Rhee. 2014. "Metabolomics of Central Carbon Metabolism in Mycobacterium Tuberculosis." *Microbiology Spectrum* 2 (3): 1–16. doi:10.1128/microbiolspec.

Brown, Henry T. 1903. *507 Mechanical Movements: Mechanisms and Devices.* New York: Brown & Seward.

Chen, Xi, Davis Goodnight, Zhenghan Gao, Ahmet H. Cavusoglu, Nina Sabharwal, Michael DeLay, Adam Driks, and Ozgur Sahin. 2015. "Scaling up Nanoscale Water-Driven Energy Conversion into Evaporation-Driven Engines and Generators." *Nature Communications* 6: 7346. doi:10.1038/ncomms8346.

Chen, Xi, L. Mahadevan, Adam Driks, and Ozgur Sahin. 2014. "Bacillus Spores as Building Blocks for Stimuli-Responsive Materials and

17 Prototype F, design of the mechanisms allows for a progressive deployment which hints at an organic behavior.

18 31A and 31B. Prototype G, working model of spore-coated elements which operate as scales, providing conditions of aperture by aggregating their individual behaviors.

Nanogenerators." *Nature Nanotechnology* 9 (2): 1–9. doi:10.1038/nnano.2013.290.

Correa, David, Athina Papadopoulou, Christophe Guberan, Nynika Jhaveri, Steffen Reichert, Achim Menges, and Skylar Tibbits. 2015. "3D-Printed Wood: Programming Hygroscopic Material Transformations." *3D Printing and Additive Manufacturing* 2 (3): 106–16.

Dynalloy, Inc. n.d. "Technical Characteristics of Flexinol Actuator Wires." Tustin, CA: Dynalloy, Inc..

Heibeck, Felix, Basheer Tome, Clark Della Silva, and Hiroshi Ishii. 2015. "uniMorph - Fabricating Thin-Film Composites for Shape-Changing Interfaces." *Proceedings of the 28th Annual ACM Symposium on User Interface Software & Technology*, edited by Celine Latulipe, Bjoern Hartmann, and Tovi Grossman. Charlotte, NC: UIST. 233–242. doi:10.1145/2807442.2807472.

Hensel, Michael, and Achim Menges. 2006. "Material and Digital Design Synthesis." *Architectural Design* 76 (2): 88–95. doi:10.1002/ad.244.

Hensel, Michael, Achim Menges, and Michael Weinstock. 2006. "Towards Self-Organisational and Multiple-Performance Capacity in Architecture." *Architectural Design* 76 (2): 5–11. doi:10.1002/ad.234.

Holstov, Artem, Ben Bridgens, and Graham Farmer. 2015. "Hygromorphic Materials for Sustainable Responsive Architecture." *Construction and Building Materials* 98: 570–82. doi:10.1016/j.conbuildmat.2015.08.136.

Holstov, Artem, Philip Morris, Graham Farmer, and Ben Bridgens. 2015. "Towards Sustainable Adaptive Building Skins with Embedded Hygromorphic Responsiveness." *Advanced Building Skins: Proceedings of the International Conference on Building Envelope Design and Technology*, edited by Oliver Englehardt. Graz, Austria: ICBEDT. 57–67.

Kumar, P. K. and D. C. Lagoudas. 2008. "Introduction to Shape Memory Alloys." In *Shape Memory Alloys: Modeling and Engineering Applications*, edited by Dimitris C. Lagoudas. 1–51. doi:10.1007/978-0-387-47685-8.

Menges, Achim and Steffen Reichert. 2015. "Performative Wood: Physically Programming the Responsive Architecture of the HygroScope and HygroSkin Projects." *Architectural Design* 85 (5): 66–73. doi:10.1002/ad.1956.

Velikov, K, and G Thun. 2012. "Responsive Building Envelopes: Characteristics and Evolving Paradigms." In *Design and Construction of High-Performance Homes: Building Envelopes, Renewable Energies and Integrated Practice*, edited by Franca Trubiano. London: Routledge. 75–92.

Wigginton, Michael and Jude Harris. 2002. *Intelligent Skins*. Oxford: Butterworth-Heinemann.

Yao, Lining, Jifei Ou, Chin-Yi Cheng, Helene Steiner, Wen Wang, Guanyun Wang, and Hiroshi Ishii. 2015. "bioLogic: Natto Cells as Nanoactuators for Shape Changing Interfaces." *Proceedings of the 33rd Annual ACM Conference on Human Factors in Computing Systems*, edited by Bo Begole, Jinwoo Kim, Kori Inkpen, and Woontack Woo. Seoul: CHI. 1–10. doi:10.1145/2702123.2702611.

Carolina Ramirez-Figueroa is a PhD researcher and designer based in the UK. Her work addresses the intersection between design and synthetic biology, especially in the way living matter challenges existing models of design and material production. In *Synthetic Morphologies* (www.syntheticmorphologies.com), she develops a number of design interventions which help her understand the impact, background and potential of living systems in the fabrication of the built environment. Carolina holds an MSc in Digital Architecture, and is originally trained in Architecture. Email: p.c.ramirez-figueroa@newcastle.ac.uk

Luis Hernan is a designer and doctoral researcher based in Newcastle University. Initially trained as an architect, Luis pursued a master's degree in Digital Architecture before embarking in his doctoral research, which explores the role of wireless infrastructure in contemporary experience of architectural space (www.digitalethereal.com). His work also extends to living technologies, investigating the interface between digital, physical and living computation. Email: j.l.hernandez-hernandez@newcastle.ac.uk

Aurelie Guyet is a research associate at Newcastle University (UK) since 2009. Aurelie holds a MSc and a Ph.D. in molecular microbiology from Denis-Diderot Paris 7 University and Institut Pasteur (Paris, France). Her Ph.D researches led to two publications as first authors on Streptomyces morphogenesis. Aurelie studied Bacillus cell envelope in Dr Richard Daniel's group at the Centre for Bacterial Cell Biology. Since October 2015, she is working on a synthetic biology project in the teams of Prof. Anil Wipat and Dr Martyn Dade-Robertson. Email: aurelie.guyet@newcastle.ac.uk

Martyn Dade-Robertson is a Reader in Design Computation at Newcastle University where he is Co-Director of the Architectural Research Collaborative (ARC), Director of the MSc in Experimental Architecture and leads a research group on Synthetic Biology and Design (www.synbio.construction). Martyn holds an MPhil and Ph.D. from Cambridge University in Architecture and Computing. He also holds a BA in Architectural Design and an MSc in Synthetic Biology from Newcastle University. Along with over 25 peer reviewed publications Martyn published the book *The Architecture of Information* with Routledge in 2011. Email: martyn.dade-robertson@newcastle.ac.uk

Knit Architecture

Exploration of Hybrid Textile Composites Through the
Activation of Integrated Material Behavior

Shahida Sharmin
Sean Ahlquist
University of Michigan,
Taubman College of Architecture
and Urban Planning

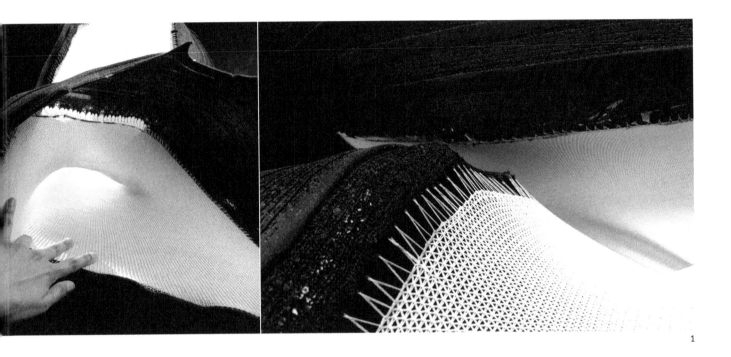

1

ABSTRACT

The hybrid system in textile composites refers to the structural logic defined by Heino Engel, which describes a system that integrates multiple structural behaviors to achieve an equilibrium state (Engel 2007). This research explores a material system that can demonstrate a hybrid material behavior defined by the differentiated tensile and bending-active forces in a single, seamless knitted composite material. These behaviors were installed during the materialization phase and activated during the composite formation process. Here, the material formation involves two interdependent processes: 1) development of the knitted textile with integrated tensile and reinforced materials and 2) development of the composite by applying pre-stress and vacuuming the localized area with reinforcements in a consistent resin-based matrix. The flat bed industrial weft knitting machine has been utilized to develop the knitted textile component of the system with a controlled knit structure. This enables us to control the material types, densities, and cross sections with integrated multiple layers/ribs and thus, the performance of the textile at the scale of fiber structure. Both of these aspects were researched in parallel, using physical and computational methods informed and shaped by the potentials and constraints of each other. A series of studies has been utilized to develop small-scale prototypes that depict the potential of the hybrid textile composite as the generator of complex form and bending active structures. Ultimately, it indicates the possibilities of hybrid textile composite materials as self-structuring lightweight components that can perform as highly articulated and differentiated seamless architectural elements that are capable of transforming the perception of light, space, and touch.

1 Hybrid Textile Composite with embedded tensile and bending-active behavior through differentiated material properties.

INTRODUCTION

The material system tries to comprehend architecture as an assembly of different interdependent variables rather than perceiving form, structure, and space as a preconceived idea or definition. The intention is to explore architecture in terms of materiality and materialization. Every element—matter, energy, environment, fabrication process, and human response—participates actively in creating the equilibrium of internal and external forces and thus, generates a dynamic formal and spatial organization.

This research intends to extend the architectural conversation on simultaneously evolving form and material formation processes, as opposed to the typical post-rationalizing process where the forms are defined first and then the materials are engineered (Ahlquist and Menges 2015). The intention is also to test the possibility of developing a lightweight but robust hybrid structural system where both the tensile and bending-active elements can be integrated in a differentiated and seamless piece of textile composite. These material behaviors were activated during and after the material formation process as a means of developing complex 3D forms from a flat initial state, as opposed to the conventional method of laying up fiber components in a three-dimensional formwork as a process of making textile reinforced composite (TRC) materials.

The research builds upon the previous studies on TRC, where knitted textile composites are utilized to produce 3D multilayer spacer fabrics suitable for high-performance composite applications (Abounaim et al. 2010). With the goal of developing 'Textile-reinforced composite components for function-integrating multi-material design in complex lightweight applications,' this particular precedent reinforces the experimentation with the performance of textile composite with varying knit structures. This, along with other precedents on the development of three-dimensional seamless multilayer textile structure generated without the use of any formwork, also defined as near-net shape preforms (Cebulla, Diestel, Offermann 2002), were particularly relevant as a way of understanding the integration of multilayer composite pre-forms. As the advanced composite manufacturing process employed in the industry demands the use of pre-assembled pre-forms, earlier research focused on developing a controlled process of manufacturing complex, lightweight TRC for its potential to achieve high flexibility.

This research also intends to extend the previous exploration of post-forming composite processes, developing a material system that internally houses the hybrid textile form and bending-active behavior (Ahlquist et al. 2014). This demonstrates the potential of hybrid textile composites, where the impact of curing time has

knitted sample

resin impregnated composite

2

a)

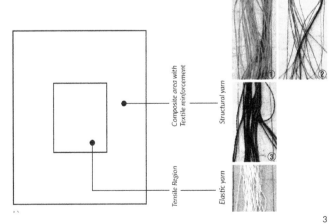

3

2 Transformation of the knitted sample after the activation of forces through applying prestress and localized composite formation.

3 a) The knitting machined used for tailoring the knit structure of the composites; b) basic material layout.

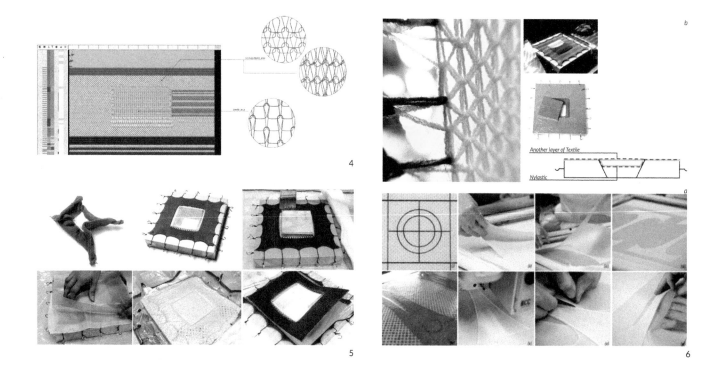

4

5

6

been examined to understand the relationship between geometry and stiffness of the material, along with the application of the pre-stress as a way of driving the three-dimensional formation. This precedent also shifts the concepts of form and bending-active structural logic achieved by deploying separately manufactured bending-active rod and tensile textiles in previous researches (Ahlquist and Menges, 2013). As this behavior is invested during the process of forming, it represents the potential of a fully materialized system embodying the aspects of material behavior.

Looking upon the precedents, this particular research tries to integrate the knowledge of developing differentiated multi-layer knitted textile, where the tensile and bending-active properties are installed during the materialization phase and activated during the composite formation process (Figure 2). The flat-bed industrial weft knitting machine has been utilized to develop the knitted textile component of the system with tailored knit structure. This enables the control of the material density and the material cross section and thus, the performance of the textile in the scale of fiber structure. Basically two types of yarn have been used. The peripheral area consists of structural yarn which, after resin impregnation, works as the bending resistant component for the system. The interior part is knitted with a type of elastic yarn with spandex core and elastic nylon sleeves around, referred to as "nylastic" yarn, which is kept isolated and utilized for providing pre-stress during the composite formation process and later, instigates three-dimensional formations upon releasing the tension. Various yarns have been tested as the reinforcing material (carbon fiber, nylon and polyester). (Figure 3).

METHODS

The series of experiments aims to explore the hybrid behavior of the knitted textile composite through the differentiated knit logics and the process of formation. It focuses mainly on two aspects: 1) development of the knitted textile with differentiated knit structure and material behavior and 2) application of pre-stress and development of the composite with resin impregnation through vacuum bagging process. Both of these aspects were researched in parallel and were informed and shaped by the potentials and constraints of one other. Thus, the research methodology was essentially based on multiple feedback loops and negotiation between several variables, including fabrication process and material behavior.

The research investigated computational and physical methods simultaneously. The computational method was mostly utilized in developing the textile component of the material system, both in design and fabrication. The knit program was developed by utilizing M1 plus software, a tool specifically dedicated for the industrial flat-bed CNC knitting machine (Figure 4). This helped to tailor the properties of the composite at the scale of fiber with a considerable number of variables, such as: 1) variation of stitch structure, 2) material types, 3) density of the material, 4) integration of yarns with different material properties, 5) location and shape of these materials/yarns, and 6) addition of layers and 3D elements in a single assembly. The development of the machine code to produce the knit requires an extensive understanding of the knitting process in the weft knitting machine, which further includes nuanced control of stitch length, machine speed, and

Knit Architecture Sharmin, Ahlquist

4   The knit program as viewed in the M1 plus software with blow-up diagrams of knit structures.

5   The fabrication process of the composite through pre-tensioning, localized resin impregnation, and vacuum bagging.

6   The formwork used to separate the integrated nylastic part, as opposed to the precedent method of cutting.

7   The prototypes with 1 x 1 alternating single jersey knit structures, before and after the resin impregnation: a) polyester, b) nylon, c) carbon fiber stretch broken yarn, and d) carbon fiber tow.

7

fabric take-down values—all of which varied according to the material types and stitch structure.

The other facet of the research regarding the development of the composite mostly involves the application of physical methods. Pre-stress was applied the textile component utilizing the integrated tensile behavior of the elastic yarns, fixing the whole piece in a flat, two-dimensional formwork. The tensile area was carefully isolated and the peripheral reinforced areas were fused in a consistently distributed resin matrix. Formworks, as well as the consolidation process, were developed to facilitate the segregation of the tensile part during the impregnation of resin without extensive post-processing or post-cutting of the masking layer, as opposed to previous studies (Figure 6). Though the different curing time in some of the initial studies showed differentiated material behavior in the cured composites, it was kept constant for most of this research to reduce the number of variables and have a better understanding of the impact of material properties. Thus, the variation in the three-dimensional outcome was essentially a result of the knit structure and the interconnection of differentiated material properties.

EXPERIMENTS

The initial experiments focused on developing a basic knit structure, integrating elastic and reinforcement yarns in the same knitted sample, where the elastic region was connected with the peripheral reinforcement areas by spaced connections. Simultaneously, different iterations were performed to develop the workable mechanism in the formwork, which can easily segregate the elastic region from the composite to prevent the bleeding of resin during the impregnation process and finally, produce a cleaner composite and allow consolidation of multiple layers in a faster and controlled way. (Figure 5).

The next iterations for the knitted samples investigated the variations in knit structure to understand and tailor the equilibrium of forces after releasing pre-tension in the cured composite. The first array of tests employed the alternating 1 x 1 knit structure, which utilized all needles in only one bed, essentially known as single jersey knit. This specific knit structure increased the amount of yarn in the sample and produced a denser knit than the regular single jersey structures. This first iteration was done using four types of yarn as reinforcement materials: polyester, unbonded soft nylon, carbon fiber stretch broken yarn, and 1.5k carbon fiber tow (Figure 7). The outcome demonstrated that heavier yarn or denser material in the reinforced area produces more bending-resistant behavior after releasing the pre-stress. This led to the next array of tests, where polyester was continued to be used as the reinforcement yarn, and the knit structure was explored as a means to tailor the material's density.

The next steps included the development of knit samples with heavier knit structures like interlock, tubular, and spacer fabric, which engaged both beds to produce the knit, known as double jersey. The 1 x 1 alternating knit structure was carried on to ensure more density in the double jersey. Simultaneously, the central elastic region was developed to achieve maximum tension during the pre-tensioning process. As the yarn at the

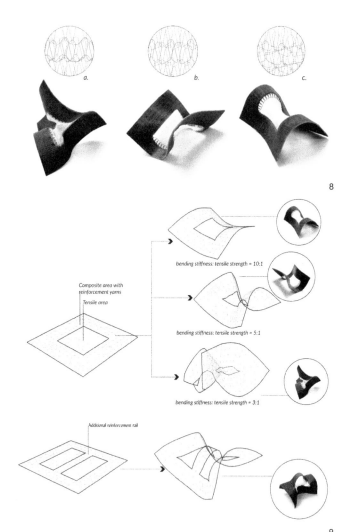

position with each other in generating variation in final outcomes from a similar initial geometric configuration. Parallel to the rigorous physical methods that have been tested, investigation has been done to simulate the interaction between the tensile and bending-active components of the hybrid composite using the particle-spring method in Grasshopper's Kangaroo physics. Since it is quite difficult to translate the knit structures accurately in terms of mesh topology in the digital model, triangulated mesh has been used in these studies to mimic the multi-directionality of the fiber connections. As the material performance is mostly dependent on the knit structure and it is still very challenging to combine the exact fiber structure in digital models, these digital studies were mainly utilized as a way of simulating differentiated surface stiffness. Here, the purpose of the computational studies is not to design any predefined form, but rather to generate a system driven by material behaviors as a way to understand the inherent interaction and simulate the formal outcome from it. This establishes the future potential for designing more complex arrangements of materials and forces, simulating the possible 3D outcome generated by the material behavior, though as basic precedents the physical models were mostly relied upon. The digital models also demonstrate the deviation from the physical models that happens because of the influence of material behavior and the fabrication process (Figure 9).

Simultaneously, other small-scale studies were done to explore the possibilities of the knitted hybrid textile composite through the insertion of additional rails above the elastic region as a separate layer to achieve topological variation. These studies demonstrate increased stiffness of the composites in the locations where multiple layers of the composites connect with each other (Figure 10a). Also, further studies demonstrate the possibilities of integrating multiple elastic regions in a single material assembly (Figure 10b). Due to the limited number of feeders in the machine, these studies focused on developing the knit structure and the knitting method using only one feeder with reinforcement yarns. The results generate a more complex interaction of forces and material behavior, altogether revealing the future potential of developing hybrid composites with complex topological variation. (Figure 10c)

CONCLUSION

Integrating differentiated material behavior in the textile composites establishes the potential of hybrid structures with controlled and tailored spatial and structural performance. Since the form and bending-active behavior is installed during the materialization phase, it compresses the involvement of separately manufactured material assembly to achieve the similar hybrid structural actions. Based on the parameters established in this research, similar results can be achieved by repeating the

8　The prototypes with double single jersey knit structures after the composite formation process with different knit structures: a) interlock, b) tubular, and c) spacer.

9　Computational studies to simulate the hybrid nature of the composite through varying stiffness and mesh topology, comparing them with the physical prototypes.

reinforcement area is non-stretchable, this required a careful determination of the ratio in terms of the number of courses being knitted for each type of yarn (1 course nylastic = 3 courses of reinforcement yarn). These experiments show how material density and behavior can be tailored through differentiated knit structures (Figure 8).

RESULT AND DISCUSSION

This research challenges the anticipated result of digitally fabricated objects as perceived while designing on a computer, since the ultimate form and structural behavior changes by the influence of material properties and how they are fabricated. Different prototypes shows the effect of material properties in terms of yarn types, fiber density, fiber orientation, and relative

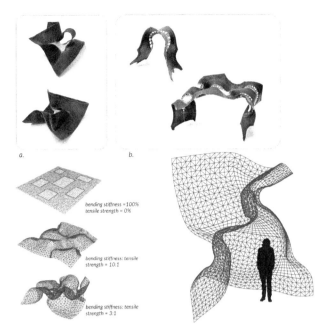

10 Studies showing the possible future trajectories of the research: a) textile hybrid prototypes with multiple reinforcement layers, b) prototypes with shaped and multiple elastic regions, and c) digital model for aggregation of multiple meso-system into one macrosystem.

experiments, as long as someone has access to a CNC knitting machine. This research also indicates the need for further studies on how different parameters (knit structures, different yarn types, composite formation process and the application of pre-stress during resin impregnation) can influence the formation of hybrid TRC in cases of more complex geometric configuration with multiple tensile regions. There were few attempts to introduce that aspect in some experiments at a very preliminary stage, but that could be researched more rigorously to explore larger systems. This ultimately can lead into developing highly articulated and differentiated full-scale architectural façade elements, designed and fabricated by inherent material logic. Additional research can also investigate the potential for a macro-system through the aggregation of mesosystems.

ACKNOWLEDGEMENTS

The research was developed as part of the capstone project under the MS in Architecture (Material Systems) program at the University of Michigan Taubman College of Architecture and Urban Planning. The authors are also thankful to the Rackham Grant for supporting the research work.

REFERENCES

Abounaim, Md., Gerald Hoffmann, Olaf Diestel, and Chokri Cherif. 2010. "Thermoplastic Composite from Innovative Flat Knitted 3D Multi-layer Spacer Fabric Using Hybrid Yarn and the Study of 2D Mechanical Properties." Composites Science and Technology 70 (2): 363–370.

Ahlquist, Sean, and Achim Menges. 2015. "Materiality and Computational Design: Emerging Material Systems and the Role of Design Computation and Digital Fabrication." In The Routledge Companion for Architecture Design and Practice, edited by Mitra Kanaani and Dak A. Kopec. New York: Routledge. 149–168.

Ahlquist, Sean, and Achim Menges. 2013. "Frameworks for Computational Design of Textile Micro-Architectures and Material Behavior in Forming Complex Force-Active Structures." In ACADIA 13: Adaptive Architecture Proceedings of the 33rd Annual Conference of the Association for Computer Aided Design in Architecture (ACADIA), edited by Philip Beesley, Omar Khan and Michael Stacey, 281–292. Cambridge: Riverside Architectural Press.

Ahlquist, Sean, Ali Askarinejad, Rizkallah Chaaraoui, Ammar Kalo, Xiang Liu, and Kavan Shah. 2014. "Post Forming Composite Morphologies: Materialization and Design Methods for Inducing Form Through Textile Material Behavior." In ACADIA 14: Design Agency—Proceedings of the 34th Annual Conference of the Association for Computer Aided Design in Architecture, edited by David Gerber, Alvin Huang, and Jose Sanchez. Los Angeles: ACADIA. 267–276.

Cebulla, H., O. Diestel, and P. Offermann. 2002. "Fully Fashioned Biaxial Weft Knitted Fabrics." AUTEX Research Journal 2 (1): 8–13.

Engel, Heino. 2007. Tragsysteme—Structure Systems, Fourth Edition. Ostfildern, Germany: Hatje Cantz.

IMAGE CREDITS

All images: Sharmin and Ahlquist, 2016

Shahida Sharmin is an architect and designer. Recently, she has graduated from the MS in Architectural Design and Research (Material System) program from Taubman College of Architecture and Urban Planning at the University of Michigan. She is working as a full-time researcher with Prof. Ahlquist and leads the development of knitting 3-dimensional (net shape) preforms with structural hybrid yarns. She is interested in advanced materials and fabrication technologies and committed in investigating the influence of material system as an agent for initiating coherent physical and social environment.

Sean Ahlquist is an Assistant Professor of Architecture at the University of Michigan. He is a part of the Cluster in Computational Media and Interactive Systems which connects Architecture with the fields of Material Science, Computer Science, Art & Design and Music. Research and course topics are centered on material computation, developing articulated material structures and modes of design which enable the study of spatial behaviors and human interaction. Ahlquist's research agendas include the design and fabrication of pre-stressed lightweight structures, innovations in textile-reinforced composite materials for aerospace and automotive design, and development of tactile sensorial environments as interfaces for physical interaction.

Bending-Active Plates

Form-Finding and Form-Conversion

Simon Schleicher
University of California, Berkeley
Department of Architecture

Riccardo La Magna
University of Stuttgart, Institute
of Building Structures and
Structural Design (ITKE)

1

ABSTRACT

With this paper, the authors aim to contribute to the discourse on bending-active structures by highlighting two different design methods, form-finding and form-conversion. The authors compare the two methods through close analysis of bending-active plate structures, discussing their advantages and disadvantages based on three built case studies. This paper introduces the core ideas behind bending-active structures, a rather new structural system that makes targeted use of large elastic deformations to generate and stabilize complex geometrical forms based on initially planar elements. Previous research has focused mainly on form-finding. As a bottom-up approach, it begins with flat plates and recreates the bending and coupling process digitally to gradually determine the final shape. Form-conversion, conversely, begins with a predefined shape that is then discretized by strategic surface tiling and informed mesh subdivision, and which in turn considers the geometrical and structural constraints given by the plates. The three built case studies exemplify how these methods integrate into the design process. The first case study applies physical and digital form-finding techniques to build a chaise lounge. The latter two convert a desired shape into wide-spanning constructions that either weave multiple strips together or connect distant layers with each other, providing additional rigidity. The presented case studies successfully prove the effectiveness of form-finding and form-conversion methods and render a newly emerging design space for the planning, fabrication, and construction of bending-active structures.

1   Berkeley Weave installed at the courtyard of UC Berkeley's College of Environmental Design (CED).

INTRODUCTION

In recent years, the architecture community has witnessed an increased availability and constant improvement of computational tools that enable not only advanced geometrical modeling, but also the integration of real-time physics-based simulations into the design process in common CAD environments. Programs like Kangaroo Physics or SOFiSTiK are used, for example, to rapidly form-find and interact with particle systems or accurately analyze structures on the bases of Finite Element Methods (Piker 2013; Lienhard et al. 2011). With the help of these programs, one can describe and evaluate the mechanical behavior and structural capacity of a model under simultaneous consideration of external forces and internal material stresses.

With the rise of these tools, architects and engineers are becoming more and more interested in structural systems whose forms and load states cannot easily be predicted, but instead result from a delicate balance between geometry, interacting forces, and material properties. It is here, in particular, where physics-based simulations that provide real-time feedback can demonstrate their strength. Bending-active structures illustrate these interrelationships and as such are chosen in this paper as a detailed example (Figure 1).

This newly established structural system is characterized by the use of large elastic deformations of initially planar building materials to generate geometrically complex constructions (Knippers et al. 2011). While the traditional maxim in engineering is to limit the amount of bending in structures, this typology actually harnesses bending for the creation of complex and extremely lightweight designs. The underlying idea of exploiting a structure's flexibility in a controlled way is rather simple and extremely versatile. It can be used, for example, as a form-giving and self-stabilizing strategy in static structures or as compliant mechanism in kinetic structures (Lienhard 2014, Schleicher 2015).

BENDING-ACTIVE PLATE STRUCTURES

Bending-active structures can be generally divided into two main categories, which relate to the geometrical dimensions of their fundamental components. One-dimensional (1D) systems can be built, for instance, by bending slender rods, while two-dimensional (2D) systems use thin plates as basic building blocks. While extensive knowledge and experience exists for 1D systems, with elastic gridshells as most prominent application, plate-dominant structures have not received much attention yet and are considered more difficult to design. One reason is certainly that plates have a limited formability, since they bend mainly along the axis of weakest inertia and thus cannot easily be forced into complicated geometries. However, what makes this subset of bending-active structures particularly interesting from

2   3

2 Buckminster Fuller's geodesic plydome in Des Moines, Iowa, 1957. The hemisphere spans 7.3 m and is made out of marine plywood sheets with a thickness of 6.4 mm.

3 ICD/ITKE Research Pavilion 2010 by the University of Stuttgart spans 10 m and consists of 80 birch plywood strips with a thickness of 6.4 mm.

a mathematical point of view is the fact that plates have a clear scale separation. They are typically very large in one dimension and progressively smaller in the other two. Their length is specified in meters, their width in centimeters, and their height only in millimeters. Having hierarchical geometrical features facilitates the further design process of bending-active plate structures and makes it easier to assess the structural behavior and accurately anticipate their deformed geometry with digital simulations.

Prominent examples for bending-active plate structures are Buckminster Fuller's plydomes and the ICD/ITKE Research Pavilion 2010 (Figures 2 and 3). While the first example follows a rational approach in which the shape of a sphere is approximated with a regular tiling of identical plates (Fuller 1959), the second example takes advantage of computational mass customization and joins 500 individual parts together (Fleischmann et al. 2012).

FORM-FINDING AND FORM-CONVERSION

A previous study identified three main design strategies for bending-active structures: a behavior-based, geometry-based, and integrative approach (Lienhard et al. 2013). According to this study, the first category refers to the traditional approach by skilled craftsmen who bend building materials intuitively on the construction site. The other two categories describe a more scientific approach, in which hands-on experiments and analytical tests were conducted beforehand and informed the further design and construction process. While the geometry-based approach relates to the idea of forcing an object to match a specific target geometry without further consideration of material properties, the integrative approach takes exactly these limiting factors into account when exploring a reachable design space. In order to best contribute to the above-mentioned classification of bending-active structures, it is the aim of this paper to further elaborate emerging design trends in the integrative approach by having a closer look at the techniques of form-finding and form-conversion.

Kangaroo Physics

SOFiSTiK in Rhinoceros

4 The form-finding approach
   starts from a flat sheet and uses
   contracting elastic cables elements
   to generate the final bent shape.
   Simulations in Kangaroo Physics
   allow for quick and interactive models while the software
   SOFiSTiK enables precise shape
   and stress analysis based on Finite
   Element Methods. (Schleicher et al.
   2015).

4

Bottom-Up Design Using Form-Finding

The term form-finding is best known for its role in the design of membranes and shell structures and refers to the concept of using physical models and numerical simulations to find an optimal geometry of a structure in static equilibrium with a design loading (Adriaenssens et al. 2014). From the 1950s onwards, architects and engineers focused on form-finding strategies that both incorporated materials and forces while enabling a systematic exploration of lightweight constructions. They became an essential part in the work of people like Buckminster Fuller, Félix Candela, Heinz Isler, and Frei Otto (Chilton 2000; Otto 2005). While these early pioneers implemented form-finding strategies to design shells and membranes determined by the shape of hanging chains and cloth, these techniques can also be applied for bending structures.

In the context of bending-active plate structures and digital simulation, form-finding is often used for a bottom-up design approach. It starts with planar sheets or strips to create the bending and coupling process in the final shape (Lienhard, Schleicher, and Knipper 2011; Fleischmann et al. 2012). By using spring-based simulations in Kangaroo Physics or finite element methods in SOFiSTiK, one can not only determine the resulting geometry of the deformed structure but also visualize the evolution of stresses within the material while the system is deforming (Figure 4) (Schleicher et al. 2015). Based on this information, one can cautiously bend a component, for instance by following an ultra-elastic cable approach until a permissible stress state is reached (Lienhard et al. 2014). The final shape and caused stresses are often unknown at the beginning, especially when multiple parts are bent and fastened together, which is a considerable drawback. A designer with a certain aim in mind would

therefore have to conduct multiple simulations with gradually changing parameters to move closer to the design objective.

Top-Down Design Using Form-Conversion

In comparison to the previous method, form-conversion pursues a different approach when integrating geometrical and material considerations into the design of bending-active plate structures. Here, the process begins with a predefined target surface or mesh, which is then discretized and further subdivided into smaller bent tiles based on the flexibility of the used plates. The main restriction in this regard is the knowledge concerning the plates' material formability. Here, it is particularly important to know that for strips and plate-like elements, the basic shapes that can be achieved by pure bending without stretching are conical and cylindrical surfaces. These shapes are also referred to as single-curved or developable surfaces. Attempting to bend a sheet of material in two directions simultaneously either results in irreversible, plastic deformations or ultimate material failure. Thus, to expand the range of achievable shapes, it is necessary to develop other methods for the induction of Gaussian curvature into the system.

To overcome the limitations related to Gaussian curvature, multidirectional bending can be induced by strategically removing material and freeing the plates from the stiffening constraint of their surroundings. This principle is illustrated in Figure 5 and a similar approach was presented by Xing et al. (2011). Here, a continuous rectangular plate is reduced to two orthogonal strips. Once again, the strips are bent using the ultra-elastic cable approach of Lienhard et al. (2014). The bending stiffness of the plate, depending proportionally on its width, results in a radical increase of stiffness in the connecting area between the strips.

Bending-active Plates Schleicher, La Magna

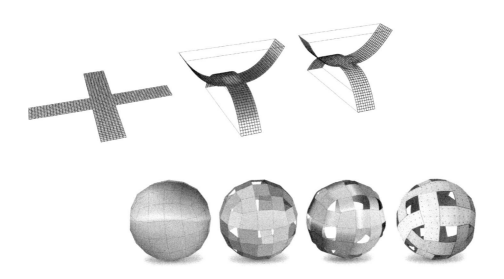

5   The form-conversion approach is informed by the mechanical characteristics of a bent material and applies these principles to the subdivision process of a given target geometry. The upper row shows the multidirectional bending of a cross-like strip based on contracting elastic cables. The center image indicates the distribution of von Mises stress. The image on the right shows the Gaussian curvature. The bottom row uses these limiting factors for a form-conversion of a target mesh into an assembly of bent components.

5

As a result, the connecting area remains almost planar and the perpendicular bending axis remains unaffected by the induced curvature. In this way, it becomes possible to bend the strips around multiple axes, spanning different directions but still maintaining the material continuity of a single element. The center image of the cross-like strips in Figure 5 depicts the resulting von Mises stresses and clearly displays an area of unstressed material at the intersection between the two strips, which supports the previous arguments. A local stress concentration appears at the junction of the strips due to the sharp connecting angle as well as the inevitable geometric stiffening in that area.

The result of multidirectional bending can be compared with an analysis of Gaussian curvature on the right. From the plot, it is clear that the discrete Gaussian curvature of the deformed mesh is zero everywhere apart from a small, localized area at the intersection of the two branches. This confirms the assumption that, for inextensible materials, most developable surfaces (or slight deviation thereof) are achievable. Based on this approach, other arbitrary freeform surfaces can be converted, following the logic of strategic material removing and defined zones of local bending and planarity, as demonstrated in the lower images of Figure 5.

CASE STUDIES

To further illustrate the design potentials of a form-finding and form-conversion approach, the following section will have a closer look at three built case studies. While the first case study takes advantage of a bottom-up, form-finding technique to design a load-bearing chaise lounge, the latter two demonstrate the form-conversion of pre-defined shapes into wide-spanning constructions that gain rigidity by either weaving multiple strips together or connecting distant layers with each other.

Bending-Active Chaise Lounge

The first case study represents the previously mentioned form-finding process and demonstrates its design possibilities in the context of a furniture-scale object. The goal of this bending-active plate structure is to meet the highest structural demands yet only use a minimum amount of material. The application chosen to address this challenge in built form was a chaise lounge for one person.

In this project, bending is primarily induced by strategically removing material from the center of a thin sheet and then pinching its naked edges together and fastening the deformed shape with rivets. This technique has multiple benefits: generating intricate forms out of a single planar surface and achieving three-dimensional shapes that perform structurally in the bent state. The general design of the chaise lounge followed a gradual form-finding approach that comprised a series of physical models and digital simulations. In a first step, the geometry of different reference chairs and the relationship with the human body was studied and a group of target angles were identified that allow for a comfortable seating position (Figure 6). Based on this information, quick sketch models were built out of paper to gain a better understanding of the interdependencies between the cutting pattern and the angles of the deformed structure after the pinching.

The second step was to turn these cutting patterns into digital models. This was done in Grasshopper and the pinching simulated in Kangaroo Physics. At this point, the main advantage of using this type of spring-based simulation was the possibility for real-time feedback and user-interactivity to further modify the cutting pattern and improve the design (Figure 7). Furthermore,

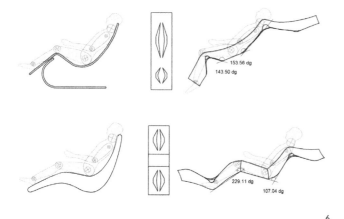

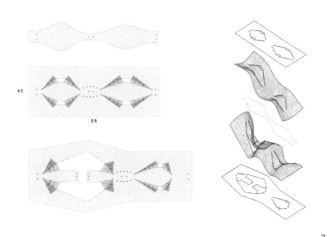

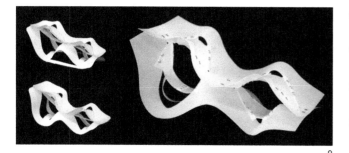

tracking the curvature of the mesh and identifying minimal bending radii allowed us to draw first conclusions if this form could be built out of a specific target material.

The following third step provided much more accurate results regarding the materialization of the chaise lounge. Here, SOFiSTiK was used to generate finite element models with defined material characteristics (Figure 8). All surfaces were given the material properties of high-density polyethylene (HDPE) with a Young's modulus of 1200 N/mm^2 and a thickness of 1.6 mm or 3.2 mm for a single or double layer. The model was then deformed using the ultra-elastic cable approach. Consulting this slightly more time-consuming method at this stage of the design process had multiple benefits. It allowed us to calculate the exact geometry of the highly deformed structure and assess the stresses within. Thus, this simulation is a much more complete description of the mechanical behavior and structural capacity of the bending-active system. Furthermore, the feedback on the structure's complex equilibrium state also allowed localizing potentially dangerous stress concentrations. And last but not least, having done the simulation in typical engineering software also made it possible to further analyze the chair's structural performance once the weight of a human body was added.

As a proof of concept, the chaise lounge was built both as a series of small-scale models as well as a full-scale chaise lounge with the dimensions of 2.44 m x 1.22 m x 1.6 mm (Figures 9–10). The construction material was HDPE, the patterns for the different sheets were cut on a Zünd blade cutter, and the pieces were connected together with steel rivets. Riveting, and in particular blind riveting, was used in this project both to permanently pinch each surface as well as to connect multiple plates with each other. Since rivets can only transmit tension and shear forces, their exact position needed to be determined carefully. In this regard, the iterative from-finding over multiple simulations

6   Different reference chairs were analyzed for their seating position and the angles therein were recreated by pinching flat sheets into a deformed shape.

7   Digital simulations in Kangaroo Physics were used to quickly test different cutting patterns and form-find the desired geometrical form.

8   Structural validation of the form-found shape based on FEM simulations and a thorough analysis of the appearing minimal bending radii.

9   A series of paper and plastic sketch models were used to gradually approach the final shape and ascertain the required cutting pattern.

10 Full-scale prototype of the bending chaise lounge is built out of 1.6 mm thin HDPE plastic and is able to carry the weight of a person.

11 Berkeley Weave installation spans over 4 m and is built out of 480 individual plywood strips with a thickness of only 3 mm.

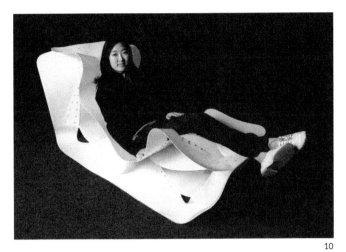

10

11

played an important role. It provided crucial information about the precise geometry of the deformed plates as well as the exact position of the rivet holes, which was needed to guarantee the alignment of all layers. As far as the actual assembly of the chaise lounge is concerned, the bending of the plastic was rather easy and could be done manually. In fact, it was surprising how rigid the structure became once all pieces were fastened together and the stored elastic energy began to pre-stress the structure. The final chaise lounge was capable of carrying the weight of a person, of course under some deflections but within permitted tolerances (Figure 10).

Berkeley Weave

In contrast to the previous example, which uses only a small number of parts, designing bending-active plate structures out of multiple components is very challenging and thus requires a different approach. For this reason, the second case study applies a different method and aims to demonstrate the design potential of form-conversion (Figure 11). It investigates an integrative approach that considers not only bending but also torsion of slender strips. The saddle-shaped design of the Berkeley Weave is based on a modified Enneper surface (Figure 12a). This particular form was chosen because it has a challenging anticlastic geometry with locally high curvature. The subsequent conversion process into a bending-active plate structure followed several steps. The first was to approximate and discretize the surface with a quad mesh (Figure 12b). A curvature analysis of the resulting mesh reveals that its individual quads are not planar but spatially curved (Figure 12c). The planarity of the quads, however, will be an important precondition in the later assembly process. In a second step, the mesh was transformed into a four-layered weave pattern with composed strips that feature pre-drilled holes. Here, each quad was turned into a crossing of two strips in one direction with two other strips at a 90-degree angle. The resulting interwoven mesh was then optimized for planarization.

However, only the regions where strips overlapped were made planar, while the quads between the intersections remained curved (Figure 12d). A second curvature analysis illustrates the procedure well and shows zero curvature at the intersections of the strips (blue areas) while the connecting arms are both bent and twisted (Figure 12e). Specific routines in the form-conversion process guaranteed that the bent zones stayed within the permissible bending radii. In the last step, this converted shape was used to generate a fabrication model that featured all the connection details and strip subdivisions (Figure 12f).

A closer look at the most extremely curved region illustrates the complexity related to this last step (Figure 13). To allow for a proper connection, bolts were only placed in the planar regions between intersecting strips. Since the strips are composed out of smaller segments, it was also important to control their position in the four-layered weave and the sequence of layers. A pattern was created which guaranteed that strip segments only ended in layer two and three and are clamped by continuous strips in layer one and four. A positive side effect of this weaving strategy is that the gaps between segments are never visible and the strips appear to be made out of one piece. The drawback, however, is that each segment has a unique length and requires individual positions for the screw holes (Figure 14).

To demonstrate proof of concept for this design approach, this case study was built in the dimensions of 4 m x 3.5 m x 1.8 m (Figure 11). The structure is assembled out of 480 geometrically different plywood strips that were fastened together with 400 bolts. The material used is 3.0 mm thick birch plywood with a Young's modulus of $Em\|= 16471$ N/mm^2 and $Em\bot = 1029$ N/mm^2. Dimensions and material specifications were employed for a finite element analysis using the software SOFiSTiK. Under consideration of self-weight and stored elastic energy, the minimal bending radii are no smaller than 0.25 m

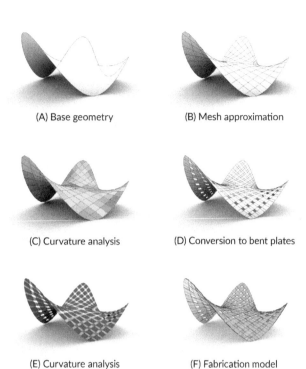

(A) Base geometry

(B) Mesh approximation

(C) Curvature analysis

(D) Conversion to bent plates

(E) Curvature analysis

(F) Fabrication model

12

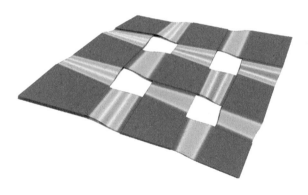

13

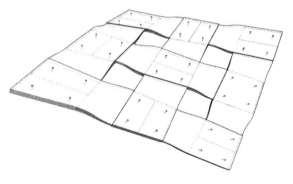

14

12 Form-conversion process and analysis of the Berkeley Weave.

13 Analysis of Gaussian curvature in the area with the highest deformation.

14 Schematic drawing of the technical details in the four-layered weave.

and the resulting stress peaks are still below 60% of permissible material utilization.

Bend9

The third case study showcases another take on form-conversion for bending-active plate structures that consists of many components. This project is a multi-layered arch that spans over 5.2 m and has a height of 3.5 m. It was built to prove the technical feasibility of using bending-active plates for larger load-bearing structures. In comparison to the previous case study, this project implements a different tiling pattern and explores the possibility of significantly increasing a shape's rigidity by cross-connecting distant layers with each other. To fully exploit the large deformations that plywood allows for, the thickness of the sheets had to be reduced to the minimum, leading once again to the radical choice of employing 3.0 mm birch plywood. Since the resulting sheets are very flexible, additional stiffness needed to be gained by giving the global shell a peculiar geometry, which seamlessly transitions from an area of positive curvature (sphere-like) to one of negative curvature (saddle-like) (Figure 15a). This pronounced double-curvature provides additional stiffness and helps avoid undesirable deformation of the structure. Despite the considerable strength achieved by the shape alone, the choice of using extremely thin sheets of plywood at that scale asked for additional reinforcement to provide further load resistance. These needs were met by a double-layered structure with two cross-connected shells.

As in the previous example, the first step of the process was to convert the base geometry into a mesh pattern (Figure 15b). In the next step, a preliminary analysis of the structure was conducted and informed the offsetting of the mesh to create a second layer. As the distance between the two layers varies to reflect the bending moment calculated from the preliminary analysis, the offset of the surfaces changes along the span of the arch (Figure 15c). The offset reflects the stress state in the individual layers, and the distance between them increases in the critical areas to increment the global resistance of the system. The following form-conversion process was once again driven by material constraints and previously determined permissible stress limits with respect to bending and torsion. The resulting tiling logic that was used for both layers affected the size of the members and guaranteed that each component could be bent into the specific shape required to construct the whole surface. More precisely, this is achieved by strategically placing voids into target positions of the master geometry, ensuring that the bending process can take place without prejudice for the individual components (Figure 15d). Although initially flat, each element undergoes multi-directional bending and gets locked into position once it is fastened to its neighbors. The flexible 3.0

(A) Base geometry

(B) Mesh approximation

(C) Layer offset

(D) Form-conversion

(E) FE analysis

(F) Fabrication model

15 The form-conversion process of Bend9 pavilion started from a base geometry (A) and approximated this shape with a mesh (B). Based on a first structural analysis, the mesh was offset and turned into a double layer. This structure was then converted into an assembly of bent plates (D). After another finite element analysis (E), a fabrication model was generated (F).

16 Detail of the assembled structure shows the layering of different components and the strategically placed voids to prevent conflicts between the bent parts.

17 Custom wood profiles were used to cross-connect the two layers together and thus increase the structural capacity of the pavilion significantly.

15

mm plywood elements achieve consistent stiffness when joint together, as the pavilion, although a discrete version of the initial shape, still retains substantial shell stiffness. This was validated in another finite element analysis that considered both self-weight as well as undesirable loading scenarios (Figure 15e).

Finally, after fabrication, the structure was assembled on site. The built structure employs 196 elements unique in shape and geometry (Figure 16). 76 square wood profiles of 4 cm x 4 cm were used to connect the two plywood skins (Figure 17). Due to the varying distance between the layers, the connectors had a total amount of 156 exclusive compound miters. The whole structure weighs only 160 kg, a characteristic that also highlights the efficiency of the system and its potential for lightweight construction. The smooth curvature transition and the overall complexity of the shape clearly emphasize the potential of the construction logic. Furthermore, the implemented form-conversion process can be applied to any kind of double-curved freeform surface, not only the one built at UC Berkeley's campus (Figure 18).

CONCLUSION

In summary, it can be concluded that the three case studies clearly illustrate the feasibility of form-finding and form-conversion techniques for the design of bending-active plate structures. All three examples showcase an integrative approach that is directly informed by the mechanical properties of the thin plastic and plywood sheets, which were employed in the different projects. Their overall geometry is therefore the result of an accurate negotiation between the mechanical limits of the materials and

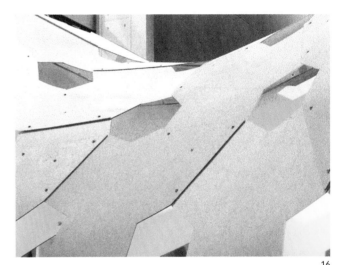

16

17

18 View of the Bend9 structure assembled out of 3 mm thin birch plywood in the courtyard at UC Berkeley's College of Environmental Design (CED).

their deformation capabilities. The very nature of all three case studies required a tight integration of design, simulation, and assessment of fabrication and assembly constraints.

Due to its small number of parts, the bending chaise lounge was a good case study to demonstrate the potential of design processes based on iterative form-finding. Depending on the simulation software used, this method can be very quick and interactive or particularly accurate and reliable regarding its results. This precision, however, comes at the expense of simulation speed. Therefore, form-finding meets its natural boundaries when the number of parts exceeds a certain limit.

The second and third case studies aimed to tackle this challenge by presenting form-conversion as an alternative design approach for bending-active plate structures that consist of many parts. Furthermore, the Berkeley Weave and the Bend9 pavilion exemplify the capacity of bending-active plate structures to be employed as larger scale, space-framing architectural interventions. For future research, the presented case studies and the underlying design routines of form-finding and form-conversion

will serve as first prototypes for the exploration of more complex surface-like shell structures that derive their shape through elastic bending.

ACKNOWLEDGEMENTS

The authors would like to thank the following student team for their amazing work on the bending chaise lounge and their indirect contribution to this paper: Cindy Hartono, Fei Du, Shima Sahebnassagh, Eleanna Panagoulia as well as their additional supervisors Prof. Kyle Steinfeld, Prof. Jonathan Bachrach, and Luis Jaggy. For the Berkeley Weave installation, the authors would particularly like to thank Sean Ostro, Andrei Nejur, and Rex Crabb for their support. Finally, the Bend9 pavilion would not have been possible without the kind support of Autodesk's Pier 9 and its entire staff.

REFERENCES

Adriaenssens, Sigrid, Philippe Block, Diederick Veenendaal, and Chris Williams, eds. 2014. *Shell Structure for Architecture: Form Finding and Optimization*. London: Routledge.

Chilton, John. 2000. *The Engineer's Contribution to Contemporary Architecture: Heinz Isler*. London: Thomas Telford.

Fleischmann, Moritz, Jan Knippers, Julian Lienhard, Achim Menges, and Simon Schleicher. 2012. "Material Behaviour: Embedding Physical Properties in Computational Design Processes." *Architectural Design* 82 (2): 44–51.

Fuller, R. Buckminster, and Robert W. Marks. 1973. *Dymaxion World of Buckminster Fuller*. New York: Anchor Books.

Fuller, R. Buckminster. 1959. Self-strutted geodesic plydome. US Patent 2,905,113, filed April 22, 1957, and issued September 22, 1959.

Knippers, Jan, Jan Cremers, Markus Gabler, and Julian Lienhard. 2011. *Construction Manual for Polymers Membranes: Materials, Semi-Finished Products, Form-Finding Design*. Basel: Birkhauser Architecture.

Lienhard, Julian, Simon Schleicher, and Jan Knippers. 2011. "Bending-Active Structures—Research Pavilion ICD/ITKE." In *Proceedings of the International Symposium of the IABSE-IASS Symposium*. London, UK: IABSE-IASS.

Lienhard, Julian, Holger Alpermann, Christoph Gengnagel, and Jan Knippers. 2013. "Active Bending, A Review on Structures Where Bending Is Used as a Self-Formation Process." *International Journal of Space Structures* 28 (3–4): 187–196. doi:10.1260/0266-3511.28.3-4.187.

Lienhard, Julian, Riccardo La Magna, and Jan Knippers. 2014. "Form-finding Bending-Active Structures with Temporary Ultra-Elastic Contraction Elements." In *Proceedings of 4th International Conference on Mobile, Adaptable and Rapidly Assembled Structures*, edited by N. De Temmerman and C. A. Brebbia. Ostend, Belgium: MARAS. 107–116. doi:10.2495/mar140091.

Lienhard, Julian. 2014. "Bending-Active Structures: Form-Finding Strategies Using Elastic Deformation in Static and Kinetic Systems and the Structural Potentials Therein." PhD Dissertation, University of Stuttgart.

Otto, Frei. 2005. *Frei Otto: Complete Works: Lightweight Construction, Natural Design*, edited by Winfried Nerdinger. Basel: Birkhäuser.

Piker, Daniel. 2013. "Kangaroo: Form Finding With Computational Physics." *Architectural Design* 83 (2): 136–137.

Schleicher, Simon. 2015. "Bio-Inspired Compliant Mechanisms for Architectural Design: Transferring Bending and Folding Principles of Plant Leaves to Flexible Kinetic Structures." PhD Dissertation, University of Stuttgart.

Schleicher, Simon, Andrew Rastetter, Riccardo La Magna, Andreas Schönbrunner, Nicola Haberbosch, and Jan Knippers. 2015.

"Form-Finding and Design Potentials of Bending-Active Plate Structures." In *Modelling Behaviour*, edited by M. Ramsgaard Thomsen, M. Tamke, C. Gengnagel, B. Faircloth, F. Scheurer. Berlin: Springer. 53–64.

Xing, Qing, Gabriel Esquivel, Ergun Akleman, Jianer Chen, and Jonathan Gross. 2011. "Band Decomposition of 2-Manifold Meshes for Physical Construction of Large Structures." In *Posters of the 38th International Conference and Exhibition on Computer Graphics and Interactive Techniques*. Vancouver, BC: SIGGRAPH.

IMAGE CREDITS

Figure 2: Marks, 1973
Figure 3: Schleicher, 2010
Figure 4: Schleicher et al. 2015
Figures 6–9: Hartono, Du, Sahebnassagh, Panagoulia, 2015
All other photography: Schleicher and La Magna, 2016

Simon Schleicher is an Assistant Professor in the Department of Architecture at the University of California, Berkeley. Simon holds a doctoral degree from the University of Stuttgart and worked for the Institute of Building Structures and Structural Design (ITKE). His transdisciplinary work draws from architecture, engineering, and biology. By cross-disciplinary pooling of knowledge he aims to transfer bending and folding mechanisms found in nature to lightweight and responsive systems in architecture.

Riccardo La Magna is a structural engineer and PhD candidate at the Institute of Building Structures and Structural Design (ITKE) at the University of Stuttgart. In his research he focuses on simulation technology, innovative structural systems, and new materials for building applications.

Bio-Inspired Kinetic Curved-Line Folding for Architectural Applications

Axel Körner
Anja Mader
Saman Saffarian
Jan Knippers
ITKE/University of Stuttgart

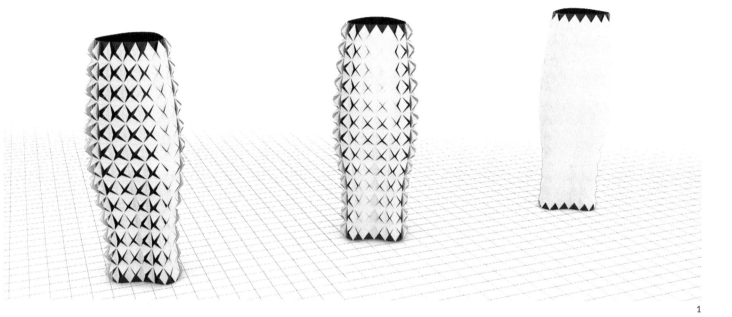

1

ABSTRACT

This paper discusses the development of a bio-inspired compliant mechanism for architectural applications and explains the methodology of investigating movements found in nature. This includes the investigation of biological compliant mechanisms, abstraction, and technical applications using computational tools such as finite element analysis (FEA). To demonstrate the possibilities for building envelopes of complex geometries, procedures are presented to translate and alter the disclosed principles to be applicable to complex architectural geometries.

The development of the kinetic façade shading device *flectofold*, based on the biological role-model *Aldrovanda vesiculosa*, is used to demonstrate the process. The following paper shows results of FEA simulations of kinetic curved-line folding mechanisms with pneumatic actuation and provides information about the relationship between varying geometric properties (e.g. curved-line fold radii) and multiple performance metrics, such as required actuation force and structural stability.

1   Application on geometry of synclastic and anticlastic double curvature.

INTRODUCTION

Buildings account for nearly 40% of the global energy consumption, and the importance of the development of energy-conscious design and construction strategies is known and well established. Building facades are the most intense energy-exchange zones and kinetic building envelopes have a high potential to improve the energy efficiency of the overall design and to perform multi-objective tasks in an integrative fashion. Solar-shading and daylight-control can be improved by controlling movement configurations, and energy-collection and lighting-integration can be integrated by embedding PV elements and light fittings into their material structure.

Past examples of kinetic facades have heavily relied on rigid body mechanics to achieve movement. These complex mechanical systems are mostly guided along straight translation or rotation axes – resulting in geometrical constraints. They are expensive to manufacture, prone to failure, hard to maintain and therefore not feasible in terms of operation economy. However, recent biomimetic research has identified strategies to achieve mobility by utilizing elastic deformation of fibrous materials to drive transformation. These kinetic elastic systems (compliant mechanisms) have the potential to dramatically reduce the mechanical complexity of kinetic elements while providing a wide range of complex yet efficient movements that are arguably more suitable for the design of freeform architectural envelopes of the near future.

BACKGROUND

Kinetic Structures

Recent research has defined a terminology for Kinetic Facades, classified them into various categories based on their system properties, and defined a methodology for their performance evaluation. Most notably, Wang defined the term Acclimated Kinetic Envelopes (AKE) as "an envelope with the aptitude to adapt itself in reversible, incremental and mobile ways" (Wang et al. 2012) and Loonen has defined Climate Adaptive Building Shells (CABS) as "a system that has the ability to repeatedly and reversibly change its features, functions or behaviour over time in response to changing performance requirements and variable boundary conditions with the aim of improving overall building performance" (Loonen et al. 2013). Furthermore CABS have been classified into multiple categories based on their (i) source of inspiration for their design (phototropism and heliotropism of plants) (ii) relevant physics of their interface with the environment (blocking, filtering, converting, collecting or storing energy), (iii) time-scales of their operation (seconds, minutes, hours, diurnal and seasons), (iv) scales of adaptation to external stimuli (micro-scale and macro-scale), and (v) control types in charge of their operation (extrinsic and intrinsic).

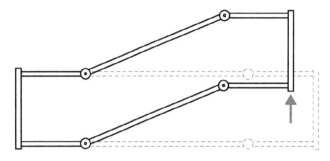

Rigid body mechanism

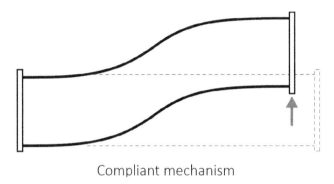

Compliant mechanism

2   Comparison between rigid body and compliant mechanism.

Adhering to the above mentioned classification, the specific Kinetic Facade Element *flectofold* discussed in this paper can be identified as a system that is bio-inspired, capable of blocking, converting and possibly storing energy, has the potential to operate on multiple time-scales, adapts on a macro-scale, and is controlled extrinsically through pneumatic actuation.

Compliant Kinetic Mechanisms

Compliant mechanisms achieve their flexibility by controlled elastic deformation of flexible members. In contrast to conventional rigid body mechanisms, they consist of only one part with locally defined stiffness. These mechanisms can transfer motion, force or energy upon deformation of the flexible parts. The main advantage of compliant mechanisms is the reduction of parts, resulting in potential economy due to simplified manufacturing and assembly. Furthermore, wear can be reduced, and unlike classical joints, there is no need for lubrication and maintenance.

Some early man-made machines, like bows, are compliant. Nowadays, compliant mechanisms known as living hinges can mainly be found on a smaller scale in medical devices or in the packaging industry, as in book covers or lids of shampoo bottles, but not on a larger scale as in architecture or building construction. (Howell et al. 2013) (figure 2)

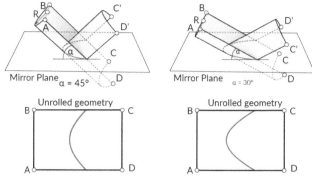

To create movements in kinetic structures, actuators are necessary to transform an energy input into motion. The input can occur as thermal, electrical or mechanical energy, allowing for a passive reaction to environmental conditions such as heat and humidity, or may demand an active trigger. Examples for passive actuation can be bimetals or anisotropic swelling. For active actuation, so called smart materials, such as shape memory materials or electrorestrictive and magnetorestrictive materials can be integrated into kinetic structures (Ham et al. 2009). Due to their high power-to-weight ratio and low material and fabrication cost, pneumatic and hydraulic systems are promising actuators for kinetic structures (Polygerinos et al. 2015).

Biomimetic Kinetic Structures

Adaptability to various geometric conditions and robustness during service is of primary importance when it comes to architectural applications. High requirements with respect to accuracy, velocity and weight can be seen in most cases compared to many other fields of kinetic technology. Many movements in biological organisms fulfill these criteria and are utilized by compliant mechanisms, triggered for example by hydraulic pressure, hygroscopic swelling and the release of stored elastic energy. This makes it valid to investigate biological compliant mechanisms as alternative solutions of adaptive building systems. Research in this field can be differentiated in two directions: Top-down and bottom-up. Biomimetic research for kinetic architectural applications has been studied intensively at the ITKE, most noteworthy in the doctoral theses of Mohammad-Reza Matini and Simon Schleicher.

Matini developed a systematic approach for the investigation of movements based on compliant mechanisms as found in biology. He abstracts the basic movement principles, transfers them into geometric models, and categorizes them into types of movements (2D or 3D, change of curvature, change of curvature direction, etc.), leading to a catalogue of compliant movements. He proposes a series of potential architectural applications using abstracted and combined principles. His work is mainly guided by technological questions, and can be seen as a top down approach. (Matini 2007).

In collaboration with biologists, Schleicher developed a methodology for abstraction and simulation of plant movements by the means of computational tools in his dissertation. His methodology includes the simulation of the actual plant movements, the disclosure of underlying geometrical principles and their variations, and analysis of the involved forces and energy. More focused on the biological role-models, it can be seen as a bottom up approach (Schleicher 2016).

METHODS

The aim of the described process is to investigate and identify basic principles of biological compliant mechanisms and the transfer into applicable technical solutions by exploring the geometrical variations.

The sequential methodology of analysis, abstraction and technical translation of biological movements builds upon the work by Simon Schleicher (Schleicher 2016), (Schleicher et al. 2015). He divided the process of modeling and simulating the bio-inspired kinetic mechanisms into three main categories: geometric, kinematic and kinetic models.

The geometrical model can be seen as a static representation of the underlying geometrical features responsible for movements. Once identified, the geometric principles can be parameterized and a catalogue of variations of topologically identical specimens can be generated.

The kinematic model investigates the actual movement. The geometrical variations can be qualitatively evaluated in terms of their influence on the movement without taking forces and needed energy into account.

The kinetic model uses non-linear finite element analysis to simulate large elastic deformations and to evaluate the involved forces. Based on exact physical material properties, it enables a sophisticated comparison of the actuation force and resultant movement for the geometrical variations and different gradients

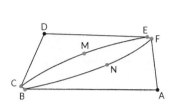
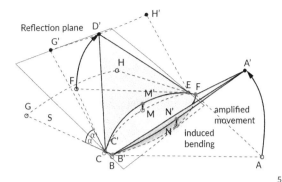

3   Biological role-model *Aldrovanda vesiculosa*.

4   Method of reflection, adapted from Mitani and Iagarashi (Mitani, Iagarshi 2011).

5   Kinematic model based on method of reflection.

5

in material and stiffness distributions.

For the abstraction of bio-inspired compliant systems, the integration of the described modeling and simulation methods and their specific application during the design process is crucial. The trade-off between speed and accuracy associated with those methods and the information exchange between them has to be taken into consideration during the design process. In particular, the simulation and evaluation of aggregations of kinetic components on an architectural scale require computational economy to gain real time feedback on the feasibility and performance of the design. Once certain mechanical and geometrical constraints and structural behaviors are determined and evaluated with the use of kinetic models, this information can be integrated into the kinematic model in a numerical manner. Thus, necessary adjustments to the underlying global geometries can be made and information of needed energy, efficiency and appearance can be extracted.

DEVELOPMENT OF KINETIC CURVED-LINE FOLDING ELEMENT

The following section will illustrate the development of the bio-inspired shading device *flectofold* starting from the investigation of the biological role-model to a functional prototype and the possibility of geometrical variations.

Biological Role-Model

The trap movement of the underwater carnivorous plant *Aldrovanda vesiculosa* served as biological role-model. The trap consists of two lobes connected with hinge zones to a mid rib. While the lobes and the central area are of higher stiffness, the hinge-zone is more flexible and describes a curved fold line (figure. 3). After being triggered by prey, a change of turgor pressure leads to a small change in the bending curvature of the mid rib which is amplified by a curved line folding mechanism and leads to a complete closure of the trap (Poppinga et al. 2016)

Abstraction and Simulation

For the translation into a technical application we based our work

on the preliminary studies done by Simon Schleicher (Schleicher 2016) which focused on the translation of the trap mechanism into a kinetic curved-line folding model where two flaps are connected to the stiffer middle part by areas of reduced bending stiffness (Figure 4). The most intriguing aspect for technical use is the extensive amplification of a small bending deformation on the central part into a large closing movement of the adjacent flaps. Geometrically, the relation between bending and opening is determined by the radius of the curved-line fold. To evaluate this correlation, Schleicher et al. developed a kinematic model using the "Rigid Origami Simulator" (Tachi 2009) which allows one to simulate a discretized simplification of the curved-line fold where the bending elements are represented by a finite number of rigid components.

To achieve a seamless integration into the design process and to reduce the computational effort, especially for the visualization of a high number of elements, we developed a geometric kinematic model based on the method of reflection (Mitani, Iagarshi 2011), applicable where the curved-line fold remains on one plane, to approximate of the geometry of curved-line folding. The plane which contains the fold line can be used to mirror the two adjacent surfaces to one fold-line. By varying the angle of this plane or by adjusting the curvature of the surface it is possible to control the folding angle (figure 4). To simulate the folding movement controlled by the induced bending, we created the bent mid-rib geometry based on the elastic-curve in relation to the displacement of the translation of the control points. The reflection plane, given by the points C', E and M', is used to mirror the lens-shaped rib surface and the initial outline of the flap surface (figure 5). With the curvature and deformed outline information, it is possible to rebuild the folded geometry. A minor displacement of C to C' and B to B' leads to an amplified deformation of A to A' and D to D'.

This method allows not only for parametric manipulation in real-time, but also provides immediate information about the relation between displacement of support points and folding angle.

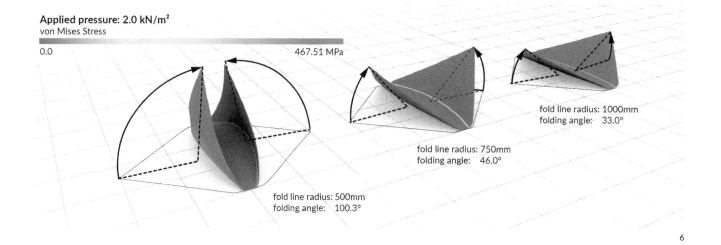

Applied pressure: 2.0 kN/m²
von Mises Stress

0.0   467.51 MPa

fold line radius: 500mm
folding angle: 100.3°

fold line radius: 750mm
folding angle: 46.0°

fold line radius: 1000mm
folding angle: 33.0°

6

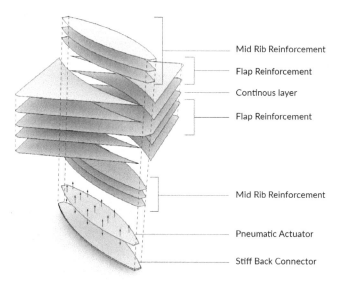

Mid Rib Reinforcement

Flap Reinforcement

Continous layer

Flap Reinforcement

Mid Rib Reinforcement

Pneumatic Actuator

Stiff Back Connector

7

The so-established geometric relationships were transferred into a kinetic model using the FEM software SOFiSTiK (SOFiSTiK AG, Oberschleißheim, Germany), taking actual material properties of glass fiber reinforced polymers (GFRP), actuation forces and resulting stresses into account. Several variations regarding the hinge zones in terms of width and stiffness gradient between hinge, flaps and mid-rib have been simulated and evaluated to establish an appropriate material gradient between the specific zones, leading to the desired closing movement. A pneumatic actuator was simulated in this study by uniformly distributed pressure onto the mid-rib (figure 7). With this method, a series of different curved-line hinges (radii of 500mm, 750mm and 1000mm) were simulated and evaluated in terms of needed pressure, resultant stresses, corresponding displacement of support points and sensitivity to the actuator (figure 6 and figure 9).

To analyze the stiffening effect of curved-line folding, we applied

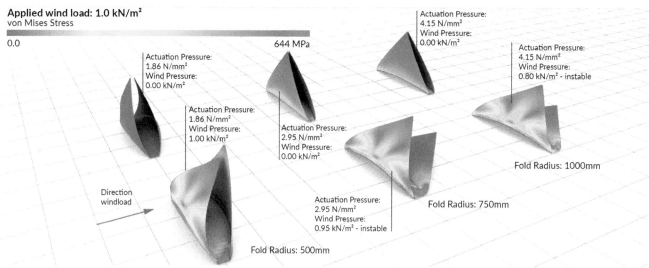

Applied wind load: 1.0 kN/m²
von Mises Stress

0.0   644 MPa

Actuation Pressure:
1.86 N/mm²
Wind Pressure:
0.00 kN/m²

Actuation Pressure:
1.86 N/mm²
Wind Pressure:
1.00 kN/m²

Actuation Pressure:
2.95 N/mm²
Wind Pressure:
0.00 kN/m²

Actuation Pressure:
4.15 N/mm²
Wind Pressure:
0.00 kN/m²

Actuation Pressure:
4.15 N/mm²
Wind Pressure:
0.80 kN/m² - instable

Direction
windload

Actuation Pressure:
2.95 N/mm²
Wind Pressure:
0.95 kN/m² - instable

Fold Radius: 500mm

Fold Radius: 750mm

Fold Radius: 1000mm

8

Bioinspired Kinetic Curved-Line Folding for Architectural Applications
Körner, Mader, Saffarian, Knippers

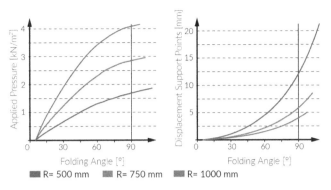

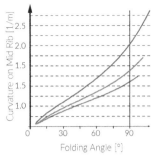

R= 500 mm   R= 750 mm   R= 1000 mm

9

wind load of 1.00 kN/m² to all models once they were fully closed. As expected, the smaller radius leads to higher curvature within the mid-rib which is transferred into the lobes, resulting in higher geometrical stiffness and therefore more stability subject to external loads (figure 8). Also a smaller radius needs less actuation energy to fold, but a higher displacement of the connection points to the substructure. A larger radius will lead to a narrower mid-rib, which would make the mechanism as a shading device more efficient since the central portion around the mid-rib will always remain closed.

Flectofold - Physical Prototype

A first physical prototype was constructed with glass fiber reinforced polymers (GFRP). The curved folding mechanism corresponds precisely to a defined stiffness gradient between the different zones—mid-rib as well as lobes must exhibit a certain stiffness while remaining flexible enough to allow for the induced bending. The hinge-zones need to be flexible and have to compensate for the bending forces in small area. The use of GFRP allows for the precise articulation of mechanical properties

in certain areas according to local demands. By adjusting the fiber orientation and varying the layer build-up, it is possible to achieve a stiffness gradient within the component that enables the curved folding mechanism (Poppinga et al. 2016).

The component is fixed to a stiff back-part. Movable connections allow sliding along the substructure to compensate the displacement induced by bending. Bending deformation of the mid-rib is generated by a pneumatic cushion which is located between the component and the stiffer back-part. The pneumatic cushion is lens shaped and fits the dimensions of the stiffer mid-rib to allow for a distributed surface actuation avoiding stress concentrations in the mid-rib as well as the hinge zone. It is fabricated from a special airbag fabric able to withstand high pressures and is laminated airtight.

By regulating the pressure according to input criteria, such as light conditions or user preferences, the folding movement can be precisely controlled (figure 10).

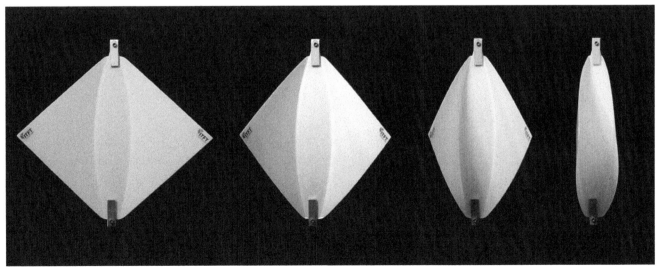

10

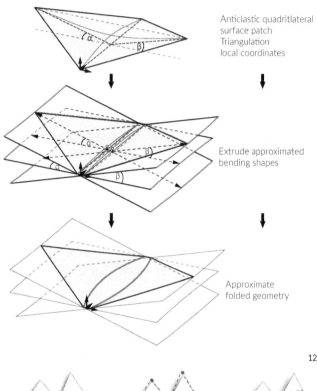

520 mm

420 mm

Actuation :
1.74 kN/m²

Actuation :
1.56 kN/m²

Actuation:
1.56 kN/m²

Actuation:
1.63 kN/m²

Actuation:
1.65 kN/m²

11

Anticlastic quadritlateral
surface patch
Triangulation
local coordinates

Extrude approximated
bending shapes

Approximate
folded geometry

12

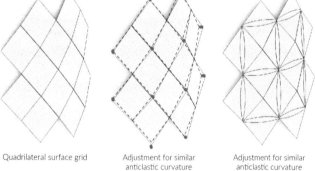

Quadrilateral surface grid

Adjustment for similar
anticlastic curvature

Adjustment for similar
anticlastic curvature

13

11 FEM simulations of asymmetric *flectofold* configurations.

12 Translation of folding mechanism into quadrilateral patches.

13 Tessellation and adjustment of global geometry.

Geometrical Adaptability

While the described simulations and the physical prototype assume an axial symmetry of the component, further investigations were carried out on the geometrical adaptability and non-symmetrical configurations. Distortions of the basic rectangular configuration have only minor influence on the folding movement and actuation energy (figure 11).

The adaptability into different boundary geometries is especially of interest for applications on double curved geometries. To translate the curved-line folding model onto geometries of positive and negative Gaussian curvature, we developed an algorithm which subdivides a given surface into quadrilateral patches of similar anticlastic curvature. A certain degree of anticlastic curvature is necessary to ensure a pre-fold in each panel while in an entirely closed configuration to give each panel the correct folding direction during actuation.

Based on the Reflection Plane Method (Mitani, Iagarshi 2011), it is possible to translate each of the quadrilateral surface patches into a curved-line folded geometry (figure 12).

The first step divides the patch into two triangles and generates the width of the mid-rib. In the next step, bending curvature is induced in the two directions of the extended mid-rib surface. By mirroring this surface and rotating it by the angle α and β respectively, the bend geometry of the flaps can be generated. The intersection of the two flap surfaces with the extended mid-rib surface gives the curved-line fold. By adjusting the bending curvature in the first step, it is possible to control the radius of the fold. Knowing the radius of each curved fold line and the degree of pre-fold, it is possible to evaluate the needed actuation force for each patch (figure 13 and figure 1).

Geometrical variations

The tessellation procedure as described in the previous section is based on a UV grid applied to the input geometry. This leads to certain geometrical constraints and limitations regarding component size variations and connections between different surface patches. Variations of the proposed mechanism using three fold-lines have been studied extensively by Aline Vargauwen (Vergauwen et al. 2014) and could be used for polygonal tessellation patterns, suitable for translating surfaces into curved-line folding elements and allowing for more design freedom (Chandra et al. 2015). To adapt the curved folding mechanism to a polygonal, preferably hexagonal, boundary condition, we imposed a triangle inside the polygon, where per definition all three vertices lie on one plane. The edges of the inner triangles can be seen as straight line representations of the curved fold lines. These straight lines can be translated into NURBs curves by using

Bioinspired Kinetic Curved-Line Folding for Architectural Applications
Körner, Mader, Saffarian, Knippers

the end points of two adjacent edges and center point of the inner triangle, and interpolation points between end points and center point as control points (figure 14). By adding an additional control point along the axis from vertices to center point the curvature of fold-lines can be adjusted.

For the geometric-kinematic model, the center point is moved in the normal direction of the plane spanned by the inner triangle. The corresponding displacement of the endpoints of each curved-line fold can be calculated by Pythagorean theorem and therefore the bend shape of actuation portion of the component can be approximated. Using the mentioned variation of the Reflection Plane Method, it is possible to approximate the folded geometry of the opening flaps. FEA simulations of GFRP panels with varying curvature in their fold lines provide first confirmation about the feasibility of the mechanism and the relation between actuation force and folding behaviour (figure 15).

RESULTS AND DISCUSSION

The process from investigation of a curved-line folding mechanism abstracted by Adlrovanda and the abstraction to a first working prototype lead to the following results:

Methodology

The development of a new approach for kinematic models for curved-line folding mechanisms using the method of reflection to simulate the movement of *flectofold* as well as the geometrical variations introduced by Vergauwen (Vargauwen et al. 2014) enables the direct integration within the design environment without external software such as rigid origami simulator or Kangaroo for Grasshopper. The established geometric rules enable the translation of surface tessellations into curved-line folding components. The geometric kinematic model reduces the computational effort compared to rigid origami simulations, which is of high advantage for the simulation of the behavior of large aggregations.

The proposed kinetic simulations are highly directed to the development of physical prototypes made of GFRP and implemented pneumatic actuators. One major aspect for the simulations was to investigate the influence of curvature of the fold line on the actuation energy required, displacement of support points and stability to external loads such as wind loads in different geometrical variations.

Geometrical Potentials and Constraints

The proposed compliant curved-line folding mechanisms are adaptable to a variety of boundary geometries. It is possible to apply the devices on double curved surfaces. Nevertheless, a central aspect of curved-line folding mechanisms is the need

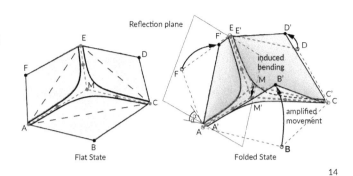

14

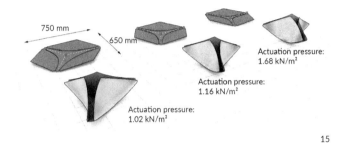

15

14 Geometric and kinematic model for curved-line folding element with polygonal boundary condition.

15 FEM simulation of geometrical variations – polygonal boundary condition.

of a certain pre-fold in the closed configuration to ensure the folding during actuation. Thus the proposed tessellation methods for freeform geometries include an adjustment of each patch to ensure the necessary pre-fold. This can require geometrical adjustments of the initial design, and can lead to an altered appearance. Also the folding components remain in this anticlastic configuration throughout the folding movement – a change of direction of curvature is not possible without complex actuation and control mechanisms. Therefore, future research will investigate biological role-models which are capable of changing from convex to concave configurations (e.g. *Adiantum peruvianum* or *Pterostylis curta*).

Material Composition and Scaling constraints

The movement of the components is triggered by bending deformation in one part of the component which is transferred to the adjacent areas. To facilitate this bending actuation, it is essential that the material is flexible enough to allow for the bending within an acceptable range of actuation force while remaining stiff enough to withstand external forces. Although the curvature induced by bending acts advantageously for the gain of geometrical stiffness, the inherent need for flexibility leads to scale limitations which need exploration in more detail.

Curved-Line Fold Radius and Efficiency

The radius of the curved-line fold influences the actuation force, and the sensitivity as well as bending deformation within the component. In addition to those mechanical parameters, it also has an impact on the width of the middle part, which will be always closed in the proposed application. A wider mid-rib will not only reduce the efficiency as shading device, but also disturb the view of users, thus it has physical and psychological impact on the performance. Trade-offs between all involved criteria must be defined from case to case.

Fabrication Advantages

One of the key advantages is the simplicity of fabrication of flat elements. The curved folded elements can be unrolled into planar elements, produced in a flat state and folded according the tessellation boundaries. Therefore, no special formwork is needed for the fabrication process.

Integrated functionality

While flexible enough to allow for the bending associated with curved-line folding, the flap surfaces provide enough stiffness and area to serve as a substrate for additional functional components such as PV cells, LED lighting, sensors, light guiding mirrors etc. This differentiates the proposed compliant systems not only from textile devices but also from roller blades or similar mechanisms, which only provide substrates with low stiffness.

CONCLUSION

This case study reveals the potential and limitations of the discussed curved-line folding principles and the use of GFRP materials for fabrication of a bio-inspired kinetic compliant mechanism to inform the design and construction of kinetic façade systems independent from linear or rotational translation movements. The curved-line folding mechanism is used to translate a one dimensional actuation force into a three dimensional deformation of the element. The study provides insight into the involved forces, actuation energy and structural behavior during the folding process.

Design and subsequent evaluation of kinetic envelopes is mainly informed and driven by quantifiable performance criteria such as energy efficiency, fabrication feasibility and operation ease, which result in strategies to optimize movement, reduce mechanical complexity and utilize compliant mechanisms. Nevertheless, there exist a number of not readily quantifiable, yet equally important factors that need to be taken into account at the concept stage of design. These include psychological and physiological effects of such systems on building occupants and methods of user control and levels of interaction as summarized by Francesco Fiorito (Fiorito et al. 2016).

In addition to technical performance aspects and user-interaction, kinetic envelopes bring forward a novel palette of time related properties that have vast design potential in composition of architectural facades and closely relate to human perception of movement. This has been well elaborated and presented by Schumacher (Schumacher et al. 2010) and includes aspects such as movement-speed, acceleration, serial repetition, complexity, weight, mystery, balance and more.

ACKNOWLEDGEMENTS

This research has been funded by the German Research Foundation (DFG) as part of the Transregional Research Centre (SFB/Transregio) 141 'Biological Design and Integrative Structures'/projects A03 and A04 and in collaboration with Larissa Born and Götz Gresser (Institute for Textile Technology, Fibre Based Materials and Textile Machinery, University of Stuttgart) Anna Westermeier, Simon Poppinga and Thomas Speck (Plant Biomechanics Group, Botanic Garden, University of Freiburg) and Renate Sachse and Manfred Bischoff (Institute for Structural Mechanics, University of Stuttgart).

REFERENCES

Chandra, Suryansh, Axel Körner, Antiopi Koronaki, Rachelle Spiteri, Radhika Amin, Samidha Kowli, and Michael Weinstock. 2015. "Computing Curved-Folded Tessellations through Straight-Folding Approximation." In *Symposium on Simulation for Architecture & Urban Design (SimAUD)* 2015. edited by H. Samuelson, S. Bhooshan, and R. Goldstein. Washington: SIMAUD. 221–228

Fiorito, Francesco, Michele Sauchelli, Diego Arroyo, Marco Pesenti, Marco Imperadori, Gabriele Masera, and Gianluca Ranzi. 2016. "Shape morphing solar shadings: A review." In *Renewable and Sustainable Energy Reviews 55*. 863–884.

Ham, Ronald van, Thomas G. Sugar, Bram Vanderborght, Kevin W. Hollander, and Dirk Lefeber. 2009. Compliant actuator designs. *In Robotics & Automation Magazine, IEEE* 16 (3): 81–94

Howell, Larry L., Spencer P. Magleby and Brian M Olsen. 2013. *Handbook of compliant mechanisms*. Chichester, West Sussex, United Kingdom: John Wiley & Sons.

Loonen, Roel C. G. M., Marija Trčka, Daniel Cóstola and Jan Hensen. 2013. "Climate adaptive building shells: State-of-the-art and future challenges." In *Renewable and Sustainable Energy Reviews 25*. 483–493.

Matini, Mohammad-Reza. 2007. *Biegsame Konstruktionen in der Architektur auf der Basis bionischer Prinzipien*. Dissertation thesis. Stuttgart: ITKE (Forschungsberichte aus dem Institut für Tragkonstruktionen und Konstruktives Entwerfen, Universität Stuttgart, 29).

Bioinspired Kinetic Curved-Line Folding for Architectural Applications
Körner, Mader, Saffarian, Knippers

Mitani, Jun, and T lagarshi. 2011. "Interactive design of Planar Curved Folding by Reflection." In *Pacific conference on computer graphics and applications*. Edited by B. Chen, J. Kautz, T. Lee, and M.C. Lin. Kaohsiung: 77–81.

Polygerinos, Panagiotis, Zheng Wang, Johannes T. B. Overvelde, Kevin C. Galloway,, Robert J. Wood, Katia Bertoldi, and Conor J. Walsh. 2015. "Modeling of Soft Fiber-Reinforced Bending Actuators." In *IEEE Transactions on Robotics 31* (3). 778–789

Poppinga, Simon, Axel Körner, Renate Sachse, Larissa Born, Anna Westermeier, Jan Knippers, Mandred Bischoff, Götz Greser, and Thomas Speck. 2016. "Compliant Mechanisms in Plants and Architecture." In *Biomimetic Research for Architecture and Building Construction: Biological Design and Integrative Structures*, edited by J. Knippers, K. Nickel and T. Speck. Dordrecht: Springer Science + Business Media B.V. (accepted)

Schleicher, Simon. 2016. *Bio-inspired Compliant Mechanisms for Architectural Design. Transferring Bending and Folding Principles of Plant Leaves to Flexible Kinetic Structures*. Dissertation thesis. Stuttgart: ITKE (Forschungsberichte aus dem Institut für Tragkonstruktionen und Konstruktives Entwerfen, Universität Stuttgart, 40).

Schleicher, Simon, Julian Lienhard, Simon Poppinga, Thomas Speck, and Jan Knippers. 2015. "A methodology for transferring principles of plant movements to elastic systems in architecture." In *Computer-Aided Design* 60. 105–117

Schumacher, Michael, Oliver Schaeffer, Michael-Marcus Vogt. 2010. *Move - Architecture in Motion - Dynamic components and elements*. Basel, Switzerland: Birkhäuser.

Tachi, Tomohiro. 2009. "Simulation of rigid origami." In: *Origami 4: the fourth international conference on origami in science, mathematics, and education*. edited by R. Lang and AK Peters. Natick: 175–187

Vergauwen, Aline, Niels De Temmerman, Lars De Laet. 2014. "Digital modelling of deployable structures based on curved-line folding." In: *Proceedings of the IASS-SLTE 2014 Symposium "Shells, Membranes and Spatial Structures: Footprints"*. edited by Brasil, R. & Pauletti, R. Brasilia.

Wang, Jialiang, Liliana O. Beltrán, and Jonghoon Kim. 2012. "From Static to Kinetic: A Review of Acclimated Kinetic Building Envelopes." In *Proceedings for American Solar Energy Society, vol. 2012*. 1–8.

IMAGE CREDITS

Figure 3: Aldrovanda vesiculosa (www.sarracenia.com)
All other image credits to authors (2016)

Axel Körner received his Diploma in Architecture at the University of Applied Sciences in Munich and his MSc. in Emergent Technologies and Design from the AA School of Architecture in London September 2013 with distinction. He worked for several architecture practices in Munich, Vienna and London, as well as for Createx and Northsails TPT in Switzerland where he was part of a multi-disciplinary team working on carbon fibre material research. Since October 2014 he has been working as research associate and PhD candidate at the Institute of Building Structures and Structural Design (ITKE), were his research is focused on bio-inspired compliant mechanisms.

Anja Mader gained a Bachelor's degree in biomimetics and a Master's degree in mechanical engineering at the University of Applied Sciences in Bremen. After working in the field of bio-based materials as research project employee at The Biological Materials Group she is now working as research associate at the Institute of Building Structures and Structural Design (ITKE). She is currently writing her PhD on bio-inspired actuation mechanisms for compliant systems in architecture as part of the Collaborative Research Center SFB-TRR 141: Biological Design and Integrative Structures.

Saman Saffarian is a Research Associate at the Institute for Building Structures and Structural Design (ITKE) at the University of Stuttgart. Previously he worked for Zaha Hadid Architects in London as a Lead Designer within the Design Cluster and was involved at the concept-stage of many projects and competitions of various scales. Additionally, in collaboration with the ZHA-CoDe group he contributed to the design and fabrication of a number of experimental and research-based installations and pavilions. Sam is currently pursuing his research interests as a PhD candidate within the Innochain research network (http://innochain. net). His research project focuses on Adaptive Building Envelopes and Material Gradient GFRP.

Jan Knippers is a structural engineer and specialises in light weight roofs and façades, as well as fibre based materials. Since 2000 Jan Knippers is head of the Institute for Building Structures and Structural Design (ITKE) at the University of Stuttgart. As such he is speaker of the Collaborative Research Centre 'Biological Design and Integrative Structures' funded by the German Research Foundation (DFG). He is also partner and co-founder of Knippers Helbig Advanced Engineering with offices in Stuttgart, New York City and Berlin. The focus of their work is on structural design for international and architecturally demanding projects.

Knit as bespoke material practice for architecture

Mette Ramsgaard Thomsen
Martin Tamke
CITA / Royal Danish
Academy of Fine Arts

Ayelet Karmon
CIRTex/Shenkar Engineering.
Design. Art

Jenny Underwood
School of Fashion and Textiles/
RMIT

Christoph Gengnagel
Department for Design and
Structural Engineering / UdK

Natalie Stranghöner
Jörg Uhlemann
Institute of Metal and
Lightweight Structure/UDE

1

1   Tower' at Danish Design Museum,
    2015.

ABSTRACT

This paper presents an inquiry into how to inform material systems that allow for a high degree
of variation and gradation of their material composition. Presenting knit as a particular system of
material fabrication, we discuss how new practices that integrate material design into the architectural design chain present new opportunities and challenges for how we understand and create
cycles of design, analysis, specification and fabrication. By tracing current interdisciplinary efforts
to establish simulation methods for knitted textiles, our aim is to question how these efforts can
be understood and extended in the context of knitted architectural textiles. The paper draws on
a number of projects that prototype methods for using simulation and sensing as grounds for
informing the design of complex, heterogeneous and performative materials. It asks how these
methods can allow feedback in the design chain and be interfaced with highly craft-based methods
of fabrication.

INTRODUCTION

Current architectural research practice is investigating the design and making of new material systems in which advanced CNC fabrication technologies allow for the precise control of material performance (Oxman 2007, Menges 2012, Palz 2009, Gramazio & Kohler 2008). This interest interfaces the design and fabrication of materials with that of buildings, allowing the conceptualisation of new structural systems that optimise material use and enable the realisation of lighter and smarter buildings (Nicholas 2013). While still experimental, the creation of new graded materials that respond to changes in site or use has become an overriding design paradigm by which the field aims to understand how we harness the potential of design for and with material performance.

Key aspects in this emerging design practice are the ability to determine the relevant design criteria and assign specifications for these new material systems while at the same time building solid models of their performance. This paper assembles conclusions across a series of CITA's interdisciplinary collaborations that examine knit as a material for architectural application. Learning from experiments using simulation and sensing, our interest is to examine how design criteria can be elicited, tested and fed back to the design system. We ask how these systems can be interfaced with knitted textile design systems and CNC fabrication technologies to understand, design, evaluate and fabricate new bespoke architectural textiles. The paper includes a discussion of the cultural interfaces that these interdisciplinary collaborations entail, querying how design, modelling and simulation are understood in the fields of architecture, textile design, technical textiles and engineering.

MATERIAL AND STRUCTURAL OPPORTUNITIES OF KNIT

The interest in knit lies with its particular ability to be both structurally and materially determined. Knit is fundamentally a highly flexible textile structure, which can be radically changed by controlling the composition of stitches. The control of stitches can happen across the entire fabric, which enables the fabrication of continuous materials with highly variegated performances. While knit is essentially a continuous yarn structure, the introduction of multiple yarns, in combination or replacing each other, enables further performance control. As such, knitted textiles embed interesting interactions between structural and material performances, thus making them highly singular and infinitely variable.

In CITA, experiments exploring the detailed fabrication of highly graded textiles—interfacing the parametric design and control of structure, material and pattern with CNC fabrication—examine knit for different classes of architectural textiles. 'Knitted Skins[i]

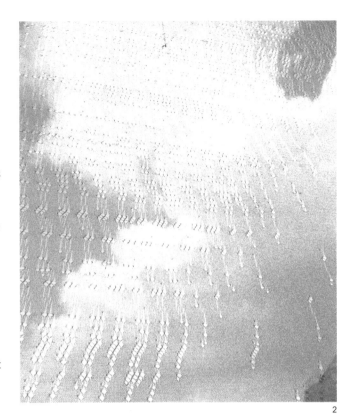

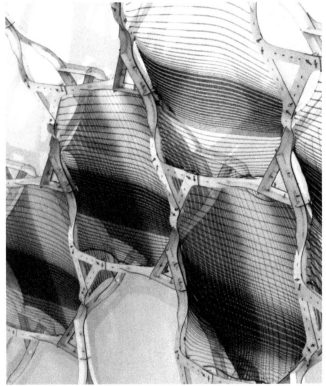

2   'Sifter' at Architecture House, Copenhagen, 2016.

3   'Derma' grading transparencies in knit-to-shape skins,, RMIT, 2013.

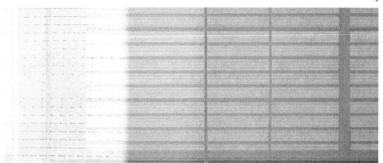

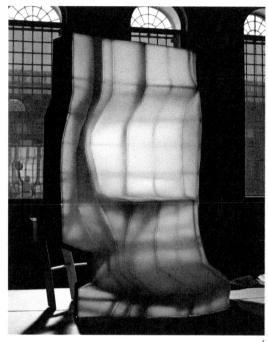

4

5

6

and 'Slow Furl' [ii] investigate the manufacture of bespoke spacer fabrics for interior wall membranes grading the structure, thereby controlling their thickness and pliability (Ramsgaard Thomsen 2008). 'Listener' [iii] explores the embedding of sensing and control into architectural membranes (Ramsgaard Thomsen 2011). 'Derma' [iv] examines the making of a building skin with bespoke transparencies that knit mass–customised textiles to shape. 'Shoji' [v] combines the interest in the interior wall with the lightweight screening device.

These investigations have led to an inquiry of how local variation can be informed and how design criteria are developed, analysed and fed back into the process. In the following we will discuss two main strategies to generate design feedback—simulation and sensing—using the two central investigations 'Tower' [v] and 'Sifter' [vi] as examples.

INFORMING MATERIAL THROUGH SIMULATION

'Tower' investigates knit as a structural membrane in which active bent GFRP rods are embedded into a bespoke knitted textile (Ramsgaard Thomsen 2015). The relationship between skin and structure is a central question in the field of architectural textiles, positioning the textile membrane either as a cladding skin, or engaged in hybrid dependencies in which membrane and scaffold act as an integrated structural system. The latter requires a high degree of control and understanding of the membrane's material behaviour. 'Tower' develops new modelling practices needed to devise hybrid behaviours. The project develops a simulation process that interfaces a projection-based relaxation

method with a finite element (FE) simulation (Ramsgaard Thomsen 2016). This approach extends the work of Lienhard (Lienhard 2015) and is able to effectively simulate the interaction of the constraining fabric, the many pretensioned actively bent GFRP rods and external forces.

The 'Tower' simulations rely on new models for testing and simulating knitted fabrics. The application of knitted fabrics on an architectural scale requires a prediction of the overall structural performance in order to guarantee stability. This requires not only an understanding of the behaviour of the knit, but also its interaction with other components and forces within the structural continuum. This complexity of these interactions and the sheer size of architecture prohibits a simple scaling up of existing techniques which are used for the simulation of smaller textile artefacts.

Cultural interfaces

There are two cultural contexts for textile design of knitted fabrics; the knitwear garment industry and the technical textiles field. Both sectors define design criteria for textiles mainly through 1:1 sampling, the 'sample' being a 2D panel that visualises and tests the surface quality (yarn, colour and stitch structure) and behaviour of the fabric. This method has its embedded problems. Theoretically, samples translate to larger fabrics directly. However, a fabric may not behave homogeneously across its entire width. When under tension, the distortion of the stitch loop varies relative to local conditions at the edge of the fabric compared to the centre. Furthermore,

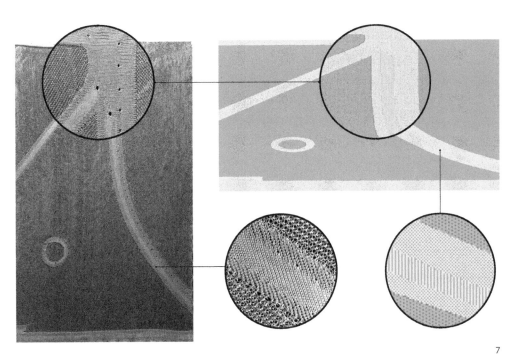

7

small variables in machine set-up, slight changes in needle movement and tensioning can lead to large performative changes across the fabric (Renkens 2010). When scaling materials up to architectural scale, these imprecisions are amplified, which makes them difficult to control or assess before production.

The engagement with knit as an architectural material questions how predictability can be established in the design chain and how alternative measures of generating material understanding can be developed.

Different kinds of simulation

For the fashion and textile industries, simulations of textiles fabrics and 3D garments are an important area of development. In the garment industry, simulation is used for design, performance evaluation such as drape, tension and pressure points associated with fit and comfort, through to supporting virtual shopping experiences. Depending on scale and area of application, different approaches for the simulation of textiles have been developed for the capture of the multiscale relations between material, structure and patch in order to simulate the overall global behaviour of the textile.

In the area of 3D virtual garment design, programs such as V-Stitcher, Marvellous Designer, and Opti Tex are developing sophisticated programing tools that work with a range of data inputs related to body dimensions, material and garment design (Kennedy 2015). All have developed extensive libraries of material knowledge based on standard fabric types commonly used in the fashion industry. However, with the material knowledge tied up in predefined software libraries intended for visualisation, there is limited possibility to directly interface with textile fabrication tools. Moreover, most software is limited in scope and only simulates textiles with highly simplified models of behaviour. The focus of these simulations is to understand the visual quality of the resulting textiles rather than their performative behaviour.

Knitting machinery companies such as Stoll and Shima Seiki have developed tools that integrate the design, visualisation and virtual prototyping of textiles; build on extensive material libraries of yarns, stitch structures and patterns; and create the technical file (G code) required for running the knitting machine. This currently applies to 2D flat fabrics in a relaxed state. As in virtual garment design software, the virtual 2D fabric can be draped across a 3D surface of an avatar, with a pattern mesh being laid across a specified garment type. Design options for 3D form are limited to predetermined garment types and the geometric tolerances (sizing and shaping) achieved in this type of simulation do not apply to architectural scale.

Yarn-level models are used for the precise simulation of knitted technical textile. They are based on complex mathematical models which simulate the behaviour of the overall fabric by analysing yarn geometry at loop scale. This includes issues such as non-circular yarn cross-sections, which affect fabric bending and draping characteristics (Kyosev 2005). FE simulation (Döbricha 2013) or specialized force models, such as bending and crossover springs in hexagonal patterns (Araújo 2004), spline

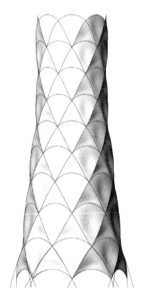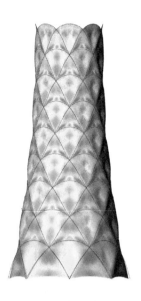

8

or volumetric representations, as in Fibre FEM by Fraunhofer, are used for the simulation of the mechanical behaviour of textiles. Volumetric representations of the knit structure are used for estimating the physical properties of textiles, such as air resistance and water absorption, through simulation (Renkens 2010, Kurback 2016). However, yarn-level models in textile research focus on small fabric sizes in controlled experiments (Cirio 2015).

A further class of textiles simulations come from CGI. Advancements of spring simulations (Stam 2009) and physics based solvers allow modelling techniques that achieve yarn-level models at the scale of complex knitted garments for virtual characters (Yuksel 2012). These cinematic effects are however decoupled from the architectural concerns of material performance, specification and fabrication.

A call for a new model of simulation

In order to create informed architectural scale design models that engage material performance, we need to develop a new class of simulation models. Where simulation in the garment industry is limited to small patches, the design and specification of textile structures in architecture has to tackle a wider geometrical range, larger scales and be able to consider the textile as part of a structural continuum. The ability to produce bespoke materials, shapes and details with CNC knit demands a tighter coupling of these currently discreet processes.

In architecture, the state of the art of textile structures can be found in membrane design. Current methods are based on material models for a limited range of laminated weaves and a two-step approach towards design and simulation that separate design and analysis. FE simulations are challenged by the inherent high degree of anisotropic and nonlinear elastic behaviour of knitted materials. These are hard to model with current methods for FE, which favour constant stress-strain ratios. It is therefore necessary to determine rough fictitious constants as an approximate simulation of materials behaviour. Finally, existing membrane materials have a relatively high degree of shear stiffness, which is not present in knit.

In 'Tower,' the challenge is to simulate the high strains and transverse contractions under load in both directions. In order to inform the simulations, we developed new testing methods to determine the material properties of the designed textiles. The testing procedure of MSAJ/M-02-1995 defines five different stress ratios (1:1, 2:1 1:2, 1:0 and 0:1) that are consecutively applied on a cruciform shaped test specimen. The result of this procedure is a stress-strain-diagram (Figures 8, 9). From this complete set of test data, ten stress-strain-paths can be extracted. A single design set of elastic constants from the

9

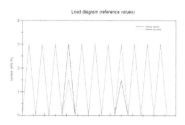

10

8   'Tower' as lightweight simulation and FE analysis.

9   Biaxial testing of the CNC 55 tex knitted fabric [© ELLF].

10   Exemplary loading sequence according to MSAJ/M-02-1995 without the 1:0 and 0:1 load ratios (left) and the stress-strain-diagram as a result of the biaxial test (right) [© ELLF].

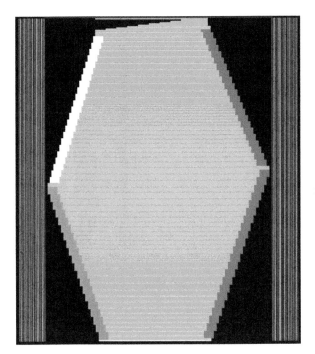 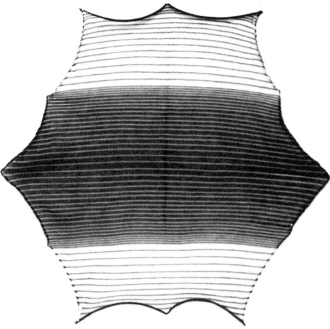

11 Derma knit to shape interface files.

extracted stress-strain-paths can be determined stepwise in a double step correlation analysis.

Existing methods are limited in terms of the scale and precision needed for architectural application. Extending knitting practice to architectural scale must therefore include the invention of new practices by which the simulation of knitted fabric can be undertaken and by which these data are fed back into the design chain. The method developed here allows us to assess the structural stability of Tower and bears the potential for further feedback into the specification of the knit, which could inform us about local requirements for stiffness or isotropy.

INFORMING MATERIAL THROUGH SENSING

The 'Tower' investigation into simulation is contrasted with the 'Sifter' (Figure 11) investigation into sensor feedback. In 'Sifter,' local sensor data, in the form of light readings, inform the making of a site specific curtain. 'Sifter' further extends the idea of directly informing material making with data by directly interfacing design environments with the physical realm.

The integration of sensors in the actual material fabric of the built environment is becoming increasingly commonplace. Embedded into materials such as concrete, steelworks and fibre glass, sensors are used for monitoring material performance and decay (Ramsgaard Thomsen 2012.) The information collected by these strategically positioned sensors generates valuable data that reflect material behaviour, environmental impact as well as

human interaction intensities.

'Sifter' aims to understand how materials could be designed directly for their context, using the scenario of a site specific curtain developed for the Architectural House, Copenhagen. The material design is informed by sense data captured while using a prototypical material in which a field of light sensors gathers the fluctuating lux values across the façade. The data is inputted into a parametric model in turn generating the perforations. The data informs both the length and the amount of perforation in areas which are mapped back onto the location of the referenced sensor. The underlying assumption is that minute changes of the lighting conditions across each window frame can be expressed by the data.

'Sifter' prototypes the means by which material design can be informed through local environmental data. By speculating that building materials belong to lifecycles in which they are replaced and renewed, 'Sifter' asks how such materials can be incrementally optimised. If 'Sifter' retained the incorporation of the sensor field as part of its own material make up, it would embody the first of multiple future cycles of material adjusting.

MATERIALISING THE DESIGN

When simulation or sensing is used to generate highly graded design criteria for material specification and fabrication, it becomes paramount to develop interfaces by which these can be transferred directly to the fabrication tools. In our practice,

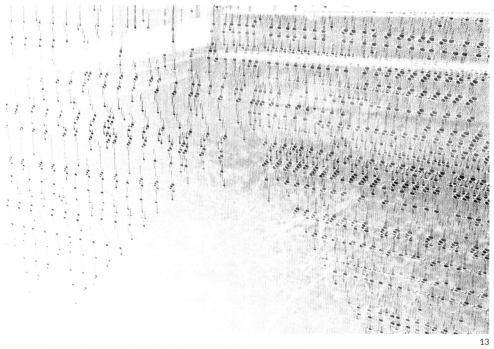

12 'Sifter' Architecture House, Copenhagen, 2016.

13 Detail 'Sifter'.

14 Generated knitting pattern for 'Sifter'.

13

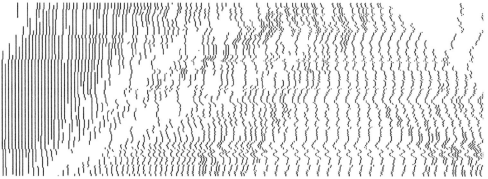

14

Knit as bespoke material practice for architecture Ramsgaard Thomsen et al.

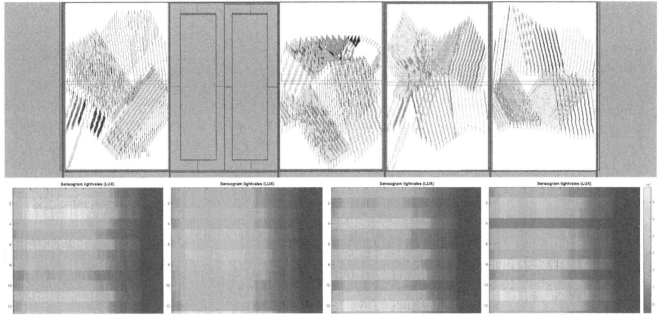

15 Correlation of generated pattern with sensor field. Sensograms visualising varying intensity in the sensor reading over time.

the interfacing between architectural design tools (rhino/grasshopper) and the CNC knitting machines has been a central focus to enable the projects' realisation. Compared to other more predictable fabrication systems such as 3D printing or CNC milling, knit necessitates a highly iterative process of prototyping. The geometrical space of design is infinitely more flexible than the material-based machine-driven knit, which depends on structurally coherent and technologically feasible combinations of knitting instructions. Therefore the creation of knitting code requires in-depth understanding, knowledge and proofing.

The development of our interfacing tools span both Stoll and Shima Seiki machines. In 'Listener,' we developed methods of directly creating the BASIC machine code consisting of patterns of letters defining the knitting beds, the yarn carries and holding patterns, and thereby controlled the formation of the knitted textile. In 'Derma,' 'Sifter' and 'Tower,' the interfacing is controlled by the generation of pixel-based files in which the controlled colour coding of each pixel denotes the particular structural and material information for each stitch. The files interface directly with the machine software. Here, tiff files are imported into the knitting software, which allows direct control of the structure, material and shape. The complexity of the design specification is controlled through bespoke definitions in Processing (Derma) or Grasshopper ('Sifter' and 'Tower') in which matrices are used to define the linear production of each knitting row (Figure 11).

In 'Tower,' this design process allows us to control sizing as well as the placement of details and reinforcements. Here,

differentiations between the base structure and its edges result in a highly differentiated knit pattern (Figure 7). In 'Sifter,' the structural pattern remains very simple. Instead, it is the drastic multiplication in the hundreds of thousands of individual perforations and their non-systematic patterning that makes up the complex and highly individualised textile.

CONCLUSION

Designing with textiles is inseparable from designing their behaviour. Architecture, engineering and textile design are struggling to find the right set of digital tools and procedures that can allow a transition from a sampling-based method towards digital processes that fully integrate and formalise unique interactions between structure and material performances. In these emergent practices, external information about material behaviour and environmental performance is fed back into the design chain, thus enabling the steering of material systems.

Knit is a valuable material technique for testing the integration of software for design, analysis, sensing and fabrication. Knitting is unique and substantially different from more commonly discussed computational techniques for material deposition such as 3D printing and enables the study of complex buildable systems with distinct properties that can be computationally steered.

The problem addressed in this paper is how to inform the design of highly detailed material systems. By presenting two strategies—design integrated simulation and sensing—our aim has been to discuss the possibilities of such strategies, their different backgrounds and the embedded complexity and limitations of their current state. However, while sensing and simulation are presented as alternative strategies, it is important to understand that these should be seen as complementary systems, where sensing informs simulation and allows an embedded and concomitant correction. This raises new questions regarding the interfacing and use of sensing for simulation. What kind of sensing can be used to understand material performance across the multiple scales that characterise architectural structures, and how do we develop new simulation types that can correlate generalised material models with highly localised data capturing?

The second aim is to understand how these highly specified designs can be interfaced with existing fabrication software and tools. Our aim is to open up the design environment by enabling a higher degree of flexibility for interfacing between multiple computationally driven environments. We have been successful in developing our own interfacing methods and establishing the necessary interdisciplinary dialogues by which the much broader possibilities of CAD and parametric design environments can be linked to the material and structural knowledge that informs CNC knitting. As such, the project spans multiple disciplinary divides combining concerns of design (architecture, textile design) with concerns of analysis (membrane engineering and virtual garment design development) and fabrication (CNC knitting).

ACKNOWLEDGEMENTS

[i] 'Knitted Wall': (listed as 'Knitted Skins' in text) CITA and Toni Hicks, Brighton University and Tilak Dias, Manchester University, 2008.

[ii] 'Slow Furl': CITA and Toni Hicks, Brighton University, 2008.

[iii] 'Listener': CITA and Shenkar College of Engineering and Design, 2010.

[iv] 'Derma': CITA and Jane Burry, Mark Burry and Jenny Underwood, RMIT, 2013.

[v] 'Shoji': CITA and KHR Architects, 2016.

[vi] 'Tower': CITA, UDK, Fibrenamics, AFF and Uni Duisburg-Essen, 2015.

[vii] 'Sifter': CITA and Shenkar College of Engineering and Design with Jan Larsen, Danish University of Technology, 2016.

REFERENCES

Araújo, M. de, R. Fangueiro, H. Hong. 2004. "Modelling and Simulation of the Mechanical Behaviour of Weft-Knitted Fabrics for Technical Applications." *AUTEX Research Journal* 4 (1): 25–32.

ASTM D2594 – 04. 2012. "Standard Test Method for Stretch Properties of Knitted Fabrics Having Low Power." http://www.astm.org/Standards/D2594.htm

Cem Yuksel, Jonathan M. Kaldor, Doug L. James, Steve Marschner. 2012.

"Stitch Meshes for Modeling Knitted Clothing with Yarn-level Detail." *ACM Transactions on Graphics (Proceedings of SIGGRAPH 2012)*, 31 (4): 37.

Deuss, Mario, Anders Holden Deleuran, Sofien Bouaziz, Bailin Deng, Daniel Piker, and Mark Pauly. 2015. "ShapeOp - A Robust and Extensible Geometric Modelling Paradigm." In *Modelling Behaviour*, edited by M. Ramsgaard Thomsen, M. Tamke, C. Gengnagel, B. Faircloth, F. Scheurer. Berlin, Heidelberg: Springer. 505–516.

Döbrich, Oliver, Thomas Gereke, Chokri Cherif, and S. Krzywinski. 2013. "Analysis and Finite Element Simulation of the Draping Process of Multilayer Knit Structures and the Effects of a Localized Fixation." *Advanced Composite Materials* 22 (3): 175–189.

Gramazio, Fabio, and Mathias Kohler. 2008. *Digital Materiality in Architecture*. Zurich: Lars Müller Publishers.

Kennedy, Kate. 2015. "Pattern Construction." In *Garment Manufacturing Technology*, edited Rajkishore Nayak and Rajiv Padhye. Cambridge, UK: Textile Institute. 205–220.

Kyosev, Yordan, Y. Angelova, and R. Kovar. 2005. "3D Modeling of Plain Weft Knitted Structures of Compressible Yarn". *Research Journal of Textile and Apparel* 9 (1): 88–97.

Lienhard, J., and J. Knippers. 2015 "Bending Active Textile Hybrids". *Journal of the International Association for Shell and Spatial Structures* 56 (1): 37–48.

MSAJ/M–02–1995. 1995. "Testing Method for Elastic Constants of Membrane Materials". Membrane Structures Association of Japan.

Menges, Achim. 2012. *Material Computation: Higher Integration in Morphogenetic Design*. Hoboken, NJ: Wiley.

Nicholas, Paul, David Stasiuk, and Tim Schork. 2014. "The Social Weavers: Considering Top-Down and Bottom-Up Design Processes as a Continuum". In *Proceedings of the 34th Annual Conference of the Association for Computer Aided Design in Architecture*, edited by David Gerber, Alvin Huang, and Jose Sanchez. Los Angeles: ACADIA. 497–506.

Oxman, Neri and Jesse Louis Rosenberg. 2007. "Material-Based Design Computation: An Inquiry Into Digital Simulation of Physical Material Properties as Design Generators." *International Journal of Architectural Computing* 5 (1): 26–44.

Palz, N., and Mette Ramsgaard Thomsen. 2009. "Computational Material: Rapid Prototyping of Knitted Structures." In *Architecture and Stages in the Experience City*, edited by Hans Kiib. Aalborg: Aalborg University.

Ramsgaard Thomsen, Mette, Martin Tamke, Anders Holden Deleuran, Michel Schmeck, Gregory Quinn, and Christoph Gengnagel. 2015. "The Tower: Modelling, Analysis and Construction of Bending Active Tensile Membrane Hybrid Structures." In *Proceedings of the International Association for Shell and Spatial Structures (IASS) Symposium 2015*. Amsterdam, The Netherlands.

Ramsgaard Thomsen, Mette, Martin Tamke, Anders Holden Deleuran, Ida

Knit as bespoke material practice for architecture Ramsgaard Thomsen et al.

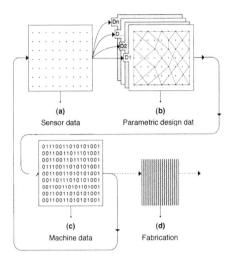

16 Flow chart for materialising the design.

(a) Sensor data
(b) Parametric design dat
(c) Machine data
(d) Fabrication

Katrine Friis Tinning, Henrik Leander Evers, Christoph Gengnagel, and Michel Schmeck. 2015. "Hybrid Tower, Designing Soft Structures." In *Modelling Behaviour*, edited by M. Ramsgaard Thomsen, M. Tamke, C. Gengnagel, B. Faircloth, F. Scheurer. Berlin, Heidelberg: Springer. 87–100.

Ramsgaard Thomsen, Mette, and Ayelet Karmon. 2012. "Informing Material Specification." In *Digital Aptitudes + Other Openings: ACSA 100th Annual Meeting*, edited by Mark Goulthorpe and Amy Murphy. Boston: ACSA. 677–683

Ramsgaard Thomsen, Mette, and Ayelet Karmon. 2011. "Listener: A Probe into Information Based Material Specification." In *Conference Proceedings Ambience 2011*, edited by Lars Hallnäs, Annika Hellström, and Hanna Landin. Borås, Sweden: Ambience. 142.

Ramsgaard Thomsen, Mette, and T. Hicks, 2008. "To Knit a Wall, Knit as Matrix for Composite Materials for Architecture." In Conference Proceedings Ambience 2008, edited by Lars Hallnäs, Pernilla Walkenström, and Lennart Wasling. Borås, Sweden: Ambience. 107–114.

Ramsgaard Thomsen, Mette, and Martin Tamke. 2013. "Digital Crafting: Performative Thinking for Material Design." In *Inside Smartgeometry: Expanding the Architectural Possibilities of Computational Design*. London: John Wiley & Sons Ltd. 242–253.

Stam, Jos. 2009. "Nucleus: Towards a Unified Dynamics Solver for Computer Graphics". In *2009 Conference Proceedings: IEEE International Conference on Computer-Aided Design and Computer Graphics*, edited by Daniel Thaimann, Jami J. Shah, and Qunsheng Peng. Yellow Mountain City, China: IEEE. 1–11.

Tamke, Martin, Anders Holden Deleuran, Christoph Gengnagel, Michel Schmeck, R. Cavalho, R. Fangueiro, F. Monteiro, N. Stranghöner, J. Uhlemann, T. Homm, and Mette Ramsgaard Thomsen. 2015. "Designing CNC Knit for Hybrid Membrane And Bending Active Structures." In *International Conference on Textile Composites and Inflatable Structures: Structural Membranes*, edited by E. Oñate, K.-U. Bletzinger, and B. Kröplin. Barcelona: CIMNE.

IMAGE CREDITS

Figures 1, 4, 6, 12, 13: Anders Ingvartsen, 2015 (photography)
Figures 9–10: © ELLF

Mette Ramsgaard Thomsen is Professor and leads the Centre for Information Technology and Architecture (CITA) in Copenhagen. Her research examines how computation is changing the material cultures of architecture. In projects such as Complex Modelling and Innochain she is exploring the infrastructures of computational modelling.

Martin Tamke is Associate Professor at the Centre for Information Technology and Architecture (CITA) in Copenhagen. He is pursuing design led research in the interface and implications of computational design and its materialization. Recent projects focus on complex modelling with interdependent materials systems and computational design.

Ayelet Karmon is a senior lecturer at Shenkar, teaching in the Master of Design program. She is the Managing Director of CIRTex – The David and Barbara Blumenthal Israel Center for Innovation and Research in Textiles. She is currently studying towards a PhD at Hebrew University Faculty of Agriculture, developing artificial textile-based materials as vertical growing substrates for plants.

Jenny Underwood is a Senior Lecturer in Textile Design and the Higher Degree by Research Coordinator for the School of Fashion and Textiles at RMIT University, Australia. Her research is practice based and explores 3D shape knitting, parametric design and embodied interaction for the design of material structures and sensory experiences.

Christoph Gengnagel is Professor at the University of Art, Berlin, School of Architecture, Chair for Construction and Structural Design. The research activities of his department focuses on design, development and analysis of innovative materials and construction systems.

Natalie Stranghöner is Full Professor at the Institute for Metal and Lightweight Structures at University of Duisburg-Essen and Director of the Essen Laboratory for Lightweight Structures. Main reserach topics are the material behaviour of textile membranes.

Jörg Uhlemann is Senior Researcher at the Institute for Metal and Lightweight Structures at University of Duisburg-Essen and Associate Director of the Essen Laboratory for Lightweight Structures. Main research topics are the determination of stiffness parameters for architectural textiles and the structural behaviour of prestressed membranea.

Pneumatic Textile System

Adam Wang
Sean Ahlquist
University of Michigan, Taubman
College of Architecture and
Urban Planning

ABSTRACT

This paper attempts to demonstrate a seamless transformable material system through an inter-dependent designed assembly of two materials with different material properties (anisotropic knit textile and isotropic silicone) but similar behaviors (stretch). The transformable system is achieved by balancing the volumetric expansion through a silicone tube, under inflation, with the controlled resistance to stretch by a custom knit fabric. The use of a CNC knitting machine allows not only an opportunity to program the stretch behavior of a knit fabric, by controlling the amount of yarn material to be deposited, but also an ability to knit multiple layers of fabric simultaneously, in order to create a space capable of accommodating an external element seamlessly.

The paper will showcase a series of experiments ranging from the initial search for compatible material combinations to the varied structures of the tube sleeve and its relationship with surrounding region. The final prototype attempts to utilize the various behavioral properties of the material system learned from the experiments to create a transformable three-dimensional structure.

INTRODUCTION

"Material System" in the context of this paper is described as an interdependent assembly of materials based on their innate properties with an intention to create a desired material behavior instead of a preconceived geometric form. A basic material system example would be a knit fabric where the process of interlock-looping of a yarn transforms the yarn's initial linear tensile nature to an expanded field condition (Fig 1).

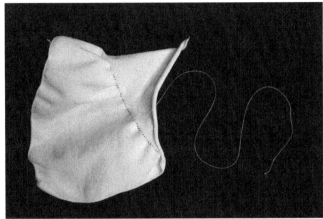

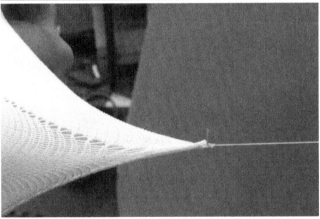

This paper focuses on the assembly of a programmable aniso-tropic knit fabric material with an isotropic silicone tube to create a deployable and 3D transformable structure. When inflated, the expansion of the silicone tube will stretch the knit textile. When taut, the knit textile will limit the degree of expansion by the silicone tube. Together, the two form an interdependent material system.

The paper hopes to contribute to the future development of textile-related design in the field of architecture by successfully demonstrating the ability of custom-knit fabric to seamlessly accommodate an external element without a secondary aggrega-tion process, such as sewing, and the ability to program a desired behavior into the textile to create a true three-dimensional structure.

Background

Similar to traditional knitting, CNC weft knitting is the process of laying a continuous piece of yarn onto a bed of crochet needles in interlocking loops. In the case of the STOLL knitting machine, there are two flat beds of needles arranged in an inversed V shape with yarn feeders running on top (Figure 2). The needles are raised to catch the yarns as the feeders move past them. The gauge of a knit refers to the number of needles required to make one inch of fabric. For example, if it takes 14 needles to knit one inch of the fabric, the full gauge of the knit is 14. However, in advanced knitting, it may sometimes be required to knit on every other needle, leaving an empty needle in between; this is called half gauging. The empty needle provides the additional space needed for transferring of needles to create a complex knit pattern. The needle activation is controlled numerically by codes generated from the graphic interface M1 Plus, where the designer can assign the exact location of needles to catch the passing yarn feeders.

Under stress, a knit fabric typically redistributes the load along one axis more than the other due to the composition of yarns and fibers. The process of CNC knitting allows an opportunity to either exaggerate or diminish the difference in force distribution through custom-knit stitch structures that either increase the stretch of fabric by more loosely arranging the yarn or increase the stretch resistance by more densely compacting the yarn. The result of localized differentiated properties within the prototype knit textile becomes more evident when activated by a uniformly expanding silicone tube, as the volume of inflation is directly affected by the willingness or resistance of the surrounding fabric to stretch.

1   Upper: Unstretched knit textile   2   Inversed V beds of needles of
    Lower: Stretched knit textile.   STOLL knitting machine.

Precedents

In "Soft Robotics Applied to Architecture," Kim et al. attempt to add a layer of intelligence to inflatable architecture by integrating soft actuated surfaces such as walls, ceilings, or floors (2015). Motion sensors are planted to detect the presence of occupants and trigger actuation of the pneumatic fixtures that, in return, create an opening in the wall. The soft surfaces are actuated through pneumatic inflation. The differentiated movement under inflation is achieved through custom ribs in the inflated bladder or varied thickness in the silicone membrane. In this context, pneumatics are an efficient means of generating movement in the silicone cell. My research prototype also hopes to demonstrate the potential for dynamic movements in an architectural surface through differentiated inflation by the custom-knit structures.

METHOD

The approach to develop a single seamless inflatable 3D structure can be divided into four categories. The first step is the selection of compatible materials for the pneumatic textile system. The second is to investigate the relationship between the knit sleeve and the enclosed silicone tube. The knit sleeve needs to accommodate the size increase of the inflated tube while maintaining enough density variation to create the desired direction of bending without overstressing the silicone material. The third aspect of the research focuses on the surface regions surrounding the planted tube. As the tube expands, it will stretch the fabric around it. Differentiated density of yarn distribution is tested to find the appropriate tensile strength to accommodate the expansion of the tube. Finally, a means of knitting multiple

3

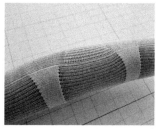

5

6

"Listener," by Thomsen et al., "examines the integration of conductive fibers within the textile matrix to enable sensing. Using capacitive sensing, the textile membrane becomes an interface for interaction (Fig 4). This is then combined with an actuation system of integrated high-pressure bladders that allow the material to inflate" (Thomsen et al. 2015). The "Listener" takes advantage of the CNC knitting process to integrate three types of yarn (polyethylene for over structure, elastomer for stretch, and silver-coated for conduction) to create a custom textile that, when paired with microprocessor, becomes a self-sensing interactive material system. Sensors were planted and inflatable bladders were inserted into individual cells to create an interface that allowed the system to respond to its environment. In this context, the fabric serves as both the interactive interface and host to the sensory system.

layers of fabric at the same time is developed to break away from the typical 2D nature of a textile.

The first stage of the research focused on the search for compatible textile sleeve and inflation bladder material. The first attempt used latex balloon and nylastic sleeve. Despite the light weight and relative thin gauge of the nylastic yarn, it produced too much friction for successful inflation of the latex balloon inside. The membrane of the latex balloon was very thin, and the weight of the fabric blocked air flow within it. Even with water-based

4

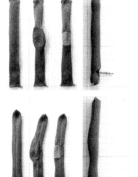

7

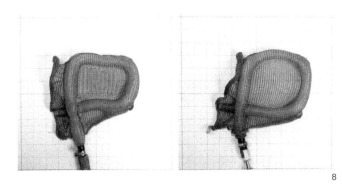

8

9

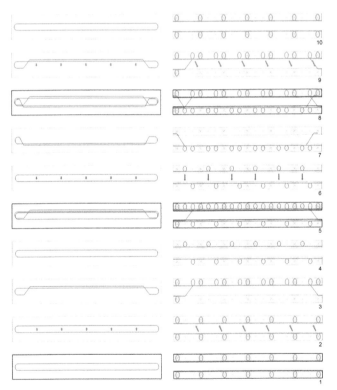

10

lubricants or soap, smooth continuous inflation was not possible. The sausage effect still persists, as shown in Figure 5. The second half of the first stage substituted the latex balloon with segmented bicycle inner tube. Continuous inflation was achieved, but the nylastic sleeve failed to provide consistent direction of planned bending due to lack of resistance to overall tube (Fig 6).

The second stage used polyester yarn as main material for the inflatable housing. It proved to be consistent in initiating the desired direction of bending. If the knit structure was loose on the top half of the sleeve and tight on the bottom half, the inflated tube would bend downward as the top half would be stretched more. The degree of bending could even be exaggerated with the introduction of nylastic yarn at selected locations. This is shown in Figure 7.

The third stage of the experiment focused on the interaction of the surrounding surface area by the inflated tube. Figure 8 shows that without a custom-knit structure, the inflated tube boundary would expand evenly in a circular manner. With the introduction of alternate miss stitches, every missed stitch reduced the loop length of the fabric to stretch and therefore limited the expansion of the tube boundary to a rectangular manner.

The fourth stage of the experiment focused on ways of ensuring the 3D quality of the future prototype. Figure 9 shows the transformative quality of the layered textile with a bridge-like tube breaking away from the 2D plane. Multiple layer knitting is done by providing additional empty needles for transferring. Figure 10 diagrams the sequencing of needle assignments to achieve multiple layers of free fabric that can share the same area on the knitting machine. Step 1 shows a loop of red yarn occupying every third needle (marked in red) on both beds. Steps 2 and 3 show how the red yarn is transferred from front to back to be deactivated. Step 4 shows the introduction a new independent blue yarn and step 5 shows the location of blue relative to the red. Steps 6 and 7 show the deactivation of the blue by transferring from back to front, while step 8 shows that the machine is now housing both red and blue yarn in four layers of fabric. Step 9 shows the transferring of red yarn from back to front again to be reactivated, and step 10 shows the start of cycle.

RESULTS

The basic prototype design is an assembly of ½ inch internal diameter (5/8 external diameter) silicone tube, as the inflatable bladder (Figure 11), inserted into a seamless custom CNC knit fabric, with the tube house dividing the textile into a 3 x 3 grid configuration (Figure 12). The diagonal rectangles of the 3 x 3 grid, marked in red in Figure 13, are high tensile zones of densely knit stitches that have limited stretch in both x and y axes. The

3   "Soft Robotics Applied to Architecture," (Kim et al. 2015).

4   "Listener," (Thomsen et al. 2009.)

5   Latex balloon and nylastic sleeve.

6   Rubber tube and nylastic sleeve.

7   Bending tests.

8   Regional tests.

8   3D bridge.

10   Needle assignment diagram.

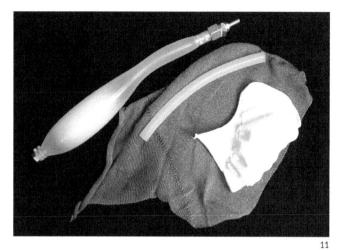

11

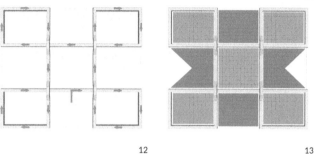

12   13

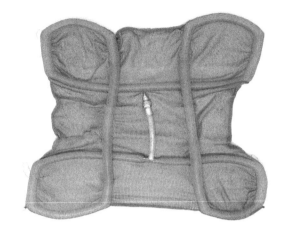

14

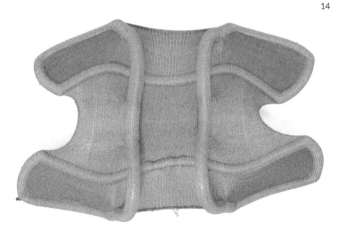

15

four middle rectangles on the outer boundary mark in blue (Figure 13) are medium-tensile zones with alternate miss stitches that create limited stretch in the x axis only. When inflated, the 3 X 3 grid boundary area of the tube is allowed for maximum expansion in volume, in order to activate the stretching of the fabric in the various zones. The resistance created by the stretching of the fabric will in return trigger a three-dimensional transformation.

There are three sets of prototypes: A.1, A.2, and B.1. Prototype A.1 is the 3 X 3 grid with emphasis on the 3D arching bridges as a means of bending to actuate the 3D transformation. Prototype A.2 is a duplicate of two sets of A.1 in one seamless textile. The lengths of the 3D bridges are varied in an attempt to differentiate the degree of deformation. Prototype B.1 is similar to A.1 but has an extended layer of fabric from the edges of the 3D bridges in an attempt to generate a pocket space between layers of fabrics.

The initial results of these inflated prototypes without the implementation of custom-knit structures reveal success in hosting the inflated bladder, but a failure to create significant 3D transformation (Fig 14). After adjusting the knit structure by reducing stitches (materials) in the webbed region to increase the tightness in the fabric, all three adjusted prototypes are able to successfully transform from the original 2D set up to a

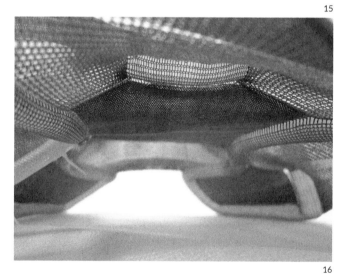

16

11 Prototype materials.

12 Tube location.

13 3 X 3 grid division.

14 Inflated prototype without successful 3D transformation.

15 Prototype A.1, top view.

16 Prototype A.1, interior view.

Pneumatic Textile System Wang, Ahlquist

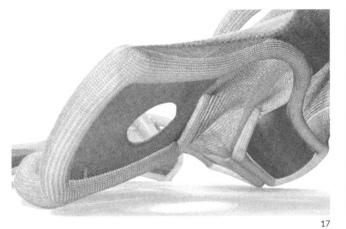

17

20

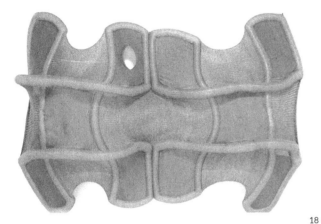

18

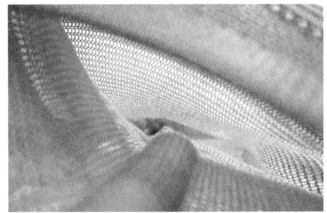

21

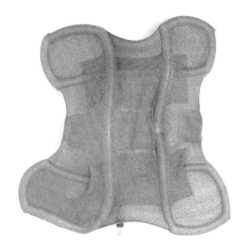

19

17 Prototype A.2, interior view.

18 Prototype A.2, top view.

19 Prototype B.1, top view.

20 Prototype B.1, external layering.

21 Prototype B.1, internal layering.

3D structure. Without the knit sleeve, the silicone tube usually shows signs of overstress at approximately 16 psi of pressure by becoming more opaque, but it maintains its integrity inside the knit sleeve to pressures of up to 40 psi without any color change. At approximately 30 psi, the inflated tube shows significant stiffness to support the knit textile lifting parts of the assembly off the ground. It is clear that the original straight orthogonal 3 X 3 grid design transforms under inflation to curvilinear forms.

Prototype A.1 and A.2 demonstrate the effect of the bridging arches in the bending of the overall structure. The longer the bridge, the more bending forces are exerted at the anchoring points.

Prototype B.1 shows how the varied knit structures not only have effects on the tensile behavior of the fabric, but also the transparency of the overall structure (Figure 20).

Computation

The study initially used Kangaroo and Maya Cloth to simulate the pneumatic textile system, but both packages focused on simulation of fabric behavior as a uniform soft body without addressing the possibility of a differentiated structural behavior within the

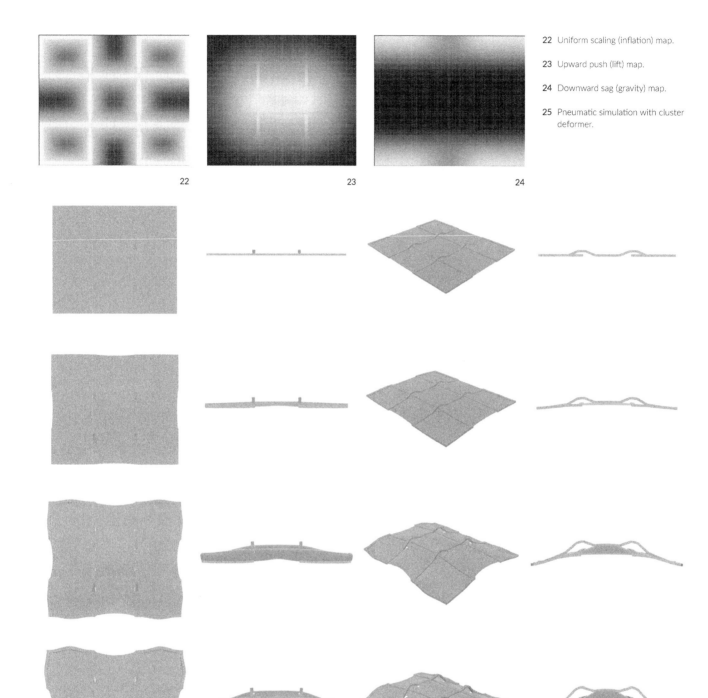

22 Uniform scaling (inflation) map.

23 Upward push (lift) map.

24 Downward sag (gravity) map.

25 Pneumatic simulation with cluster deformer.

22

23

24

25

Pneumatic Textile System Wang, Ahlquist

fabric and the continuity of the original linear yarn. Therefore, the project decided to mimic the behavior of the design structure in Maya through the systematic use of cluster deformers.

A geometric model is created in Maya and a cluster deformer is later applied. The cluster deformers generate uniform scaling similar to inflation and effect vertical movement similar to gravitational force. Assigning varied weights to the individual vertices in the geometric model, differentiated mesh movements are generated in response to the same uniform scaling or vertical movements by the cluster deformer. The weight of the deformer is scaled 0.000 to 1.000 and is applied to an individual vertex through a graphic interface of "painting" that has 255 levels of grey (white to black) to mimic the dissipation of the tensile forces. Three clusters are used to simulate inflation (uniform scaling), upward movement by expanding tube (+ Z axis translation), and gravitational pull (- Z axis translation). Figure 22 shows the inflation map were areas of the tube location are at the 100% effective range (white color) of the cluster. The tube area then gradually dissipates toward the center of each rectangle into shades of grey. Figure 23 shows areas of the prototype that will be propped upward during the expansion of the inflating tube. Figure 24 shows the downward drag around the outer edges due to weight of the prototype.

CONCLUSION
The prototypes demonstrate the ability of custom knitting to integrate external elements to form a transformative material system. However, the process of textile design requires many rounds of trial and error until the desired behavior is achieved. The knit textile design process is actually suited for computational design because either the "knit" or "miss" conditions of knitting are similar to the binary conditions of 1 or 0. Computing will resolve the different shades of grey between black and white similar to the way that knit fabric redistributes its applied forces. Figure 22 attempts to show how areas of different fabric density respond differently to the stretch caused by the same inflated tube. The tedious task of measuring the individual stitch spacing will eventually lead to the rendering of mathematical equations that describe the force dissipation by the linear yarn of the fabric. Data gathered from the analogue model can feed into the design of a more accurate computational model.

Immediate advancements in the pneumatic textile system can be obtained with more experiments with different yarn materials, different geometric patterns of bladder inflation implementation, or even the use of the custom textile as soft formworks, since casting plaster or concrete can lead to stretching in a manner similar to inflation.

REFERENCES
Kim, Simon, Mark Yim, Kevin Alcedo, Michael Chung, Billy Wang, and Hyeji Yang. 2015. "Soft Robotics Applied to Architecture." In *Computational Ecologies: Design in the Anthropocene—Proceedings of the 35th Annual Conference of the Association for Computer Aided Design in Architecture* edited by Lonn Combs and Chris Perry. Cincinnati, OH: ACADIA. 232–242.

Thomsen, Mette Ramsgaard, Martin Tamke, Phil Ayres, and Paul Nicholas. 2015. *CITA Works*. Toronto: Riverside Architectural Press. 178.

IMAGE CREDITS
Figure 3: Soft Robotics Applied to Architecture, Kim et al., 2015
Figure 4: Listener, Thomsen et al., 2009
All other figures: Wang and Alquist, 2016

Adam Wang graduated from the University of Michigan with an MS in Architecture, with a Material Systems concentration, in 2016. Prior to Michigan, Adam received a B.Arch from the University of Kentucky and has worked in the field of architecture both in the United States and Asia for the past 10 years.

Sean Ahlquist is an Assistant Professor of Architecture at the University of Michigan. He is a part of the Cluster in Computational Media and Interactive Systems which connects Architecture with the fields of Material Science, Computer Science, Art & Design and Music. Research and course topics are centered on material computation, developing articulated material structures and modes of design which enable the study of spatial behaviors and human interaction. Ahlquist's research agendas include the design and fabrication of pre-stressed lightweight structures, innovations in textile-reinforced composite materials for aerospace and automotive design, and development of tactile sensorial environments as interfaces for physical interaction.

Highly Informed Robotic 3D Printed Polygon Mesh

A Novel Strategy of 3D Spatial Printing

Lei Yu
Tsinghua University

Yijiang Huang
University of Science and
Technology of China

Zhongyuan Liu

Sai Xiao
Archi-solution Workshop
(Beijing)

Ligang Liu

Guoxian Song
University of Science and
Technology of China

Yanxin Wang
Archi-solution Workshop
(Beijing)

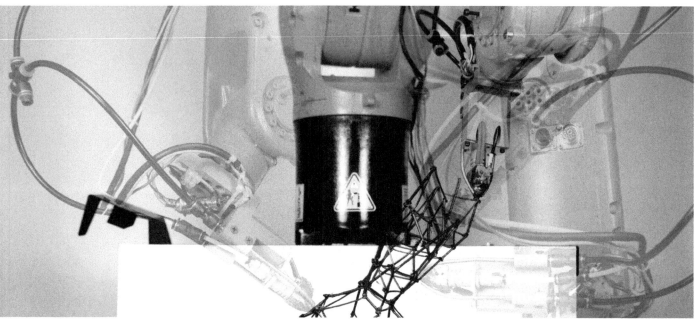

1

ABSTRACT

Though robotic 3D printing technology is currently undergoing rapid development, most of the research and experiments are still based on a bottom up layering process. This paper addresses long term research into a robotic 3D printed polygon mesh whose struts are directly built up and joined together as rapidly generated physical wireframes. This paper presents a novel "multi-threaded" robotic extruder, as well as a technical strategy to create a "printable" polygon mesh that is collision-free during robotic operation. Compared to standard 3D printing, architectural applications demand much larger dimensions at human scale, geometrically lower resolution and faster production speed. Taking these features into consideration, 3D printed frameworks have huge potential in the building industry by combining robot arm technology together with FDM 3D printing technology. Currently, this methodology of rapid prototyping could potentially be applied on pre-fabricated building components, especially ones with uniform parabolic features. Owing to the mechanical features of the robot arm, the most crucial challenge of this research is the consistency of non-stop automated control. Here, an algorithm is employed not only to predict and solve problems, but also to optimize for a highly efficient construction process in coordination of the robotic 3D printing system. Since every stroke of the wireframe contains many parameters and calculations in order to reflect its native organization and structure, this robotic 3D printing process requires processing an intensive amount of data in the back stage.

3D PRINTED FRAMEWORK

A polygon mesh is often known as a collection of vertices, edges and faces that define the shape of a polyhedral object in 3D computer graphics and solid modeling. Typical 3D printing techniques slice the polygon mesh with evenly spaced horizontal planes and rebuild the model through the layering of contour crafted slices. But this method of fabrication is too slow to satisfy the scale of production for the building industry. When Solid Doodler came out a few years ago, its spatial thermoplastic extrusion drew much attention. Especially with its powerful and flexible maneuvering that could potentially simulate the performance of a human arm, the industrial robot arm has proven to be a perfect choice for the task.

Several research teams have had tremendous achievements in spatial 3D printing by robot arms, for instance, Freeform Printing (Oxman et al. 2013), Iridescence Print (Helm et al. 2015), 'TN-01' by Branch Technology, and Mesh Mold (Norman et al.2015, Figure 2). These precursors mainly use a 6-axis industrial robot mounted with an FDM extruder as the end effector to produce a matrix of wireframes. The outputs could either be used as a mesh mold that is an alternative to conventional formwork or as a self-supporting structure. The features of the framework evoke a truss system that extends beyond the traditional notion of polygon mesh that is only defined by geometric surfaces.

All the projects above share certain common characteristics as the construction processes are all based on a layering process from the ground up. The nature of the process would result in the following shortcomings:

- The polygon mesh is rebuilt after being sliced into layers with a certain thickness, which in turn causes loss of geometric detail and resolution.
- These layers are different from the typical 3D printing thin slices, but no more than the thicken "strata" composed with triangle strut groups that are seen in the example above.
- The inherent weakness between the layers in a structure fabricated through the contouring process could cause fractures between layers when the structure is under abrupt force and loads. Thus this feature would render the process improper for fabrication of large-scale building components that require additional infill.

Based on the current research of robotic 3D printing, there are three potential aspects that are worth investigation.

- How to maintain a proper resolution of the original mesh in the definition used for physical output.

2

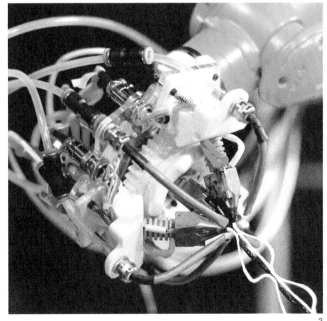

3

2   Mesh Mould by Norman Hack and Willi Viktor Lauer.

3   Hyperbolic Extruder on Robot, a 2014 Workshop in School of Architecture and Urban Planning, Tongji University, Shanghai China.

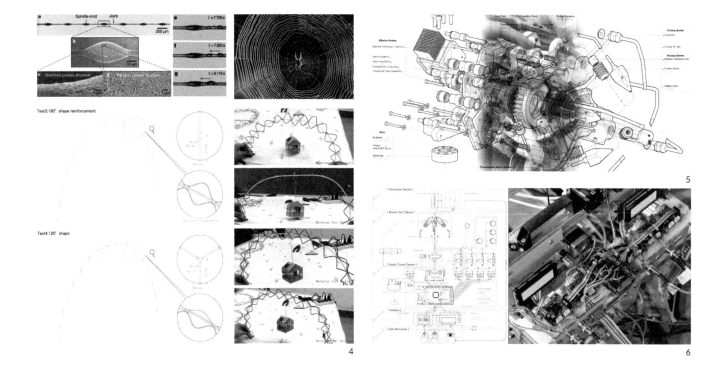

4

5

6

- How to utilize the potential of the robot arm's dexterity which is much more flexible than the generic 3D printing CNC machine.
- How to take advantage of the geometrical features of the original polygon mesh to generate a 3D spatial wireframe and reinforce its strength in all directions.

ROBOTIC 3D SPATIAL PRINTED POLYGON MESH

In the past two years, we have dedicated a constant effort on how to employ an industrial robot arm in the rapid construction of frameworks. This combination has shown great potential that could fundamentally alter the conventional method of concrete mold-work, which has always faced criticism for overly high cost and being labor intensive, especially for customized building parts. Our research focuses on a novel direction that mesh framework could be strategically 3D printed according to its native topographies. The key focus of the research is the spatial configuration of the 3D printing process and its strength, with particular attention to the span, deformation, connection and generation sequence. It contains two stages that involve different attitudes toward material configuration. The former is based on multi-threaded extrusion, while the latter is single-threaded extrusion.

Multi-threaded Extrusion and Form Definition

The goal of this research stage is to create standalone lineal overhang, which is capable of rising up as a self-supporting structure. In order to increase the thin plastic filament's capacity for

a larger span, this research concentrated heavily on the design of the robot arm's end effector, which could produce significant composite extrusions embedded with more structural freedom.

Spider silk was investigated under a microscope for more evidence. We were surprised to discover that the profile of spider silk appears as a series of spindle knots joined together under the microscope at 200 μm. After a few tests with hand-made models lifting the same block, including a model of one center tread surrounded by two or three corrugated ABS filaments (Figure 3&4), we discovered a model that is highly efficient in increasing toughness of the extrusion through the spatial configurations. Design decisions for the robot's 'hand' were based on creating these regulating knots along the motion path of the robot. There are four nozzles required for this mechanism. One center nozzle extruding a fat straight thread, and the other three outer nozzles generating undulating curves that meet periodically with the center thread (Figure 5). The outer nozzles are fixed onto a sweeping mechanism driven by stepper motors to ensure their open-close movements. Extruder and robot arm are synchronized by an Arduino system (Figure 6). There are many essential driving parameters, such as the robot effector's moving speed, extruding speeds, rotation speed of the center sweeper, nozzle temperature etc. All of the parameters need to be coordinated in order to create a smooth lineal profile similar to the hand-made test model.

The robotic 3D printer generates a hyperbolic curve that addresses extra geometric and structural challenges (Figure 7).

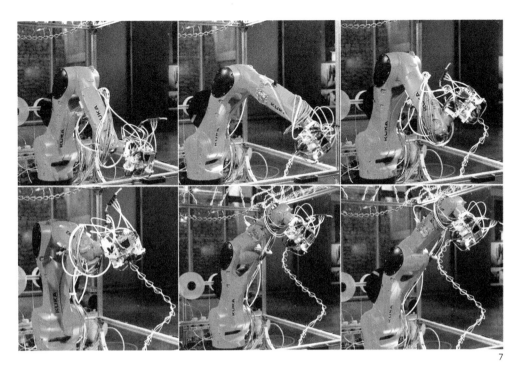

7

Unlike the arch model with two anchor points, this hyperbolic helix has only one standing point while supporting itself by its own rigidity derived from its biomimetic configuration. We are also aware that the ABS filament, a common thermo-plastic, has a significant shrinkage rate, which means that after the melted plastic extrusion cools down, the shape will deform slightly, which could cause joint dispositions under some circumstances. Even though the multi-threaded extrusion gains certain achievement with its attractive performance, three main concerns drive the research into the second stage – a massive structural frame:

- Deformation due to material shrinkage and gravity will cause joints to dislocate from the adjacent connecting struts.
- The hyperbolic configuration is hard to connect with other components other without extra supports.
- The multi-threaded robot end effector is too big to avoid collision with the existing 3D printed elements, and always causes the robot to run out of working range.

Single-threaded Extrusion and Form Definition

From the previous study, we have come to a conclusion that in order to achieve a better-defined 3D printed mesh, the robot effector should be as small as possible for the process to be collision free. Therefore, we started to reconsider single-threaded extrusion. Even though we abandoned the multi-threaded extrusion approach, much of the experience and parameters are were transferred into the next stage.

Industry robots have inborn constraints which have to be considered at each step of motion to plan a valid working path. The constraints of motion planning include existing obstacles in the working space, limits from articulated singularities and articulated robot joint angles. According to the American National Standard for Industrial Robots and Robot Systems, singularity is defined as a condition caused by the collinear alignment of two or more robot axes resulting in unpredictable robot motion and velocities.

Singularities occur as:

- Wrist singularities: occur when the axes of Joints 4 and 6 are aligned;
- Alignment singularities: occur when Joint 6 (wrist) and Joint 1 axes are aligned;
- Elbow singularities: occur when the arm is fully extended. In this case, as the elbow joint becomes further extended, higher joint speeds are required to maintain constant Cartesian speed. The robot cannot extend beyond its reach.

These constraints affect a tiling style construction process much less than an 'orientation style' construction technique, which is the method involved in this paper. If the effector's orientation and the space angle of mesh edges are to be matched within a certain angle for making straight and strong struts, there are many more parameters that need to be systematically managed. The data requires careful calculation and simulation for validation and optimization beforehand, testing for any possibility of halting during fabrication.

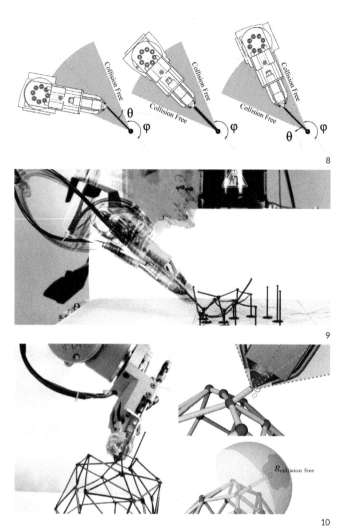

two considerations, the sequence that the mesh edges are built needs to be specified, where every step of operation from the first edge to the last one is ensured to be valid regarding these hard and soft constraints.

The Collision Area (Figure 10) illustrates the virtual working zone, which tightly relates to a few angles that contribute to the development of the mesh construction sequence and avoid the singularities of robot. The process of pursuing this magical zone consists of three steps that almost entirely depend on the algorithm.

- Rebuild mesh polygon: explode mesh into edges and reorganize them in an adaptable queue for robotic 3D spatial printing.
- Validation of the rebuilt mesh: each re-connection has to be qualified not only to avoiding the existing printed struts, but also to be responsible for the construction of the remaining mesh edges which will be potentially influenced by the former process.
- Validation of robot control: singularity or self-lock cases need to be entirely removed in the whole motion planning to avoid accidental disruption to the fabrication process.

The algorithm is meticulously driven by this comprehensive solution. Compared to the generic tiling 3D printing method, this method has to be pre-calculated repetitively to guarantee the correct joint-to-joint plastic extrusion and the fluency of the overall fabrication process. Therefore, each stroke has a set of information including both the basic point XYZ positions and their spatial vectors. This process generates a massive amount of data. The data is filtered, optimized and organized into the most appropriate groups, becoming the data package to drive the actual working path.

Re-composition of Polygon Mesh

The reason the Stanford Bunny (Figure 11) was selected as the test object is that this typical case of computer graphics bears most of the geometrical features of a polygon mesh. The result of the test could potentially be applied within the construction of most architecture forms.

To reduce the calculation, the edges of the polygon mesh are sorted into different groups based on their geometrical features. As in the Stanford Bunny, the portion of the ear color coded red and the top skull portion in colored in green represent grouped sub-entities. Then each sub-entity is calculated for sequence configuration separately. In this way, each edge's fabrication data only needs to be considered and validated within a smaller set instead of the whole. Even though it still performs like the

8   Diagrams of Nozzle Orientation, Thermoplastic Aggregation and Collision Free zone.

9   Process of One Single Extrusion.

10 Relationship of Working Angles.

The mechanism of the robot end effector design and the proper algorithm are both key elements to this research that rigorously rely on each other. First of all, the cone volume of the extruding effector should be sharp enough on angle α (Figure 8) to avoid collision with existing static obstacles. Furthermore, only when angle θ is limited within a certain degree will the extruded thermoplastic be aggregated into a straight line(- Figure 9). However, the cone volume of the heating block has to contain a heater, heating nozzle, dual cooling nozzles and a thermistor. All these necessary parts together force the cone to be above a certain volume in size. On the other hand, each spatial stroke has to consider its neighborhood condition so that it will not touch or collapse onto its adjacent parts. This process requires the management of the φ angles. On the top of these

Highly Informed Robotic 3D Printed Polygon Mesh Yu et. al.

layering concept in a certain sense, this sub-entities concept results in a much larger volumetric structure than flat slices. This methodology takes not only point-to-point definition into consideration, but also point-to-point orientation.

Specifically, the strut set S is decomposed into two layers, B and T, each time, where T is the top layer and B the bottom layer. The decomposition methodology has to guarantee that:

- The size of upper part T is small enough so that the sequence-searching algorithm can be applied efficiently within the capacity of the computer.
- The lower part B is in a structurally stable state, i.e. each node has limited deformation so that the robot can locate it in the subsequent printing process.

The decomposition algorithm can be recursively applied on the lower part B until the termination criteria are satisfied.

Physical constraint: At any intermediate fabrication state, the fabricated frame shape should keep stable with limited deformation for each node. Thus, for the deformation vector:

$$||d_i^{\mathrm{transf}}|| < \epsilon, \qquad \text{where } i = 1, 2, ..., N \qquad (1)$$

Before introducing the optimized model, we will define the notations that are used. As the basic fabrication element in this solution is the strut, layer decomposition divides the set of frame strut ε into two strut sets every time, or in mathematical context, finds a cut on ε. The division has the relation of ε = B∪T, where B and T share common nodes. If strut e_i and e_j share one common node, we then say they are connected, and then define the weight between them as w_{ij}. Thus, the cut would be:

$$w(S,T) = \sum_{e_i \in B, e_j \in T} w_{ij} \qquad (2)$$

Optimization model: The objective of the layer decomposition is to limit the size of the upper part S. The constraint is that the lower part must have limited deformation. We need to verify the constraint by confirming the deformation vector computed by the stiffness equation (Hughes 2012, Kassimali 2012) satisfies the physical constraint. To achieve this goal, we define a weight graph on the current frame structure. The aim of the optimization model is to find a minimum cut on the weighted graph with a stable constraint:

$$\min_{\mathcal{B},\mathcal{T}} \; w(\mathcal{B},\mathcal{T}) \qquad (3)$$
$$\text{s.t. } \mathbf{K}(\mathcal{B})\mathbf{d}(\mathcal{B}) = \mathbf{f}(B).$$
$$||\mathbf{d}_i^t(\mathcal{B})||_2 < \varepsilon.$$
$$S^t \subset \mathcal{T}.$$
$$S^b \subset \mathcal{B}.$$

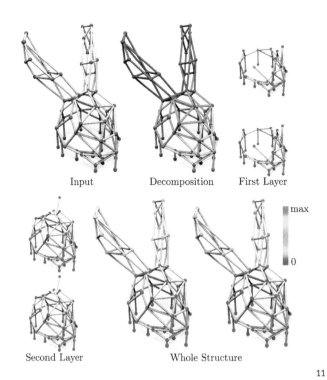

Input   Decomposition   First Layer

Second Layer   Whole Structure

11

Algorithm 1: Layer Decomposition Algorithm

1 **while** $|T| >$ *tolerance for sub-part size* **do**
2   G = T;
3   Define the weight w_{ij} according to Equation (1);
4   Construct stiffness matrix;
5   Set boundary value;
6   $iter = 0$;
7   **while** *Reweighing termination criteria = False* **do**
8   ADMM solver;
9   Update reweighting factor;
10    $iter = iter+1$;
11   **if** *Solution is feasible* **then**
12   Cut G into Set T and B according to the computation result;
13   **else**
14   Return False (no feasible decomposition can be found);

12

Algorithm 2: SequenceSearching(k, s_{i_k})

1 Add s_{i_k} into the printed strut set;
2 Delete s_{i_k} from the unprinted strut set;
3 **while** *candidate set is not empty* **do**
4   Set the strut s' with the minimal cost $E(s')$ as $s_{i_{k+1}}$;
5   **if** *SequenceSearching$(k + 1, s_{i_{k+1}})$ = True* **then**
6   Print strut $s_{i_{k+1}}$;
7   Return True;
8   **else**
9   Delete s' from the candidate set;
10 **if** *Candidate Set is not empty* **then**
11   Return True;
12 **else**
13   Return False;

13

11   Polygon Mesh Decomposition in Grouped Layer.

12   Algorithm 1: Layer Decomposition.

13   Algorithm 2: Sequence Searching.

Algorithm 3: Fabrication sequence design (overall)

1 Input frame shape G is decomposed into layers of $G_1, ..., G_L$ by
 iteratively applying Algorithm 2;
2 **for** *each layer G_l* **do**
3   **if** *SequenceSearching$(1, s_{i_1})$ = False* **then**
4   Return False (No valid printing sequence found);
5 Return True (a valid fabrication sequence is obtained);

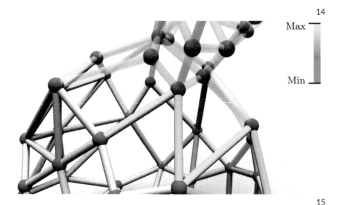

14

Max

Min

15

14   Algorithm 3 for Fabrication Sequence Design.

15   Extruding Sequence Probing Simulation.

Here S^t and S^b are respectively the top and bottom boundary struts of the input wireframe. K is the stiffness matrix of sub-structure B and f is the gravity force associated with the bending moment of the nodes (Hughes 2012, Kassimali 2012). To make the size of set S small, the cut should be close to the top boundary. To achieve this purpose, we define weight w_{ij} as:

$$w_{ij} = \begin{cases} exp(-\beta h^2(v_k)), & e_i \text{ and } e_j \text{ share node } v_k \\ 0, & \text{otherwise.} \end{cases} \quad (4)$$

Here, parameter β is used to control the weight distribution and $h^2(v_k)$ is the normalized height of the node v_k. In this way, the weights of the nodes close to the top boundary are small, which makes the cut boundary as high as possible and thus the top sub-part T is as small as possible.

Numerical computation of the decomposition algorithm: This

is a constrained discrete optimization problem. We first simplify the binary optimization problem into a continuous optimization problem and then use an iterative re-weighting scheme (Lai et al. 2013) to solve this constrained optimization problem. In each iteration re-weighing loops, we adopt the Alternating Direction Method of Multipliers (ADMM) (Boyd et al. 2011) to solve sub-problems, as shown in Algorithm 1 (Figure 12).

Fabrication Sequence Finding

Collision is the most critical concern in this phase of the fabrication sequence. Quite different from the tiling method, in which the nozzle is strictly perpendicular to the build plate, the

'orientation style' extrusion in our approach puts the extruder tip inside the existing wireframe in most of situations. Hence, collision possibility has to be evaluated carefully for every single stroke. After many experiments, when the extruder nozzle extrudes along the direction of each polygon edge mesh, the cone volume, defined by angle θ (Figure 8), provides a partial solution to avoid the obstacles during extrusion.

To compute the complete fabrication sequence, the first step is to consider how to choose the next strut to print at each state. Assume we have already printed k struts, which form sub-structure S_k. Now we have to find an optimal strut with minimal cost to use for subsequent printing. We construct a candidate set of eligible struts for the current decision state, where each eligible strut s in the set satisfies:

- s is connected to the current printed structure.
- There are collision-free extruder orientations eligible for the printing s.

There might be several eligible struts in the candidate set. We then introduce cost evaluation for each candidate to help us select the most appropriate one. The sequence-searching algorithm 2 (Figure 13) is a kind of searching strategy with quick pruning, using the evaluated cost as a guide, to quickly prune off unnecessary items to avoid searching an unnecessary printing sequence.

Overall algorithm

The re-composition is also considered as a reconstruction of an architectural framework. Each sub-entity contains a structure-like framework by means of its triangle formation. A C++ program was chosen to work with Grasshopper in the Rhinoceros environment. A genetic algorithm, one type of Heuristic algorithm, is employed in this case for a suitable solution to the decomposition and re-organization. (Latombe 2012; Choset 2005). A similar algorithm is applied in the fabrication sequence finding as well.

However, more collision free zones are defined through study of the kinetic features of the industrial robot. This six axis robot performs in a way similar to human limbs but with a lot more strength and precision. Through calculation, there are multiple angles free from collision, but there are only a few candidates that are collision free and compatible with the next strokes. So the algorithm of sequence design attempts to optimize not only the next single struts, but the rest of its group. After tapping all related information into the algorithm (Figure 14), the result is quite unexpected. The optimal sequence is not a continuous polyline-making motion, but a discrete order. In Figure 15, the struts are color coded from yellow to orange according to their

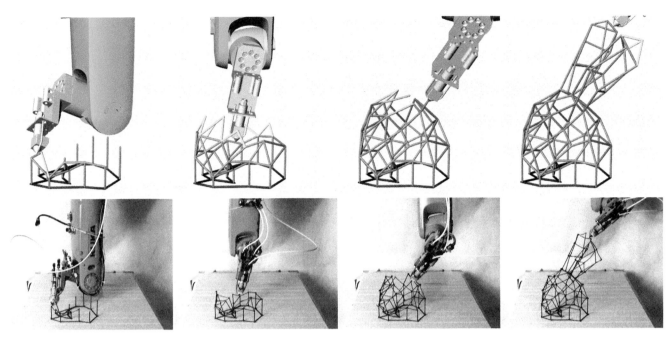

16 Validation Between Virtual Simulation and Physical Fabrication.

order in the construction sequence. The yellow struts have more priority than the orange. In the optimal sequence, the extruder will generate the struts with similar spatial directions first. Then the robot changes its gesture to work on another direction to close the polygon mesh. During the construction process, there are lots of transitional moments that the robot turns its axis drastically from one gesture to another. These transitions actually permit the machine to 3D print in a fluent sequence by skipping any cessation caused by the robot's over-range and singularities.

Spatial Printing Material

Almost all of the spatial thermoplastic extrusion projects confront the same trouble: shrinkage. In the multi-threaded extrusion project, the outer nozzles have to swing a little more toward the center tip to allow all four threads to run into each other. This is because ABS thread has a certain shrinkage during the cooling phase. This phenomenon brings more drawbacks in single-threaded frame printing. Because of shrinkage and self-weight, the single standalone strut which does not have any support from adjacent elements, will have physical deviation causing the end point to lose its position for the next member to pick up. Consequentially, the remaining struts will not keep the relay and the end result is a disastrous mess.

Many thermoplastic filaments, such as ABS and PLA, were tested and proven to have similar deviation problems. PLA has a little more advantage over ABS which has a higher melting temperature and longer cooling time. A custom-made PLA mixed with certain amount of carbon powder is applied to fix this problem.

As expected, the strut extruded by this composite PLA filament reduces the deviation into an acceptable range.

Robot Control Based on Mathematics

Robots only works in a certain range due to mechanical constraints. If it goes beyond this range, it will self-lock and cause the whole process to halt. There is also a singularity problem that the alignment of 4th axis and 6th axis causes robot to behave in an unpredictable manner, followed again by an immediate halt. It means that the sets of working sequences figured out from the sequence design have to be re-selected to avoid these issues. KUKA|prc, a Grasshopper plug-in used for generating KUKA robot code, is taken into consideration in this phase. KUKA|prc simulates the whole printing process in Rhinoceros and eliminate the bad key frames, and helps to pipe out the final robotic G-code.

The custom-made extruder is installed on the robot's 6th axis plate. To reduce the extruder's size, a stepper motor is installed on the top of the 4th axis. This extruder design takes the reference of a Bowden extruder, which is popularly used in a lot of desktop 3D printers In this way, the robot effector is lighter and has less interference with other elements. The effector's length and tip angle also heavily influence the validation phase, which has been improved many times based on both calculation and mechanical optimization (Figure 16).

Final Spatial Printing

The Stanford Bunny mesh model was reduced to 179 edges. It

17 Concrete Cement Spray on Metal Wireframe.

18 Voronoi Tessellated 3D Polygon Mesh of Stanford Bunny.

19 Soap Bubbles, New York, 1945-6 by Berenice Abbott.

took 14.09 seconds to finish the calculation of mesh decomposition (section 2.3), 36.61 seconds for sequence searching (section 2.4) with a powerful CPU working at 4G frequency. The overall printing time was 91 minutes for a model that is 371 mm in height. However, it took almost one hour to calculate the valid robot working path which was free of collision and self-lock; this was because there were interpreting processes between Rhino and C++, and the most time was wasted in the middle of transition. Since the whole process was validated through computer simulation, the physical printing process was non-stop. All edges were joined together in good condition, especially in the ears area, which is the most difficult part with large overhangs (Figure 20).

CONCLUSION

The method to generate a physical wireframe model based on FDM and robotic technology involved in this paper is divided into two consecutive stages based on different strategies and applications. The latter strategy would have significant potential on the rapid industrial fabrication of large scale building components. The involvement of robotic technology allows architects to gain more control on the fabrication of spatial structures through utilizing mathematical algorithms. The spatial printing methodology based on the original mesh's topological logic confronts many technological challenges on one hand. On the other hand, it is genuinely superior to the layering method of

fabrication. Thus this research proposed a novel direction in the field of robotic spatial printing. This direction distinguishes itself from its predecessor in ways such as its geometrical optimization is embedded with many conditions of robotic control, and the process itself involves an intensive amount of complicated data collection and filtration.

However, the current method of thermal melting extrusion places heavy restrictions on the physical quality of the printable materials. This is also the bottleneck of the 3D printing industry on the whole. Only with more adequate materials, can robotic 3D printing technology be transformed into a more adaptable fabrication method that could be wildly applied in the architecture industry.

FUTURE DEVELOPMENT

This research is still currently restricted to a lab environment. A desktop level 3D printer extruder was modified and installed onto a desktop level industrial robot (KUKA KR10 R1100). Thus the operation volume of the machine limits the ability to make architecture components in 1:1 scale. In the upcoming stage we will adopt a heavy-load industrial robot with professional heat melting glue gun, such as a BAK heat gun, installed for fabrication and installation in an architectural dimension.

The Stanford Bunny serves only as a test prototype at this stage. Next, we will adapt this process to the fabrication of complicated surfaces and use reinforced fiber concrete spray casting techniques (Figure 17) for fabricating a series of composite products. We are also aiming to complete a 'solid' mesh (Figure 18&19) with an internal cellular structure after optimizing the algorithms for calculation of robotic motion path, and print a truly spatial physical mesh mold.

REFERENCES

Boyd Stephen, Neal Parikh, Eric Chu, et al. "Distributed optimization and statistical learning via the alternating direction method of multipliers." *Foundations and Trends in Machine Learning*. 3(1): 1–122.

Choset, Howie; Kevin M. Lynch, Seth Hutchinson, George A. Kantor, Wolfram Burgard, Lydia E. Kavraki and Sebastian Thrun. 2005. *Principles of Robot Motion: Theory, Algorithms, and Implementation*. Cambridge: MIT press.

Hack, Norman; Willi Viktor Lauer, Fabio Gramazio and Matthias Kohler. 2014. "Mesh Mould: Differentiation for Enhanced Performance".. Kyoto: CAADRIA. 139–148

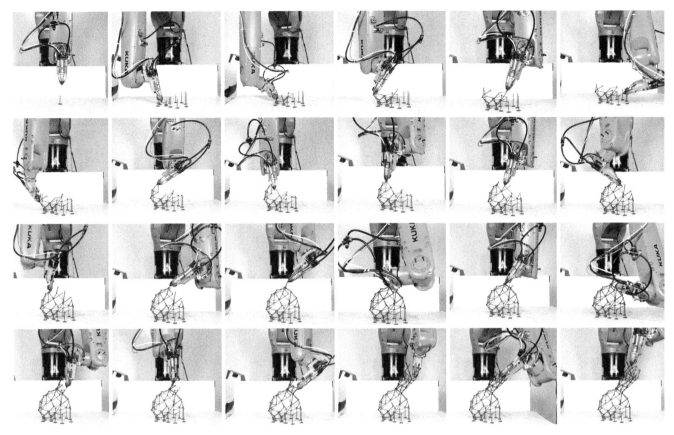

20 Entire Process of 3D Robotic Spatial Printing.

Hughes T J R. 2012. *The Finite Element Method: Linear Static and Dynamic Finite Element Analysis*. Mineola: Dover Publications.

Helm, Volker; Jan Willmann, Andreas Thoma, Luka Piškorec, Norman Hack, Fabio Gramazio, and Matthias Kohler 2015. "Iridescence print: Robotically printed lightweight mesh structures." *3D Printing and Additive Manufacturing* 2(3): 117–122.

Lai, Ming-Jun; Yangyang Xu, And Wotao Yin. "Improved Iteratively Reweighted Least Squares for Unconstrained Smoothed l_q Minimization." *SIAM Journal on Numerical Analysis*.51(2): 927–957.

Latombe, Jean-Claude. 2012. *Robot motion planning*. New York: Springer Science & Business Media.

Kassimali Aslam. 2012. *Matrix Analysis of Structures SI Version*. Indenpendence, KY: Cengage Learning.

Oxman, Neri; Jared Laucks, Marcus Kayser, Elizabeth Tsai, and Michal Firstenberg. "Freeform 3d printing: Towards a sustainable approach to additive manufacturing." *Green Design, Materials and Manufacturing Processes*. Edited by H. Bartolo et. al. London: Taylor & Francis. 479–484.

IMAGE CREDITS

Figures 1, 8, 9, 17, 20: Lei Yu, 2016
Figure 2: Norman Hack, Willi Viktor Lauer, Fabio Gramazio and Matthias Kohler, 2014
Figures 3, 4, 5, 6, 7, 8,: Lei Yu, 2014
Figures 10, 16: Yijiang Huang, Lei Yu, 2016
Figure 11, 12, 13, 14, 15: Yijiang Huang, 2016
Figure 18: Ligang Liu, Lei Yu, 2016
Figure 19: Bernice Abbott, New York, 1945–6

Lei Yu is a Ph.D Candidate, School of Architecture, Tsinghua University. He got the MArch degree from GSD Harvard in 2004. He is the founder of Archi-solution Workshop (ASW,Beijing) , the co-founder of Laboratory for Creative Design (LCD, Beijing), and the co-founder of DADA (Digital Architecture Design Association). His teaching experience includes ETH Zurich, Tsinghua University, and Tongji University. He was featured as Beijing Design Week Elite of 2015.

Concepts and Methodologies for Multiscale Modeling

A Mesh-Based Approach for Bi-Directional Information Flows

Paul Nicholas
Mateusz Zwierzycki
David Stasiuk
Esben Nørgaard
Mette Ramsgaard Thomsen
CITA / Royal Danish
Academy of Fine Arts

1

ABSTRACT

This paper introduces concepts and methodologies for multiscale modeling in architecture, and demonstrates their application to support bi-directional information flows in the design of a panelized, thin skinned metal structure. Parameters linked to the incremental sheet forming fabrication process, rigidisation, panelization, and global structural performance are included in this information flow. The term multiscale refers to the decomposition of a design problem into distinct but interdependent models according to scales or frameworks, and to the techniques that support the transfer of information between these models.

We describe information flows between the scales of structure, panel element, and material via two mesh-based approaches. The first approach demonstrates the use of adaptive meshing to efficiently and sequentially increase resolution to support structural analysis, panelization, local geometric formation, connectivity, and the calculation of forming strains and material thinning. A second approach shows how dynamically coupling adaptive meshing with a tree structure supports efficient refinement and coarsening of information. The multiscale modeling approaches are substantiated through the production of structures and prototypes.

1   The installation 'StressedSkins,' at the Danish Design Museum 2015.

INTRODUCTION

Thin panelized metallic skins play an important role in contemporary architecture, often as a non-structural cladding system. Strategically increasing the structural capacity—particularly the rigidity—of this cladding layer could offer significant savings for secondary and primary structural systems. Achievable through the specification of geometric and material properties, the development of skin-stiffening techniques marked the early history of metallic aircraft manufactuing (Hirschel, Prem, and Madelung 2012), and are currently applied within the automotive industry, where selective local differentiation of sheet thickness and yield strength combine with locally specific rigidizing geometries that increase structural depth.

To improve the rigidity of thin skinned metal structures requires a modeling approach that guards against instabilities due to buckling at three distinct scales: buckling of the structure, buckling within panel elements which have to carry compressive load, and also buckling and tearing that can occur during the sheet forming process itself (Nicholas et al. 2015). This necessitates a multiscale perspective. In this research, much of the multiscale challenge is related to the fabrication technique used to form the steel sheet—robotic incremental sheet forming (ISF)—and the desire to connect information regarding localized material change that results from this process to the design and finite element analysis of the larger structure. This is accomplished through a transition between multiple mesh resolutions, and an approach to meshing that supports effective flows of information about both geometric and material properties. In this paper, we introduce these modeling frameworks through a description of the installation 'StressedSkins' (Figures 1–3).

The paper is organized as follows: section one describes a conceptual background for multiscale modeling, the ISF process, and the geometric and material transformations that it implicates. Section two describes our application of multiscale modeling, and presents two adaptive mesh-based approaches. The first supports predominantly unidirectional information flow and the second implements bidirectional information flow through a coupled meshing/tree traversal.

MULTISCALE MODELING

Most physical and social phenomena are multiscale, and exhibit what Cyril S. Smith has described as the "deep entanglement of macro and micro" (1981). We organize time into days, months, and years as a result of the multiscale dynamics of the solar system (E 2011). We understand materials to combine "macrocosm and microcosm consist[ing] of innumerable material objects... each material object capable of supporting and transmitting forces" (Otto 1992). Architectural structures can be

2

3

2   The installation 'StressedSkins,' at the Danish Design Museum 2015.

3   Forming of connection and rigidisation geometries on the inner and outer skins enables stability and force transfer without a frame.

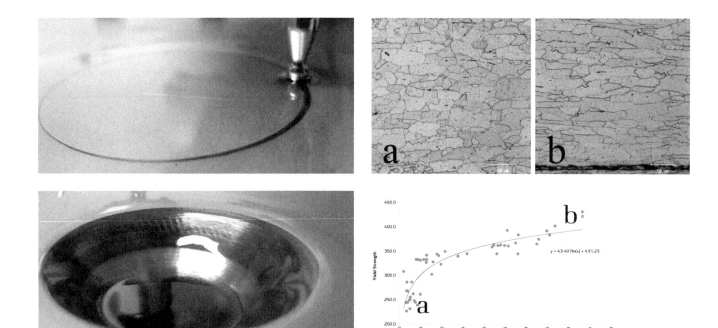

thought of similarly, as nested organizations from which features, behaviors, and properties emerge on the basis of interactions across scales and systems. Macroscopic domains—often concerned with territories, topologies, and structures—provide environmental constraints for the micro-scale concerns—material distributions, loads, limits—that also inform them.

Modeling approaches in architecture typically follow traditional drawing practice: a focus on one scale at a time, and a gradual refinement from greater to smaller scales. Relations between scales work under the assumption that processes at any other scale are homogenous, or can be described via highly simplified linear relationships. But in other fields, including materials science, economics, and meteorology, alternate modeling approaches that support a different and less linear set of relations and flow of information have developed.

These modeling approaches—termed multiscale—simulate underlying phenomena that span a sequence of scales or, more accurately, frameworks (E 2011). They have developed on the basis of several realizations: 1) that no single model or framework is adequate on its own to capture the full behavior of a system, since the information and models that we have about the world are partial and bounded; 2) that modeling efficiencies can be gained by exploiting different levels of resolution; and 3) that high-resolution models quickly becomes intractable at larger scales. For example, molecular dynamics and quantum mechanics models can capture differentiation at the smallest scales, but because of computational issues, these simulations are currently

constrained to approximately 107–108 molecules, or about fifty nanometers. The problem of modeling larger collections is not simply computational; the mathematical complexities are so great that it is impossible to apply them directly to common problems (E 2011). Given that architectural models—when attempting to model differentiation within the bounds of a single scale and a single model—are similarly constrained in computing dynamic information flows between large numbers of entities, multiscalar approaches become a promising architectural tool.

Instead of attempting a complete description within a single scale or model, multiscale approaches assemble a multiplicity of models, each capable of describing an important feature using a particular framework. These models are connected together, so that the output of a given model becomes the input for another. Multiscale modeling is therefore the identification and construction of suitable models and frameworks, together with the application of modeling techniques that relate or 'bridge' these models and frameworks (Elliot 2011) by coupling together different kinds of description.

Within architecture and engineering, one approach to multi-scale modeling is to link a macro-scale structural domain with a micro-scale material domain. With either design generation or optimization as a goal, each level is varied so as to achieve a specific global effect. In the simplest case, this involves the iterative solution of one problem at the macro level (stability, for example), and several problems (which together inform the best local configuration) at the material level (Coelho et al. 2008).

310

Concepts and Methodologies for Multi-scale Modelling Nicholas et al.

4   The start and end states of the incremental sheet forming process, which induces 3D form through the application of a continuous localised plastic deformation.

5   Above left: The material implications of the forming process. Above right: the elongation of grain geometry under strains induced by forming is observable via optical microscopy. Lower: Graph of yield strength as a function of strain, derived from Vickers hardness testing.

6   The machine setup at CITA used for single- and double-sided forming.

6

Some multiscale models, including the approach described in this paper, include an intermediate meso-scale level, in this case related to an architectural component and its detailing. But because the type and level of detail of information is different for the different levels of description, multiscale models can easily be constrained by the need to translate information. For this reason, bridging or 'handshaking' techniques (Winsberg 2010)—which translate, coarsen, or refine information as it passes it between models—are central to the multiscale modeling process. The mesh-based techniques described in this paper directly address this issue.

Considering the Fabrication Process as a Site For Localised Material Property Variation

The modeling process addresses the design of a thin-sheet steel structure fabricated via a specific fabrication method—robotic incremental sheet forming. Incremental sheet forming (ISF) is an innovative fabrication method for imparting 3D form on a 2D metal sheet, directly informed by a 3D CAD model. In the ISF process, a simple tool moves over the surface of a sheet (Figures 4 and 6) to cause localized plastic deformation (Jeswiet et al. 2005). The primary advantage of ISF is to remove the need for complex molds and dies, which only become economically feasible with large quantities (Wallner and Pottmann 2011). For this reason, in contexts such as automotive fabrication, ISF is explored for its potential to dramatically reduce the costs of prototyping.

Transferred into architecture, ISF moves from a prototyping technology to a production technology. Within the context of

mass customization, it provides an alternate technology through which to incorporate, exploit, and vary material capacities within the elements that make up a building system. Potential architectural applications have been identified in folded plate thin metal sheet structures (Trautz and Herkrath 2009) and customized load-adapted architectural designs (Nicholas et al. 2015; Kalo and Newsum 2014; Brüninghaus et al. 2013). Recent research has established ISF as structurally feasible at architectural scale (Nicholas et al. 2015; Bailly et al. 2015).

The Transformative Implications Of ISF

The ISF process has effects that are both geometric and materially transformative. Geometric features can be introduced by locally stretching the planar sheet out of plane. These increase structural depth and therefore increase rigidization, and can also provide architectural opportunities for connection and surface expression.

As the steel is formed, there is an increase in surface area, and a corresponding local thinning of the material. This change in thickness is important to calculate so that the material is not stretched too far, and does not tear or buckle as the thickness approaches zero. Forming also activates a process of work hardening—a deliberate application of deformation that helps resist further deformation—with the effect of raising the yield strength of the steel. Depending on the geometric transformation, the effects of the material transformation are locally introduced into the material to differing degrees, depending on the depth and angle attained through the ISF process. At an extreme, yield strength for steel can almost double (Figure 5), while material

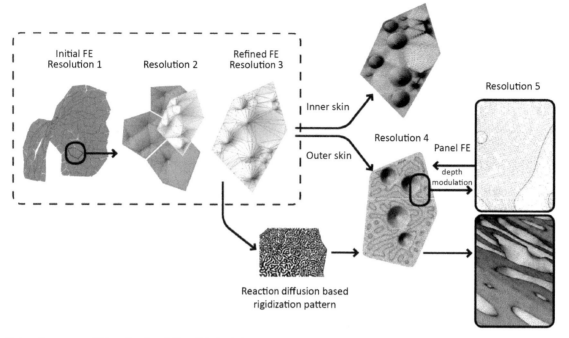

7   Flow of information across multiple scales of resolution within the design process.

thickness can reduce to zero. Because the transformative implications of ISF fabrication are significant, it is very important to incorporate them into the design phase.

DESIGN APPLICATION

The context of this research is the application of ISF to the forming of panels within unframed, panelized, stressed skin structures. Stressed skins are lightweight, thin sheet structures in which the skin is structurally active, and bears tensile, compressive, and shear loads as well as providing rigidity. ISF is particularly suited to this application, as it provides a method for customizing each panel so that it can be informed by local, performance driven requirements for rigidization and connection, as well as by the geometries needed to negotiate these conditions in a seamless manner. In our design application, ISF is used to make all out of plane geometric features within a panel, including connections between the inner and outer skin, as well as the rigidization geometries that are applied to the outer panels.

A full-scale demonstrator was installed at the Danish Design Museum in May 2015, and prototype panels that also test the meshing methods described in this paper have been produced afterwards. The basis of the customized tool-pathing algorithm is the established method of a spiral descent (Jeswiet et al. 2005), which can be run on different levels of mesh resolution to achieve different aesthetic effects, but extended to vary stepping and tooling speed in relation to wall angle, measured from the normal of the mesh face.

METHOD

One of the main problems in the design of thin-skinned metal structures is to ensure rigidity, and to guard against instabilities due to buckling at three distinct scales: buckling of the structure, bucking within panels which have to carry compressive load, and also buckling that can occur within the sheet-forming process itself. This design context necessitates a multiscale approach and the development of techniques that enable the information generated within models to flow to others.

The modeling framework for StressedSkins defines three scales—macro, meso and micro—that coincide with the considerations regarding rigidity outlined above. In addition, the macro scale encompasses the resolution of global design goals, overall geometric configurations, a full-scale understanding of structural performance and discretization, and is informed by the available scale of production. The meso scale considers the project at an assembly and sub-assembly level, and is concerned with material behaviors tied to geometric transformation, detailing, and component-level tectonic expression. The micro scale is concerned with relevant material characteristics at the most discretized level. To act as a communicative substrate and efficiently bridge between different levels of resolution to capture the required dynamics, small-scale geometry, and scale-sensitive calculations, the adaptation of a non-structured grid is pursued. This mesh supports all relevant outputs for form-finding, analysis, fabrication, and representation.

Concepts and Methodologies for Multi-scale Modelling Nicholas et al.

Communication Across Scales Through Half-Edge Mesh Structure

The first approach focuses on incrementally refining a mesh subdivision so that one mesh can support understandings of coarser topological relationships between individual panels, granular understandings of local material behaviors, and refined geometries for defining digital fabrication drivers and toolpaths. The basis of the approach is a half-edge (or directed-edge) mesh data structure. Half-edge meshes enable the deployment of N-gon faces, rather than more standard triangulated or quadrilateral faces. This opens up the possibility for designing with more complex topologies.

The sequential increase in resolution is shown in Figure 7. Initial increases in resolution are achieved through node insertions related to specific geometries, and later refinements by Loop subdivision (Loop 1987). The refinement of the mesh maintains anchored nodes, seams, and creases as they are established at different levels of resolution. At a first resolution, a generative pentagonal tiling algorithm arrays a double skin of pentagonal tiles across a base surface. The nodes of this base mesh are positioned so that edges are oriented to minimize any global hinge effects using constraint-based form finding. At a second resolution, nodes describing low-resolution details related to connection are added to the mesh. The conical geometries are integrated with the panels and connective faces—with inherited data structures—into a coarse triangulated mesh. An iterative process of finite element analysis performed upon this mesh refines the number and distribution of connection elements, which are located in as great a number as possible near high-shear forces, and aligned perpendicular to them.

A third resolution introduces new nodes that more accurately describe all connection geometries, and the mesh is then subjected to finite element analysis. The results of this analysis—utilization and bending energy—directly drive the tectonic patterning of the skins, which introduces a fourth resolution. For this, utilization forces within each panel are used to drive the depth of either oriented dimples or a non-oriented pattern within the structure (Figure 9). The complex geometries that result are informed by the calculation of thinning (Figure 8) and increased yield strength, on the basis of strain measurement via circle projection and numeric models generated from Vickers hardness testing. Empirical testing provided a means to accurately inform the model at this scale, as available theoretical models such as the sine law do not yet provide accurate models (Ambrogio et al. 2005). A final skin fabrication model at a fifth scale of resolution is synthesized, and each panel systematically arrayed for extracting toolpaths.

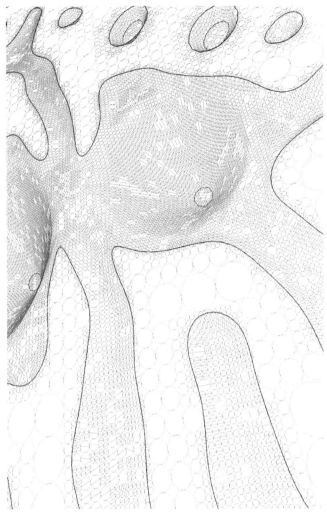

8   Calculation of strains and thinning are achieved using circle projection and a measure of deformation.

Communication Across Scales Through Coupled Meshing/Tree Traversal

The second communication approach is focused on refining two phases of the modeling process: mesh subdivision and data transmission between different scales. As experienced with the first modeling workflow, the geometries produced by subdivision can become computationally expensive, whereas their high resolution is necessary only locally within each panel, specifically where the out-of-plane deflection occurs. To reduce the mesh density without coarsening the geometry, an adaptive Loop Subdivision algorithm (Pakdel and Samavati 2004) was implemented and further developed to incorporate additional constraints. The subdivision method was extended to support creases (chains of edges which break the curvature continuity) and anchor points (points which stay in place during the process), which are utilized to efficiently and precisely model the deformation. Using this adaptive subdivision strategy, the resolution of a typical mesh used in the first demonstrator can be reduced

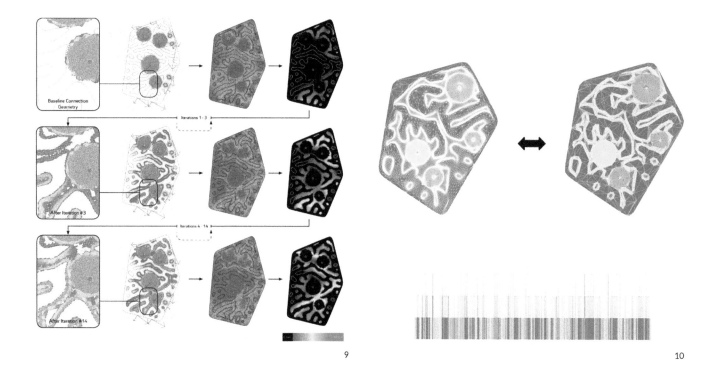

by up to 30%, yet still maintain the shape (Figure 11). Structural analysis occurs at different mesh resolutions/scales: the structural efficiency of the global shape is optimized at the macro level, where the low resolution mesh is sufficient. On the other hand, the plastic deformation is computed at the micro level, being analyzed for a single panel at a time. The meso-level information accounts for joinery and analysis of relationships between panels. It is highly desirable to tie the analysis information with the discrete model produced by the subdivision algorithm, as that way, the efforts to transition data back and forth between different models/scales should be made much less noticeable. The ultimate goal is to consider multiple various scale representations as a single model.

The HNode Class

The HNode Class is developed to support continuity of information between different resolutions. The modeling framework is based on Grasshopper, where the principal collection type is called Data Tree. Contrary to its name, this object is not a proper tree-like collection (rather a dictionary), as it doesn't have a query method for parent and child nodes. A custom-tailored class provides a better foundation to accomplish geometry-data coupling through a recursive tree object. The HNode Class (Hierarchy Node), is a type of a tree data structure which can be traversed efficiently. As with tree structures, all of the data is stored in the root level node. In our case, the root represents the complete demonstrator structure composed of multiple panels, which are stored separately as the second level of the tree. The third level represents the initial low-resolution mesh, where each

node keeps information for each mesh face. To keep track of different resolutions, the subdivision algorithm introduces new layers to the tree: for each subdivided face, multiple children are added (2–4 for adaptive Loop Subdivision), and to keep the tree easy to read and manipulate, the nodes of the faces that are not subdivided are given a singular child. Additionally to storing information about its children, an HNode collection can store and/or convey some more information just like a binary tree. Contrary to that kind of structure, the values are decoupled from the topology of the tree (in our case the topology is derived from the subdivision process) and come from structural analysis at various levels. As the analysis can be done for any of the levels of the tree at any time, various upstream and downstream methods of propagation have been implemented. One of the examples of upstream data propagation is the minimal wall thickness information gained from strains calculation. This process happens at the lowest level of the tree (the highest density mesh), and to visually inspect the results it is easiest to recursively query each top-level parent to get the lowest value of each of its children. At this highest level, this results in an easy-to-verify visualization (Figures 10).

Two major ways of keeping the data up-to-date within the tree have been tested: active and passive. The active way means that the value of dependent nodes (both parents and children) is updated automatically each time any value in the tree is changed. The passive method requires the user to manually trigger the upstream or downstream propagation from a selected level of the tree. During the tests, it became clear that for the sake of

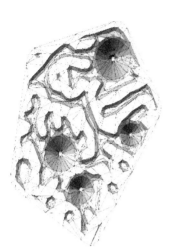 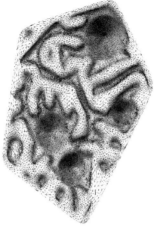 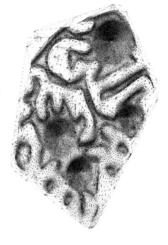

9   Multi-directional data propogation
    to improve panel performance.
    From left: Base panel with
    translation (blue) and rotation
    (green) vectors at connection
    nodes; Calculation of local material
    proerties; Utilisation calculated via
    structural analysis; Change to depth
    of rigidisation geometry; Continue
    loop.

10 Bidirectional data propagation
    between low and high resolution.

11 Face count comparison. From top
    left: original mesh; Loop subdivi-
    sion; adaptive Loop subdivision.

11

computational efficiency and clarity, the passive method seems more appropriate.

The HNode library is written in .NET, and our implementation wraps it up as a data type compatible with Grasshopper. The generic nature of this collection type likely makes it useful in other applications, where keeping track of dependencies and relationships might not be as easy to achieve with the native to Grasshopper Data Tree collection because of the previously stated dictionary-like characteristics.

REFLECTION & CONCLUSION

This paper examines adaptive mesh-based modeling as a means to support the computational design of panelized thin sheet structures built using the ISF fabrication process. Fabrication parameters are not usually included within architectural modeling or simulation even when, as is the case with ISF, they have significant impacts on material properties. A greater awareness of these impacts, together with a greater capacity to include them within simulation models, provides just one motivation for the greater use of multiscale approaches within architectural design.

Two approaches have been described in this paper: the first is characterized as unidirectional and the second as bidirectional. The context of the research exemplifies the need for a back and forth between fabrication, design, and analysis. With multiple scales of material organization—multiple parts, highly hetero-geneous in terms of their shape, their surface geometry, and

their material properties—modeling necessitates a discretization for reasons of control, accuracy, and workability. However, a successful discretization relies on retaining as many possibilities for information flow as possible, and on an efficient and effective organization of that information flow.

One could ask why it is necessary to have multiple scales of resolution, and not simply compute every aspect at the highest level of resolution. Beyond pragmatic reasons, which include limitations of any given model, computation time, and work-ability, there is a greater issue of simplicity. The generation of unnecessary data can render a design workflow unusable, or can generate subsequent filtering activities that displace effort.

The first approach sequentially varies a single mesh topology to manage the complexity of bridging scales and functions while maintaining continuity of information flows down scale. However, a realization of this approach is that for each scale, there is some data that we want to pass up or down. This is because a model does not necessarily have the possibility to recognize or even correct a problem within the model itself. Instead, geometry needs to be passed to another level of resolution for its implica-tions to be tested accurately. Equally, something can be learnt on a lower level that forces adjustment on the upper level, which cannot be tested for at the resolution of prior levels. This cannot be well addressed by a unidirectional model.

In the second described approach, the bidirectional workflow ties multiple scales together in a more consistent and manageable

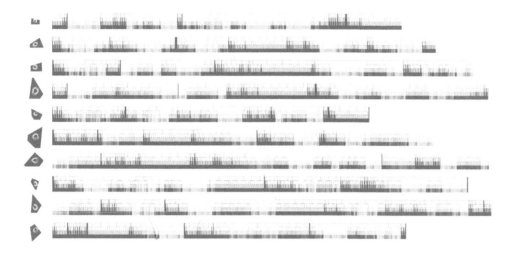

12

12 Example of upstream data propagation. From lower left: original mesh, subdivided mesh, strain calculation, results propagated up the subdivision tree, colorizing the panels with respect to the maximal strain value. Above: results propagated up the subdivision trees.

way compared with the previous method. Ability to reference the data through common interface to other levels makes an element on one level aware of information at any other level of the tree. This enables adaptation of any particular element based on higher- or lower-level information. Future research will connect this bidirectional workflow with an automated feedback loop, and develop visualization techniques that allow analysis and comparison at different resolution levels.

ACKNOWLEDGEMENTS

This project was undertaken as part of the Sapere Aude Advanced Grant research project 'Complex Modeling,' supported by The Danish Council for Independent Research (DFF). The authors want to acknowledge the collaboration of Bollinger Grohmann consulting engineers, KET at UdK, Daniel Piker and Will Pearson, the research departments DTU Mekanik and Monash Materials Science and Engineering, and the robot command and control software HAL.

REFERENCES

Ambrogio, Giuseppina, Luigino Filice, Francesco Gagliardi, and Fabrizio Micari. 2005. "Sheet Thinning Prediction in Single Point Incremental Forming." *Advanced Materials Research* 6–8: 479–486.

Bailly, David, Markus Bambach, Gerhard Hirt, Thorsten Pofahl, Giovanni Puppa Della, and Martin Trautz. 2015. "Flexible Manufacturing of Double-Curved Sheet Metal Panels for the Realization of Self-Supporting Freeform Structures." *Key Engineering Materials* 639: 41–48.

Brüninghaus, Jan, Carsten Krewet, and Bernd Kuhlenkötter. 2013. "Robot Assisted Asymmetric Incremental Sheet Forming: Surface Quality and Path Planning." In *Rob | Arch 2012: Robotic Fabrication in Architecture, Art and Design*, edited by Sigrid Brell-Çokcan and Johannes Braumann. Vienna: Springer. 155–160.

Coelho, P. G., P. R. Fernandes, J. M. Guedes, and H. C. Rodrigues. 2008. "A Hierarchical Model For Concurrent Material And Topology Optimisation Of Three Dimensional Structures" *Structural and Multidisciplinary Optimization* 35 (2): 107–115

E, Weinan. 2011. *Principles of Multiscale Modeling*. Cambridge, UK: Cambridge University Press.

Elliot, James A. 2011. "Novel Approaches to Multiscale Modeling in Materials Science." *International Materials Reviews* 56 (4): 207–225

Hirschel, Ernst H., Horst Prem, and Gero Madelung. 2012. *Aeronautical Research in Germany: From Lilienthal Until Today*. Berlin: Springer.

Jeswiet, J., F. Micari, G. Hirt, A. Bramley, J. Duflou, and J. Allwood. 2005. "Asymmetric Single Point Incremental Forming of Sheet Metal." *CIRP Annals - Manufacturing Technology* 54 (2): 88–114.

Kalo, Ammar, and Michael J. Newsum. 2014. "An Investigation of Robotic Incremental Sheet Metal Forming as a Method for Prototyping Parametric Architectural Skins." In *Rob | Arch 2014: Robotic Fabrication in Architecture, Art and Design*, edited by Wes McGee and Monica Ponce de Leon. Cham, Switzerland: Springer. 33–49.

316

Concepts and Methodologies for Multi-scale Modelling Nicholas et al.

Loop, Charles Teorell. 1987. "Smooth Subdivision Surfaces Based on Triangles." MS Thesis, University of Utah. http://research.microsoft.com/apps/pubs/default.aspx?id=68540.

Nicholas, Paul, David Stasiuk, Esben Clausen Nørgaard, Christopher Hutchinson, and Mette Ramsgaard Thomsen. 2015 "A Multiscale Adaptive Mesh Refinement Approach to Architectured Steel Specification in the Design of a Frameless Stressed Skin Structure." In *Modelling Behaviour*, edited by M. Ramsgaard Thomsen, M. Tamke, C. Gengnagel, B. Faircloth, F. Scheurer. Berlin: Springer. 17–34.

Pakdel, Hamid-Reza, and Faramarz Samavati. 2004. "Incremental Adaptive Loop Subdivision". In *Computational Science and Its Applications: Part III*, edited by Antonio Laganà, Marina L Gavrilova, Vipin Kumar, Youngsong Mun, C. J. Kenneth Tan, and Osvaldo Gervasi. Assisi, Italy: ICCSA. 237–246.

Smith, Cyril S. 1981. *A Search for Structure: Selected Essays on Science, Art, and History*. Cambridge, MA: MIT Press

Trautz, Martin, and Ralf Herkrath. 2009. "The Application of Folded Plate Principles on Spatial Structures With Regular, Irregular and Free-Form Geometries." In *Proceedings of the 50th Jubilee Symposium of the International Association for Shell and Spatial Structures*. Valencia, Spain: IASS.

Wallner, Johannes, and Helmut Pottmann. 2011. "Geometric Computing for Freeform Architecture." *Journal of Mathematics in Industry* 1 (1): 4.

Winsberg, Eric. 2010. *Science in the Age of Computer Simulation*. Chicago: University of Chicago Press.

IMAGE CREDITS

Figures 1–2: Ingvartsen, 2015, © Centre for Information Technology and Architecture

Figures 3–5: Nicholas, 2015, © Centre for Information Technology and Architecture

Figure 6: Leinweber, 2016, © Centre for Information Technology and Architecture

Figures 7–9: Stasiuk, 2015, © Centre for Information Technology and Architecture

Figures 10–12: Zwierzycki, 2016, © Centre for Information Technology and Architecture

Paul Nicholas holds a PhD in Architecture from RMIT University, Melbourne, Australia. After a period of practice with Arup Engineers from 2005 and AECOM/Edaw from 2009, Paul joined the Centre for Information Technology and Architecture (CITA) in 2011. Paul's particular interest is the development of innovative computational approaches that extend architecture's scope for design by establishing new bridges between design, structure, and materiality. His recent research explores sensor-enabled robotic fabrication, multiscale modeling, and the idea that designed materials such as composites necessitate new relationships between material, representation, simulation, and making.

Mateusz Zwierzycki is an architect, designer, Grasshopper user, and co-author of the projektowanieparametryczne.pl (the first Polish website about parametric tools in architectural design). He is also the author of the Starling, Squid, Anemone, and Mesh Tools plugins for Grasshopper, and many more disassociated scripts scattered all over the Grasshopper community, as well as the founder of the Milkbox group, a long time workshop tutor, teacher, and a parametric design populariser.

David Stasiuk's academic research exists within the larger framework of CITA's 'Complex Modeling' project, which investigates the digital infrastructures of design models, examining concerns of feedback and scale across the expanded digital design chain. His work discusses adaptive reparameterisation, focusing on the dynamic activation of data structures that allow for model networks to operate holistically as representational engines in the realisation of complex material assemblies.

He is currently the Director of Applied Research at Proving Ground, a technology consultancy for architects, engineers, and manufacturers, which focuses on the development of advanced computational tools that facilitate data-driven design and project collaboration.

Esben Clausen Nørgaard is an educated civil engineer with a specialty in architectural design from Aalborg University in 2014, and joined CITA after graduation. His primary research and interest lies within prototyping, fabrication, and rationalization. Since joining CITA, his primary focus has been on fabrication with industrial robots and how this can be used to create relationships between traditional craftsmanship and digital environments.

Mette Ramsgaard Thomsen is head of the Centre for Information Technology and Architecture (CITA). Her research centres on the intersection between architecture and computer science. During the last 15 years, her focus has been on the profound changes that digital technologies instigate in the way architecture is thought, designed and built.

At CITA, she has piloted a special research focus on the new digital-material relations that digital technologies bring forth. Investigating advanced computer modeling, digital fabrication, and material specification, CITA has been central in the formation of an international research field examining the changes to material practice in architecture.

From Bones to Bricks

Designing the 3D Printed Durotaxis Chair and La Burbuja Lamp

Alvin Huang
University of Southern California
School of Architecture /
Synthesis Design and
Architecture

ABSTRACT

Drawing inspiration from the variable density structures of bones and the self-supported cantilvers of corbelled brick arches, the Durotaxis Chair and the La Burbuja lamp explore a material-based design process by responding to the challenge of designing a 3D print, rather than 3D printing a design. As such, the fabrication method and materiality of 3D printing define the generative design constraints that inform the geometry of each. Both projects are seen as experiments in the design of 3D printed three-dimensional space packing structures that have been designed specifically for the machines by which they are manufactured. The geometry of each project has been carefully calibrated to capitalize on a selection of specific design opportunities enabled by the capabilities and constraints of additive manufacturing.

The prototypes: The Durotaxis
Chair (left) & the La Burbuja Lamp
(right).

The Durotaxis Chair is a half-scale prototype of a fully 3D printed multi-material rocking chair that is defined by a densely packed, variable density three-dimensional wire mesh that gradates in size, scale, density, color, and rigidity. Inspired by the variable density structure of bones, the design utilizes principal stress analysis, asymptotic stability, and ergonomics to drive the logics of the various gradient conditions.

The La Burbuja Lamp is a full scale prototype for a zero-waste fully 3D printed pendant lamp. The geometric articulation of the project is defined by a cellular 3D space packing structure that is constrained to the angles of repose and back-spans required to produce un-supported 3D printing.

INTRODUCTION

3D printing has been heralded as the next industrial revolution (Anderson 2010). It is projected that in the near future, our homes, schools, and small businesses will all have the potential to become micro-factories for personal fabrication (Ratto and Ree 2012). Currently, the prevalence of 3D printing within the worlds of architectural practice and architectural academia has been evidence of the evolution of 3D printing from a novel and exclusive prototyping tool to a ubiquitous and accessible fabrication medium. However, this prevalence has led to the widespread overuse of 3D printing for the production of forms and geometries that have not been designed for the constraints of the machine, and in some cases could be better manufactured by an alternate process. Alternatively, a material-based design process utilizes computation to integrate the logic of fabrication technologies with structure, material, and form (Oxman 2010). With a material-based design process, the capabilities and limitations of 3D printing become latent design opportunities which can drive the design process.

This paper will explore the potentials provided by the constraints of additive manufacturing to provide design opportunities that are informed both by its capabilities as well as its limitations through the design and production of the Durotaxis Chair and La Burbuja Lamp. Both projects are fully 3D printed design artefacts that are designed specifically for the 3D printers which manufactured them, and are thus the projects of a design process informed by fabrication.

Designing a 3D print, as opposed to 3D printing a design, is the fundamental challenge we are confronting in both projects. Our goal was to produce structures which could not be manufactured by any other process. By prioritizing the chosen fabrication method and materiality as the generative design constraints that inform geometry, both projects are experiments in the design of 3D printed three-dimensional space-packing structures that have been designed specifically for these machines and materials by which they are manufactured. They have each been pre-calibrated to capitalize on specific design opportunities enabled by the capabilities and constraints of additive manufacturing.

In this capacity, both projects are experiments in Structuring—defined as the process whereby the elements of architecture develop a unique logic of parts-to-whole relationships where the static pattern of structural order (tessellations, configurations, etc.) can be mediated into a system of both generative and differentiated potential (Oxman 2010).

THE DUROTAXIS CHAIR

The Durotaxis Chair is a half-scale prototype of a fully 3D printed

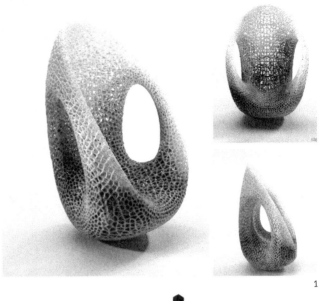

1   The fully 3D printed prototype of the Durotaxis Chair.

2   3D printed samples illustrating the multi-material capabilities of the Objet 500 Connex3 3D printer manufactured by Stratasys.

multi-material rocking chair that is defined by a densely packed, variable three-dimensional wire density mesh that gradates in size, scale, density, color, and rigidity (Figure 1). The chair is inspired by the biological process of the same name, which refers to the migration of cells guided by gradients in substrate rigidity (Plotnikov et al. 2012). The project also takes inspiration from the variable density structure of bones, utilizing principal stress analysis of its geometry to inform the distribution of matter.

The chair is an ovoid rocking chair which has two positions, as an upright rocking chair and as a horizontal rocking lounge. The design of the chair was commissioned by Stratasys to showcase the capabilities of their new Objet 500 Connex3 3D printer. As a response to the recent ubiquity (and perceived overuse) of 3D printing within the design industry—just because you can 3D print something, doesn't mean you should—the challenge that our design team established for ourselves was to produce something which could not be manufactured without 3D printing, and more specifically, which capitalizes on the multi-material

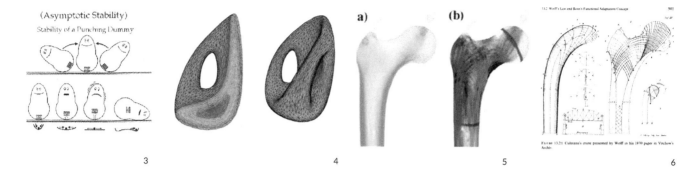

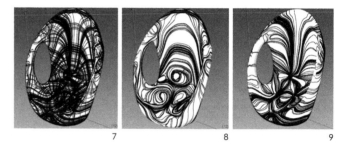

printing capabilities of the Objet 500 Connex3 3D printer to produce gradients of material performance (Figure 2).

Surface vs. Solid vs. Lattice

Due to the fundamental nature of 3D modeling tools and the process of preparing files for 3D printing, most 3D printed models are either defined as surface models (with thickness) or as solid models (with mass). All of these modeling techniques can be materialized (albeit with varying degrees of difficulty) through existing fabrication methods of molding, casting, or milling. As such, a decision was made to explore a design which was articulated not as a surface or a mass, but rather as a variable density three-dimensional lattice—similar to a sponge. For the design team, this was a fundamental paradigm shift in the way in which we thought about the chair, which enabled us to explore the gradient not only as a condition of materiality, but also as a condition of density (spacing) and thickness (dimension). The combination of these three gradient conditions allowed us an opportunity to explore gradients that were informed by three guiding performative criteria: ergonomics (rigidity of material), stability (distribution of mass), and structure (density of members).

As such, the starting point for the design was a manually modeled closed mesh (produced through low-poly subdivision modeling in Maya), which was then articulated as three-dimensional lattice. In short, the goal was to manually dictate global form, while computationally generating the localized articulation of that form.

Gradient Stability (Distribution of Mass)

The multivalent form of the Durotaxis chair allows for multiple readings and uses of its unique geometry. It is defined by an ovoid form, which has two principle resting positions: as an upright rocking chair, and as a horizontal lounge chair. The single curved belly and tighter radius of the upright position allows for a more dynamic rocking condition of rest, while the shallow curvature of the horizontal position provides for a more static resting condition. In both cases, the chair is able to roll from one position to the next. At a global scale, the geometry of the chair is defined

3   Conceptual diagram illustrating the concept of Asymptotic Stability in an inflatable punching dummy.

4   Sectional diagram of the redistributed center of gravity of the densely packed three-dimensional mesh structure.

5   (a) von Mises stress distribution of the femur under tension and compression loading; (b) principal stress lines: green—compression, purple—tension.

6   Diagram of Wolff's Law of Bone Adaption, illustrating the trabecular bone and the way its structure grows most where needed, ending up following principal stress trajectories.

7   Composite principal stress analysis of the chair done in the Karamba structural analysis solver for Rhino.

8   Tensile principal stress analysis of the chair done in the Karamba structural analysis solver for Rhino.

9   Compressive principal stress analysis of the chair done in the Karamba structural analysis solver for Rhino.

by an ergonomic desire for multiple positions, an aesthetic desire to express continuity between those positions, and a structural desire to produce a closed system for force distribution.

However, the irregular form and multiple positions of the chair introduced a fundamental problem: balance. One of the primary concerns of the chair's design was how to redistribute its center of gravity so that it would allow the chair to consistently return to an upright position while rocking. To achieve this, the concept of "Asymptotic Stability" was applied (Figures 3 and 4). This is a strategy for using counterweighting forces through the re-distribution of mass to give a non-linear system a gravitation towards a particular orientation (Bender and Orszag 1999).

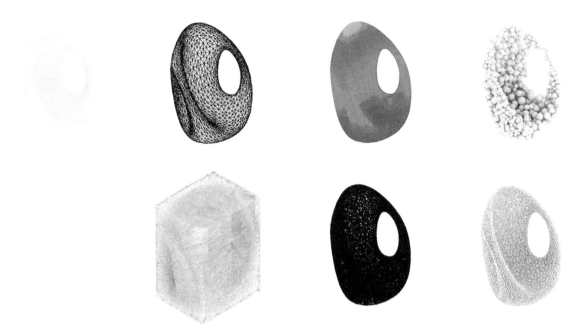

10 The generative sphere packing process.

By conceiving of the chair neither as a homogenous solid (with a consistent distribution of mass through an irregular volume) or as a hollow skin (with a consistent surface mass distributed by an irregular surface area distribution), we are able to re-distribute the mass of the chair through a variable density and variable thickness three-dimensional mesh matrix. The resulting hierarchical cellular structure allowed us to place the center of gravity of the chair (both pre- and post-occupant) in a position which naturally gravitates towards a vertical position.

Gradient Structure

As 3D printing is a layer-based additive manufacturing process, where each subsequent layer is placed upon the previous one, a fundamental paradigm shift in the way we conceived of the chair was to relate it towards the growth and performance of bones. Bones are thought of as solid homogenous elements, yet at a microscopic level they are what is known as hierarchical structures. They are actually a variable density structure of spongy cellular tissue, with increased material and density placed in areas of the greatest principal stress (Figures 5 and 6). The structural density of bone adapts over time to the forces and loads that are applied to it (Wolf 1892).

We approached the analysis of the Durotaxis chair as an opportunity to take a similar strategy. Through the use of the Karamba structural analysis plug-in, we were able to analyze the mesh geometry of the chair to discover its principal stress flow lines (Figures 7, 8, and 9).

These principal stress lines illustrate the flow of vectors through the geometry in both pure tension and pure compression, and serve as the basis for generating a gradient heat map that informs the redistribution of mass, and reorientation of structure to accommodate and respond to structural force.

Generative Sphere Packing

In order to achieve the desired effects of a variable density three-dimensional space frame, a computational process for generating the articulation of the sponge-like structure was proposed. The workflow for generating the variable densely packed 3D mesh jumped among a number of software packages and oscillated from the manual and intuitive modeling of the global form of the chair, to the analytical study of structural forces, to the generative articulation of gradient conditions into a cellular matrix (Figure 10).

- The chair geometry is developed through low-poly subdivision modeling in Maya.
- The subdivision mesh is exported to Rhino, and the mesh topology is processed through a plug-in called Weaverbird.
- The refined mesh is analyzed in the structural analysis plug-in Karamba to discover principal stress distributions.
- The structural analysis is converted into a gradient heat map through the associative modeling plug-in Grasshopper.
- A dynamic springs & nodes is utilized to "sphere pack" a variable density cloud of points within the boundary of the mesh, utilizing the gradient heat map as scalar distribution map.
- Each of the nodes of the network (the centers of the spheres)

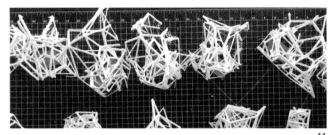

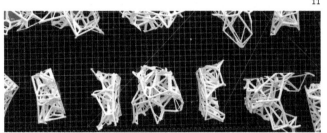

11 Breakdown of the chair into 5 build volumes, and each build's respective gradient steps.

12 Mapping the gradient steps against a material and color gradient.

13 The 3D printing and post-production process.

is extracted as a point cloud, including the centroids of all boundary mesh faces.

- The composite point cloud of both sphere-packed volume and the boundary mesh topology are used to generate a three-dimensional voronoi network.
- All voronoi cells outside of the original boundary condition are removed, and the wireframe network of the cells is extracted to create the underlying topology of the new mesh.
- The wireframe network is given a variable thickness and color gradient in relation to the heat map of its principal stress analysis through the plug-in Exoskeleton.

The Connex 500 is able to print in a smooth gradient, but the process of providing gradient information to the machine to print was a much more painful process than expected. At the time of production, the only process for printing the multi-material conditions of the chair was to break the file into a series of stepped gradient conditions as separate closed meshes. Grasshopper was used to break the color and scale gradient into precisely ten color/scale steps, each resulting in a single closed mesh (Figure 11). The resulting ten-step file is combined into a single STL file, which is broken down into exactly 4 pieces: a back rest, seat, and two arm rests (Figure 12). These pieces are split to be able to fit the bed of the Objet 500 Connex3.

Fabrication

Though the translation of information to matter is an automated translation of computational protocols, the end result involves an intense amount of manual labor and a high level of precision with hand craft. The post-production process of the model far exceeded our expectations for the amount of time and labor required to produce the part. This process can be defined by the following steps:

- The individual pieces are 3D printed with full support material on the Objet 500 Connex3. You can see in Figure 13 that volume of the chair is printed as a solid object with the support material filling the voids of the model.
- The raw 3D prints are placed in a chemical bath which dissolves the support structure. Nearly 70% of the printed material is wasted. This waste is non-recyclable and cannot be repurposed or recovered.
- The individual pre-fabricated pieces of the chair are removed from the bath and post-processed to clean off any remaining support structure. The individual pieces are carefully and precisely welded together by hand using chemical adhesives.

LA BURBUJA LAMP

The La Burbuja Lamp is a bespoke 3D printed pendant luminaire articulated as a densely packed three-dimensional field of virtual

14 The La Burbuja Lamp.

soap bubbles. The intricately articulated lattice of the pendant is the result of a design research experiment in three-dimensional space packing structures that are constrained by the limitations of the angles of repose required to support extruded PLA plastic in a supportless FDM 3D printing process with zero waste. The object can be read as a series of multiple micro figures embedded within a monolithic macro figure via the constitution of over 1200 virtual bubbles computationally packed into its spherical bubble-like mass. The juxtaposition of two primitive and platonic global forms is defined by an internal decahedron skin and a spherical exterior skin, with the volume between them defined by variably localized articulations, including lace-like forms which produce multiple readings of figures within figures.

Following the completion of the Durotaxis Chair prototype, we began to question some clear redundancies in the fabrication process for the chair. The cost of the support material for the build was extremely high. Over 75% of the printed material was support material, which was removed and wasted. The process of post-production (removal of support material) was highly laborious and manual. We proposed a second challenge to the design team, which was to produce another similarly intricate 3D printed prototype, but this time, with zero waste.

To achieve this, we were inspired by three basic ideas:

- The fact that most 3D printing software such as MeshMixer currently utilize various algorithms for the generation of scaffolding as support structure for cantilevered elements.
- The fact that 3D printing is currently being used to fabricate

15 The filligree effect of the La Burbuja Lamp.

scaffolds with complex internal structures for synthetic bone replacement materials (Leukers et al. 2005).

- The concept of corbelling bricks, which consists of incrementally cantilvered layers of purely horizontal brickwork that meet at the top (Rovero and Tonnieti 2014).

In short, the goal was to prototype a complex structure that was actually composed of scaffolding as opposed to supported by scaffolding, thus omitting the requirement to print or waste support material. Furthermore, we wanted to propose a prototype which could capitalize on a secondary purpose for the intricacy of the geometry, thus deciding to produce a pendant lamp which would also profit from the richly varied pattern of light and shadow produced by the lattice network.

Experiments in Unsupported 3D Printing

The hypothesis of our experiment proposes that there is a potential connection between the angle of inclination and the amount of back span within each horizontal layer of a 3D print, which works similar to the concept of corbelling in brick structures. The restrictions of corbelling are based on the horizontal length and vertical depth of each brick, and the resultant center of gravity of each building unit. This position of the center of gravity in the horizontal axis defines the maximum unsupported cantilver length, which in turn defines the maximum angle of inclination (Figure 16). In short, the greater the back span, the longer the unsupported cantilever can extend until it reaches a ceiling where it must be intercepted by another angle, thus allowing us to discover the extreme non-linear relational constraints that can inform the design of the mesh.

Utilizing a Makerbot 2.0, a fused deposition desktop 3D printer that prints layers of PLA plastic, a series of material experiments were conducted to test over 50 combinations of various angles of repose against various thicknesses (Figure 17).

Geometric Rationalization

In order for the build process to work, it is required that each build had to start from a flat construction plane. Given the build constraints of the 3D printer, the spherical form of the out skin of the lamp was constrained against the planar faces of a polyhedron. A decahedron was used to create 20 build faces in order to build the organic growth from a two-dimensional plane, thus producing 20 individual 3D printed components (Figure 19). This polygonal-faceted form is then assembled into a continual composite sphere.

Fabrication

Each of these segmented components was printed through the Makerbot Replicator 2, with all support structure options turned off. Each component had an average build time of 12 hours, with no post-production for the removal of support material. Subsequently, all of the components are welded together to form the complete entity of the La Burbuja lamp.

CONCLUSION

Both the Durotaxis Chair and the La Burbuja Lamp challenged us as designers to consider both the constraints and opportunities offered by 3D printing. By employing a material-based design process paired with a strategy of structuring as a form of design exploration, both projects were able to be materialized as 3D printed prototypes.

By simultaneously considering the structural, material, and fabrication constraints of 3D printing, our hope is to further pursue these discoveries at larger scales and further explore the benefits of 3D printing for producing highly precise complex geometries. These are especially notable for capitalizing on the material and geometric opportunities afforded by 3D printing without the costly drawbacks of wasted material and unnecessary post-production.

REFERENCES

Baerlecken, Daniel, and Sabri Gokmen. 2015. "Osteotectonics-Trabecular Bone Structures and Their Adaptation for Customized Structural Nodes Using Additive Manufacturing Techniques." In Real Time: Proceedings of the 33rd eCAADe Conference, vol. 2, edited by B. Martens, G. Wurzer, T. Grasl, W. E. Lorenz, and R. Schaffranek. Vienna, Austria: eCAADe. 439–448.

Bender, Carl M., and Steven A. Orszag. 1999. Advanced Mathematical Methods for Scientists and Engineers. New York: Springer. 549–568.

Bonswetch, Tobias, Daniel Kobel, Fabio Gramazio, and Matthias

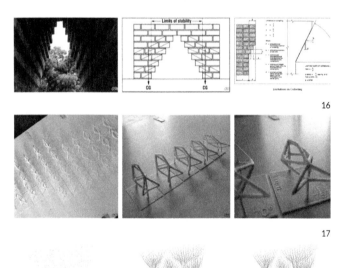

16 The concept of corbelling bricks.

17 3D printed experiments in unsupported cantilvered angles.

18 Branching growth algorithms constrained by maximum angles of repose.

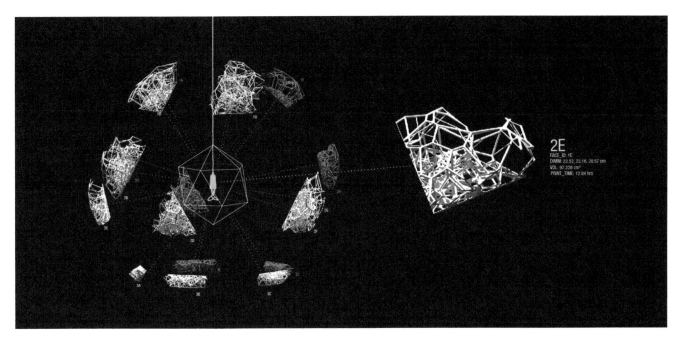

19 The rationalized components of the inner decahedron.

Kohler. 2006. "The Informed Wall: Applying Additive Digital Fabrication Techniques on Architecture." In Synthetic Landscapes: Proceedings of the 25th Annual Conference of the Association for Computer-Aided Design in Architecture. Lexington, KY: ACADIA. 489–495.

Leukers, Barbara, Hülya Gülkan, Stephan H. Irsen, Stefan Milz, Carsten Tille, Matthias Schieker, and Hermann Seitz. 2005. "Hydroxyapatite Scaffolds for Bone Tissue Engineering Made by 3D Printing." Journal of Materials Science: Materials in Medicine 16 (12): 1121–1124.

Oxman, Rivka. 2010. "The New Structuralism: Conceptual Mapping of Emerging Key Concepts in Theory and Praxis." International Journal of Architectural Computing 8 (4): 419–438.

———. 2012. "Informed Tectonics in Material-Based Design." Design Studies 33 (5): 427–455.

Oxman, Neri, and Jesse L. Rosenberg. 2007. "Material Computation." International Journal of Architectural Computing 5 (1): 21–44.

Plotnikov, Sergey V., Ana M. Pasapera, Benedikt Sabass, and Clare M. Waterman. 2012. "Force Fluctuations Within Focal Adhesions Mediate ECM-Rigidity Sensing to Guide Directed Cell Migration." Cell 151 (7): 1513–1527.

Ratto, Matt, and Robert Ree. 2012. "Materializing Information: 3D Printing and Social Change." First Monday 17 (7).

Rovero, L., and U. Tonietti. 2014. "A Modified Corbelling Theory for Domes with Horizontal Layers." Construction and Building Materials 50: 50–61.

Wolff, Julius. 1892. Das Gesetz der Transformation der Knochen. Berlin: August Hirschwald.

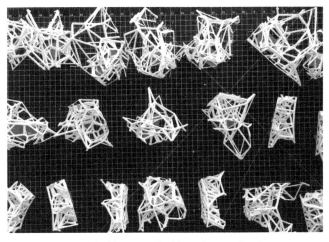

20 Branching growth algorithms constrained by maximum angles of repose.

IMAGE CREDITS

Figure 1: Imstepf., 2015

Figures 2–20: Synthesis Design & Architecture, 2015

Alvin Huang, AIA is the Founder and Design Principal of Synthesis Design. He is an award-winning designer and educator specializing in the integrated application of material performance, emergent design technologies and digital fabrication in contemporary design practice. This exploration of "digital craft" is identified as the territory where the exchange between the technology of the digitally conceived and the artisanry of the handmade is explored. His wide ranging international experience includes significant projects of all scales ranging from hi-rise towers and mixed-use developments to bespoke furnishings.

Composite Systems for Lightweight Architectures

Case studies in large-scale CFRP winding

Andrew Wit
Rashida Ng
Cheng Zhang
Temple University

Simon Kim
University of Pennsylvania

1

ABSTRACT

The introduction of lightweight Carbon Fiber Reinforced Polymer (CFRP) based systems into the discipline of architecture and design has created new opportunities for form, fabrication methodologies and material efficiencies that were previously difficult if not impossible to achieve through the utilization of traditional standardized building materials. No longer constrained by predefined material shapes, nominal dimensions, and conventional construction techniques, individual building components or entire structures can now be fabricated from a single continuous material through a means that best accomplishes the desired formal and structural objectives while creating minimal amounts of construction waste and disposable formwork. This paper investigates the design, fabrication and structural potentials of wound, pre-impregnated CFRP composites in architectural-scale applications through the lens of numeric and craft based composite winding implemented in two unique research projects (rolyPOLY + Cloud Magnet). Fitting into the larger research agenda for the CFRP-based robotic housing prototype currently underway in the "One Day House" initiative, these two projects also function as a proof of concept for CFRP monocoque and gridshell based structural systems. Through a rigorous investigation of these case studies, this paper strives to answer several questions about the integration of pre-impregnated CFRP in future full-scale interventions: What form-finding methodologies lend themselves to working with CFRP? What are the advantages and disadvantages of working with pre-impregnated CFRP tow in large-scale applications? What are efficient methods for the placement of CFRP fiber on-site? As well as how scalable is CFRP?

1   CFRP winding detail from the rolyPOLY prototype.

2   roboWINDER CFRP winding prototype.

3   rolyPOLY on display at the Tyler School of Art.

INTRODUCTION

As global populations continue to rise and environmental conditions become increasingly unpredictable, material and resource availability will become an ever greater problem affecting the discipline of architecture, especially following times of disaster. With the vast amounts of resources required to create current building typologies, the formulation of a more robust, intelligent, adaptable, yet affordable housing system will become an ever more significant issue. Existing housing typologies in the U.S. find themselves lagging behind other industries such as aerospace and automotive, lacking advanced lightweight materials, novel fabrication methodologies and adaptability, as well as embedded intelligent systems allowing for more efficient building processes, minimization of waste and a reduction of life-cycle energy consumption.

2.

In 2013, the "One Day House" initiative commenced, which is focused around the creation of a new typology of housing. Based around a wound composite structural system, adaptable robotic skin and intelligent living systems, the project called into question our understanding of the dwelling and it's role in society. To realize the project, a series of short-term prototypical projects were initiated, each solving a single aspect associated with this newly proposed building typology. Initial research focused around material processes, adaptive systems and the utilization of industrial robotics for CFRP placement (Figure 2).

This paper investigates two recent projects (rolyPOLY + Cloud Magnet) to understand how novel material and fabrication methodologies, along with formal adaptability and environmental responsiveness, could be integrated into the larger research question centered around the creation of the "One Day House".

CFRP

The introduction of CFRP into the discipline of architecture

has opened up new opportunities in the design, engineering and fabrication of large-scale, lightweight structures as found in recent research completed at the University of Stuttgart (Dörstelmann, et al 2015, Reichert, et al 2014). CFRP's ability to effortlessly vary density, patterning and overall shape allows for the creation of forms, efficiencies and visual/tactile qualities that were previously impossible to achieve through traditional materials and processes. Additionally, the introduction of robotic coreless FRP winding has removed previous constraints requiring time- and money-intensive formwork that limit design freedom while creating vast amounts of waste (La Magna, et al 2016).

Building upon CFRP's robust properties, in addition to previously completed projects in robotic coreless winding (Wit, et al 2016), both of the discussed projects utilized CFRP for their overall design and fabrication. A pre-impregnated resin system was chosen for consistency, stability, low-temperature curing, structural attributes and minimal toxicity. Two CFRP strand counts with identical resin systems (±27.51% resin content) were utilized. rolyPOLY was wound from a 12K strand CFRP tow while a 24k strand tow was chosen for Cloud Magnet. The resins required a firing duration of 4 hours at 260°F for curing.

CASE STUDY 1: ROLYPOLY

Following the completion of several small-scale projects focused around robotically wound CFRP, rolyPOLY was initiated as the first large-scale intervention. Designed as a traveling exhibition (Figure 3), the artifact functions as a reconfigurable shelter for a single occupant while remaining structurally stable in all orientations. Through the operation of assisted tumbling, unique spaces, opacities and textures are created.

3

Based on a sphenoid hendecahedron with two open faces, each 7' x 4.5' x 4.5' module easily aggregates with adjacent modules creating larger, more complex, self-supporting structures. Wound from a single strand, each module contains over 100,000 linear feet of CFRP tow with a minimal self-weight of ±20 lbs. Additionally, the design and implementation of variable methods for fiber winding created varying skin thicknesses ranging between 1/16"–1/4" while simultaneously allowing unique winding patterns to emerge. Initially requiring 24 hours of winding, subsequent projects have substantially reduced winding times making the process more feasible for large-scale architectural applications.

Methodologies

Rather than allowing aesthetics to dictate the artifact's design, material processes were utilized for formal definition. The processes and structural characteristics tested throughout the design defined the overall aesthetics and were as follows: 1. Part minimization through self-supporting monocoque shells; 2. Material/structural optimization through tensile form-finding; 3. Waste elimination through coreless CFRP winding; 4. Seamless fabrication of unique elements through reconfigurable/reusable framework; 5. Unique structural, visual and tactile effects through variable winding patterns.

To facilitate a robust form able to handle stresses from inhabitation and tumbling, a monocoque structure was chosen. In addition to strength, the single structural unit minimized complexity during fabrication as no assembly was required; allowed for continuous winding of the structure as a whole; enabled consistent structural properties by firing the entire artifact at once; and showed promise for easy scalability in our move towards an architectural scale.

For the fabrication of a successfully wound shell, it was necessary for all surfaces to maintain double curvature, creating maximum tensioned fibrous overlap. As the base form contained no double curvature, tensile cable/membrane simulation was utilized through the MPanel and Kangaroo interfaces, converting the form's linear edges into catenary cables stressed under predefined loads. Simultaneously, doubly-curved tensile membranes were created between the newly redefined edges. Lastly, the new membranes were deflated by a factor of 10%, simulating surface deformation encountered during fiber winding.

To facilitate CFRP tow winding, a reconfigurable steel frame was created. The frame consisted of 8 unique elements of 3–4 curved, welded edges, each bolted together on-site with conduit hangers forming the 10 doubly curved surfaces of the artifact. Upon assembly of the steel frame, plywood CFRP grippers were laser cut and attached to the frame though a similar system (Figure 4).

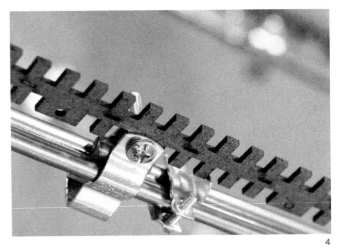

4

5

The flex in the gripper material allowed for seamless flowing along the steel frame. Upon completion of baking, the steel frame simply unbolted and separated from the CFRP shell.

Winding over the reconfigurable frame was an important aspect of the fabrication process (Figure 5). Initially simulated digitally in Rhino 3D, then reproduced through hand winding, the process was completed in three stages: Peak winding, valley winding and spiral winding. Through these operations, several attributes were achieved: Structural rigidity, varying opacities, varied thickness and most importantly, minimal fiber creep and delamination during the winding process. Peak winding was the first stage of winding. Connecting the high points of each frame, this operation created high levels of positive fiber offset. Following, valley winding connected the frame's low points. Pulled with 5 lbs more force, this operation helped pull initial layers into their nominal positions while also equalizing internal tensions. To eliminate delamination, spirals were wound around interconnecting faces drawing in stray fibers. Through a layering of these operations, variation in opacity and texture was also achieved.

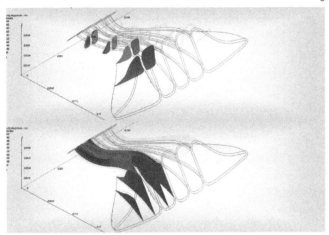

4　Frame to CFRP gripper detail.

5　String winding study.

6　rolyPOLY Confessional on display at the Tyler School of Art.

7　Cloud Magnet CFD simulations.

Results

rolyPOLY showed the potential of harnessing CFRP in craft-based monocoque shell structures while maintaining a high level of structural rigidity and winding repeatability (Figure 6). The doubly curved surfaces created through tensile modeling laid a robust structural framework rigid in all orientations. Additionally, rolyPOLY was produced in a minimal amount of time from a continuous material, allowing for the rapid production of both identical and one-off modules. The fabrication process itself also produced virtually no waste. Although many positive attributes are visible through this project, there are several areas that must be further addressed for larger-scale architectural interventions.

Delamination of layers as tension mounts within the overall structure must be further resolved to insure structural and formal consistency. Additionally, further research in methods of

robotic and hand winding will aid in the definition of more robust patterning while ensuring proper fiber placement orientation.

CASE STUDY 2: CLOUD MAGNET

Similar to rolyPOLY, Cloud Magnet also offered the opportunity for one-to-one testing of wound CFRP systems in lightweight, large-scale construction typologies. Cloud Magnet researches the co-dependencies between material, form, energy, and environment through the design and fabrication of a series of environmentally performative kites within the cloud forest in Monteverde, Costa Rica. Cloud forests have been rapidly disappearing due to climate change and deforestation. Rising global temperatures cause a cloud-lifting effect raising the cloud cover above the tree canopy and forest ecosystem that depends on constant moisture and humidity to support its life. The impetus for this project is to explore the ways by which design can contribute to the stabilization of the atmosphere and the restoration of the forest.

In the early stages of research, prototypes focused on the CFRP winding of the kite frames. Prototypes were fabricated at 1/4 scale to test the suitability of the winding patterns in fabricating a strong, yet flexible frame. The full-scale artifacts will measure approximately 12' wide by 8.5' long by 3.1' high. CFRP was selected for its strength to weight characteristics, which meet the specific needs of the Cloud Magnet project while also informing the larger set of research questions previously described.

Methodologies

Initial forms for the Cloud Magnet prototypes were based on Bernoulli's principle of pressure, correlating reduced pressure with increased speed of fluids, such as air. Their forms were derived through, one, the production of venturi tubes directing air from wider into narrower cross-sections; and, two, the aerodynamic lift produced by the flow of air around airfoils, similar to the wings of an airplane, which creates increased airspeed and decreased upper surface pressure on the form. Prototypical designs were evaluated using computational fluid dynamic (CFD) software based on the average and maximum wind speeds. The CFD simulations confirmed the performance of the proposed forms allowing for the development of a fabrication system (Figure 7).

The forces acting upon Cloud Magnet in flight are similar to those acting upon buildings in extreme weather conditions. To deal with these forces, several directions were explored including monocoque shells, rigid diagrid frames (Figure 8), hybridly wound diagrid frames and finally, flexible variable density gridshell structures. Although the monocoque and rigid diagrid structures had an extremely high strength to weight ratio, their lack of flexibility

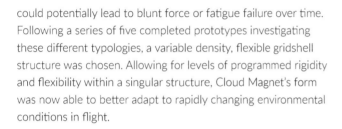

8   Clound Magnet module V.4.0 winding detail.

9   Cloud Magnet complete winding study V.3.0.

could potentially lead to blunt force or fatigue failure over time. Following a series of five completed prototypes investigating these different typologies, a variable density, flexible gridshell structure was chosen. Allowing for levels of programmed rigidity and flexibility within a singular structure, Cloud Magnet's form was now able to better adapt to rapidly changing environmental conditions in flight.

Unlike rolyPOLY, the creation of a vast number of unique prototypes was necessary to satisfy the requirements for flight, stability, the generation of desired internal compression ratios as well as a robust yet flexible structure. For these reasons, a rapidly adaptable, constructible, and disposable formwork was created for 1/4 & 1/2 scale prototyping. Utilizing 1/4" cardboard or chipboard, a unique slicing algorithm and a laser cutter, the project's form was sliced into 15 unique sections with eight additional CFRP grippers, nested and sent for cutting. This workflow allowed for the rapid assembly and testing of a large number of prototypes in a short span of time and for little upfront cost.

Winding of the prototype frame was rather straightforward. Each gripper was numbered between 1–8 beginning on the top of a given section and moving clockwise. Winding began at S1G1 (section 1 gripper 1) and moved clockwise to S2G2 and so on. Upon reaching the final section S15 in one direction, the CFRP shifts one gripper clockwise and continues back towards S1 now shifted one gripper below the original pass. Once eight full rotations in a single direction were complete and all grippers had been filled with a single layer of CFRP, the same process was repeated in a counterclockwise direction, intersecting all previous layers. Variations in this process (i.e. additional passes in one direction, or doubling up every other pass) allowed for the tuning of the artifacts' flexibility. The 1/4 scale prototype iteration of Cloud Magnet consisted of two clockwise layers sandwiching a single counterclockwise layer (Figure 9). Upon completion of the winding process, the artifact was fired and the cardboard was

simply soaked in water and removed via pliers.

Results
Cloud Magnet showed that composite structures have the ability to be both extremely rigid or flexible depending on winding patterns, amount of material applied, overlap and the desired outcome. The use of simplistic winding patterns helped ensure easy repeatability, constant fiber tension throughout the entire winding process and zero CFRP waste during the entire fabrication process. Additionally, the disposable formwork allowed for the rapid testing of a large number of variables in a short duration of time with minimal costs.

At the same time, several issues need to be addressed before implementation on full-scale architectural artifacts. For small scale prototyping, the disposable formwork is efficient, but in large scale, high-output situations, this method will produce vast amounts of waste. Different methods must be investigated. Additionally, further variable winding investigations should be carried out in conjunction with structural testing to verify its feasibility in long term, high stress environments.

CONCLUSION
Although much more research and prototyping is necessary to realize an architectural-scale prototype, the completion of these two projects within the framework of the "One Day House" illustrates a potential for the further integration of wound pre-impregnated CFRP into the architectural workflow. As demonstrated in rolyPOLY, the ease of use and workability of CFRP allow for not only the creation of complex forms and robust structures, but also extremely tailored aesthetic properties. Initial research in Cloud Magnet also suggests that robust structures could remain flexible, creating the potential for building structures with the ability to slightly adapt to internal and external forces.

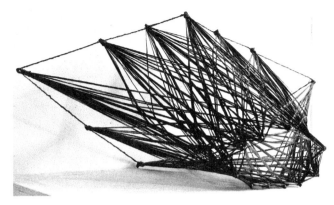

10 Cloud Magnet 1/4 scale winding study V.7.0.

Current research has combined the fabrication-process projects into larger prototypes for the Cloud Magnet projects. Rather than working with form first, form is now derived from forces applied during the winding process on a coreless frame (Figure 10). This new series of prototypes has also eliminated most waste while creating robust, flexible and large-scale (±5') prototypes which could be translated to larger architectural installations.

Further transformation of hand winding typologies into numeric winding and robotic programming for robotic coreless winding will also be an important step in bringing this research into the realm of future architectural-scale interventions.

ACKNOWLEDGEMENTS

The authors would like to thank the following researchers, institutions and partners for their continued assistance throughout this ongoing research: John Williams & Chad Curtis (CFRP Baking); Tim Rusterholz (Steel Fabrication); Sneha Patel (Schematic Design and Consultation); Kerry Hohenstein (Modeling and CFD Simulations); Aidan Kim, Daniel Lau, Han Kwon, Junghyo Lee, Yue Chen, Lyly Huyen, Joseph Giampietro, Gary Polk (Documentation & Production); Temple University Tyler School of Art: Division of Architecture and Environmental Design; University of Pennsylvania School of Design; TCR Composites; MPanel Software Solutions.

REFERENCES

Dörstelmann, Moritz, Jan Knippers, Valentin Koslowski, and Lauren Vasey. "ICD/ITKE Research Pavilion 2014–15: Fibre Placement on a Pneumatic Body Based on a Water Spider Web." *Architectural Design* 85(5): 60–65.

La Magna, Ricardo, Frédéric Waimer and Jan Knippers. "Coreless winding and assembled core - Novel fabrication approaches for FRP based components in building construction," *Construction and Building Materials* (2016): accessed May 01, 2016, doi: 10.1016/j.conbuildmat.01.015.

Reichert, Steffen, Tobias Schwinn, Ricardo La Magna, Frédéric Waimer, Jan Knippers, and Achim Menges, "Fibrous structures: An integrative approach to design computation, simulation and fabrication for light-weight, glass and carbon fiber composite structures in architecture based on biomimetic design principles," *Computer-Aided Design*. 52: 27–39.

Wit, Andrew, Simon Kim, Mariana Ibañez, and Daniel Eisinger, "Craft Driven Robotic Composites." *3D Printing and Additive Manufacturing* 3(1): 2–9.

IMAGE CREDITS

Figures 1–5, 8–10: Andrew Wit, 2015–2016.
Figure 6: Joseph Giampietro, 2016.

Andrew John Wit is the Assistant Professor of Digital Practice within Temple University's Tyler School of Art in Philadelphia, PA. Additionally he is a co-founder of WITO*, "Laboratory for Intelligent Environments" where he creates projects that fringe design, technology and robotics. Prior to his current appointment, Prof. Wit taught courses and led workshops in architecture, urbanism and robotics in both in the U.S. and Japan.

Professor Wit earned his Bachelors of Science in Architecture from the University of Texas at San Antonio, and his Masters in Architecture from M.I.T where he also researched in the Media Lab's "Smart Cities Lab".

Rashida Ng is an Associate Professor and Chair of the Architecture Department at Temple University's Tyler School of Art. Her research and creative activities broadly explore the measurable and experiential characteristics of new material technologies and performative architecture. She has authored numerous papers on these topics and co-edited the book, *Performative Materials in Architecture and Design*, published by Intellect in 2013. Professor Ng is a registered architect in Pennsylvania and the Co-founder and President of SEAM*Lab*, a collaborative think tank dedicated to research and the dissemination of design-based knowledge focused on materiality within the built environment.

Cheng Zhang is a designer in Philadelphia, PA exploring the fecund intersection between performative materials, human experience and the architectural environment. His academic projects and research have been highly recognized receiving awards such as the DaVinci Prize & Temple University's Alumni Thesis Prize. Cheng received his Bachelors and his Masters in Architecture from Temple University.

Simon Kim is a registered architect, and an assistant professor at the University of Pennsylvania, where he directs the Immersive Kinematics research group. As principal of Ibañez Kim, Simon is interested in the augmentation of architecture and urbanism given agency and behavior by new media and material. In support of this project, Simon has worked in both the domains of science and art. He has published numerous papers on robotics and multiagent interaction and is funded by an NSF grant. Simon has also worked with leaders in theater, opera, and dance to develop and test prototypes in the production of culture.

Discrete Computational Methods for Robotic Additive Manufacturing

Combinatorial Toolpaths

Gilles Retsin
Manuel Jiménez García
The Bartlett School of
Architecture / UCL

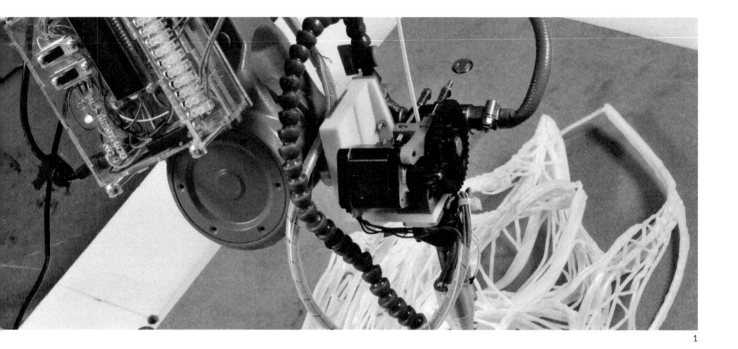

1

ABSTRACT

The research presented in this paper is part of a larger, emerging body of research into large-scale 3D printing. The research attempts to develop a computational design method specifically for large-scale 3D printing of architecture. Influenced by the concept of Digital Materials, this research is situated within a critical discussion of what fundamentally constitutes a digital object and process. This requires a holistic understanding, taking into account both computational design and fabrication. The intrinsic constraints of the fabrication process are used as opportunities and generative drivers in the design process. The paper argues that a design method specifically for 3D printing should revolve around the question of how to organize toolpaths for the continuous addition or layering of material.

Two case-study projects advance discrete methods as efficient ways to compute a continuous printing process. In contrast to continuous models, discrete models allow users to serialize problems and errors in toolpaths. This allows a local optimization of the structure, avoiding the use of global, computationally expensive, problem-solving algorithms. Both projects make use of a voxel-based approach, where a design is generated directly from the combination of thousands of serialized toolpath fragments. The understanding that serially repeated elements can be assembled into highly complex and heterogeneous structures has implications stretching beyond 3D printing. This combinatorial approach for example also becomes highly valuable for construction systems based on modularity and prefabrication.

1   Robotic plastic extrusion of continuous systems. The Bartlett AD-RC4,
2013–14. Project: SpaceWires.
Team Filamentrics.

LARGE-SCALE 3D PRINTING

The research presented in this paper is part of a larger, emerging body of research into large-scale 3D printing for architecture. Large-scale 3D printing is often associated with engineer and innovator Behrokh Khoshnevis' Contour Crafting method. Developed at the University of South California, contour crafting is a process where concrete is extruded from a nozzle that is mounted on a large, gantry-like structure. A similar process developed by WinSun in Shanghai has entered the commercial market, producing a number of full-scale prototypes in the past few years. Enrico Dini's D-Shape printer is also based on a large gantry, but it uses a binder to solidify stone dust into a sandstone-like material. Some research makes use of existing, commercially available 3D printers. Architects Ronald Rael and Virginia San Fratello use, for example, ZCorp machines to develop a wide range of printable materials, such as ceramics. Their company, Emergent Objects, has successfully produced a series of larger-scale architectural prototypes. There are a number of important precedents where robots are used as large 3D printers. Research led by Marta Malé-Alemany at IAAC was first to focus on robotic processes for additive manufacturing in an architectural context. Gramazio and Kohler's research at the Future Cities Laboratory in Singapore introduced spatial plastic extrusion with a robot arm (Hack and Lauer 2014). Spatial plastic extrusion is a process where a robot arm extrudes plastic in the air, rather than in horizontal layers. Outside of architecture, the aerospace industry has been investigating metal sintering processes with robots.

While these precedents successfully innovate with the development of the machine or material, their aim is not to innovate with the design methods themselves. Their main focus is the fabrication process itself. Although highly innovative, the knowledge produced in terms of architectural design is rather limited. On the other hand, research by people like Benjamin Dillenburger and Michael Hansmeyer is specifically focused only on design methods, and not on fabrication (Dillenburger and Hansmeyer 2014). The designers assume the existence of a large-scale 3D printer, using commercially available printers such as the Voxeljet sand printers. The same argument applies for SoftKill Design, which produced a complete proposal for a 3D printed dwelling, but does not address the actual printing process. In this context, the theorist Neil Leach argues, "while there is clearly a practice of designing that involves the use of digital tools, there is no product as such that might be described as digital" (2015). Digital design and fabrication tools merely allow a specific type of design to be realized, but they can as well be used for objects which are not "digital"—for example, a replica of an old Indian temple. The research projects mentioned before affirm this argument. Research focused only on fabrication can be used as effectively to 3D print a computationally generated model as a

Type 1
Example of connection

All possibilities to connect

Rotation 1   Rotation 2   Rotation 3   Rotation 4

Example of connection

All possibilities to connect

Rotation 5   Rotation 6   Rotation 7   Rotation 8

Type 2
Example of connection

All possibilities to connect

Rotation 1   Rotation 2   Rotation 3   Rotation 4
(Type 1's R1 + R2) (Type 1's R3 + R4) (Type 1's R5 + R6) (Type 1's R7 + R8)

2

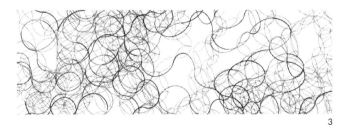

3

2   Combinatorics diagram. The Bartlett AD-RC4, 2014–15. Team Curvoxels.

3   Combined Voxels into continous toolpaths. The Bartlett AD-RC4, 2014–15. Team Curvoxels.

4   Screenshot of Processing application for toolpaths generation. The Bartlett AD-RC4, 2014–15. Project: SpatialCurves. Team: Curvoxels.

a historic replica, or a modernist column. The machine process is unrelated and ignorant of the object it fabricates. Hansmeyer and Dillenburger's highly detailed prints also comply with Leach's theory, as these are merely made possible through the affordance of the machine.

Mario Carpo, architect and historian, provides the beginning of a counter argument to Neil Leach's statement. Carpo identifies the digital character of a design method, arguing for the intrinsically discrete nature of computational processes (2014). The research presented in this paper continues asking this question, and looks at whether there is a design method specific to large-scale 3D printing. This requires a more holistic understanding, taking into account both computational design and fabrication. The intrinsic constraints of the fabrication process are used as opportunities or drivers in the design process.

3D PRINTING AND CONTINUITY

3D printing processes are fundamentally continuous in nature. A 3D printer is a machine which continuously extrudes material, or continuously glues or melts material particles together. This process happens additively, as the machine deposits material layer by layer. To materialize the complex structures generated in a digital environment, it has to be reduced to a series of slices, contours, or layered toolpaths. The translation to physical form reduces the complexity of the structures, removing information. With high-resolution, small-scale printers, the issue of the layer is maybe not relevant, but for large-scale additive manufacturing,

it becomes crucial. The resolution of large-scale printing is much lower, and layers are clearly distinguishable. The actual organization of these layers is often not computed or designed. In that case, it can be argued that these objects are post-rationalized, or merely representational. They bear no relation to their materialization. A design method specifically for 3D printing should revolve around the question of how to organize toolpaths for the continuous addition or layering of material.

As layers in architectural-scale printing are clearly visible, they should be the main subject of the digital design process. The method presented in this paper proposes one possible approach to the problem of the layer. The design process allows for heterogeneous toolpaths, which are continuous, non-intersecting, and printable with specific material properties.

The continuous character of 3D printing initially gave rise to algorithmic work that also had a continuous character. Designers mostly used algorithmic processes inspired by nature, referencing growth and morphogenesis. The formal complexity and continuous differentiation of these processes implied a certain need for 3D printing, as they would be very hard to fabricate in any other way. An example of a continuous algorithm is, for example, an agent-based system. Although it makes use of discrete entities, it has a continuous search space and makes possible continuous variation. However, continuous systems prove to be very hard to optimize for large-scale, robotic 3D printing. They usually require a significant amount of post-rationalization. Project

SpaceWires by Filamentrics, for example, is based on the use of agents to deposit material. Interpolating a vector field, an agent is programmed to create a toolpath trajectory for the robot. The agent gets a series of constraints that relate to the constraints of the fabrication process. A minimum and maximum distance between trajectories is also constrained. In a subsequent stage, a second set of agents connects the previously generated lines together. The generative process takes a number of constraints into account, but the resultant structures still require a significant effort to solve errors, intersections. Due to the heterogeneity and large amount of variation in the generated toolpaths, it's difficult to automate the post-rationalization. (Figure 1)

Problems and errors are, just like the structure itself, continuously different and require unique solutions. The nature of this problem finds its origin in the continuous character of the generative process. Errors can't be serially solved, and large amounts of time or computational power are needed to prevent them from occurring. To incorporate all the constraints from the printing process in a continuous toolpath requires a lot of computing and a large amount of memory. This renders the process increasingly inaccessible, requiring expensive computing equipment or hardware such as sensors to detect unique errors in the toolpath.

Based on the experience with continuous models, the research in this paper looks into design methods based on discreteness. Discrete models allow users to serialize problems and errors in toolpaths. This allows a local optimization of the structure, avoiding the use of global, computationally expensive, problem-solving algorithms. All problems in the toolpath can be isolated locally at the level of the cell or the cluster of cells. Toolpaths are first generated in one voxel, where all the constraints are optimized and tested. In a second stage, many voxels are combined together into one continuous path. This method only requires local computation, and as such, is computationally inexpensive and quick. The prototyping aspect also becomes much quicker, as only one voxel and its immediate neighbors have to be checked for potential problems. Rather than continuous differentiation, heterogeneous structures are then achieved by always rotating the piece of toolpath contained in the voxel into different positions.

The main argument of the research is that the most efficient way how to calculate a continuous printing process is through a discrete method. To illustrate the wide applicability and efficiency of the method, the two projects are based on different materials; one makes use of plastics, the other of concrete.

PROJECTS

The projects described in this paper are produced in a

5

6

5 Robotically fabricated Chair Prototype v3.0. The Bartlett AD-RC4, 2014–15. Project: SpatialCurves. Team: Curvoxels.

6 Robotic plastic extrusion of continuous systems. The Bartlett AD-RC4, 2014–15. Project: SpatialCurves. Team: Curvoxels.

research-through-teaching context in the Bartlett Architectural Design (AD) Research Cluster 4 (RC4). AD is a part of BPro, an umbrella of post-graduate programs in architecture at the Bartlett School of Architecture at University College London. The research is led by Manuel Jiménez García and Gilles Retsin, and started out in 2013. Since the start of the research, there has been a close collaboration with Vicente Soler. For two years, the research agenda of RC4 has focused on large-scale additive manufacturing for architecture. The research makes use of industrial robots, which are turned into 3D printers by attaching custom-designed end-effectors. The research, carried out by teams of postgraduate students, has a holistic character. Students develop material studies, as well as the mechanical and electrical design for the end-effectors, robotic programming, and design prototypes. The student work is set within the framework of an architectural research lab. This means that data and knowledge from previous years of research is accessible.

This paper will discuss two projects that implement a computational design and fabrication method based on discreteness. Both projects make use of a voxel-based approach, where a design is generated from the combination of thousands of serialized toolpath fragments. A continuous toolpath is then generated by combining lots of different toolpath fragments together into the most optimal option. An optimal toolpath can be understood as being continuous: the less the extruder has to stop, the better. It should also prevent singularities or intersections where the robot would disrupt previously printed parts. The toolpaths are computed in Processing, and then passed on to Rhino/Grasshopper. The Grasshopper plugin HAL is used to translate the toolpaths to the ABB IRB 1600 robot.

Project A: SpatialCurves by Curvoxels—Curved Spatial Plastic Extrusion

StatialCurves is based on robotic spatial printing, which was mentioned before. A custom developed nozzle with a print width of 6 mm was used. This nozzle makes use of ABS filament, and has an embedded cooling system that blows cold air under high pressure. The design method developed in this project is based on the idea of combining a curved toolpath segment with a voxel-based data structure. A combinatorics algorithm is used to then aggregate this single curvilinear element into a continuous, kilometers-long extrusion. This single line is folded into a complex and heterogeneous structure, with multiple hierarchies of scale and density (Figures 2–3).

The research uses a Panton chair as test case. The Panton chair has a complex geometry, with concave and convex surfaces—as such, a good 3D test model. Within the research of RC4, it has become the equivalent of 3D test models like the Stanford

7

Bunny or Utah Teapot. It was also the first industrially produced plastic moulded chair, and as such forms an interesting precedent for a 3D printed plastic chair. The original chair is voxelized, and then structurally analyzed in Rhino/Grasshopper using the Millipede plugin. The resulting data, in this case, maximum stress levels, is then transferred to Processing (Figure 4), and then used to redistribute voxels, adding new ones where necessary. The size of the voxels also varies in response to the amount of stress. This is done through an OcTree voxelization, with 4 different scales of voxels. Once printed, the smallest voxel becomes the densest one. The largest voxels become the most porous, occupying areas with a low level of stress.

Every voxel is embedded with a single Bezier curve, which has been tested on printability. Specific curvatures are not printable, for example, under specific angles where the nozzle could intersect with the already printed curve. Extensive tests were undertaken to see if a curve was printable or not. The scale of the voxel also influences the parameters for printing. Larger voxels, which contain longer curves, require different settings for extrusion speed than smaller ones. When voxels are very small, the embedded spatial curve effectively becomes no more than a line when printed. What in the final object appears to be two different formal syntaxes, curvilinear or linear, is the product of a single spatial curve on different scales.

The combinatorial logic works by calculating tangents and points of connectivity between voxel curves. Every voxel curve attempts to find a continuous connection to a neighboring voxel curve.

7 Robot End Effector: Three-nozzle
   PLA filament extruder. The Bartlett
   AD-RC4, 2014–15. Project:
   SpatialCurves. Team: Curvoxels.

8 Robotically fabricated Chairs
   displayed during the BPro
   Exhibition. The Bartlett AD-RC4,
   2014–15. Project: SpatialCurves.
   Team: Curvoxels.

8

Each discrete voxel unit has twenty-four possible rotations. In a number of these rotations, the curve becomes non-printable, so these are left out. To generate the toolpath, the code starts from the bottom and generates layers of curves going up along the z axis (Figure 6). Every layer needs to be connected to a previous layer, so curves need to touch across layers. A series of printable combinations or patterns was developed. These patterns also produce a certain level of non-repetition and heterogeneity.

The fundamental advantage of this serial approach is that a toolpath only has to be optimized and tested for one voxel, in twenty-four different rotations. Afterwards, thousands of these voxels can be aggregated, but the connection problems remain finite and manageable. The process proves to be very efficient and easy to prototype. Three chairs were printed (Figure 8), requiring a minimum of interventions to correct mistakes. During the printing process, there were no errors or singularities where the robot intersected previously printed material. The third and final chair is structurally stable, and can withstand the significant load of a human body (Figure 5).

However, it also proves inefficient to print curves in space rather than straight lines. To achieve a curve, the robot has to move slower and stop at multiple points. Moreover, curves are harder to control and deform a lot compared to the original. This raises the question of incorporating material behavior.

The premise of continuously printing curves in mid-air would require more thorough insight into the behavior of the printing material, or the use of a material that is easier to control.

Project B: Fossilized by Amalgama—Supported Extrusion in Concrete

Amalgama develops a project based on heavy compression-based structures and materials, such as concrete. The proposed fabrication method combines two already existing concrete 3D printing methods: extrusion and printing. This combination of techniques can be described as supported extrusion. Concrete is extruded layer by layer over a bed of granular support material. The supports allow for more formal freedom; for example, large cantilevers become achievable. Traditionally, concrete printing is heavily constrained by a limited degree of possible overhangs. These constraints defy the initial promise of 3D printing to achieve more complexity and formal freedom. As the production system makes use of support material, a bounding box is needed. This bounding box is based on the maximum reach of the ABB 1600 robots used for printing. The bounding box is designed to allow for easy extraction of the model after printing. ABS pellets were originally intended as support material, but at a later stage rock salt was chosen as a more economical option. Two types of nozzles were used: one for concrete extrusion and one for gluing the support material. The glue nozzle is a relatively simple extruder, based on compressed air with a valve to regulate the amount of glue. The concrete nozzle was entirely 3D printed on a Makerbot (Figure 10). It is connected to a peristaltic pump, which is able to pump thick material with a low consumption of energy. The pump is made of CNC-routed aluminum components, and uses skateboard wheels as pressure rollers.

The team developed a combinatorics-based code, where a voxel has a specific type of pattern inscribed on its faces. Every face has a different pattern, but can be combined with other faces and voxels (Figure 11). The pattern also indicates what has to be printed in concrete, and what in salt. There are four types of patterns: directional structural ribs, nondirectional infills, semi-directional infills and directional emphasized edges. Structural edges create edge-conditions or boundaries for the design. These patterns are also referred to as "structural skeletons." The voxels rotate initially with the goal of translating a series of principal stress lines into continuous structural ribbons. The process starts with the voxelization of a generic base model, with a resolution of approximately 200 mm. This model is then structurally analyzed. The data from the structural analysis is stored as a cloud of environmental conditions that can be accessed by the voxels. This includes stress levels, direction, and rotation of neighbors. In a first instance, the voxels are programmed to organize their structural patterns to align with stress and densify structurally weak areas. Subsequently, other voxels adapt to complete the patterns set out by the initial voxels. In a second stage, these two-dimensional patterns or skeletons are translated into a volumetric organization of smaller-scale voxels of 8 x 8 x 8 mm. The 8 mm height of the voxel is roughly the thickness of one layer of concrete extrusion.

In contrast to SpatialCurves, the weight of the concrete used in Project Fossilize requires a layered printing method. This requires a meaningful translation of the three-dimensional voxel structure into a linear two-dimensional organization of layers. To avoid interrupting the extrusion, the toolpaths should also be as long as possible. Cellular Automata-like logics are used to generate a continuous toolpath from the voxels. Each voxel checks its immediate neighbors, to see which ones share the same materiality. Voxels with only one neighbor are identified as the beginning of a line. To create a continuous line, every voxel has to connect to two neighbors. When only a limited number of points without any connections are left, the code moves to the next layer of voxels.

As a test case, Amalgama printed a small coffee table (Figures 9 and 13) and a two-meter-high column. Both prototypes were assembled out of larger pieces, limited to the bounding box of the support material. These pieces have printed male-female joints, which help to connect them seamlessly together.

TOWARDS DISCRETE FABRICATION

The proposed workflow aims to embed fabrication constraints into the design process from the very start. The resulting physical objects are therefore deeply related to the digital design process and vice versa. However, it can still be argued that the physical object is not digital in its material organization. In his concept of "Digital Materials," MIT professor Neil Gershenfeld distinguishes between analogue and digital organizations (Gershenfeld et al. 2015). He draws a parallel to the way data is organized. Analogue data is continuous, digital data is discrete. In an analogue or continuous system, a piece of matter has infinite connection possibilities, whereas a discrete or digital system only has a

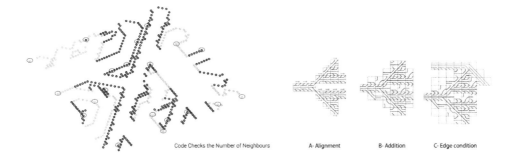

Code Checks the Number of Neighbours   A- Alignment   B- Addition   C- Edge condition

9   Robotically 3D printed Concrete
    Table Prototype. The Bartlett
    AD-RC4, 2014–15. Project
    Fossilized. Team: Amalgama.

10 Robotic Concrete 3D printing. The
    Bartlett AD-RC4, 2014–15. Project
    Fossilized. Team: Amalgama.

11 Combinatorics diagram. The
    Bartlett AD-RC4, 2014–15. Project
    Fossilized. Team: Amalgama.

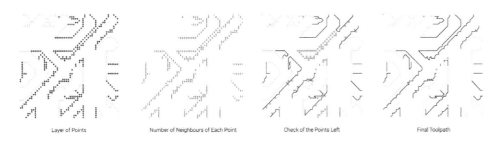

Layer of Points   Number of Neighbours of Each Point   Check of the Points Left   Final Toolpath

11

limited number (Ward 2010). In that sense, a 3D printed object is necessarily always analogue, as it doesn't have a limited connection scheme. According to these definitions, the objects described in this paper would not be considered digital. As a result of the continuous character of the printing method, there is also a fundamental gap between computation and fabrication. The object is first generated, simulated, and optimized, and then passed on to be fabricated. There is no interaction with the design during the fabrication process; as such, there are effectively two separate processes.

The understanding that serially repeated elements can be assembled into highly complex and heterogeneous structures has implications stretching beyond 3D printing. This combinatorial approach is highly valuable for construction systems based on modularity and prefabrication. This insight gave rise to a new research agenda, which investigates discrete additive assembly methods and robotic modular assembly. Continuous fabrication processes such as 3D printing have intrinsic problems with fundamental issues such as speed, structural performance, multi-materiality, and reversibility. Discrete fabrication has the same type of advantages in terms of problem solving as discrete computation; problems are serialized and solutions therefore become repeatable and cheap. Rather than using robots as 3D printers, this next phase of research uses robots as voxel assemblers or voxel printers. Robots quickly pick and place discrete bits of matter, assembling it into heterogeneous aggregations.

Aligning discrete computation with discrete fabrication enables the designer to bridge the gap between the digital and the physical. Digital data becomes the same as physical data. Computation and fabrication can happen in parallel, with the robot computing and evaluating the assembled structures while building them. The possibility of re-assembling enables the robot to correct mistakes, or go back and rebuild parts of the structure. In this case, the physical organization of matter is "digital," along Gerschenfeld's definition.

CONCLUSION

These discrete approaches prove to be successful and robust. The serialization of discrete toolpath patterns reduces the amount of unique problems to solve. One fragment of the toolpath can be optimized, and then serially repeated and combined into a larger toolpath. Continuously generated toolpaths have a large number of unique connection problems, each of them requiring a different solution to create a printable structure.

To overcome the risk of generating repetitive and homogenous structures due to the serial repetition of voxels, the concept of combinatorics was used. By always combining the discrete element in different positions, highly heterogeneous and differentiated structures become feasible. This is a fundamental shift in digital design thinking from mass-customization and continuous differentiation to discrete, serially repeated systems that still maintain a high degree of heterogeneity. This approach not only brings the feasibility of printing digitally intelligent structures a step closer to reality, but also makes 3D printing more accessible and robust. As problems are serialized and easy to solve, there is no need for expensive problem solving equipment such as advanced sensors, camera trackers, or supercomputers. The projects described challenge Neil Leach's assumption about the relation between digital tools and their outcomes. They are intrinsically based on the specific operation of a robotic 3D printing workflow. The physical organization of the printed object is a result of a negotiation with machine- and process-specific constraints. There is no post-rationalization, where the digitally designed object is sliced into layers for printing. Instead, the design is directly developed at the level of the toolpath itself (Figure 12).

The design strategy, based on combinatorial assembly of serially repeating elements with a voxel-based structure, is also valuable beyond the field of large-scale 3D printing. This design strategy can be applied to discrete fabrication strategies such as the concept of digital materials, as described by Neil Gerschenfeld. This opens up interesting territory where digital and physical discreteness are aligned, bridging the gap between digital design and fabrication.

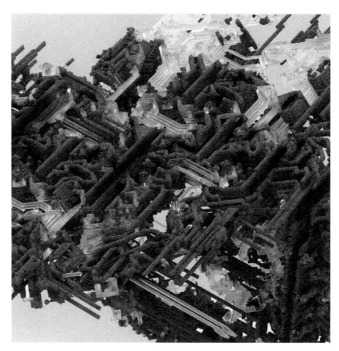

12 Rendering of horizontal formation achieved through combinatorial process. The Bartlett AD-RC4, 2014–15. Project Fossilized. Team: Amalgama.

ACKNOWLEDGEMENTS
The Bartlett AD Research Cluster 4 (RC4) Studio Masters: Manuel Jimenez Garcia, Gilles Retsin. Technical support: Vicente Soler; Team Filamentrics: Nan Jiang, Yiwei Wang, Zheeshan Ahmed, Yichao Chen; Team Curvoxels: Hyunchul Kwon, Amreen Kaleel, Xiaolin Li; Team Amalgama: Francesca Camilleri, Nadia Doukhi, Alvaro Lopez Rodriguez and Roman Strukov

The Bartlett AD Director: Alisa Andrasek; BPro Director: Professor Frédéric Migayrou.

Main Software and Libraries used: Processing by Casey Reas and Ben Fry, Toxiclibs library by Karsten Schmidt, McNeel Rhinoceros + Grasshopper, Robots fro Grasshopper by Vicente Soler, HAL by Thibault Schwartz.

REFERENCES
Carpo, Mario. 2014. "Breaking the Curve: Big Data and Design." *Artforum*, February. 169–173.

Dillenburger, Benjamin, and Michael Hansmeyer. 2014. "Printing Architecture: Castles Made of Sand." In *Fabricate: Negotiating Design and Making*, edited by Fabio Gramazio, Matthias Kohler, and Silke Langenberg. Zurich: ETH. 92–97.

Gershenfeld, Neil, Matthew Carney, Benjamin Jenett, Sam Calisch, and Spencer Wilson. 2015. "Macrofabrication with Digital Materials: Robotic Assembly." *Architectural Design* 85 (5): 122–127.

Hack, Norman, and Willi V. Lauer. 2014. "Mesh-Mould: Robotically Fabricated Spatial Meshes as Reinforced Concrete." *Architectural Design* 84 (3): 44–53.

Leach, Neil. 2015. "There is No Such Thing as a Political Architecture; There is no Such Thing as Digital Architecture." In *The Politics of Parametricism: Digital Technologies in Architecture*, edited by Matthew Poole and Manuel Shvartzberg. New York: Bloomsbury. 58–78.

Ward, Jonathan. 2010. "Additive Assembly of Digital Materials." PhD Thesis, Massachusetts Institute of Technology.

IMAGE CREDITS
Figure 1: Jiang, Wang, Ahmed, Chen, 2014, © UCL The Bartlett AD-RC4
Figures 2–8: Kwon, Kaleel, Li, 2015, © UCL The Bartlett AD-RC4
Figures 9–13: Camilleri, Doukhi, Lopez, Strukov, 2015, © UCL The Bartlett AD-RC4

13 Robotically 3D Printed Concrete Table Prototype displayed during the BPro Exhibition. The Bartlett AD-RC4, 2014–15. Project Fossilized. Team: Amalgama.

Gilles Retsin is the founder of Gilles Retsin Architecture, a young
award-winning London-based architecture and design practice, investigating new architectural models that engage with the potential of
increased computational power and fabrication to generate buildings
and objects with a previously unseen structure, detail, and materiality.
He graduated from the Architectural Association Design Research Lab
in London. Alongside his practice, Gilles directs a research cluster at the
Bartlett School of Architecture, UCL, investigating robotic manufacturing
and large-scale 3D printing, and he is a senior lecturer at UEL.

Manuel Jiménez García, a registered architect in UK and Spain, is the
founder and director of MadMDesign, a London-based research practice,
which mainly focuses on the integration of computational methods and
digital fabrication. He is also co-director of Nanami Design, a robotic
manufacturing startup based in Madrid and London. Alongside his
practice, Manuel is currently Course Master of AD Research Cluster 4,
as well as Unit Master of MArch Unit 19, both at The Bartlett School of
Architecture, UCL. He is also curator of the Bartlett Computational Plexus
and Programme Director at the Architectural Association's Visiting School
in Madrid (AAVSM).

Posthuman Engagements

Kathy Velikov
University of Michigan, Taubman College of Architecture and Urban Planning

Central to posthumanist theory is the interrogation and eventual elimination of boundaries, particularly through the transgression of the biological with the technological. The human body ceases to be understood as a complete, homeostatic entity, but is—and has been since the very first use of tools—a body prosthetically expanded and articulated through tools and media technologies.[1] This alternative configuration of embodiment produces the possibility of new identities and ontologies. The decentering of the human subject from a position of privilege, moreover, recognizes the agency of other living and nonliving entities. Here, the paper by Leach approaches this question by interrogating materialist philosophies of contemporary digital tools and objects. The agency of nonhuman entities is further explored by the other papers in this section through modalities of expanded embodiment and interaction between humans and intelligent machines.

In *How We Became Posthuman*, N. Katherine Hayles traces a genealogy of posthumanist theory as one closely linked to the history of cybernetics, the science of communication, feedback, and control in both living and machine systems that emerged as a paradigm and transdisciplinary field of study during the Macy Conferences on Cybernetics held between 1943 and 1954.[2] The ideas that emerged from these conferences not only began to "dismantle the liberal humanist subject" (by looking at humans as "information-processing entities" and models for machine intelligence), but also laid the foundations for thinking about the behavior and language of first- and second-order cybernetics, which define interactive protocols and open systems.[3]

Second-order—or "learning, conversing" systems[4]—are the operational basis for Beesley's "Hybrid Sentient Canopy." The synthetic system achieves life-like behavior through the combination of a broad spectrum of distributed, cooperative sensor technologies, curiosity-based learning algorithms, and form-found digitally fabricated spatial structures capable of kinetic movement. Farahi explores sentient behavior at a more intimate scale in a "body architecture" that combines smart material actuators and multimaterial 3D printing with sensing and interactive capabilities. Both explore the design of life-like behaviors and experiment with the production of emotional and cognitive relations between humans and synthetic systems. Costa Maia and Meyboom, whose paper describes experiments with models of interaction that aim to empower inhabitants, explore the question of participant agency in the field of Interactive Architecture (IA). In different ways, López, Pinochet, and Eisinger and Putt also experiment with the novel creative agencies afforded to "humans" in posthuman collaboration with digital machines.

In aggregate, the papers in this section explore tools, objects, and modes of existence beyond human-defined consciousness-based life, and thus contribute to growing conversations regarding the posthuman. Instead of amplifying anxieties about conditions of alienation produced by advanced technologies and machines, these projects suggest that computation and technology might be used to discover material and "thing languages"[5] that allow for newfound intimacies and relations between ourselves and the world around us.

1. Marshall McLuhan, *Understanding Media: The Extensions of Man* (New York: McGraw Hill, 1964).

2. N. Katherine Hayles, *How We Became Posthuman: Virtual Bodies in Cybernetics, Literature, and Informatics* (Chicago: University of Chicago Press, 1999), 7.

3. Ibid.

4. Hugh Dubberly, Usman Haque, and Paul Pangaro, "What is Interaction? Are There Different Types?" *Interactions* 16, vol. 1 (2009): 75.

5. Kyla Anderson, "Object Intermediaries: How New Media Artists Translate the Language of Things," *Leonardo* 47, vol. 4 (2014): 352.

Posthuman Engagements

Digital Tool Thinking: Object Oriented Ontology versus New Materialism

Neil Leach
EGS/Tongji/FIU

ABSTRACT

Within contemporary philosophy, two apparently similar movements have gained attention recently, New Materialism and Object Oriented Ontology. Although these movements have quite distinct genealogies, they overlap on one key issue: they are both realist movements that focus on the object. In contrast to much twentieth-century thinking centered on the subject, these two movements address the seemingly overlooked question of the object. In shifting attention away from the anthropocentrism of Humanism, both movements can be seen to subscribe to the broad principles of Posthumanism.

Are these two movements, however, as similar as they first appear? And how might they be seen to differ in their approach to digital design? This paper is an attempt to evaluate and critique the recent strain of Object Oriented Ontology and question its validity. It does so by tracing the differences between OOO and New Materialism, specifically through the work of the neo-Heideggerian philosopher Graham Harman and the post-Deleuzian philosopher Manuel DeLanda, and by focusing on the question of the 'tool' in particular. The paper opens up towards the question of the digital tool, questioning the connection between Object Oriented Ontology and Object Oriented Programming, and introducing the theory of affordances as an alternative to the stylistic logic of 'parametricism' as a way of understanding the impact of digital tools on architectural production.

The paper concludes that we need to recognize the crucial differences between the work of DeLanda and Harman, and that—if nothing else—within progressive digital design circles, we should be cautious of Harman's brand of Object Oriented Ontology, not least because of its heavy reliance on the work of the German philosopher, Martin Heidegger.

INTRODUCTION

In the 1990s, two philosophers emanating from the Deleuzian tradition, Manuel DeLanda and Rosi Braidotti, independently introduced the term New Materialism. It is unclear who used the term first and whether they were referring to the same issues. Since then a number of other thinkers—such as Karen Barad—have followed suit.[1] This paper focuses on DeLanda's version of New Materialism [NM], not least because DeLanda has had such a significant impact on architectural discourse.[2]

Broadly speaking, NM is an attempt to counter the emphasis on not only the subject, but also representation and interpretation in twentieth-century thinking in general, and Postmodernism in particular, by focusing instead on the object, material processes, and expression. It is not that we should discount the previous tradition, in that—from a Deleuzian perspective—representation and process are locked into a mechanism of reciprocal presupposition. Rather it is strategically important to redress the balance and counter the scenographic tendencies of Postmodernism. While NM emerges out of a Deleuzian tradition, precedents can be found in the work of biologist D'Arcy Wentworth Thompson and philosopher Henri Bergson. It challenges not only the linguistic turn in twentieth-century philosophy, but also the Dialectic Materialism of Karl Marx.[3]

From an architectural perspective, it is Deleuze's distinction between the Romanesque and the Gothic spirit that highlights the key difference between Postmodernism and NM. Whereas Postmodernism privileged the Romanesque logic of representation and symbolism, NM focuses on the Gothic logic of process and material performance. (Deleuze and Guattari 1988, 394) Important precursors of NM within an architectural tradition include Antoni Gaudí and Frei Otto, while perhaps Achim Menges best articulates the tradition today.

In 1999, the philosopher Graham Harman coined the term Object Oriented Philosophy in his doctoral dissertation on Martin Heidegger's thinking about tools, *Tool-Being: Elements in a Theory of Objects*. Subsequently, Levi Bryant introduced the term Object Oriented Ontology, which has now been adopted as the name of the movement.

Broadly speaking, Object Oriented Ontology [OOO] seeks to challenge the hegemony of the previously dominant anthropocentric outlook—traceable back to Immanuel Kant—that privileged human beings over objects, and viewed objects primarily through the mind of the subject. The movement has attracted a range of followers—such as Levi Bryant, whose intellectual formation is indebted largely to the work of Gilles Deleuze—and falls under the umbrella of the somewhat disparate movement of Speculative Realism, which encompasses a broader range of thinkers, including Ray Brassier, Iain Hamilton Grant, and Quentin Meillasssoux. Harman himself has a background in the work of Martin Heidegger, and his current thinking could be described as neo-Heideggerian. The key difference between Harman and Heidegger, is that while Harman adopts the fourfold structure drawn from Heidegger, he does not adopt his "distinction between 'object' (which he uses negatively) and 'thing' (which he uses positively)" (Harman 2011, 5). In short, Harman places a greater emphasis on the "object."

From an architectural perspective, it is perhaps too early to detect any significant impact of OOO, although Harman himself has a number of followers in architectural circles. However, we can trace the antecedents of OOO in architectural thinkers from the phenomenological tradition, such as Alberto Pérez-Gómez, Steven Holl, Dalibor Vesely, Juhani Pallasmaa and Christian Norberg-Schulz.

Significantly, Harman sees overlaps between his thinking and that of DeLanda, although the sentiment is not necessarily reciprocated[4] (Harman 2011, 170).

TOOL THINKING

Given that Harman's doctoral thesis was an attempt to elucidate and explain Heidegger's thinking on tools, it could be assumed that Harman himself buys into the basic principles of Heidegger's approach to tools. It therefore makes sense to introduce Harman's position by outlining Heidegger's thinking on tools.

In his famous early analysis of tools, Heidegger makes a distinction between 'ready-to-hand' and 'present-at-hand'. (Heidegger 1962). As Harman puts it, "[Heidegger's] famous tool analysis in Being in Time shows that our usual way of dealing with things is not observing them as present-at-hand (*vorhanden*) in consciousness, but silently relying on them as ready-to-hand (*zuhanden*)" (Harman 2010, 36). Jonathan Hale provides a helpful summary of this distinction:

> Heidegger's now famous example describes how a piece of equipment like a hammer can be approached in two distinct

ways: we can either pick it up and use it, or we can contemplate it from a distance. When we pick up the hammer and use it, it becomes what Heidegger calls 'ready-to-hand,' the hammer is ready to be put to work, assuming we know how to wield it. In the second case, what Heidegger calls 'present-at-hand,' we simply stare at the hammer as an object, trying to make sense of it by some kind of intellectual analysis. In this case Heidegger claims that we never uncover the true being of the hammer as a tool, we are simply confronted with a curious lump of inert physical stuff. (Hale 2013)

As Hale astutely points out, however, there are problems with Heidegger's overly binary analysis. In short, it seems very 'black and white'. The tool is either 'ready-to-hand' or 'present-to-hand,' and there appears to be no space in between. What is clearly lacking in this account is our gradual proprioception of tools over time. For a subtler understanding of how we come to accommodate the tool, and 'think through' it in an unselfconscious way through use so that it gradually becomes a prosthesis to the motility of the body, we should perhaps turn to Maurice Merleau-Ponty, who, as Hale notes, "describes the tool becoming incorporated (literally) into an extended 'body-schema', as a kind of prosthetic bodily extension that allows me to experience the world through it" (Hale 2013).

This problem is repeated on a larger scale in Heidegger's approach to technology in general, articulated most explicitly in his essay, "The Question Concerning Technology."[5] Here, Heidegger draws upon the notion of 'standing–reserve' that seemingly echoes the notion of 'present-at-hand' used in the context of tools: "Everywhere everything is ordered to stand by, to be immediately on hand, indeed to stand there just so that it may be on call for a further ordering. Whatever is ordered about in this way has its own standing. We call is the standing-reserve [Bestand]" (Heidegger 1993, 322). And it is this sense of 'standing-reserve' that lies at the heart of modern technology: "The essence of modern technology shows itself in what we call Enframing… It is the way in which the real reveals itself as standing-reserve' (Heidegger 1993, 328–329). The problem is not so much of nature being devalued as standing-reserve, but humankind finding itself in the same condition: "As soon as

what is concealed no longer concerns man even as object, but exclusively as standing-reserve, and man in the midst of object-lessness is nothing but the orderer of the standing-reserve, then he comes to the very brink of a precipitous fall; that is, he comes to the point where he himself will have to be taken as standing-reserve" (Heidegger 1993, 332).

Heidegger illustrates his notion of 'standing-reserve' with the example of an airliner. The airliner on the runway, for Heidegger, "stands on the taxi strip only as standing-reserve, inasmuch as it is ordered to insure the possibility of transportation" (Heidegger 1993, 322). The problem here is that the airliner disappears into the abjection of its 'standing reserve' (Heidegger, 1993: 324). The possibility that the airliner might be appropriated and become part of our horizon of consciousness is not entertained.

By contrast, German philosopher Theodor Adorno—an outspoken critic of Heidegger in his book, *The Jargon of Authenticity* (1973), and elsewhere—refers to the airliner as an object, much like a tool in Merleau-Ponty's terms, that can be appropriated: "According to Freud, symbolic intention quickly allies itself to technical forms, like the airplane, and according to contemporary American research in mass psychology, even to the car" (Adorno 1997, 10).

What Heidegger fails to address, then, is the progressive way that we come to appropriate technology in general, and tools in particular, and absorb them within our horizon of consciousness. It is the difference between Heidegger's decidedly static notion of 'Being'—or indeed the 'being of Being'—and Deleuze and Guattari's more fluid, dynamic concept of 'becoming.' This contrast could be extended into their understanding of the 'machine,' which—for Deleuze and Guattari—is far more than merely a piece of mechanistic technology. Whereas for Heidegger, technology is perceived as antithetical to what it is to be human, Deleuze and Guattari celebrate a more empathetic relationship between humans and technology. As John Marks observes, "Everything is a machine, and everywhere there is production. For Deleuze and Guattari, the machine is not a metaphor; reality is literally 'machinic'. The concept of the machinic is set against the traditional opposition between vitalism and

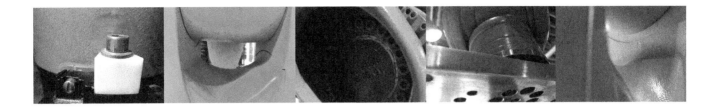

mechanism. . . In short, there is no difference between categories of living and the machine" (Marks 1998, 98). Importantly for Deleuze and Guattari, the machinic is also associated with desire: "A direct link is perceived between the machine and desire, the machine passes into the heart of desire, the machine is desiring and desired, machined" (Deleuze and Guattari 1983, 285).

This difference extends into the equally problematic distinction between Heidegger's somewhat instrumental approach to technology and the more poetic approach that he refers to as *techné*, that echoes the distinction that he makes between 'authenticity' and 'inauthenticity.' On what authority does Heidegger make these distinctions? What is 'inauthentic' for some can easily be 'authentic' for others. In short, we must move away from Heidegger's universalizing claims to 'truth' – as though in our postmodern world we could claim that there could be any universalizing, singular 'truth'– just as we need to move away from Heidegger's homogeneous view of the world towards a more open-minded Deleuzian understanding of multiplicities. As Guattari comments on the subject of technology: "Far from apprehending a univocal truth of Being through *techné*, as Heideggerian ontology would have it, it is a plurality of beings as machines that give themselves to us once we acquire the pathic or cartographic means of access to them" (Guattari 1993, 26).

DeLanda, by contrast, finds Merleau-Ponty more useful in understanding our relationship with tools, and is also sympathetic to J. Gibson, whose concept of 'invariance' forms a central part of his thesis in *Intensive Science, Virtual Philosophy* (DeLanda 2002). For DeLanda, invariances are transmitted by structured light and inform the organism about the 'affordances' of objects and tools, which opens up an entirely different way of thinking about tools. Let us turn, then, to Gibson's 'theory of affordances.'[6] This theory suggests that there is a particular set of actions 'afforded' by a tool or object. Thus a knob might afford pulling—or possibly pushing—while a cord might afford pulling. It is not that the tool or object has agency as such, or the capacity to 'invite' or 'prevent' certain actions. Rather, it merely 'affords' certain operations that it is incumbent on the user to recognize, dependent on pre-existing associations with that tool or object. Likewise, those operations are also dependent upon our capacity to undertake them. Thus certain operations might not be afforded to those without the height or strength to perform them. Moreover, certain tools afford certain operations, but do not preclude others. For example, we could affix a nail with a screwdriver—albeit less efficiently—if we do not have a hammer at hand. Similarly, it is easier to cut wood with a saw than a hammer.

In his own discussion of tools, however, DeLanda refers not to 'affordances' so much as to 'capacities,' whereby the 'virtual'—in the Deleuzian sense—has the capacity to become 'actual' through use:

> The causal capacity of the knife to cut is not necessarily actual if the knife is not actually being used. In fact, the capacity to cut may never be actual if the knife is never used. And when that capacity is actualized it is always as a double event: to cut—to be cut. In other words, when a knife exercises its capacity to be cut it is interacting with a different entity that has the capacity to be cut. This implies a realist commitment not only to the mind-independence of actual properties but also of causal capacities that are real but not necessarily actual. (DeLanda 2011, 385)

DeLanda offers a further gloss: "Although there are many species that use tools (crows, chimps, even insects) we humans certainly excel at inventing new ones. But to conceive of tools as autonomous we need to redefine mind-independence because tools would not exist if human minds did not exist. Hence, the expression should signify independence from the content of our minds, not from their existence" (DeLanda and Harman, 2016).

This highlights a crucial difference between the two, in that Harman still clings to the notion that the world is dependent on our minds, whereas for DeLanda the world exists independently of our minds. Another important difference is that Harman—following Heidegger—believes in essences, whereas DeLanda—along with the entire poststructuralist tradition—is highly critical of essences (Harman 2009b, 204–206; DeLanda 2002, 9–10). In short, despite the initial appearances, Harman and DeLanda are clearly at odds, and are caught between the basic differences between OOO and NM.

DIGITAL TOOL THINKING

Why then are some architects so fascinated by OOO? In part, it might be because architects design 'objects.' At first sight, there also appears to be a connection between Object Oriented Ontology and Object Oriented Programming. However, this connection is highly problematic. Ian Bogost, for example, who is both a philosopher and a game designer, explicitly distances himself from the term 'object' precisely because of the confusion that the apparent similarity of names might cause, and uses the term 'unit' instead:

> Then and since, I've been secretly bothered by "object-oriented philosophy" (the name, not the idea). I was reminded of this concern when I saw that Harman had shorthanded his term with the acronym OOP, one also commonly used to refer to the programming paradigm. My worry arose not from the perception that Harman had absconded with the appellation without giving it proper credit (he has never to my knowledge noted the similarity of the terms in his writing, although I know he is aware of it) but because I feared the sense of "object-oriented" native to computer science didn't mesh well with that of speculative realism. (Bogost 2009)[7]

In the end, however, there would appear to be little difference between 'tools' and 'digital tools.' They are all ultimately 'tools,' and instructions given to a construction laborer are not so dissimilar to code inputted in computational operations. Both are 'algorithms.' The theory of affordances can therefore also be applied to digital tools to refer to the progressive instantiation of certain operations that eventually become hegemonic. An obvious example would be the progressive adoption of curvilinear forms within architectural design. Rather than subscribing to Patrik Schumacher's controversial theory of 'parametricism' as a new style, we might gain a better understanding of the growing popularity of curvilinear forms through the theory of affordances (Schumacher 2009). In the days of the parallel motion, it was very easy to draw straight lines, and also possible to draw curved lines using 'French curves.' However, there was no way to define the precise nature of the curves drawn, let alone to reproduce them. With the advent of the computer, however, it became possible to define the curve very precisely. Not only that, but

when simulating the bottom up logic of multi-agent systems, it is actually difficult to generate a straight line. As these new operations afforded by the computer are repeated, they become the norm. Thus, although some regard curvilinear forms as a style, they are more likely the result of the 'emergence' of certain practices based on the affordances of the digital tools deployed.

FORGET HEIDEGGER

The main problem with Harman's neo-Heideggerian version of OOO is that it returns us to some of the key debates between advocates of Poststructuralism and Phenomenology during the 1990s. The challenge, however, is that much depends on the starting point adopted, in that Poststructuralism and Phenomenology seem to be fundamentally incommensurable. Either we buy into a Phenomenological position of collapsing the subject into the object and therefore assume that we are all at one with culture—and therefore have no problem understanding culture—or we adopt a more skeptical Poststructuralist position that always offers an epistemological check and constantly problematizes the relationship between subject and object. It is simply a question of position taking.

One key problem for Poststructuralists is that Phenomenologists are often guilty of ascribing 'agency' to the object. This returns with a vengeance if we consider the controversial Actor Network Theory [ANT] of Bruno Latour, an influence on OOO. ANT assumes that objects 'act' in social networks. Latour himself uses the example of a door-closer 'on strike' to illustrate his point (Latour 1988). However, the most well known architectural example of ascribing agency to objects is perhaps Louis Kahn asking a brick what it wants to do, as though the brick has the capacity to think and speak.[8] This—for the Poststructuralist—is simply a question of ventriloquism, of projecting onto the object a form of anthropomorphic agency. Of course, there is always a natural tendency to think in this way. Many of us give our cars names, and perhaps even speak to them. And, as Lacan has observed, there is a 'primordial anthropomorphism' that underpins knowledge, and he therefore questions "whether all knowledge is not originally knowledge of a person before being knowledge of an object, and even whether the knowledge of an object is not, for humanity, a secondary acquisition" (Lacan 1975,

392). However, from a Poststructuralist viewpoint, the problem is that we end up 'appropriating' the object—be it door-closer or brick—as though we understand it. Jacques Derrida has offered a powerful critique of the potential relativism of phenomenological hermeneutics in his analysis of Heidegger's interpretation of a painting of shoes by Van Gogh (Derrida 1987).

There are further problems with Heidegger. For example, we should not forget his disturbing anti-Semitism and affiliation with the National Socialists—for which Heidegger never apologized—issues foregrounded following the publication of his 'Black Notebooks' (Farías 1989; Wolin). While the question of whether this should disqualify Heidegger's philosophy remains unclear, it is possible to see connections between his right wing political agenda and conservative philosophical agenda. There is a 'dark side' to his 'philosophy of the soil,' which can lead inexorably to a disturbing celebration of the *heimat* and the homeland (Leach 1999).

Nor should we overlook the issue of 'forgetting' that is central to Heideggerian philosophy. As Jean-François Lyotard points out, the crime in Heidegger is the *forgetting*—and the *forgetting* of the *forgetting*—that leads him to conveniently overlook his involvement in National Socialism (Lyotard 1990). However, the real problem in accepting Harman's seemingly uncritical acceptance of Heidegger is that we too end up forgetting the very powerful critiques of Heidegger by Derrida, Lyotard, Deleuze, Guattari, and others, alongside the clearly unpalatable politics to which Heidegger ascribed. Several years ago, calls were made within the architectural community to forget Heidegger (Leach 2006). His thought was seen to be overly conservative, essentializing, anti-technological, and out of touch with contemporary concerns, such as capitalism and global warming.

OH, OH, OH, NO!

OOO has become something of a trendy movement in some design circles, and has even been linked—erroneously—to digital design thinking. Given that DeLanda, as a former programmer, has far greater command of digital thinking, and exerted much greater influence on architectural culture, it is surprising that Harman is even taken into consideration.

With Harman we have a highly Postmodern rebranding of a relatively unreconstructed version of Heidegger's thinking, complete with appropriately trendy Postmodern acronyms—OOO and so on. It is as though the previous sixty years of sustained critique has been *forgotten* and Heidegger—like some Teflon-coated politician—has effectively escaped to re-emerge as some post-apocalyptic figure rising from the dead. Not only that, but Harman has even been linked to the domain of digital computation. For a thinker who invests so much in a philosopher such as Heidegger, who has such a pejorative view of technology, this seems remarkable.

There are, of course, many other variations of OOO that have not been covered here—variations that do not subscribe to the conservative neo-Heideggerianism of Harman himself—that merit further consideration. But as far as Harman is concerned, especially in terms of digital design, it is surely time to *forget* OOO.

ACKNOWLEDGEMENTS

I am grateful for the feedback and advice of Manuel DeLanda, Gray Read, Hélène Frichot and Brian Cantrell in the preparation of this article.

NOTES

1. For an overview of NM, see Dolphijn; van der Tuin (2012) .
2. DeLanda has taught at Columbia GSAPP, USC, Pratt, UPenn and Princeton.
3. As Leach observes, "New Materialism can be compared and contrasted with the old Historical Materialism of Karl Marx. Famously, Marx had turned Hegelian dialectics 'on its head', and - against Hegel's idealistic theory of the dialectic - had stressed the primacy of the material world. Equally there are echoes of Marxist famous dictum from his Theses on Feuerbach, 'The philosophers have only interpreted the world, in various ways; the point is to change it', in relation to DeLanda's critique of postmodern hermeneutics. There are echoes too of Marx's basic premise that what we see on the surface of cultural phenomena is the product of deeper underlying forces. But New Materialism extends the range of Historical Materialism. For Marx the only form of economic production considered was labour, whereas for New Materialism any cultural expression – social, economic or political – can be understood in terms of the forces that produce it." (Leach 2015).

4. Harman and DeLanda are currently writing on a book about their respective positions. I am grateful to DeLanda for allowing me to view a draft of that book.

5. Heidegger was not opposed to technology as such. But rather he saw in technology a mode of 'revealing', and it was here that the danger lay. "The essence of modern technology," as he puts it, "lies in enframing. Enframing belongs within the destining of revealing" (Heidegger, 1993: 330). And this form of 'revealing' is an impoverished one as it denies the possibility of a deeper ontological engagement: "Above all, enframing conceals that revealing which, in the sense of *poiesis*, lets what presences come forth into appearance" (Heidegger 1993, 332). Rather than opening up to the human it therefore constitutes a form of resistance or challenge to the human, in that it 'blocks' our access to truth: "Enframing blocks the shining-forth and holding sway of truth." (Heidegger 1993, 332).

6. The theory of affordances was introduced by Gibson in an article, "The Theory of Affordances" (Gibson 1977), and later elaborated in *The Ecological Approach to Visual Perception* (Gibson 1979). It was later developed by Eleanor Gibson and Anne Pick (2000).

7. Bogost goes on to list his understanding of the 'object' as used in Object Oriented Programming:

   "To wit, an object in the computational sense:
   - describes a pattern, not a thing
   - exists in stable relation to its properties
   - exists in stable relation to its abilities.
   - has direct access to other objects via their properties and abilities
   - is not a real object (but can be made real, e.g. on magnetic tape or as a series of instructions on a processor stack)
   - always relates to an intentional object (both because it is a designed object and because it strives to embody and enact direct modeling of the world)

   Many—perhaps all—of the aspects above conflict with Harman's understanding of objects and what it means to be oriented toward them." (Bogost 2009)

8. "You say to a brick, 'What do you want, brick?' And brick says to you, 'I like an arch.' And you say to brick, 'Look, I want one, too, but arches are expensive and I can use a concrete lintel.' And then you say: 'What do you think of that, brick?' Brick says: 'I like an arch'" (Wainwright 2013).

REFERENCES

Adorno, Theodor. 1973 *The Jargon of Authenticity*. London: Routledge.

Adorno, Theodor. 1997. "Functionalism Today." In *Rethinking Architecture: A Reader in Critical Theory*, edited by Neil Leach. London: Routledge. 5–18.

Bogost, Ian. 2009. "Units and Objects: Two Notes Apropos of Graham Harman." Bogost January 11. http://bogost.com/blog/units_and_objects/

Bogost, Ian. 2012. *Alien Phenomenology, or, What it is Like to be a Thing*. Minneapolis: University of Minnesota.

Borch-Jacobsen, Mikkel. 1991. *Lacan: The Absolute Master*, translated by Douglas Brick. Stanford, CA: Stanford University Press.

Bryant, Levi. 2008. *Difference and Givenness: Deleuze's Transcendental Empiricism and the Ontology of Immanence*. Evanston: Northwestern.

Dallmayr, Fred. 1993. *The Other Heidegger*. Ithaca, NY: Cornell University Press.

DeLanda, Manuel. 2002. *Intensive Science, Virtual Philosophy*. London: Continuum.

———. 2011. "Emergence, Causality and Realism." In *The Speculative Turn: Continental Materialism and Realism*, edited by Levi Bryant, Nick Srnicek, and Graham Harman. Melbourne: re.press. 381–392.

DeLanda, Manuel and Graham Harman. 2016. *Conversation*. Unpublished

Deleuze, Gilles and Félix Guattari. 1983. *Anti-Oedipus: Capitalism and Schizophrenia*, translated by Robert Hurley, Mark Seem, and Helen R. Lane. Minneapolis: University of Minnesota Press.

Deleuze, Gilles and Félix Guattari. 1988. *A Thousand Plateaus: Capitalism and Schizophrenia*, translated by Brian Massumi. London: Athlone Press.

Derrida, Jacques. 1987. *Truth in Painting*, translated by Geoff Bennington and Ian McLeod. Chicago. University of Chicago Press.

Dolphijn, Rick and Iris van der Tuin. 2012. *New Materialism: Interviews and Cartographies*. Ann Arbor, MI: Open Humanities Press.

Farías, Victor. 1989. *Heidegger and Nazism*, translated by Paul Burrell et al. Philadelphia: Temple University Press.

Gibson, Eleanor J. and Anne D. Pick. 2000. *An Ecological Approach to Perceptual Learning and Development*. New York: Oxford University Press.

Gibson, James. 1977. "The Theory of Affordances." In *Perceiving, Acting, and Knowing: Toward an Ecological Psychology*, edited by Robert Shaw and John Bransford. London: Wiley.

Gibson, James. 1979. *The Ecological Approach to Visual Perception*, Hove: Psychology Press.

Guattari, Félix. 1993. "Machinic Heterogenesis." In *Rethinking Technologies*, edited by Verena Andermatt Conley. Minneapolis: University of Minnesota Press. 13–27.

Hale, Jonathan. 2013. "Harman on Heidegger: 'Buildings as Tool-Beings." *Body of Theory*, May 29. https://bodyoftheory.com/2013/05/29/harman-on-heidegger-buildings-as-tool-beings/

Harman, Graham. 2002. *Tool-Being: Heidegger and the Metaphysics of Objects*. Chicago: Open Court.

———. 2009a. *Towards Speculative Realism: Essays and Lectures*. New York: Zero Books.

———. 2009b. *Prince of Networks: Bruno Latour and Metaphysics*. Melbourne: re.press.

———. 2011. *The Quadruple Object*. New York: Zero Books.

Heidegger, Martin. 1962. *Being and Time*. London: SCM Press.

———. 1993. *Basic Writings*, edited by David Farrell Krell. New York: Harper Collins.

Lacan, Jacques. 1975. *De la psychose paranoïague dans ses rapports avec la personnalité*. Paris: Éditions du Seuil.

Lacoue-Labarthe, Philippe. 1990. *Heidegger, Art and Politics: The Fiction of the Political*, translated by Chris Turner. Oxford: Blackwell.

Latour, Bruno [Jim Johnson]. 1988. "Mixing Humans with Non-Humans: Sociology of a Door-Closer." *Social Problems* 35 (3): 298–310.

Leach, Neil. 1999. "The Dark Side of the Domus: The Redomestication of Central and Eastern Europe." In *Architecture and Revolution: Contemporary Perspectives on Central and Eastern Europe*, edited by Neil Leach. London: Routledge. 150–162.

———. 2006. *Forget Heidegger*. Bucharest: Paideia.

Leach, Neil. 2015. "Design and New Materialism." In *De-signing Design: Cartographies of Theory and Practice*, edited by Elizabeth Grierson, Harriet Edquist, and Hélène Frichot. London: Lexington Books. 205–216.

Lyotard, Jean-François. 1990. *Heidegger and "The Jews,"* translated by Andreas Michel and Mark S. Roberts. Minneapolis: University of Minnesota Press.

Marks, John. 1998. *Gilles Deleuze: Vitalism and Multiplicity*. London: Pluto.

Scheibler, Ingrid. 1993. "Heidegger and the Rhetoric of Submission." In *Rethinking Technologies*, edited by Verena Andermatt Conley. Minneapolis: University of Minnesota Press. 115–139.

Schumacher, Patrik. 2009. *"Parametricism: A New Global Style for Architecture."* Architectural Design 79 (4): 14–23.

Sluga, Hans. 1993. *Heidegger's Crisis: Philosophy and Politics in Nazi Germany*. Cambridge, MA: Harvard University Press.

Wainwright, Oliver. 2013. "Louis Khan: The Brick Whisperer." *Guardian*, February 26. http://www.theguardian.com/artanddesign/2013/feb/26/louis-kahn-brick-whisperer-architect

Wolin, Richard, ed. 1993. *The Heidegger Controversy: A Critical Reader*. Cambridge, MA: MIT Press.

IMAGE CREDITS

All photographs by Neil Leach, 2016

Neil Leach teaches at the European Graduate School, Tongji University and Florida International University, and is a member of the Academia Europaea. He has been working on a NASA funded research project to develop robotic fabrication technologies to print structures on the Moon and Mars. He has published 27 books including *Designing for a Digital World* (Wiley, 2002), *Digital Tectonics* (Wiley, 2004), *Digital Cities* (Wiley, 2009), *Machinic Processes* (CABP, 2010), *Fabricating the Future* (Tongji UP, 2012), *Scripting the Future* (Tongji UP, 2012), Robotic Futures (Tongji UP, 2015) and *Digital Factory: Advanced Computational Research* (CABP, 2016).

Caress of the Gaze: A Gaze Actuated 3D Printed Body Architecture

Behnaz Farahi
University of Southern California

1

ABSTRACT

This paper describes the design process behind Caress of the Gaze, a project that represents a new approach to the design of a gaze-actuated, 3D printed body architecture—as a form of proto-architectural study—providing a framework for an interactive dynamic design. The design process engages with three main issues. Firstly, it aims to look at form or geometry as a means of controlling material behavior by exploring the tectonic properties of multi-material 3D printing technologies. Secondly, it addresses novel actuation systems by using Shape Memory Alloy (SMA) in order to achieve life-like behavior. Thirdly, it explores the possibility of engaging with interactive systems by investigating how our clothing could interact with other people as a primary interface, using vision-based eye-gaze tracking technologies.

In so doing, this paper describes a radically alternative approach not only to the production of garments but also to the ways we interact with the world around us. Therefore, the paper addresses the emerging field of shape-changing 3D printed structures and interactive systems that bridge the worlds of robotics, architecture, technology, and design.

1   Caress of the Gaze: 3D printed garment, which moves in response to the onlooker's gaze by Behnaz Farahi. Photographers: Charlie Nordstrom, Elena Kulikov. 2015.

INTRODUCTION

3D printing and interactive design have revolutionized approaches to design in many different industries, including fashion, and wearable and architectural design. The relationship between the body and architecture has a long history. From Vitruvius onwards, there have been attempts to relate buildings to the proportions of the human figure. More recently, the connection between the body and architecture has developed into an interest in the fashion industry among architects, as in the "Skin + Bones: Parallel Practices in Fashion and Architecture" exhibition (Museum of Contemporary Art, Los Angeles, 2007), which drew extensively on architects designing wearables. Since then, there has been a veritable explosion of architects working in the realm of fashion—largely, but not exclusively operating through 3D printing technologies. Architects such as Neri Oxman, Philip Beesley, Julia Korner, and Daniel Widrig have collaborated with leading figures within the fashion industry, such as Iris van Herpen.

The term "Body Architecture" has been used by the avant-garde media artist, Stelarc, who notes that "altering the architecture of the body results in adjusting and extending its awareness of the world." Later, the term was used by artist Lucy McRae to refer to practices which aim to transform conceptions of the human body. In the context of this paper, Body Architecture is a form of proto-architecture where there is an attempt to engage with the human body and its relationship with the space immediately around it.

The intention is to explore the domain of the body as the site of architecturally inspired investigations. The main attempt in such a work, is to engage with geometries and forms in order to understand dynamic material behaviors in relation to the human body. Inevitably, these lines of thought draw our attention to some groundbreaking works in the realm of material intelligence, including 4D printing by Skylar Tibbits of the MIT Self-Assembly Lab. In these attempts, material intelligence is informed not only by geometry but also by real-time changes in the material properties that inform matter.

As Neri Oxman states, "A bio-inspired fabrication approach calls for a shift from shape-centric virtual and physical prototyping to material-centric fabrication processes" (2011). Advances in 3D printing technologies have enabled us to design shape-shifting objects by distributing a range of different materials in various compositions, so as to open up a variety of mechanical behaviors. This provides us with the opportunity to design objects that are becoming closer to natural systems in terms of their morphologies as well as their behaviors. As researchers from MIT Assembly Lab explain, "3D printing enables us to arrange the stretching

2

3

2   Removing the print from the Polyjet Connex 500 3D Printers.

3   Using multi-material 3D Printing by Objet Connex500. This prototype is demonstrating varying flexibilities and densities from stiff to soft.

and folding primitives in different orientations, which allows stretching and folding to happen at the desired position in the 3D vector space". (Raviv et al. 2014)

Moreover, advances in interactive design technologies are changing the way that we interact with the world around us by influencing our perception, ways of communication, and awareness. These technologies are also changing the perception of our bodies by allowing them to become augmented, enhanced, and expanded in terms of their functionality. This will fundamentally change what it means to be human and make us closer to being posthuman. As Parks puts it, "ubiquitous and embedded technologies are allowing our devices to become more and more part of us with increasing mobility and pervasiveness" (2008). Hansen explains that the human body does not end at the boundaries of its own skin, but rather constructs intimate relationships with digital information flows and data spaces (2006: 191). Gray believes that when every part of human life, including birth, education, sex, work, and death are transformed by intimate connections with technologies, then "the language of technology will begin to 'invade' the ways we express and perceive these experiences" (1995, 6). But these technologies are equally influential in this coupling. For instance, computer technologies as "non-organic, clear and functioning prostheses" are transformed by their coupling with the human body. Likewise, the human body is perceived as "augmented and enhanced by its attachment to technology" (Clark 2002, 36) This suggests that the notion of the "cyborg" or "posthuman" should not be limited to early speculations from the world of either space exploration

or sci-fi about the potential extensions of the human body. Rather, it implies that our understanding of the term has itself evolved to include the use of everyday technological devices. Thanks to embedded computing and bio-sensing, our bodies can now operate as primarily interactive interfaces with the world through the use of interactive systems such as "vision-based eye-gaze tracking" for Human Computer Interface (HCI) that can be attached to the body.

The necessity for such developments is unquestionable. Firstly, they allow us in the process of design to go beyond simply replicating the human body as a static form in order to truly track and understand the dynamic nature of the body as a living and changing entity, which is able to respond to various internal and external stimuli. Secondly, by exploring material intelligence, they allow us to move on from traditional robots to "soft robotics" that have become an increasingly popular area of research as a sub-branch of biomimicry.[1] This is important given that "traditional robots have rigid underlying structures that limit their ability to interact with their environment"[2] (Trivedi et al. 2008). Thirdly, they illustrate how our garments could behave as a second skin capable of changing their shape and operating as an interface with the world, thus influencing social issues such as intimacy, gender, and even personal identities. Lastly, and more importantly, this approach can provide a framework that can contribute to the field of interactive architecture by addressing the emerging field of shape-changing 3D printed structures and interactive systems.

4 Study of Auxetic behavior of 3D
   Printed prototypes with cellular
   structures using two materials.

5 Various topologies and taxonomy
   of geometries and material
   combinations.

5

CARESS OF THE GAZE: A GAZE ACTUATED 3D PRINTED BODY ARCHITECTURE

What if our outfits could recognize and respond to the gaze of the other? Caress of the Gaze is an interactive 3D printed garment, which can detect the gaze of other people and respond accordingly with life-like behavior. This project offers a vision of the future by exploring the possibility of designing a second skin, fabricated using multi-material 3D printing and enabling the wearer to experience and "feel" the most subtle aspect of social interaction: people's gaze (Figure 1).

The design process behind this project can be broken down into three different sections: morphology (shape), actuations (shape-changing mechanism) and interaction (control). As such, the task is to understand the relationship between morphology, behavior, and interactive systems of control.

Morphology: Multi-Material 3D Printing

Many natural materials have exceptional mechanical properties through the organization of their cells or fibers. (Ashby et al. 1995) As such, natural materials are often superior to man-made materials. The study of natural systems can therefore show how materials can be distributed to allow for a desired mechanical behavior. In this sense, the design of forms capable of move-ment based on their material composition can be inspired by the elastic mechanisms in biological systems. In the human body, there are many examples of flexible structures made of springy, soft tissues to be found both in active elastic elements, such as muscles, and in passive tissue elements, such as skin. As Roberts

and Azizi put it, "Elastic mechanisms are likely to play a significant role in all vertebrate locomotors systems" (2011). Understanding the morphology and functionality of elastic structures in nature can lead to new concepts in the realm of design. Ultimately, these designs can be fabricated using 3D printing technologies.

3D printing includes a range of techniques, such as selective laser sintering (SLS), selective laser melting (SLM), stereolithography (SLA), fused deposition modeling (FDM), multi met modeling 3D printing (MJM), electron beam melting (EBM), laminated object manufacturing (LOM), and PolyJet Connex500 printing. For the fabrication of Caress of the Gaze, Polyjet Connex 500 printing was used (Figure 2). This technology has made it possible to generate composite materials with varying flexibilities and colors, and to combine materials in several ways, which enables the simultaneous use of two different materials (flexible and stiff) or any combination of them known as "digital material." [3] This has allowed forms to be fabricated with elastic properties by creating digital materials with full Shore scale.[4] Besides its comparatively large bed size, this high-resolution machine has an accuracy of up to 16 microns. After being extruded, each photopolymer layer is cured by UV light. Gel-like support material can be removed with pressurized water.

The design of Caress of the Gaze was inspired by the system of scaling in animal skins, which is composed of both stiff and flexible tissues (Figure 3). However, the material composition of the Caress of the Gaze was simplified to (a) a cellular mesh varying in its properties in various nodes ranging from stiff to

6

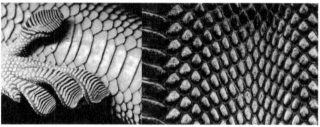

7

architecture, "with a hard, mineralized outer layer and a softer inner layer composed of collagen fibers arranged in a Bouligand, or plywood-like pattern" (2011). What is very interesting about fish scales is their capacity to protect against predator penetration while also allowing for the uniquely flexible mechanical behavior of fish locomotion. However, in this project, the main focus has simply been on the characteristics and flexible behavior of natural skin systems.

For the design of the elastic component of Caress of the Gaze, a cellular mesh, which can hold scale-like members, was deployed. The cellular mesh is able to provide sufficient porosity and therefore desirable contraction/expansion behaviors. Initially, a series of experiments was conducted in order to understand the flexibility of the mesh and its scales by changing parameters such as density, as well as material properties of the mesh (particularly with respect to Shore values and their influence on curvilinear motion). This allowed the mechanical behavior of the mesh to be fine-tuned. The cellular mesh also controlled the scale position and overlap and thus the scale-scale interactions. Meanwhile, control of scale sizes and their softness/stiffness was also essential in terms of the distribution of forces as well as aesthetic expression (Figure 5). As Browning et al. demonstrate, the overlapping of the scales in fish skin is crucial for the distribution of forces across the material, making it possible to tailor the overall stiffness of the material by adjusting structural parameters such as the scale overlap, orientation, aspect ratio, and volumetric filling within the substrate. (2013) Therefore, a series of experiments was also conducted exploring the different

soft, which holds (b) semi-rigid, stiff scales in order to provide maximum curvilinear contraction and expansion. This was a highly iterative and hands-on process whereby printed samples were continuously tested and catalogued (Figure 4). Various geometrical and material combinations were also explored in order to enhance the aesthetic expression of the form as well as to control the types of motion it could afford. It is important to note that the composition of natural biological skin morphologies varies across the body, with varying densities and porosities providing different behaviors. For example, the microstructure of the skin on our face is different to the skin on our feet. It is also worth mentioning that a particular cell composition and distributed density will inform a specific behavior in biological skin. For example, fish scales—although quite hard—are located on a semi-flexible mesh, which provides a certain flexibility for the fish to move and bend its body in various directions. Therefore, the main intention here was to study natural skin systems in order to develop a formal language that exhibits different behaviors ranging from stiff to flexible in various parts of the body.

As discussed by Lin et al., fish scales display a hierarchical

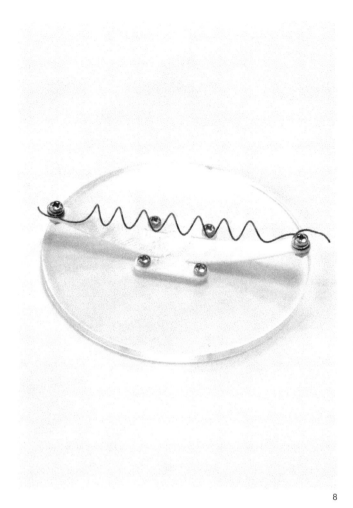

ratios of flexible to stiff materials on the scale-like members. While the flexible material (Shore 60 Black in this case) provides flexibility to the entire structure, the stiff material (Vero White in this case) provides structural rigidity and stability.

The lessons from the aforementioned experiments were then applied to the form of the garment, so as to allow it to move, open, close, and change its shape based on stimuli from the onlooker's gaze (Figure 6).

Actuation System: Shape Memory Alloy

The actuation system for Caress of the Gaze was inspired by animal/human skin and its complex architecture—the interplay of muscles, hair, feathers, quills, scales, etc.—and endowed with life-like behavior by incorporating Shape Memory Alloy (SMA) actuators as a main shape-changing mechanism (Figure 7). This project therefore explored the potential of a shape-changing actuation system assembled as a form of muscle system using SMA that informed the motion of the 3D printed quills.

Muscle functions have inspired numerous researchers and scientists to explore the potential of developing smart materials embedded with soft and compliant actuators such as EAPs (Electro-Active Polymers), SMAs, and so on. We refer to materials that can be significantly changed in a controlled fashion by external stimuli—such as stress, temperature, moisture, electric, or magnetic fields—as "smart materials." As Coelho notes, "Smart

8

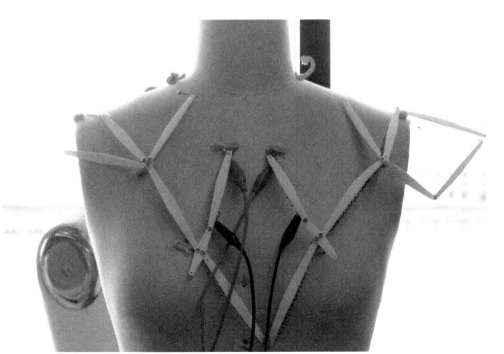

6 Morphologies applied to the form of the garment.

7 Inspiration from natural skin systems such as fish scales and geco skin.

8 SMA assembled to the PLA 3D Printed prototype to develop a compliant actuator. Exploring material properties as a mean to compensate for the nonequal cooling and heating curvature.

9 Study of the motion using SMAs and cataloguing behaviors.

9

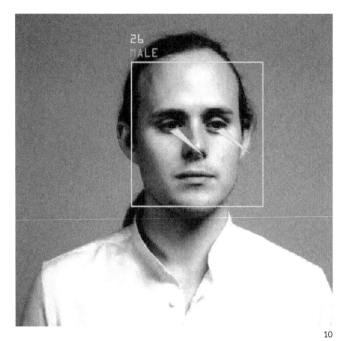

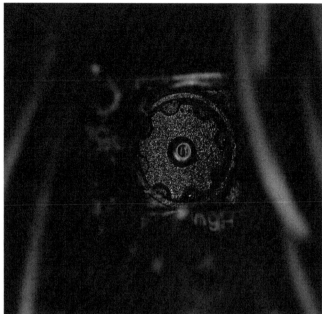

10

11

materials and their composites are strategically positioned to fulfill this desire by transforming input stimuli into controlled material responses, while presenting a wide range of material properties and behaviors" (2008, 23).

The intention here behind deploying smart materials was to embed them within a 3D printed structure whose inherent properties could be changed to meet dynamic external conditions. One possible solution for creating a living system is to develop prototypes using "smart materials" and define their behavior and form based on different stimuli. For this research, SMAs were chosen. SMAs change their shape when heated to the required activation temperature and subsequently return to their initial state. In other words, SMAs can be deformed and then "remember" their original shape when triggered by a specific activation temperature. SMAs are often manufactured as wires. Since these are small, strong, durable, and easy to trigger, they can be used in many products. And since they move silently and organically, they could be a good medium for developing life-like movements. These small diameter wires contract typically 2% to 10% of their length. The diameter of the wire is one of the really important factors in terms of their actuation. Higher-diameter wires have lower resistance and need more power. Thus, they are more likely to overheat and lose their original contraction abilities. Wires that are 0.006" or smaller can be charged constantly without fear of overheating. High-temperature (90°C) wires take longer to cool than low-temperature (70°C) wires.

One of the main challenges in working with SMA wires is that their cooling and heating curvatures are not equal. As Paik,

Hawkes, and Wood explain, when working with SMA actuators, there is no standard model that prescribes temperature, load, and material geometry for a desired performance; the convention therefore is to "derive an actuator model's thermo-mechanical properties experimentally, either partially or fully" (2010). A considerable amount of time in this process was dedicated to fine-tuning the desired motions, including bristling, swirling, and changing shapes (Figure 8) in various sections of the garment and cataloging their behaviors (Figure 9). Please note that although in this project the actuation system was embedded in a post assembly process, the possibility of 4D Printing and printing with smart materials could be explored in the future.[5]

Interaction: Vision-Based Eye Gaze Tracking

Thirdly, Caress of the Gaze explored how our clothing could interact with other people as a primary interface using computer vision technologies.

While HCI researchers have focused on complex mechanisms for acquiring precise data from the eye tracking technologies as a mean of developing human computer interaction, designers can now take advantage of these technologies to design systems that go beyond simple 2D digital interfaces. Therefore, this section is less concerned with adding new scientific findings per se to the field of computer science, and more with introducing a new design application into the field of interactive dynamic design. This might also provide a great opportunity to link recent research in HCI to developments in interactive wearable design.

Caress of the Gaze aimed to create an interactive system which

12

10 Caress of the Gaze equipped with facial tracking camera, detecting age, gender and gaze orientation of the onlookers.

11 Small tracking camera -with lens smaller than 3mm embedded underneath the quills.

12 Caress of the Gaze: A gaze actuated 3D printed garment.

would make the garment become an extension of the wearer's skin and respond to external stimuli in a manner not so dissimilar to some of our involuntary skin responses such as pupil dilation or goose bumps. Our skin is dynamic and constantly in motion. It expands, contracts, and changes its shape based on various internal/external stimuli. In animals, we can see more or less similar responses, particularly for mating or defense mechanisms. For instance, a porcupine feeling threatened might bristle its quills, or a bird trying to find the right mate might puff up its feathers, or fish trying to camouflage themselves from predators might change their skin colors.

The aim of Caress of the Gaze was to establish an interactive autonomous system enabling the wearer to "feel" the invisible and the most subtle aspects of our social interaction: the gaze of other people. This process is not dissimilar to way that we develop goose bumps on our skin as a natural involuntary response to the cold or even to emotions such as fear, nostalgia, sexual arousal, and admiration.

For this purpose, an image sensing camera with a lens of less than 3 mm, capable of detecting gender, age, and orientation of the onlooker's gaze, was positioned underneath the garment quills. The garment then responds autonomously to a front-mounted camera as it detects the orientation of an onlooker's gaze. Basically, the data related to the orientation of the onlookers' gaze is relayed to a Teensy microcontroller on the back of garment, which is capable of actuating and controlling up to eight various nodes of SMAs in the garment (Figures 10 and 11). Technically speaking, this was achieved by mapping the yaw and pitch values

of the onlooker's eyes to the garment actuators and allowing the garment to move in response to the onlooker's gaze.[6]

In a sense, by using facial tracking algorithms, this outfit, with its responsive skin of 3D printed spikes, opens up numerous possibilities for the interaction of the wearer to the environment and people around. First of all, it increases the awareness of wearers to their social context by being able to literally "feel" the garment on their bodies as they move according to the onlookers' gaze. Although, at the moment, this garment is equipped with one front-mounted camera, further developments could also incorporate back-mounted cameras in order to increase security and awareness for the wearer. Secondly, it allows the wearer to communicate to onlookers that their actions have been perceived, thereby allowing them to ward off any further disturbing actions. However, the full potential of such a system is not solely to prevent onlookers from staring, but also in some cases to attract desired onlookers. As mentioned above, information on age and gender can also be gathered. As a result, if one is willing to use this interactive garment as a mechanism of attraction, one can possibly define the desired age and gender and the garment would just respond by attracting the attention of that desired group.

CONCLUSION

Caress of the Gaze addresses the challenge of developing an interactive shape-changing body architecture through the design of a 3D printed dynamic garment operated by a vision-based eye-gaze tracking system that bridges the worlds of fashion, architecture, technology, and design.

Traditional static, single-state 3D printing has now given way to printing flexible, multi-state technologies employing a range of different materials, which can provide opportunities for physical shape changing through the distribution of those materials. The project has also demonstrated the novel ways of interacting with our environment through the use of eye-gaze tracking technologies. Just as our skin can respond to various internal and external stimuli, so too our outfits can now do the same and also respond to various social issues such as intimacy, privacy, gender, and identity (Figure 12).

Therefore, by implementing 3D Printing, SMA, and gaze-tracking technologies, this project has served as a provocation, encouraging us to rethink both the production of the garment and also the way that we interact with the world around us. It has also been an attempt to demonstrate a new coupling of our biological body with a nonorganic garment as an extension of our skin. It has done so in the belief that by implementing design/motion principles inspired by natural systems, both in terms of morphology and behavior, we might be able to rethink the relationship between our bodies and the surrounding environment. Even though this research is still speculative, it opens up the possibility of a radical new approach to interactive dynamic design, particularly in the field of interactive clothing, interactive architecture, and soft robotics.

ACKNOWLEDGEMENTS

This project was made with support from Autodesk/ Pier9 and the MADWORKSHOP. Special thanks to Paolo Salvagione (mechanical consultant), Sebastian Morales and Julian Ceipek (coding consultant).

NOTES

1. "Soft robotics" is a new field of research in robotics. As Verl, Albu-Schäffer, Brock and Raatz explain, thanks to bio-inspired technology design, instead of implementing the rigid mechanical structures of the past, a new robotics paradigm is now starting to focus on soft, pliable sensitive, organic representations on soft robotic.

2. The use of smart material and compliant actuators in this research was an attempt to move away from a mechanical system of actuators such as servos and motors toward soft robotics.

3. For further information on digital materials see: http://www.stratasys.com/materials/polyjet/digitalmaterials

4. The Shore Scale was invented by Albert Ferdinand Shore, who developed a device to measure Shore hardness in the 1920s. The term 'shore durometer' is often used as a measure of hardness in polymers, elastomers, and rubbers. http://www.matweb.com/reference/shore-hardness.asp

5. For 4D Printing and Self-Assembly see the research of the Self-Assembly Lab at MIT. http://www.selfassemblylab.net/

6. Yaw value refers to horizontal angel between onlookers gaze to the center of camera lanes. This value can be measured in degrees. Meanwhile, pitch value refers to the vertical angle between the onlookers' gaze and the center of the camera lens.

REFERENCES

Ashby, M. F., L. J. Gibson, U. Wegst, and R. Olive. 1995. "The Mechanical Properties of Natural Materials. I. Material Property Charts." *Proceedings: Mathematical and Physical Sciences* 450 (1938): 123–140.

Bar-Cohen, Yoseph, ed. 2006. *Biomimetics: Biologically Inspired Technologies.* Boca Raton, FL: CRC Press.

Browning, Ashley, Christine Ortiz, and Mary C. Boyce. 2013. "Mechanics of Composite Elasmoid Fish Scale Assemblies and Their Bio-Inspired Analogues." *Journal of the Mechanical Behavior of Biomedical Materials* 19: 75–86.

Clark, Andy. 2003. *Natural-Born Cyborgs: Mind, Technologies, and The Future of Human Intelligence.* New York: Oxford University Press.

Hansen, Mark B. N. 2006. *Bodies in Code: Interfaces with Digital Media.* New York: Routledge.

Haraway, Donna J., Chris Hables-Gray, Ron Eglash, Manfred E. Clynes, Alfred Meyers, Robert W. Driscoll, Sarah Williams, Motokazu Hori, Monica J. Casper, and Philip K. Dick. 1995. The Cyborg Handbook. 1st edition. New York: Routledge.

Le, Chan Hoang, Quang Sang, and Hoon Park. 2012. "A SMA-Based Actuation System for a Fish Robot." *Smart Structure and Systems* 10 (6): 501–515.

Levine, Simon P., Jane E. Huggins, Spencer L. BeMent, Ramesh K. Kushwaha, Lori A. Schuh, Mitchell M. Rohde, Erasmo A. Passaro, Donald A. Ross, Kost V. Elisevich, and Brien J. Smith. 2000. "A Direct Brain Interface Based on Event-Related Potentials." *IEEE Transactions on Rehabilitation Engineering* 8 (2): 180–5.

Lin, Y. S., C. T. Wei, E. A. Olevsky, and Marc A. Meyers. 2011. "Mechanical Properties and the Laminate Structure of Arapaima Gigas Scales." *Journal of the Mechanical Behavior of Biomedical Materials* 4 (7): 1145–1156.

Norman, Donald. 1998. *The Invisible Computer: Why Good Products Can Fail, the Personal Computer Is So Complex, and Information Appliances Are the Solution.* Cambridge, MA: MIT Press.

Oxman, Neri. 2011. "Variable Property Rapid Prototyping." *Journal of Virtual and Physical Prototyping* 6 (1) 3–31.

Paik, Jamie K., Elliot Hawkes, and Robert J Wood. 2010. "A Novel Low-Profile Shape Memory Alloy Torsional Actuator." *Smart Materials and Structures* 19 (12). http://dx.doi.org/10.1088/0964–1726/19/12/125014

Parkes, Amanda. 2008. "Phrases of the Kinetic: Dynamic Physicality as a Construct of Interaction Design." PhD Thesis Proposal, Massachusetts Institute of Technology.

Raviv, Dan, Wei Zhao, Carrie McKnelly, Athina Papadopoulou, Achuta Kadambi, Boxin Shi, Shai Hirsch, Daniel Dikovsky, Michael Zyracki, Carlos Olguin, Ramesh Raskar, and Skylar Tibbits. 2014. "Active Printed Materials for Complex Self-Evolving Deformations." *Scientific Reports* 4. doi:10.1038/srep07422

Roberts, Thomas J. and Emanuel Azizi. 2011. "Flexible Mechanisms: The Diverse Roles of Biological Springs in Vertebrate Movement." *Journal of Experimental Biology* 214: 353–361.

Trivedi, Deepak, Christopher D. Rahn, William M. Kier, and Ian D. Walker. 2008. "Soft Robotics: Biological Inspiration, State of the Art, and Future Research." *Applied Bionics and Biomechanics* 5 (3): 99–117.

Verl, Alexander, Alin Albu-Schäffer, Oliver Brock, and Annika Raatz, eds. 2015. *Soft Robotics: Transferring Theory to Application.* Berlin: Springer.

IMAGE CREDITS

Figure 1,10,11,12: Charlie Nordstrom and Elena Kulikov, 2015.
Figures 2–10: Behnaz Farahi, 2015.

Behnaz Farahi is a designer exploring the potential of interactive environments and their relationship to the human body working in the intersection of architecture, fashion, and interaction design. She also is an Annenberg Fellow and PhD candidate in Interdisciplinary Media Arts and Practice at USC School of Cinematic Arts. She has an Undergraduate and two Masters degrees in Architecture.

Hybrid Sentient Canopy

An implementation and visualization of proprioceptive curiosity-based machine learning[1]

Philip Beesley
Living Architecture Systems Group, University of Waterloo

Zeliha Asya Ilgün

Giselle Bouron
CITA / Royal Danish Academy of Fine Arts

David Kadish

Jordan Prosser
Philip Beesley Architect Inc.

Rob Gorbet

Dana Kulić
Living Architecture Systems Group, University of Waterloo

Paul Nicholas

Mateusz Zwierzycki
CITA / Royal Danish Academy of Fine Arts

1

ABSTRACT

This paper describes the development of a sentient canopy that interacts with human visitors by using its own internal motivation. Modular curiosity-based machine learning behaviour is supported by a highly distributed system of microprocessor hardware integrated within interlinked cellular arrays of sound, light, kinetic actuators and proprioceptive sensors in a resilient physical scaffolding system. The curiosity-based system involves exploration by employing an expert system composed of archives of information from preceding behaviours, calculating potential behaviours together with locations and applications, executing behaviour and comparing result to prediction. Prototype architectural structures entitled Sentient Canopy and Sentient Chamber developed during 2015 and 2016 were developed to support this interactive behaviour, integrating new communications protocols and firmware, and a hybrid proprioceptive system that configured new electronics with sound, light, and motion sensing capable of internal machine sensing and externally-oriented sensing for human interaction. Proprioception was implemented by producing custom electronics serving photoresistors, pitch-sensing microphones, and accelerometers for motion and position, coupled to sound, light and motion-based actuators and additional infrared sensors designed for sensing of human gestures. This configuration provided the machine system with the ability to calculate and detect real-time behaviour and to compare this to models of behaviour predicted within scripted routines. Testbeds located at the Living Architecture Systems Group/Philip Beesley Architect Inc. (LASG/PBAI, Waterloo/Toronto), Centre for Information Technology (CITA, Copenhagen) National Academy of Sciences (NAS) in Washington DC are illustrated.

1   Produced at CITA, this experimental testbed focuses on modeling and form-finding through laser-cut acrylic thermoforming, integrated with computational controls and proprioceptive functions. The latter functions support cycles of sensor and actuator data which enables it as a physical environment for machine learning.

INTRODUCTION

This paper provides a detailed exposition of physical testbeds, electronics, and software developed as proof-of-concept prototypes for curiosity-based responsive architecture. An interaction algorithm is implemented using distributed machine learning, supported by a distributed system of custom microprocessor electronics hardware integrated within interlinked cellular arrays of sound, light, and kinetic actuators and proprioreceptive sensors in a resilient physical scaffolding system. The sentient canopy system is supported by network communications and real-time visualizations of complex system dynamics. Specialized craft and problems associated with implementation of machine-based curiosity, framed within a general context of complex systems research are described here.

The Living Architecture Systems Group/Philip Beesley Architect Inc. (LASG/PBAI, Waterloo/Toronto) and Centre for Information Technology (CITA, Copenhagen cooperated during 2015 and 2016 to explore machine-based motivation for responsive systems housed within architecture-scale prototype structures. Machine-based motivation for the responsive architecture prototypes was formulated as an internal drive to satisfy curiosity. Workshops staged in November 2015 and February 2016 were staged for development of firmware, software, communications, and coupled dynamic visualizations. Fundamental questions were explored within this cooperative development: how can we design kinetic, living architecture that engages with visitors during extended interactions and enhances human experience in an immersive environment? How do humans respond to these evolving interactions, in a process of mutual adaptation? Such environments present novel challenges of finding new methods of presentation and representation supporting further exploration and study.

Responsive kinetic architecture requires support scaffolds, control systems applicable to distributed organizations, mechanism and hardware configurations, communication and networking systems, software-based controls, and algorithms for learning, adaptation and human-machine interaction (Spiller 2008; Fox and Kemp 2009; Bullivant 2006). The hybrid physical environments and control systems being engineered by LASG with CITA are accompanied by conceptual models rooted in non-equilibrium systems (Nicolis and Prigogine 1977; England 2012). This emerging field is characterized by systems inde-terminacy, requiring cycles of development that test potential combinations of assemblies and interdisciplinary working methods in practical implementations with public occupants.

The current CITA and LASG testbeds are related to a new generation of prototypical spaces developed by LASG containing large arrays of actuators and sensors that are linked together by networks of nodes (Beesley 2010) Specialized development of individual systems within this research includes the iterative whole-systems development of an evolving physical testbed, serving as the integrative platform for individual elements. These testbeds mature into full-scale public prototypes, usually housed in museums or galleries, where observation and data collection allow the validation of prototype function and resilience, responsive intelligent behaviour models, and hypotheses around occupant interaction and reaction.

The testbeds integrate arrays of interconnected, interactive components within lightweight kinetic scaffolds and massively distributed proprioreceptive sensor networks. The vital aspect of proprioreception in these sensor networks is a particular feature of new LASG work providing feedback between sensing and mechanisms and setting the stage for machine learning. Computational functions include layered communication between nodes and interactive curiosity-based learning. The combination of these computational and physical systems creates substantial complexity and unpredictability. Interactions with sensors organized within individual nodes influence the behaviours of actuators within the local node, adjacent nodes, and also within global organizations of the system . Sensors vary from simple range detectors to a vision system with image processing and pattern recognition. Actuators vary from simple LED lights to shape memory alloys and pneumatic actuators with complex dynamics that challenge normal methods of modeling. Interaction between human occupants and neighbouring nodes increase the unpredictability and complexity of behaviours. By studying the technologies and implications of these layered, interdependent systems, insight can be gained that contributes to new discourse examining complex systems and interconnectedness.

A key question when designing interactive installations and environments is the design of the interactive behaviours. The use of learning algorithms for automatically generating behaviours is an active research topic in developmental robotics (Clune et al. 2009) but is still at early stages of implementation within architectural construction systems. In early work by the LASG group, interactive behaviours were designed as reactive programs organized by coupling between distributed nodes. Stochastic variations were included, but those machine-generated random signals were not configured to feed back into creation of new behaviours (Beesley 2010). Those programs included capture of system-wide behaviours, including rapid responsiveness and diffusion over space but they did not include the capability to change over time. The pursuit of adaptation is now a priority for test-beds installed within locations permitting long durations of occupation and study. Interactive behaviours which have the ability to change over time and adapt to users could lead

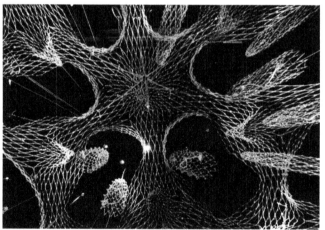

to hybrid couplings (Cairns-Smith 1982). To this end, we are investigating the use of curiosity-based learning algorithms to automatically generate interaction behaviours.

Curiosity-based learning algorithms were originally proposed in the field of developmental robotics, to enable robots to learn about their capabilities and the environment, emulating childhood development (Oudeyer, Kaplan, and Hafner 2007). We have adapted these algorithms to distributed interactive installations (Chan et al. 2015). The interactive system initiates interaction with and responds to human visitors by using its own internal motivation, formulated as an internal drive to satisfy curiosity. This enables the system to generate continuously evolving interactive behaviours and move beyond purely reactive interaction towards shared-initiative interaction.

CONTEXT

A particular focus of the research is the algorithmic development of intelligent behaviour controlling kinetic, light and sound-based responses embedded within these environments. The use of machine learning for autonomous robot behaviour generation has been an active topic in the developmental robotics community (Peteiro-Barral and Guijarro-Berdiñas 2013). For example, the Playground Experiment performed by Oudeyer, Kaplan, Hafner, and Whyte (Oudeyer, Kaplan, and Hafner 2005) was based on a curiosity-driven learning algorithm which tries to select action that can potentially minimize the prediction error in the same sensorimotor context. They implemented and validated the algorithm on a Sony AIBO robot which only had three sensors and allowed three types of actions. They showed that the robot was able to discover complex behaviours which lead to improved knowledge and sensorimotor ability.

Curiosity, in this context, is formulated as a driving force for the learning algorithm to explore motor functions that offer the highest potential for increased knowledge. This means that the learning algorithm is incentivised to explore regions of sensorimotor space that are neither well-known nor seemingly random. One of LASG's research challenges is to develop a similar learning algorithm that uses this sense of curiosity and apply it to a distributed system with a large number of sensors and actuators. Due to the distributed nature of the system, the algorithm must also deal with the sharing of information among different nodes in the system. Distributed machine learning techniques enable learning from multiple sites while avoiding the need of transferring a large amount of data to be processed centrally in one processor.

CANOPY PROTOTYPE

The canopy prototype consists of four interdependent layers: a physical structure that forms the space as a forest of deeply

2   Elevation view of interactive canopy from below, showing scaffold and LED cluster adjacent to vibrating sensor mechanisms.

3   LASG installations are supported by a hybrid triangular flexible space-grid structure that provides flexibility and stability through tensile and compressive support. (Musée des Beaux Arts, Montreal, 2007).

4a Interface between the laptop-based CBLA learning module and the Teensy microcontroller board.

4b Modular software configuration for curiosity-based learning; expert systems associated with the CBLA are overlaid with flexible occupancy mapping and with pre-scripted behaviours.

5   Overview of the Series 3 hardware systems.

Hybrid Sentient Canopy Beesley et al.

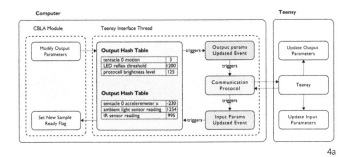

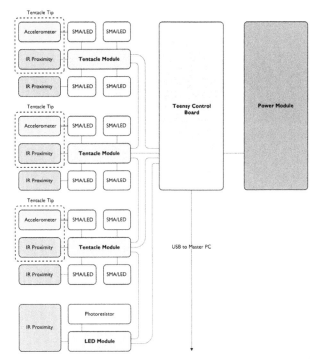

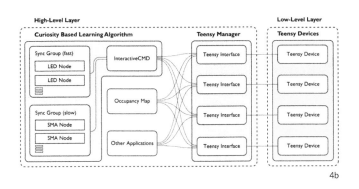

4a

4b

5

interwoven material; an electronic system that includes sensors, actuators, and microcontrollers; firmware that provides low-level functionality at the microcontroller stage; and software that is executed on a remote computer and provides higher-level intelligence. Together, these layers form a meshwork characterized by resilience and hybrid coupling (Beesley and Macy 2010). The new structural system is organized by a hybrid triangular flexible space-grid, stiffened by expanded-mesh hexapods that support telescoping posts and spires contacting the floor and ceiling for stability. Tensegrity coupling is featured, employing metal rod cores that stabilize the system surrounded by expanded meshwork hyperbolic shells that provide alternating tensile and compressive support. This structure offers minimal material consumption, achieved through efficient digital manufacturing employing laser machining and thermal forming of expanded meshwork.

The electronics form the basis of the interactive system that imbues the canopy with a prototypical sentience. A portion of the canopy's primitive intelligence is distributed throughout its structure in a series of connected nodes. These nodes are each built around a Teensy 3.1/3.2 microcontroller, an Arduino-compatible platform that provides the low-level interface to an array of sensors and actuators. These sensors and actuators form the canopy's perceptive and reactive body and are configured to allow the canopy to perceive and influence its own physiology as well as its environment. This ability to self-sense is called proprioception and is an essential capacity for an intelligent system that can understand itself within the context of its environment.

Proprioceptive signals—both sensing and actuation—are first produced at the firmware level. Firmware refers to the portion of code that is written for and executed on the Teensy microcontroller. It is installed onto every node and performs low-level processing of actuation commands and sensor data. Simple behaviours, such as the generation of fades and activation envelopes for the various actuators, are produced at the firmware level. Furthermore, sound production is handled entirely by the firmware, though the actual triggering of sound can be communicated from the software layer. The firmware also plays the role of translating raw sensor data into filtered and error-corrected signals that can be used by the software as representations of the sensed environment.

The supervisory layer is executed on a remote computer and implements the learning algorithm, named the curiosity-based learning algorithm (CBLA) . The CBLA engine aims to learn about the relationship between its actions and its sensory observations. At each time step, the engine considers a set of behaviours and predicts the outcome of each of the behaviours. It then chooses one of the behaviours to execute, observes the outcome and compares its prediction to its observations. The agent's curiosity is highest for those actions which are likely to lead to decreasing prediction error. This means that it specifically selects actions that are outside of its past experience and for which its predictions of the outcome are mostly likely to be incorrect. This configuration ensures that the canopy and its nodes do not simply settle into routines of recurrent action and response, but continue to find and explore unfamiliar territory.

6

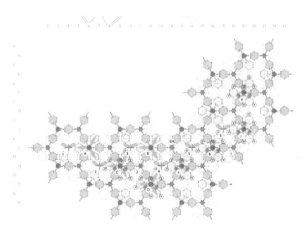

7

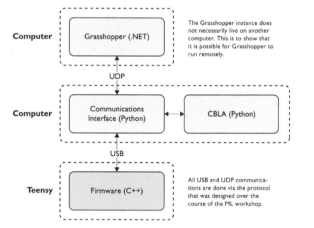

| Computer | Grasshopper (.NET) | The Grasshopper instance does not necessarily live on another computer. This is to show that it is possible for Grasshopper to run remotely. |

UDP

| Computer | Communications Interface (Python) | CBLA (Python) | |

USB

| Teensy | Firmware (C++) | All USB and UDP communications are done via the protocol that was designed over the course of the ML workshop. |

8

Each node learns and explores its own subset of the system. The distributed nodes are coupled to each other both physically, by sharing sensors, and virtually, by considering node outputs as inputs to other nodes in the system. This coupling creates the potential for neighbour and group behaviour by connecting the perceptive spaces of each node. In addition to running the CBLA learning process, the Python-based software also passes signals from the sculpture over the Internet and local networks via UDP. This UDP connection exposes the data from the canopy for analysis and visualization outside of the CBLA software and opens the canopy's actuators to external control. The data streaming capability is used to drive a Grasshopper-based parametric model for the canopy's 3D Rhino model. This dynamic 3D visualization of the testbed's live operations is a valuable tool for the analysis and study of the canopy's behaviour and its evolution over time.

TESTBEDS

Three new testbeds based on this new control framework are currently installed in the PBAI and CITA studios, and at the National Academy of Sciences in Washington, D.C. In these prototypes, the pre-programmed behaviours seen in the preceding Series 2 (Beesley 2014) have been replaced with

supervisory adaptive behavioural algorithms based on the work of Oudeyer et al. (2007). The aim is to improve the behavioural and perceptual capabilities of the interactive sculpture systems through developing learning algorithms that can acquire novel and engaging behaviours through their interactions with the users.

These algorithms require that the system is able to detect and monitor its own activity as well as changes in the environment. This ability, equivalent to proprioception in animals, plays an important role in enabling the sculpture to generate a model of itself and the world and thereby learn. The current design includes three main types of actuators: high power LEDs, Shape-Memory Alloy (SMA) fin mechanisms, and custom-made speakers. Actuators are paired with proprioceptive sensors in order to provide feedback on their behaviour for the CBLA algorithm. The kinetic SMA actuator is equipped with an accelerometer which provides feedback on movement caused by either the SMAs or external interactions such as a person touching the fin. The LED is paired with a phototransistor that provides feedback on ambient and actuated brightness. The sound modules are paired with nearby microphones that pick up ambient noise as well as sounds produced by the sculpture.

Hybrid Sentient Canopy Beesley et al.

In the case of all of these sensor-actuator pairs, the sensors are able to register a change in their state when their actuator is triggered by the CBLA system. At the same time, they can register environmental changes that move the CBLA system into a new sensorimotor zone that it is able to then explore.

In addition to the sensors that provide feedback on the actuators, there are sensors that provide information about the environment that the system is in. The Infrared (IR) proximity sensors are currently the only type of sensor in this category and are located on the sound modules as well as the tentacle nodes. These sensors provide feedback on the observers in the vicinity of the system. This system offers a unique, physical kind of machine vision that offers complex responsive kinetic functions. For example, occupants interacting with this system could find arrays of individual fronds following their motions accompanied by outward-rippling motions. Increased complexity approaching peer-like playful kinetic responses could, with further development, result from this arrangement. Proprioception necessitates a vast expansion of the sculpture's sensing capabilities in comparison to previous generations of sculpture hardware, with a corresponding radical increase in the amount of data that is generated in each control cycle. The communication system plays a central role in ensuring that all of the data generated by these expanded sensor systems are able to be used for modeling and visualization.

DIGITAL COMMUNICATION

Previous generations of this work within LASG have established a communications protocol designed to facilitate the transmittal of sensor information and control signals between the Teensy-based nodes and the laptop-based CBLA software. The inclusion of a model-based graphical visualization of the system necessitated a re-design of this protocol and the addition of a system for forwarding signals to a remote computer to perform visualization. The new communications protocol uses a predefined message format to execute a set of designated commands on specific end-points within the canopy. For example, the CBLA software may request the value of a sensor at a particular address on a particular node, or it might direct a specific LED to fade out over a given timespan. These commands are purposely not limited to being used in the USB communication system that connects the hardware nodes to the computer running CBLA. Using the protocol it is possible to extend the communications system over other channels.

A new event-based communication protocol was created for this prototype, following a query-response schema. In order to efficiently organize large exchanges of individual messages required by this approach, pre-programmed behaviours are configured

9

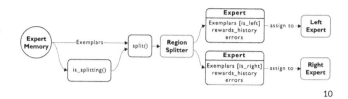

10

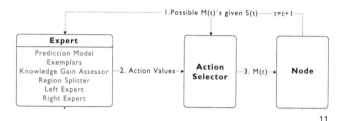

11

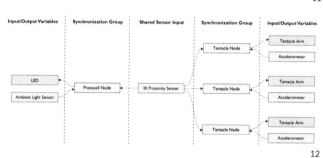

12

6 View of Sentient Canopy scaffold with distributed interactive system within tensegrity coupled expanded-meshwork structure.

7 Organization of devices and electronic systems within Sentient Chamber.

8 User Datagram Protocol ('UDP') provides the interface between the model/visualization space and the CBLA software.

9 S(t+1) and S'(t+1) are the actual and predicted sensor input variables. (1) The KGA computes the error as the root-mean-square of their difference. (2) After that, it computes the mean error over a window of previously calculated errors. Note that the mean error is only calculated based on errors associated with this particular region. (3) The metaM predicts the error by taking the mean error over time. (4) Finally, the KGA outputs the Reward by subtracting the actual mean error from the predicted mean error.

10 When the function "split()" is called, it first checks if the split criteria are met by calling "is_splitting()". If the split criteria are met, it forwards the exemplars to the Region Splitter. The Region Splitter then splits the exemplars into two and assigns them to the Left and Right Expert. Other properties such as the previous errors and rewards are passed on as well.

11 At the start of each time step, the Node outputs the current possible actions M(t)'s given S(t). (2) Then, the Expert provides the Action Values associated with those possible M(t) to the Action Selector. (3) After that, the Action Selector selects an action and actuates the Node. (4) After the Node has performed the action, it returns the actual resultant state S(t+1) to the Expert. The Expert can then improve its internal prediction model and update the Action Value associated with sensorimotor context SM(t).

12 Nodes' connections and groupings.

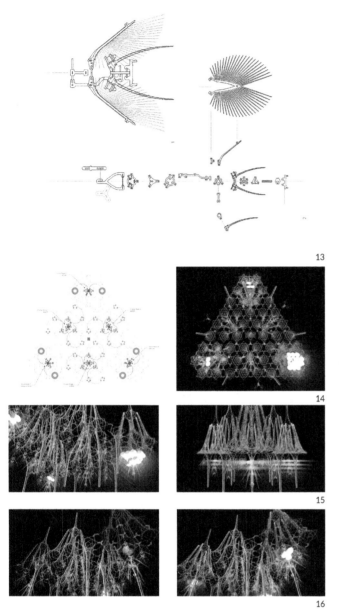

13 Design details of vibrating actuator and sensor assembly, Sentient Canopy.

14 The basis for the visualization (right) is a spatially explicit 3D Rhino model of the canopy (shown in plan on the left). Grasshopper is able to access real-time information about sensors, actuators, and the state of the learning system and translate them into a rich visual representation of the canopy's current state.

15 The colouration of the visualization reflects the learning system's mean error over time in the current sensorimotor context where bluer tones indicate a lower mean error, while redder tones reflect a higher mean error. Higher mean error means that there is more knowledge to be gained, meaning that the system is more curious in that context and will reflect that in its behaviour.

16 A view of the visualization with (right) and without (left) the particle system enabled. The colour spectrum notation for mean error (left) is extended with a particle system to visualize the relationship between mean error, curiosity, and the reward for an action. When the mean error over time (colour spectrum) differs from the mean error calculated for a specific action, a high reward is associated with that action.

for each hardware node within the installation. This nested hierarchical organization shifts a portion of response to local controllers, reducing the necessity for high-frequency messaging with the central CBLA engine and easing the load on the USB communication system. User Datagram Protocol (UDP) provides the interface between the model/visualization space and the CBLA software. UDP is formulated as a low-latency transport layer, making it well suited for near-real time connections and supporting rapid development of new interfaces for the canopy system. The UDP component for Grasshopper is used to link the firmware-software messaging system with the parametric controls that Grasshopper provides for the 3D Rhino model of the system. A custom message parsing component processes messages in the communications protocol format and then uses those values to drive the visualization in the model space. The initial steps described here anticipate significant further development exploring complex interactions and learning processes within these environments.

SOFTWARE COMPONENTS

The software is structured as a modular hierarchy, consisting of a low-level layer and a high-level layer. The two layers are connected physically through USB. A low-level layer of firmware written in C++ runs on the Teensy 3.1/3.2 USB-based development boards which interface with the peripherals that connect with the actuators and sensors. High-level software written in Python runs on a central computer. The use of the central computer as a development platform provides flexibility for development free from the limited processing power and specialized functions inherent to the Teensy microcontroller hardware. Moreover, Python is cross-platform and supports multi-threading, permitting operation within many operating systems and allowing multiple sets of software instructions to be executed in parallel. In this configuration, it is possible to use the connection to the sculpture for multiple simultaneous but independent applications. For example, a CBLA learning engine may read the sculpture data and send control signals, while a visualization is run using the same sculpture data stream.

In previous installations, the interactive behaviours of the sculptures have been pre-scripted. Each node responded to the occupants and influences the behaviours of its neighbouring nodes in deterministic ways. As described above, the software and hardware platforms were updated in the current series in order to allow the more demanding CBLA to be implemented. However, the new systems also support the design and implementation of pre-scripted behaviours that can run alongside or in place of the CBLA learning engine, depending on the desired mode of operation.

COUPLED VISUALISATION

The real-time visualisation of information provides opportunities to observe, analyse or even feed back into and affect a dynamic system. The system-level view that one can attain in the virtual space provides a fundamentally different way of experiencing the canopy from a sensorially immersed presence within it. Through the UDP communication system, Grasshopper is able to access information about sensors, actuators, and the state of the learning system and translate them into a rich visual representation of the canopy's current state. The basis for the visualization is a spatially explicit 3D model of the canopy. As real-time sensor and actuator data are passed to the model, Grasshopper allows for the recording and visualization of the canopy's behaviour.

To enable real-time linking between the physical and virtual, a custom 'listener' software module links a 3D Rhino model to the communications protocol described in a previous section. As events occur in the physical installation, relevant information propagates into the virtual model through a series of interlinked parametric data filters configured within Grasshopper software. In the case of light visualisation, for example, values received from photoresistors positioned adjacent to each LED actuator are decoded. These values are assigned to spatially explicit 3D points corresponding to the location of the particular LED. The model-space below the installation is discretized into a matrix of points, and fall-off from the LED input values is calculated across this point matrix.

With more development, the parsing and spatial mapping process described here can provide a means for the predictive simulation and testing of different behaviours. The eventual aim of a digital representation is not only to visualize the physical effects and learning status but to support an informational space that can act as an interface capable of controlling the physical system. The real-time visualization of phenomena such as light, sound or the learning process can feed back into the physical installation and modify its learning pattern or trigger actions based on emerging patterns discovered within the digital visualization. This can provide an alternate pathway for exploration of the canopy for both researchers and participants and open exciting new modes of understanding the drivers behind the canopy's actions.

DISCUSSION

The responsive functions implemented within the testbeds described here are organized by both prescripted and curiosity-based behaviours. The prescripted system is deterministic, involving predictable behaviours employing actuation tightly coupled to sensor stimulus, linear translations of sensed data into actuation response, propagating into neighbour and global

responses, modified by timings and filtered damping. In addition to sensor stimulus, background behaviours provide orchestrated cycles of actuation. Prescripted behaviours include combinations that include highly responsive behaviours akin to reflexes, directly associated with gestures and positions of the viewer, and also intermittent events occurring within the surrounding environment. Some background behaviours are coupled to viewer activations, with long echoing cycles of response associated with prior action. Complex actuations occur as a result of the interaction of multiple overlapping viewer stimulations in combination with these background behaviours.

In contrast, the curiosity-based system involves exploration by employing an expert system composed of archives of information from preceding behaviours, calculating potential behaviours together with locations and applications, executing behaviour and comparing result to prediction. Machinic decision processes involves searches for new information with reward structures implemented for behaviours yielding new information. When new actions become reliable with stable stimulus, the reward recedes and the system narrows its response, stabilizing its searching functions. It then eventually shifts to searching for new behaviour. Changes in sensor values tend to introduce new sensorimotor context placing the system in new states for exploration. When a viewer triggers a sensor, initial responses involve diverse possible reactions within the system at widely varying levels. If a viewer maintains a stable behaviour with relatively consistent sensor stimulus, activity levels associated with that stimulus will eventually drop, corresponding to machinic understanding of its new environment. If a viewer continues to explore with widely varying stimulus, the system may continue to explore the context with a wide spectrum of potential behaviours.

While both prescripted and curiosity-based responsive systems offer open-ended exploration for viewers, experience of primary implementations of these two states tends to be widely polarized. Notwithstanding the prescripted behaviours embedded within these environments, the open-ended spatial field offered to viewers appears to support highly involved responses by viewers, who tend to associate the environments with empathetic, mutual responses.. In contrast, the behaviour manifested by machinic curiosity can be perceived by viewers as random, even alienating. However, when viewers interact with sustained attention to this implementation of curiosity-based machinic interaction, certain patterns can become apparent. When viewer behaviours are held to highly predictable states, the expert system guiding machine selection of behaviours is configured to connote relatively comprehensive 'understanding' and in turn actuations tend to cease. Memory functions are configured within the expert system, imparting states of 'forgetting' that,

17 Detail view of actuated prototype cells and reflex sensors within Sentient Chamber testbed, National Academy of Sciences, Washington.

over time, allow the environment to re-explore and re-engage viewers with interactive cycles.

Next stages of development would involve hybrid coupling in which prescripted behaviours could be combined and adapted by curiosity-based learning. We anticipate that this combination would impart visibly coherent nuanced action to the system, employing reflex-based responses enriched by machinic exploration.

CONCLUSION

The Sentient Canopy implementation described here integrated the new features rendering the system capable of both self-reflexive internal sensing and externally-oriented sensing for human interaction. A number of specialized technical refinements were developed during this phase of research covering key parts of the system including communication, firmware, electronics, and proprioceptive sensing. These innovations help support precise manipulation of this relatively unstable software routine and open the system to new methods for visualization and analysis. The communication system was re-implemented to use an extensible protocol with data transmitted over a USB connection.

A modular, object-oriented framework was developed for firmware that allows for rapid extension and configuration of nodes. An Arduino-compatible node-based electronics system was designed and implemented that exposes dense sensing and actuation capabilities to a flexible array of sub-node (device) level modules allowing for the inclusion a wide range of low-voltage peripheral devices including shape-memory alloy and DC motor-based mechanisms, amplified sound, and high-power LED lighting. Proprioception was implemented by producing custom electronics serving phototransistors, microphones, and accelerometers for motion and position, each coupled to sound, light and motion-based actuators and additional infrared sensors designed for sensing of human gestures. This configuration provides the machine system with the ability to calculate and detect actual behaviour, and to compare this to behaviours predicted by the system's knowledge to date, allowing the system to explore and experiment with new behaviours. The configuration supports machinic introspection.

Physical interconnection of light, sound and motion-based sensor and actuator systems can result in hybrid relationships. Within these testbeds, coupled relationships have been observed in which vibration from sound emissions have resulted in stimulation of accelerometers originally designed to track the motions of adjacent kinetic devices. Similarly, motion-based sequences designed to act in response to sensor stimulus tracking gestures of human occupants have been discovered also resulting from self-stimulation resulting from stray mechanical movements in adjacent mechanisms. Cycling patterns of feedback result, creating emergent behaviour.

Discussion of hybrid couplings has significance for study of evolving life. Canonical discussions such as Cairns-Smith's Genetic Takeover and the Mineral Origins of Life (Cairns-Smith 1982) have suggested that highly circumstantial couplings could well explain key relationships within the complex systems of organisms. By observing and analyzing the patterns of behaviour seen within the sentient canopy prototypes illustrated here, insight might be gained in ways that responsive architectural environments might relate to living systems evolved by natural processes. By coupling the kind of intelligent virtual models described here with their corresponding proprioceptive dynamic physical environments, and by visualizing and analyzing the behaviours that they contain, increasingly complex hybrid relationships can become legible. These visualizations offer practical means for working with the systems indeterminacy that tends to challenge control of interactive systems by designers, and could help support skills for the rapidly emerging field of design of near-living systems.

Hybrid Sentient Canopy Beesley et al.

NOTES

1. The findings presented here are the result of collaborations between the Living Architecture Systems Group (LASG) at the University of Waterloo, the Center for Information Technology and Architecture (CITA) at The Royal Danish Academy of Fine Arts (KADK) and Philip Beesley Architect Inc. (PBAI), drawn from a workshop series in late 2015 and early 2016.

REFERENCES

Beesley, P. et al. 2014. Near-Living Architecture: Work in Progress from the Hylozoic Ground Collaboration. Toronto: Riverside Architectural Press.

Beesley, P. et al. 2010. Kinetic Architectures &Geotextile Installations. Toronto: Riverside Architectural Press.

Beesley, P. et al. 2010. Hylozoic Ground: Liminal Responsive Architecture. Toronto: Riverside Architectural Press.

Bullivant, L. 2006. Responsive Environments: Architecture, Art and Design. Victoria & Albert Museum.

Cairns-Smith, A. G. 1982. Genetic Takeover and the Mineral Origins of Life. Cambridge University Press.

Chan, M.; Gorbet, R.; Beesley, P.; Kulic, D. 2015. "Curiosity-Based Learning Algorithm for Distributed Interactive Sculptural Systems." In 2015 IEEE/RSJ International Conference on Intelligent Robots and Systems (IROS), 3435–41. IEEE.

Clune, J.; Beckmann, B.; Ofria, C.; Pennock, R. 2009. "Evolving Coordinated Quadruped Gaits with the HyperNEAT Generative Encoding." In 2009 IEEE Congress on Evolutionary Computation, 2764–71. IEEE.

Fox, M; Kemp, M. 2009. Interactive Architecture. NJ: Princeton Architectural Press.

Nicolis, G; Prigogine, G. 1977. Self-Organization in Nonequilibrium Systems. CA: University of California.

Oudeyer, P-Y.; Kaplan, F.; Hafner, VV. 2007. "Intrinsic Motivation Systems for Autonomous Mental Development." IEEE Transactions on Evolutionary Computation 11 (2): 265–86.

Oudeyer, PY.; Kaplan, F.; Hafner, VV. 2005. "The Playground Experiment: Task-Independent Development of a Curious Robot." Proceedings of the AAAI.

Peteiro-Barral, D, and B Guijarro-Berdiñas. 2013. "A Survey of Methods for Distributed Machine Learning." In Progress in Artificial Intelligence, Vol. 2, Issue 1. 1–11.

Spiller, N. 2008. Digital Architecture Now: A Global Survey of Emerging Talent. Thames & Hudson.

IMAGE CREDITS

Figure 1, 2: Anders Ingvartsen, 2016
Figure 3–13,17: PBAI, 2015
Figure 14–16: CITA, 2016

CITA is an innovative research environment exploring the intersections between architecture and digital technologies. Identifying core research questions into how space and technology can be probed, CITA investigates how the current forming of a digital culture impacts on architectural thinking and practice. Using design and practice based research methods, CITA works through the conceptualisation, design and realisation of working prototypes. CITA is highly collaborative with both industry and practice creating new collaborations with interdisciplinary partners from the fields of computer graphics, human computer interaction, robotics, artificial intelligence as well as the practice based fields of furniture design, fashion and textiles, industrial design, film, dance and interactive arts.

Living Architecture Systems Group, associated with the University of Waterloo, is an international consortium of pioneering academic, institutional and industry partners in a multidisciplinary research cluster dedicated to developing built environments with qualities that come close to life—environments that can move, respond, and learn, and which are adaptive and empathic towards their inhabitants. The LASG partnership is focused on developing innovative technologies, new critical aesthetics, and integrative design working methods for working with complex environments. LASG's work spans a broad field of specialized disciplines, but is firmly situated within the context of responsive architecture, an emerging sub-discipline, that responds to building occupants. Specializations within the group include advanced structures, mechanisms, control systems, machine learning, human-machine interaction, synthetic biology, and physiological testing.

Philip Beesley Architect Inc. is an interdisciplinary design company located in Toronto, Canada. PBAI specializes in architectural design of public buildings, public art and experimental installations. The group is closely associated with the multi-partner Living Architecture Systems Group, the School of Architecture and Faculty of Engineering at the University of Waterloo, the architectural practice of Pucher Seifert, and Riverside Architectural Press. Sculptural work in the past two decades has focused on immersive textile environments, landscape installations and intricate geometric structures. The most recent generations of these works feature interactive sound, light and kinetic mechanisms with distributed control systems. Studio research focuses on aesthetics, technology and craft of responsive envelope systems including digital fabrication of extremely light-weight, flexible component arrays containing embedded sensors and actuators.

Researching Inhabitant Agency in Interactive Architecture

Sara Costa Maia
AnnaLisa Meyboom
University of British Columbia

1

ABSTRACT

The study of Interactive Architecture (IA) spans over several decades and appears to be gaining increasing momentum in recent years. Yet, inhabitant-centered approaches towards research and design in the field still have a long way ahead to explore. Particularly, we observed that the examination of IA's social relevance in literature is still incipient and ill supported by evidence. The study discussed in this paper is attempting to remediate this gap by exploring one of the first socio-political arguments around the relevance of IA, namely inhabitant empowerment and agency. It investigates whether an inhabitant's relation and experience with interactive spaces, conceived according to different interaction strategies, increases the participants' perception of their own agency in the space. In this paper, we briefly explain the prototyping of an interactive space-plan designed to emulate the behavior of four basic models of interaction. Finally, the paper presents an experimental study set to test inhabitant agency in IA. It concludes that IA has the potential to increase inhabitant agency, but that this is very dependable on the system's design regarding behavior and interaction.

1   Interactive space-plan prototype.

INTRODUCTION: A SOLUTION LOOKING FOR A PROBLEM

Interactive Architecture (IA) can be broadly defined as an architectural setting computationally enabled to sense its environment and respond accordingly in a dynamic feedback system. It is still a field that relates more closely to science fiction than to the mainstream production of architecture. Yet its study spans over several decades and has gained substantial momentum in the last few years, possibly due to the increasing availability of inexpensive and easy-to-use electronic components.

In a previous study (Costa Maia and Meyboom 2015), it was argued that most recent investigations in the field are centered on technological availability, with ad hoc discussion regarding its context in the built environment. Little research exists to date that tries to identify whether IA can be an adequate solution for real-world problems and demands, specially regarding inhabitants' needs. In fact, very little research has been done on inhabitant experience of interactive spaces in general, hindering our ability to justify its use or to properly ground design decisions.

The literature on IA abounds with arguments regarding the social relevance of data-driven adaptable environments (Costa Maia and Meyboom 2015). Each of the several rationales for IA presented in literature require further scrutiny. They address important topics, but they do not empirically demonstrate that IA is an adequate response for them.

This research begins the task of investigating IA's possibility of fulfilling one of its many untested claims. We explore one of the main assumptions which we deem highly pertinent to the domain of IA: inhabitant empowerment and agency. Its selection among others is justified by: 1) its critical participation in early debates in the field, 2) its conceptual relation with fundamental notions of IA, 3) a recent upsurge of interest on (architectural) production democratization, and 4) a personal interest on the political implications of the topic.

IA's initial relation with the problem of inhabitant agency is explained, to large extent, by its foundation in cybernetics. Gordon Pask, a main proponent of the second generation of cyberneticians, introduced the relevance of feedback, systems design, and underspecification in architecture. Through systems thinking, IA would allow for buildings to become components of empowering environments by integrating the human user as part of a larger control loop.

Pask explicitly claimed that, in IA, "the designer is no longer conceived as the authoritative controller of the final product;" instead, "an environment should allow users to take a bottom up

role in configuring their surroundings in a malleable way." Haque (2007) also argued that applying Pask's ideas to architecture "is about designing tools that people themselves may use to construct — in the widest sense of the word — their environments and as a result build their own sense of agency." Authors such as Negroponte (1975) have also extensively framed IA, or computing enabled environments, as primarily concerned with freeing the user from the paternalistic figure of the architect by instead providing agency and responsiveness.

More broadly, interactivity has often been intrinsically associated with the idea of user empowerment. Andrejevic (2009) explains that early advocacy for interactivity in media studies sought to subvert traditional structures that helped reproduce power and social relations.

The extensive discussion around participatory design in architecture since the 1960s proves that this question of who should control the formation of the built environment is an important concern of architecture. Instead of focusing on anticipatory demands, exploring inhabitant agency and empowerment in IA may help address immediate challenges faced by designers.

As already argued, there is a very scant number of studies generated by the IA community which address specific current demands in architecture and which generate evidence regarding the adequacy of specific strategies towards specific problems. Academic work to provide support for decision-making in the design of such systems will become increasingly relevant and potentially determinant on the success of following projects. The lack of such supporting information might be a key reason why IA is largely still limited to speculative manifestations, even several decades after the concept was popularized.

This research is therefore framed to provide an understanding of IA with regard to one specific relevant problem, and in testing the adequacy of IA as a solution for that problem. Only through proper scrutiny and data support may we begin to understand the relevance of IA towards inhabitant agency, or any other rationale of interest.

This paper presents a study of how different models of interaction might influence inhabitants' perception of agency and related concepts. It describes the process of prototyping interaction models and conducting a user experience study on four different forms of system behaviour and user interaction.

WHAT IS INHABITANT AGENCY?

The topic of agency is not a simple one and there is much to say in relation to the subject and its study in social sciences and other fields. However, in the context of this research, a few concepts will be highlighted. Firstly, in its most fundamental meaning, agency can be broadly defined as the capacity of an actor to act on the world. It is also important to clarify the distinction between the ideas of inhabitant agency and architectural agency in IA. In architectural agency it is the architecture itself which must have internal goals and be able to act in their pursuit, and these goals may or may not be aligned with goals of the inhabitants. In such instances, IA is typically approached as intelligent and/or autonomous machines. Several authors have directly or indirectly discussed an idea of agency in IA that refers to architectural agency primarily (Calderon 2009; Adi and Roberts 2010; Jaskiewicz 2013). In this research, however, we focus on inhabitant agency—that is, on how the environment enables people to accomplish goals of their interest.

This research adopts the concept of human agency as it is proposed by the Capability theory and most centrally to the work of the Nobel prize-winning economist Amartya Sen. Sen's definition of "human agency" stands for people's freedom to act in pursuit of whatever goals or values they regard as important (Alkire 2005). It allows them a "systematic ability to achieve high levels of functioning (which may or may not be realised in practice)" (Johnstone 2007). Systematic is the key word in this definition because, contrary to allowing a person accidental access to wellbeing, it ensures that wellbeing may be sustainable and robust across time and context (Johnstone 2007).

Therefore, this research is interested in the social sciences' use of the term agency, which transcends a subject's basic capability of action on the world and studies it with regard to the socio-political structures in which it is inserted. How might an IA apparatus provide the spatial/environmental support to improve an inhabitant's agency with regard to the processes that shape his/her built environment? This kind of argument questions the architect's often paternalist role in designing environments and spatial configurations deemed best for the inhabitant. It also draws an immediate connection with IA, in IA's ability to offer inhabitants a sustained say in the shaping of space, thus potentially allowing systematic access to environmental wellbeing.

In order to assess agency in inhabitants' experiences, this research used measures of empowerment, autonomy (from self-determination theory) and capability (from self-efficacy theory) as proxies for human agency (Alkire 2005).

HOW CAN WE TEST INTERACTIVE ARCHITECTURE?

This paper briefly discussed how interaction could potentially be

an instrument to bring about inhabitant agency within architecture. To validate these types of claims, we need to be able to empirically test the interaction aspect of IA concepts. After all, it is its interaction component which qualifies IA differently than other, better-known instances of architecture. This work based itself in well-tested methods from the fields of user-centered design and interaction design, creating a "microcosm" that could be used to test ideas out and to evaluate specific user experiences of interest, such as that of human agency.

The mental model is a central concept to the field of human-computer interaction (HCI). Upon exposure to a system, users can develop a mental model of that system; that is, a conceptual understanding of what the system is and how it works. Mental models were also important concepts in early discussions around IA (e.g. Negroponte 1975; Wellesley-Miller 1975), as they were very relevant to the artificial intelligence debates of the time. When addressing IA as intelligent participants in a conversation that manifests understanding, users will have not only a mental model of the building, but also a mental model of themselves, and a mental model of the building's model of themselves. All of these are important factors influencing the interaction.

Considerations of mental models in the context of IA raises some challenges. The first one is that people typically already have strong mental models of what buildings are and how they behave. Also, given the fact that the built environment is composed of different juxtaposed layers, which extend themselves in a continual fabric, it might be difficult for users to build unified mental models for IA—or for an ecosystem of IAs. Scholtz and Consolvo (2004) point out a similar question regarding Ubiquitous Computing. They ask: "how do users know when they are in a smart room, and how will they know how to interact with such a room?" In fact, IA can easily go from empowering to oppressive if inhabitants of the space cannot understand the system.

Given these difficulties, we propose that metaphors provide a useful way to study different forms of IA interaction, especially with regard to empowerment. The issue of metaphors is directly related to that of mental models. Metaphors are intended to encapsulate an understanding of how a system is expected to work a priori, as part of a deliberate design effort. That is, they seek to start with the desired mental model users should formulate—one that is familiar and pertinent to the problem domain—and allow designers to conceptualize a system design based on that information, not the other way around.

In general, we observed that the descriptions and analysis of interaction that can be found in IA publications may be roughly classified in three groups of metaphors. The groups are hardly isolated categories, but rather sequent marks in a continuum.

The first group of authors assume that IA will not only have human-like intelligence, it will interact as such (e.g. Adi and Roberts 2010; Achten 2013) by following a metaphor of human-human interaction (see Waugh and Taylor 1995). Interaction happens via natural speech as well as deictic and representational gestures, following the "put-that-there" paradigm, and possibly extending to cover the whole breadth of human communication forms. Only with human-like intelligence and an ability to maintain a dialogue with users can a truly interactive architecture be achieved within this paradigm. This group of authors explicitly focus on concerns of intelligence (mostly around cybernetic concepts) and speculate on its implications.

The second group of authors is by far the most numerous (e.g. Fox and Kemp 2009; Cetkovic 2012), although its approach to IA seems to avoid the discussion of interaction in detail. They are the group that a "black box" critique most suitably fits. In these cases, IA is the perfect machine that can identify needs when these needs are reasonably identifiable, provide logical spatial support to activities it understands, and allocate/manage spatial and functional resources in ways to better support specific goals. The machine is efficient and invisible, and the inhabitant wants to be disburdened of as many tasks as possible.

Authors like Jaskiewicz (2008) argue that achieving such goals will only be possible if IA has an ability to reason and learn, although it can also be argued that other means exist (e.g. Gajos et al. 2002). However, the biggest difference between the first and the second groups is that of approach, each comprising very different behavior/interaction metaphors. In one case, IA behaves and communicates as a human agent. In the other case, the agent ethos is transparent, with communication being mostly limited to context interpretation and adequate physical response.

The third group of authors, perhaps of smallest representation, describe the emergent behavior of distributed intelligence and/or the organic-like constant adaptation of a building to the ecosystem it is inserted in (e.g. Fox 2009). Interaction with inhabitants is undefined, possibly occurring in different ways, and it is non-deliberate. This kind of IA is perhaps the one that most closely approximates the natural environment around us.

In the interest of studying inhabitant agency, and after reviewing the literature on empowerment in human-computer interaction (see Weller and Hartson 1993), one more type of behavior was also included in this research. This fourth group can be described as one where inhabitants are given the ability to directly control a space's outcome, and where contextual data is used to assist inhabitants in their decision-making instead of making decisions for them.

Not a single publication was found by the authors in the field of IA describing such a system, and it can be argued that fully controllable systems would fail to fulfill the most basic definitions of IA. However, it is possible to conceive of systems where inhabitants simply assume more roles in the feedback control loops, conceptually sustaining the pertinence of this form of interaction and system behavior.

Sterk, for instance, discusses the concept of dynamic stability (2006), which can also be referred to as shared or assisted control. It defines systems that allow inhabitants to focus on symbolic actions of control, while leaving the system to provide solutions for lower level components of that action. Yona Friedman also describes assistive systems that provide inhabitants with feedback on the consequence of their design decisions (information), although all the decisions are still the inhabitants' to take (Negroponte 1975).

These four groups defined the basis for four interaction models/metaphors that this research prototyped and tested. Testing inhabitant agency considering only one approach to interaction would greatly limit our ability to assess IA more integrally in relation to its claims. Instead, by testing a set of popular but different approaches to interaction, we can achieve results that are more meaningful and generalizable. Respectively, we refer to the four models developed as: human-like intelligence model, self-adjustment interaction model, emergent behavior interaction model, and direct manipulation interaction model.

HOW WAS INTERACTION PROTOTYPED?
Overview
An interactive space was assembled in a classroom in the School of Architecture and Landscape Architecture at UBC. It was conceived as an interactive space-plan (i.e. only one of Brand's [1995] building layers), where internal partitions could transform according to inhabitants' input. In order to allow for unrestricted arrangements of the space-plan, the internal partitions were proposed in a mixed reality environment: the internal walls and partitions are virtual, and projectors were used to create visual representations of these partitions directly onto the real world. The complete design process has been described in Costa Maia (2016). Figure 2 illustrates the physical setup, including the location of the two short throw projectors.

The behavior of virtual internal partitions was the focus of this project, and their prototyping was crucial for testing IA in an

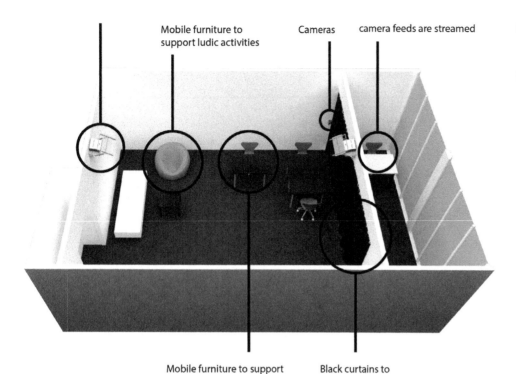

Mobile furniture to
support ludic activities

Cameras

camera feeds are streamed

2   Model illustrating physical set-up of
    the interactive space-plan in mixed
    reality.

3   Example of an inhabitant interacting
    with the space-plan via the direct
    manipulation interaction model.

Mobile furniture to support
task-oriented activities
(e.g. study, having lunch)

Black curtains to
shut out light and
to hide control space

2

experimental setup. The main technique used for the interaction prototypes in three of the four experiments is one known in Interaction Design as Wizard of OZ, which consists of an operator effectively controlling the behavior of the space without the knowledge of inhabitants. This methodology allows seamless interaction between the 'room' and inhabitants without the necessity to program the room's intelligence. This is important as any technical problems with the behavior would affect people's response to the experiment and their perception of agency. An application was developed in JavaScript that allowed for the easy creation and modification of the representation graphics in these cases. The final interaction model operation, on the other hand, was completely automated (described further on).

The following subsections describe the behavior of each interaction model as it was developed and employed in the interactive space prototype.

Direct manipulation interaction model
Users control final disposition through gestures and by directly interacting with boundaries. The computer checks if output is coherent and gives informational feedback on problematic dispositions, e.g. lack of access. Users can directly 'push' any boundary to make it move, create or delete spaces, merge adjacent spaces, create spotlights, or control color and lighting. A total of eight intuitive gestures must be learned. This way, users

can easily redraw the spaces according to need, and they have an ample power to define the final internal partition of a building. Figure 3 illustrates in a set of frames how an inhabitant can directly modify the virtual partitions and spaces using the inputs supported by this interaction model.

Human-like communication and intelligence model
Similar functioning as the previous condition. However, users do not manipulate the system directly and they do not need to remember any of the command gestures. Instead, an intelligent entity controls the system. The intelligent entity can learn about users and might not need instructions or requests for many of the procedures. Otherwise, the intelligent entity communicates with users using natural language and high-level symbolism. Participants can talk to the room to request any changes they may want or to socially engage with the room. The room may also engage with inhabitants proactively, according to its own goals and its personality. Figure 4 illustrates a verbal dialog between the inhabitant and the intelligent room with a purpose of modifying the space.

Self-adjustment interaction model
This model offers fewer possibilities for users to directly control the output. A standard, base phenotype exists and is adapted depending on use patterns, in a context-aware manner. That is, as users engage in activities, the system recognizes each context

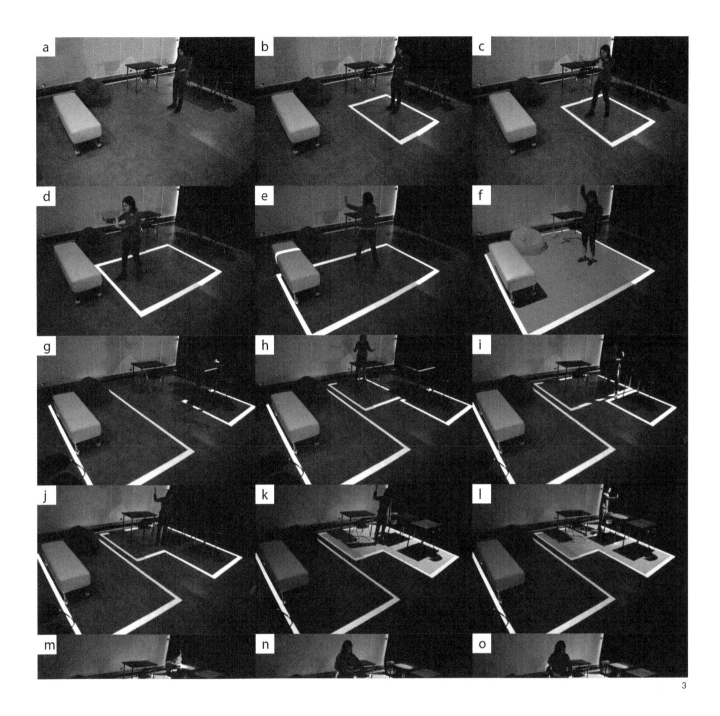

3

and provides the adequate enclosure layout, as well as adequate lighting levels, for that specific activity and context. Adequate layout for any activity is pre-established (although evolving) but parametric, so it is responsive to a set of parameters (such as number of people participating). The logic of the system is primarily concerned with functional arrangement depending on activities taking place. The output is always based on the drawing of the perimeter with confined properties. Users can only give correctional feedback. Thus, this interaction strategy follows the following logic flow: (1) to identify which activity is taking place, (2) to identify contextual parameters for that activity, such as

number of people involved, (3) to define adequate area, lighting, and color based on activity and parameters, (4) to constantly check for changes in activity and parameters, adapting accordingly, (5) to constantly check whether the last change performed received correctional feedback, and (6) if correction is flagged, undo changes, record learning data, and re-run logic flow. Any change in the room always follows these steps (see figure 5).

Emergent behavior interaction model
The different bounded spaces interact and respond to each other in a constantly evolving configuration. Users' actions in the room

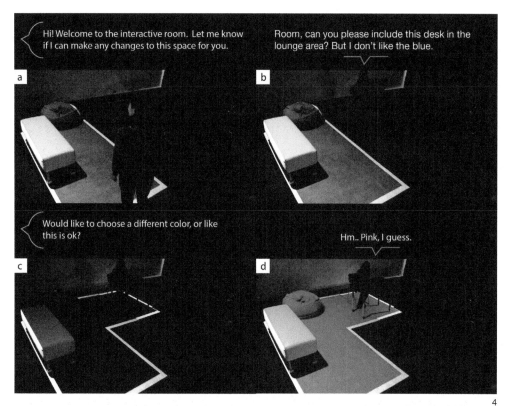

4　Example of an inhabitant interacting with the space-plan via the human-like intelligence interaction model.

5　Example of an inhabitant interacting with the space-plan via the self-adjustment interaction model.

6　Example of an inhabitant interacting with the space-plan via the emergent behavior interaction model.

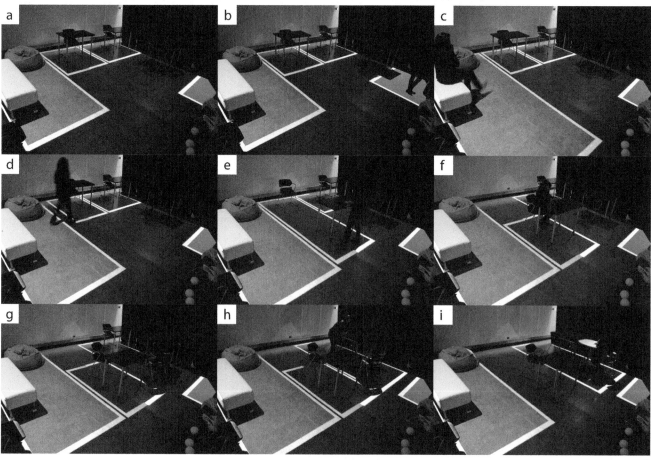

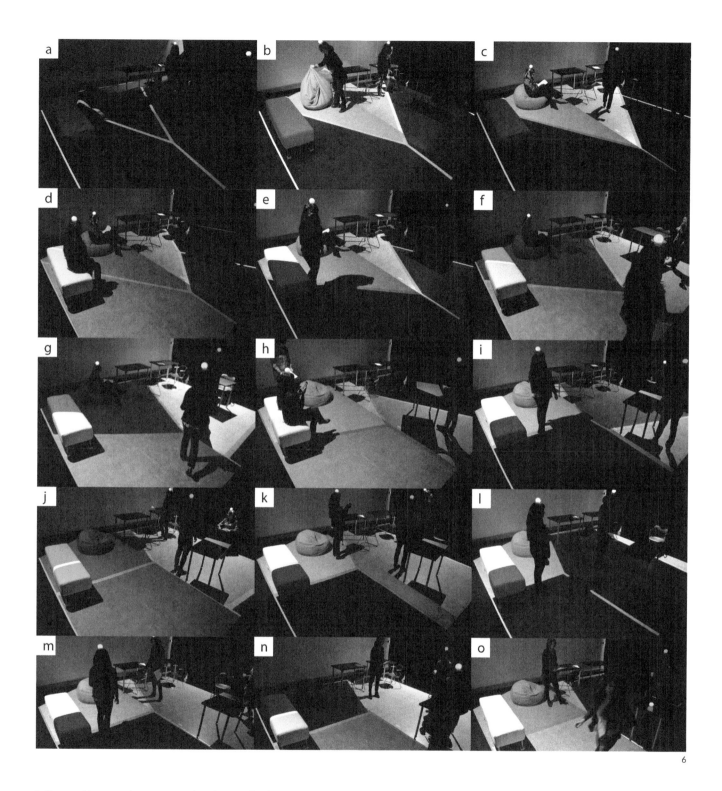

6

influence the way the space evolves by causing local changes and disturbances that may propagate and influence the system at a larger scale. The rules of the system's behavior, however, do not refer to functions or spatial fitness, nor do they have a higher order goal governing to whole. Instead, the whole is simply the emergent result of smaller, local interactions. The ways partici- pants may influence the evolution of the room are: crossing of boundaries, persistent occupation, amount of movement inside a bounded space, entering or leaving the room.

Because of the complexity of the emergent behavior, this model did not use the Wizard of Oz method. The entirety of the behavior was executed by a computer program custom written in Python, with a tracking system using glowing markers, cameras

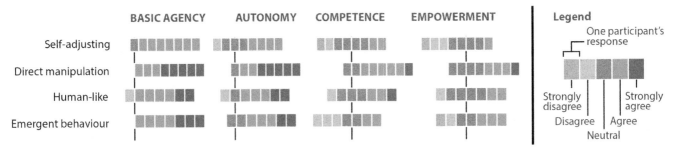

7 Comparison of measures of basic agency, autonomy, competence and empowerment: summary diagram.

and the OpenCV library. Figure 6 illustrates the room's behavior, showing frames taken 5 minutes apart.

HOW DID WE TEST AGENCY IN IA?

After the interactive space was assembled, participants were invited to occupy it. The study ran for four days, and in each day the space emulated one of the interaction models.

Participants filled questionnaires after participating in the study. The questionnaire asked participants about their experience of the space (using established UX survey material), plus specific questions designed to inquire about participants' perception of personal agency towards the space (based on self-efficacy and self-determination theories). Self-reported perceptions of autonomy, empowerment, and competence were measured as instrumental proxies for agency. Related concepts, such as perception of control, ownership, authorship, and attachment were also measured in the survey.

Additionally, participants were observed during their use of the space. Aspects such as whether participants chose to use the system features, and the situations when they chose to do so, were recorded in written form. Thirty participants completed all the stages of this study.

DID IA RESULT IN INHABITANTS EXPERIENCING INCREASING AGENCY?

The main results of the study are summarized in graphical form in Figure 7, which shows the variation of different indicators on simple agency, autonomy, empowerment, and competence for each interaction model. Extensive analysis can be found in Costa Maia (2016).

All models show a significant perceived increase in agency and it is clear that the indicators vary considerably depending on the model of interaction. The self-adjusting model has the lowest levels in all measures of agency, whereas direct manipulation had the highest. Statistical analysis of the results, however, does not bear out a significant relation between the higher levels of

inhabitant agency experienced and aspects such as perception of control, autonomy, attachment, authorship, and ownership. On the other hand, when jointly analyzing the models that allow for direct input (Direct Manipulation + Human-like Intelligence interaction models) and indirect input (Self-Adjusting + Emergent Behaviour interaction models), the results show that there is a statistically significant difference between the groups, suggesting that in order to promote human agency IA must allow inhabitants direct input on the interaction outcome.

CONCLUSION

This study supports the plausibility of Negroponte and others' thesis that inhabitant agency can be produced by IA. In taking this first step, we can start to rely on evidence to support the different claims that have been put forward regarding the social relevance of Interactive Architecture. The testing and analysis methodology put forward by this paper is intended to be a proto-type for future experiments on the efficacy and relevance of IA. Despite the fact that further studies are necessary to provide more robust and statistically defensible evidence, this exploratory research is an important step forward.

This study also presents a way to study IA through different models of interaction. The results show how experiences such as agency can vary significantly according to different forms of interaction and behavior. It suggests that interaction models have a significant impact on inhabitants' perceptions of agency and empowerment. As such, careful attention should be paid to the models of interaction that are adopted and it is likely that the most effective interaction model may vary depending on the design intent of the interactive architecture being developed.

LIMITATIONS AND FUTURE RESEARCH

The short periods of time that participants spent inside the inter-active room, alongside the high density of interaction (average of one interaction every 1.28 minutes), portrays a pattern that suggests that most participants came to the interactive room to explore the space as a primary intent. This observation defines a significant limitation of this study. Future studies must consider surveying the use of interactive spaces on a daily basis for an extended period of time, allowing inhabitants to interact with the architecture and IA systems more thoroughly and habitually, for beyond the effect of novelty. Future research should also test interaction metaphors and options within those metaphors more thoroughly, as different design decisions are possible inside each metaphor.

REFERENCES

Achten, Henri. 2013. "Buildings with an Attitude: Personality Traits for the Design of Interactive Architecture." In *eCAADe 2013: Computation and Performance–Proceedings of the 31st International Conference on Education and research in Computer Aided Architectural Design in Europe*, edited by Rudi Stouffs and Sevil Sariyildiz. Delft, The Netherlands: eCCADe. 477–486.

Adi, Mohamad Nadim, and David Roberts. 2010. "Can You Help Me Concentrate Room?" In *2010 IEEE Virtual Reality Conference (VR)*, edited by Benjamin Lok, Gudrun Klinker, and Ryohei Nakatsu. Waltham, MA: IEEE. 131–134.

Alkire, Sabina. 2005. "Subjective Quantitative Studies of Human Agency." *Social Indicators Research* 74 (1): 217–260.

Andrejevic, Mark. 2009. "Critical Media Studies 2.0: An Interactive Upgrade." *Interactions: Studies in Communication & Culture* 1 (1): 35–51.

Calderon, Roberto. 2009. "Socio-Political Communication Enabled Spaces." Master's thesis, University of British Columbia.

Cetkovic, Alexander. 2012. "Unconscious Perception in a Responsive Architectural Environment." In *Proceedings of MutaMorphosis Conference*. Prague: CIANT.

Costa Maia, Sara, and AnnaLisa Meyboom. 2015. "Interrogating Interactive and Responsive Architecture: The Quest of a Technological Solution Looking for an Architectural Problem." In *International Conference on Computer-Aided Architectural Design Futures*, edited by Gabriela Celani, David Moreno Sperling, and Juarez Moara Santos Franco. Berlin, Heidelberg: Springer. 93–112.

Costa Maia, Sara. 2016. "Testing Inhabitant Agency in Interactive Architecture: A User-Centered Design and Research Approach." Master's thesis, University of British Columbia.

Fox, Michael. "Flockwall: A Full-Scale Spatial Environment With Discrete Collaborative Modules." In *reForm( ) - Building a Better Tomorrow: Proceedings of the 29th Annual Conference of the Association for Computer Aided Design in Architecture (ACADIA)*, edited by Tristan de Estree Sterk, Russell Loveridge and Douglas Pancoast. Chicago, IL: ACADIA. 90–97. .

Fox, Michael and Miles Kemp. 2009. *Interactive Architecture.* New York: Princeton Architectural Press.

Gajos, Krzysztof, Harold Fox, and Howard Shrobe. 2002. "End User Empowerment in Human Centered Pervasive Computing." In *Proceedings of Pervasive*, edited by Friedemann Mattern and Mahmoud Naghshineh. Zürich, Switzerland: Pervasive Computing. 1–7.

Haque, Usman. 2007. "The Architectural Relevance of Gordon Pask." *Architectural Design* 77 (4): 54–61.

Jaskiewicz, Tomasz. 2008. "Dynamic Design Matter [s]: Practical Considerations for Interactive Architecture." In *First international conference on critical digital: What Matter(s)?*, edited by Kostas Terzidid. Cambridge, MA: Harvard University Graduate School of Design. 49–50.

Jaskiewicz, Tomasz. 2013. "Towards a Methodology for Complex Adaptive Interactive Architecture." Doctoral Thesis, Delft University of Technology.

Johnstone, Justine. 2007. "Technology as Empowerment: A Capability Approach to Computer Ethics." *Ethics and Information Technology* 9 (1): 73–87.

Negroponte, Nicholas. 1975. *Soft Architecture Machines*. Cambridge, MA: MIT press, .

Scholtz, J., and S. Consolvo. 2004. "Toward a Framework for Evaluating Ubiquitous Computing Applications." *Pervasive Computing* 3 (2): 82–88.

Sterk, Tristan d'Estree. 2006. "Responsive Architecture: User-Centered Interactions Within the Hybridized Model of Control." In *Game Set and Match II: On Computer Games, Advanced Geometries, and Digital Technologies*, edited by Kas Oosterhuis and Lukas Feireiss. Rotterdam, The Netherlands: Episode Publishers. 494–501.

Weller, Herman G., and H. Rex Hartson. 1992. "Metaphors for the Nature of Human-Computer Interaction in an Empowering Environment: Interaction Style Influences the Manner of Human Accomplishment." *Computers in Human Behavior* 8 (4): 313–333.

IMAGE CREDITS

Figures 1, 3–6: Nair, 2016

Sara C. Maia is a graduate student at the School of Architecture and Landscape Architecture in the University of British Columbia. Her research focuses on social aspects of emerging technologies in architecture and urban spaces.

AnnaLisa Meyboom is a associate professor at the School of Architecture and Landscape Architecture in the University of British Columbia. Her research and teaching looks at applications of technology in our environment where the highly technical meets the human environment. AnnaLisa's areas of teaching emphasize the ability to integrate the technical, the beautiful, and the environmental simultaneously and seamlessly into a built form. Her research area examines the forces of advanced technologies on architecture and infrastructures.

Human Touch in Digital Fabrication

Déborah López
Hadin Charbel
Yusuke Obuchi
Jun Sato
Takeo Igarashi
Yosuke Takami
Toshikatsu Kiuchi
University of Tokyo

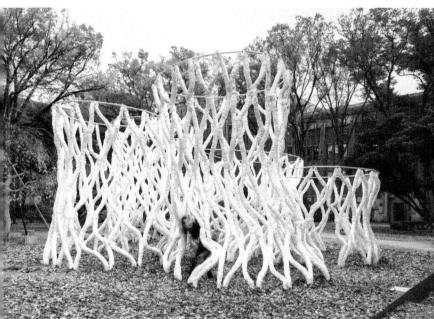

1

ABSTRACT

Human capabilities in architecture-scaled fabrication have the potential of being a driving force in both design and construction processes. However, while intuitive and flexible, humans are still often seen as being relatively slow, weak, and lacking the exacting precision necessary for structurally stable large-scale outputs—thus, hands-on involvement in on-site fabrication is typically kept at a minimum. Moreover, with increasingly advanced computational tools and robots in architectural contexts, the perfection and speed of production cannot be rivaled. Yet, these methods are generally non-engaging and do not necessarily require a skilled labor workforce, bringing to question the role of the craftsman in the digital age. This paper was developed with the focus of leveraging human adaptability and tendencies in the design and fabrication process, while using computational tools as a means of support. The presented setup consists of (i) a networked scanning and application of human movements and human on-site positioning, (ii) a lightweight and fast-drying extruded composite material, (iii) a handheld "smart" tool, and (iv) a structurally optimized generative form via an iterative feedback system. By redistributing the roles and interactions of humans and machines, the hybridized method makes use of the inherently intuitive yet imprecise qualities of humans, while maximizing the precision and optimization capabilities afforded by computational tools—thus incorporating what is traditionally seen as "human error" into a dynamically engaging and evolving design and fabrication process. The interdisciplinary approach was realized through the collaboration of structural engineering, architecture, and computer science laboratories.

1　The constructed pavilion as an outcome of the case study.

INTRODUCTION

Where is the human?

The development and proliferation of digital tools in design have increased precision in performance and the fabrication process. Construction sites have become highly networked, necessitating an extreme degree of accuracy, and redefining the role of human involvement.

However, fluctuating environmental conditions often make construction sites unpredictable, requiring on-the-spot decisions—a characteristic that is inherently human. Could human improvisations, though inaccurate, be integrated into the design and construction process with the support of digital technologies?

Human capabilities have been the source of design and fabrication processes of different architectures at various scales; hence the production and assembly of form emerged as a hybrid of human capacity and a symbiotic relationship with materials, tools, and systems.

Currently, predetermined outcomes are secured by the use of standardized premade elements, formworks, and machinery, which also increase efficiency. Similarly, in-house fabrication can control production as a result of a controlled environment, further reducing the margin of error. Yet these methods are not necessarily novel, in that features of contemporary construction methods and techniques often have undetected precedents (Rudofsky 1964). Moreover, while errors can be reduced, the search for increasing precision generally entails breaking away from means of mitigating them (Hughes 2014), bypassing what Pye (1968) refers to as the workman's ability to approximate during an integrated design and fabrication process.

There is a recent growing interest in the relationship between human-driven methods, technological integration, and architectural output. Aesthetically, this could be due to a quality of craft that can be read in an object, something Marble (2010) attributes to the detection of human input, and which is achieved by a mediated relationship between humans and tools. Similarly, according to Hight (2008), these interests are due to an understanding of the human body as a hybrid site of mediation, and thus, the human can be considered in terms of all its capacities and lack thereof.

As interactions among human actors and technical networks are already supported (Carpo 2011), human limitations could otherwise be understood as unused potential, whereby constraints can serve as catalysts for design solutions as every design move creates additional constraints, thus triggering further contextual responses (Killian 2006).

In understanding what is inherent in machines and humans, Obuchi (2015) observes that while robots are highly precise, they are not particularly adaptive and do not integrate changes with ease, whereas humans, though highly adaptive, lack the precision of robots—raising the question of whether tools can be developed with traditional forms of human engagement in mind.

This research speculates that unskilled humans, embedded with some degree of intuition, have the capacity to engage in architectural fabrication via computational support. Three primary technical aspects were addressed and developed to explore this idea, each entailing their own sub-developments:

1) A catalog of human movements: this provided a library from which possible geometric combinations could be sourced. Research explored human movements that came instinctively and were replicable.

2) A material and tool: to materialize human movement, a combination of fast drying foam and porous steel mesh was used with a modified spray gun as a "smart" tool.

3) A feedback system to detect deviations in materialized human movements via a recursive correction process according to structural stability, therefore integrating human imprecision and decision-making.

Tested in the form of a pavilion (Figure 1), this research envisioned that with digital technology, such an approach could see a new form of craftsmanship and motivation reintroduced to on-site fabrication—incorporating the human touch as a design element and tool.

BACKGROUND

The multifaceted nature of the research prompted investigations into different fields of study—examining historical and contemporary methods of fabrication, as well as human-computer interaction, and feedback systems.

Production models based in vernacular tradition indicated that unique constraints result in the development of specific techniques; making use of inherent qualities in human physiology and intuition. At the architectural scale, human physical capabilities can be incorporated as a defining element in the construction process, as seen in the mud hut *tolek* structures produced by the Musgum people in Cameroon, where perceived patterns on the exterior façade are hand formed during construction. While ornamental, these patterns serve primarily as footholds for climbing the structure—resolving the need for conventional scaffolding (May 2010). In craftsmanship, basket weaving is performed as a sequence of repeated movements that are initially taught and guided through instruction, but can over time culminate into a skill—demonstrating the combination of using acquired knowledge and human intuition in smaller scaled production.

The above examples demonstrate how human involvement at physical and mental levels can be integrated into a generative production process. Yet current trends and the advancement of digital fabrication have seen the increasing disappearance of human production-based models in favor of automated computationally driven methods.

Still, many common everyday devices that are designed specifically to be used by humans have been subtly instilled with computation in accordance with human tendencies. For instance, predictive text and auto-correct functions in mobile phones allow users to type in an imprecise manner, but because the system recognizes the sequence of letters it can compute which word is most likely to come next while also correcting misspellings—allowing inaccuracy while maintaining a precise result. This example, though simple, suggests that computation could be used to absorb human imprecision. Could such an approach reposition the human as an integral part of the digital fabrication process?

In an effort to better understand how acquired knowledge can be supported through human-computer interactive processes, precedents in architectural and non-architectural fields were investigated. *Possessed Hands* introduces the acquirement of a skill (such as playing an instrument) through muscle memory by triggering muscle movements with electrical stimulus that are timed with computational precision (Tamaki et al. 2011). *Becoming Knowledge* proposes a new dance as a result of engaging with a "virtual dancer" programmed to grow and evolve in response to simulated mechanical (human) constraints and to a database of film material (Leach 2015). These studies helped provide an understanding of skill development through the engagement of bodily movements and improvisation.

In digital fabrication-based design, different research has been developed around integrating specific human tendencies. A real-time feedback system was developed in *Pteromys* for the design and fabrication of paper planes—allowing users to draw them free-hand, while also visualizing optimal solutions for aerodynamic performance through minor variations (Umetani et al. 2014). *FreeD* combined a handheld tool with a three-dimensional guidance system, permitting complex carving tasks to be executed by unskilled makers, which are tracked and controlled with reference to a virtual 3D model (Zoran et al. 2013). Motion tracking allowed a more bodily-driven design process, which was used in *Sketch Furniture* to materialize three-dimensional air-drawn furniture

scale sketches into objects through rapid prototyping (Lagerkvist et al. 2005). At the tabletop scale, the *3Doodler* materialized three-dimensional drawings in real-time with a plastic extruding pen (Bogue et al. 2012). Most recently, *Making Gestures* imbued fabrication machines with behavior using artificial intelligence, creating a real-time relationship between body gestures and the control of machine movement (Pinochet et al. 2015).

Lastly, *STIK Pavilion* (Yoshida et al. 2014) was examined as an instance of human integration as part of architectural production. Used as mobile 3D printers, humans guided by a projection system used handheld tools to deposit material over target areas, using their physical capabilities as a substitute for an otherwise unfeasible large-scale onsite 3D printer. A scanning and feedback system was used to compare and match the built form as closely to the original target as possible—marking a distinction between human error-correcting (as used in the aforementioned research) and human error-integrating (as speculated in this research).

The above models demonstrate the potentials and shortcomings in human-machine hybridized fabrication. By incorporating the different parts mentioned (feedback system, handheld tool, guidance, and human movements) into an entire coherent system, the present research seeks to evolve the dialogue between human patterns, machine learning, and communication in on-site three-dimensional fabrication.

METHODS

To examine how human capabilities and computational precision might be combined in an on-site fabrication process, the devised system sought to integrate: 1) human movement, based on inherent human tendencies, 2) an appropriate material able to materialize those tendencies, 3) a tool that would ergonomically deploy the material, and 4) a feedback system that could embrace the difference between target form and actual form.

Human Movements

As the human body can be understood as being composed of fixed and rotational joints, a person drawing a line from the ground up with their arm fully extended tends to result in arc-like motions. These movements were analyzed through two parameters that are contingent on the person producing the movement: 1) the arc's radius and length, which is derived from arm length, and 2) the type of movement, which was determined by each person's starting and ending points relative to their body's disposition. Five different and intuitively replicable movements, which created five different arcs, were tested by four people, scanned three-dimensionally using ARToolKit (Kato 1999) visual markers, and transferred to Rhinoceros. It was observed that the same person repeating the same movement resulted in

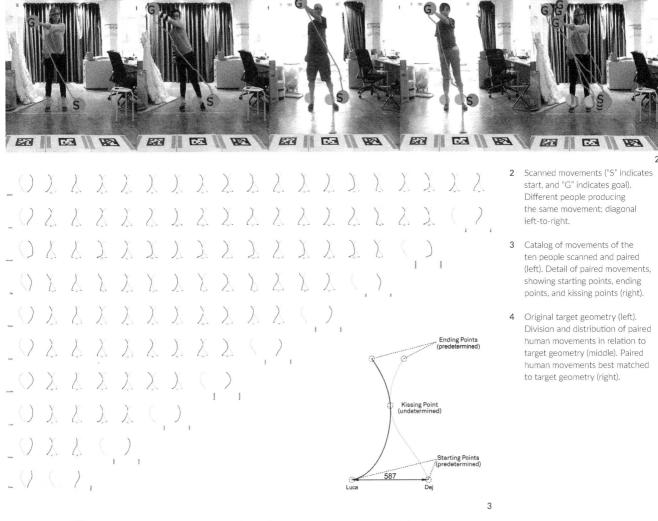

2

   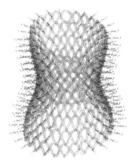

3

2 Scanned movements ("S" indicates start, and "G" indicates goal). Different people producing the same movement; diagonal left-to-right.

3 Catalog of movements of the ten people scanned and paired (left). Detail of paired movements, showing starting points, ending points, and kissing points (right).

4 Original target geometry (left). Division and distribution of paired human movements in relation to target geometry (middle). Paired human movements best matched to target geometry (right).

4

minor but noticeable deviations. Furthermore, the same movement produced by different people demonstrated a variation in tendencies, similar to differences in writing. For instance, asked to repeat a diagonal movement from left to right, some demonstrated a dipping tendency, while others demonstrated a bulging tendency (Figure 2).

Catalog of Movements

Two kinds of movements, 1) a right-bending arc and 2) a left-bending arc, were selected for application in the research. Ten people were scanned producing the two movements, and the resulting arcs were then paired to produce the catalog (Figure 3). The two paired arcs met somewhere tangentially between the two people—this point came to be referred to as the *kissing point*.

Using Rhinoceros and Grasshopper, the archived curves are used to pair one person's left arc with another person's right arc. The target geometry is divided into the number of paired movements that best completes it, whereafter the paired movements are then selected according to those that most closely match it (Figure 4).

Material

To materialize these human movements, a light and fast drying material is required. The selected material for research was a combination of 1) a common spray polyurethane foam (SPF) composed of equal parts polyol blend and polymeric methylene diphenyl diisocyanate, and 2) a flexible and porous woven stainless steel mesh tube 70 mm in diameter with 2 mm spacing, and 0.1 mm thickness. The steel mesh served as reinforcement and physical guide for the foam. The foam's quasi-unpredictable expansion maintained the concept and the practice of human imprecision (Figure 5, 6, 7).

To determine the most appropriate SPF for use, the material was compared structurally against soy-based SPF and water-based SPF in controlled conditions, revealing that the tack-free time of all three foams was 30 seconds, the time to completely solidify was 3 minutes, and the expansion ratio once the foam completely dried was 300%. Compression and bending tests suggested that common SPF was the most suitable for construction because of its superior structural capacity. However, the material is commonly used as insulation for buildings and is significantly weaker than traditional architectural materials, having a Young's modulus of 1.9 MPa (19.40 kgf/cm², equivalent to a 1.16 cm square section of cedar).

In order to find the most suitable reinforcement option, three aspects of several materials were evaluated: foam adhesion, structural strength, and ease of manipulation. Mesh fabrics of various materials and porosities were tested on two, three, and four sides of sprayed foam (Figure 8).

The results of these tests, however, were deemed structurally ineffective, while they also limited possible angles in movement as foam tended to droop and fall from the open edges. Ultimately, a custom-ordered woven stainless steel mesh in tube form performed the best, increasing the Young's modulus to 12.8 MPa (131 kgf/cm²) (Figure 9) and allowing unrestricted movement. To secure structural stability and facilitate calculations, the radius of each element was fixed at 8.5 cm.

Once foam components are sprayed, they require a method of being connected and assembled. In this research, horizontal components were connected at the aforementioned *kissing point* and at vertical connections with H-shaped joints consisting of two 300 mm U-shaped aluminum profiles. A base jig and a tripod jig were used to hold the joints in place at the starting and

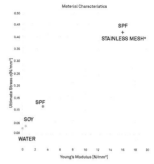

5

9

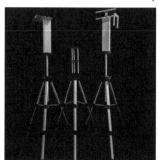

6

10

7

11

8

12

5   Common SPF foam.

6   Woven metal mesh.

7   Sprayed component with combined materials.

8   Material tests. A stress test and bending test indicated SPF is much stronger than water-based and soy-based foam.

9   The material's Young's modulus increased significantly when sprayed inside a stainless steel mesh.

10   Tripod jig.

11   "H" joint.

12   Base jig.

13   Early kissing point test.

     Human Touch in Digital Fabrication López et al.

ending points of each component (Figure 10, 11, 12) (explained further in Handheld smart tool).

The jigs were also used as part of the overall construction method of the project case study; once a completed layer of components had been sprayed in a ring formation, each layer could be lifted and held in place with the tripods. After locating the new base jig positions, workers could then spray the next connecting layer. This process eliminated the need for scaffolding and allowed workers to perform all tasks at ground level.

Handheld Smart Tool

To deploy the material as an extrusion made from human movement (Figure 13, 14), a handheld "smart" tool was developed around five key design criteria: 1) augmentation of standard spray gun, 2) secured injection of SPF inside of mesh tube for guided expansion, 3) minimal size and weight to facilitate use by pairs of humans, 4) controlled amount of foam sprayed, and 5) guided movement speed to achieve consistent radii in sprayed elements (Figure 15).

Once the trigger is pulled, the foam begins filling the stainless steel mesh. The speed of human movement is guided with a beeping device connected to a rotary encoder and an Arduino, which sounded only when the movement was in excess of the ideal 7 cm/s. The advantage of using an audible device is that workers can remain visually engaged while intuitively adjusting their speed (Figure 16).

In previous prototypes (Figure 17), a stepping motor was used to deploy the mesh, which controlled the human's speed. However, in the interest of pursuing adaptation to human inaccuracy, a system that distinguishes itself by guiding, rather than controlling, was preferred.

Feedback Loop

To integrate human imprecision as a part of the design process, a feedback loop was developed between five intercommunicative aspects: 1) target, 2) guidance, 3) scanning, 4) structural re-calibration, and 5) structural validation (Figure 18). The loop connected humans on site, material, and construction. The creation of the feedback loop was facilitated by collaboration between three fields of study; structure, architecture, and computer science (described further in the following section).

Target Geometry

A target geometry is defined as a model for production. Due to the effects of wind on such a light material, the target geometry is modified using Kangaroo simulation software in order to reduce the maximum amount of deformation, thereby reducing

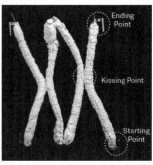

13

14

15

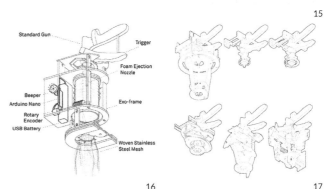

16

17

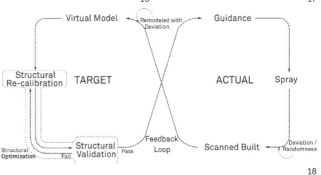

18

14　Two components sprayed and joined at kissing points and starting and ending point joints (right).

15　Handheld smart tool: Final prototype.

16　Handheld smart tool diagram.

17　Smart tool prototype evolution: different sized exo-frames, two-sided and three-sided mesh, passive and automated mesh deployment.

18　Feedback loop diagram.

·Camera

·Screen

·Jig Position

·AR codes

19

Ending Point (predetermined)

Kissing Point (undetermined)

20

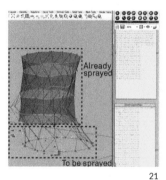

Already sprayed

To be sprayed

21

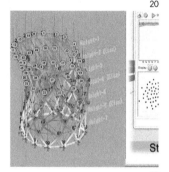

22

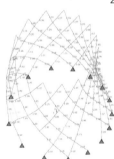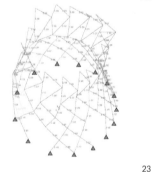

23

19 (Left) Guidance system setup: base jig location (top right), and tripod jig location (bottom right). (Right) Display setup and view on screen.

20 (Left) AR markers placed on joint/vertex locations. (Right) Red outlines on screen around AR markers indicate successful detection.

21 Actually sprayed layers (indicated in red) are used to update the target model by replacing the previously

modeled vertices with newly scanned ones, while the bottom layer is ready to be re-calibrated.

22 Genetic algorithm for structural re-calibration. Blue labels on the points indicate materialized and scanned layers. Unlabeled points indicate newly calculated points of proceeding layers.

23 Structural validation comparison between original target geometry.

risks of structural failure and avoiding the need for additional foreign supports.

Guidance

To indicate on site where each person's movement must be performed in order to achieve the target lattice geometry, an augmented reality (AR) guidance system shows the starting and ending points of each movement. Initially, it was tested by displaying the exact arc to be produced in its entirety. However, providing only the minimum amount of information required—starting and ending points—allows the workers to move freely between those two points.

ARToolKit was selected as the guidance system because of its user-friendliness and minimal setup of AR codes, web cam, and monitor. The AR codes are placed in a grid on the ground, and the camera must detect a minimum of three for accurate positioning. This facilitates visualization on the monitor of 1) starting point, indicated by the rectangular footprint of the base jig, and 2) ending point, indicated by three circles defining the tripod's legs (Figure 19). The flexibility in the minimal guidance method allowed for in-situ fabrication decisions, facilitating the intuitive aspect and making deviation from the target geometry expected.

Scanning

Upon completing a layer, a scanning system is used to compare the target model with the actual sprayed geometry. Scanning was done at the two types of vertices: (i) controlled vertices, which are the starting and ending points of the movements, and (ii) uncontrolled vertices, which are the varying middle (kissing) points where two different humans' movements meet. Using ground markers as reference points, the vertices of newly sprayed members can be digitally located in three-dimensional space, and used to update the original model (Figure 20, 21).

Structural Re-calibration

Once the sprayed geometry is scanned, a newly optimized target geometry that absorbs materialized deviations is calculated via Karamba and Galapagos (plug-ins for Grasshopper) with a genetic algorithm (GA), minimizing the maximum deformation of the whole structure (Figure 22). The GA works by moving the vertices of the layer to be built horizontally along a circular path contained in the target geometry—negotiating the tolerance of human imprecision with respect to structural integrity.

Structural Validation

A structural analysis (Figure 23) of the optimized model is performed in Hogan (Sato 1993), which uses load, cross-section, and boundary conditions as input parameters, in addition to calculating Young's modulus and yield point as material properties. If the newly optimized model does not pass the structural validation, it returns to the re-calibration stage, where a new solution is generated and tested again. This step is repeated until the validation is passed.

The feedback loop described in this section is repeated throughout the entire construction, producing a different possible outcome with each added layer and optimization (Figure 24).

CASE STUDY

To validate the hypothesis of an adaptable, hybridized human-machine on-site fabrication system, a pavilion was designed and constructed using the methods described in Section 3.

Design Overview

Five towers were bundled together to create the overall geometry. Each tower varied from 2.3 to 3.0 m in diameter and ranged in height from 2.4 to 4.2 m. The towers were translated into a lattice and divided vertically into 1.2 m layers, each with 15 horizontal divisions—the dimensions of which were decided in relation to human spraying heights and spacing, as well as structural integrity.

Site and Construction Prep

The use of AR codes as the guidance method necessitated a perfectly flat surface, so a platform was constructed. 121 AR codes were placed in a grid of 11 by 11 with 950 mm spacing.

A laptop PC with Intel 1.86 GHz core 2 Duo CPU and 2 GB memory for the computer monitor and a Logicool HD Pro Webcam C920r were mounted on a small trolley to facilitate on-site relocation of the scanning and guidance system.

On-site Construction Flow

Five people were needed on site for production: two sprayers, responsible for spraying components; one person responsible for ensuring a successful "kiss"; and two supporters, to maintain a smooth process.

The two sprayers, at their respective starting points and aware of their ending points, inferred a general area where the kissing would take place. Sprayers counted down from 3 together and created their arcs, communicating and adjusting as necessary. The two supporters aided the process in the event of a malfunction, and the fifth person ensured the appropriate amount of contact area at kissing points by applying pressure to both sides until the foam elements sufficiently bonded together (Figure 25).

When a full ring of lattice components was completed, the layer was lifted to a 1.2 m height (requiring eight people), and held in

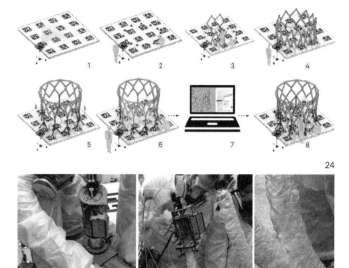

24

25

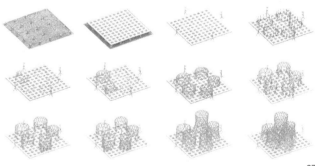

26

27

24 Feedback loop workflow: 1) scanning of AR codes on the ground, 2) guided placement of jigs, 3) spraying of members, 4) completion of layer, 5) lifting of layer, 6) scanning of layer, 7) GA structural re-calibration and structural validation of subsequent layer, 8) the process is repeated for all layers.

25 On-site paired spraying (from left to right): respective starting points, meeting in the middle, parting to ending points, ensuring kissing point.

26 Lifting of sprayed layers of two towers (background), and completed single layer (foreground).

27 Construction overview diagram.

place with tripod-jigs. The initial starting points of the first layer became the ending points of the subsequent layer (Figure 26).

Next, the geometry was scanned, revealing deviations between the target and the actual sprayed geometry. After updating the virtual model by replacing the original vertices with the newly scanned ones, the GA then created a new target, which was tested for structural validity. After passing structural validation, the base and tripod jigs (start and end points) were positioned on site according to the newly generated model, followed by the spraying process—this was repeated for each new layer until all the towers were formed (Figure 27) When all five towers were complete, the AR system was used to determine each of their originally intended final locations. Each tower was manually lifted, relocated to its target position, and screwed to the platform at the base joints.

Three different possible final geometries had been generated from the start of the construction to its completion (Figure 28).

RESULTS AND DISCUSSION

The proposed hypothesis made two assumptions—the first was that human intuition and tendencies could become a generative part of architectural fabrication, and the second was that the integration of humans in an on-site fabrication process would inevitably produce "errors," which with the support of computational tools, could become essential parts of the design and fabrication process.

Evolving Outcomes

A comparison of the original target geometry and the final built geometry shows there is a difference between a precisely computed geometry and a human made one, but structurally speaking, they are both viable (Figure 29). This demonstrated that flexible, human-based production could embrace unpredictability, and improvisation could become a part of the global design agenda.

A Glitch in Tower 2

The first layer sprayed was that of Tower 2. An error in the AR code caused a significant deviation from where the ending points were intended to be and where the guidance system was actually indicating them to be. This was only discovered once the layer had been sprayed, scanned, and compared against the original target. However, because the overall system was designed around the intention of incorporating and resolving similar types of error throughout the entire process, the layer was preserved and used as part of the final outcome (Figure 30).

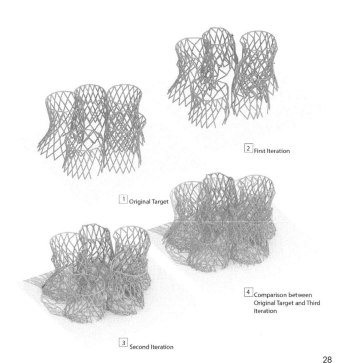

2 First Iteration

1 Original Target

4 Comparison between Original Target and Third Iteration

3 Second Iteration

28

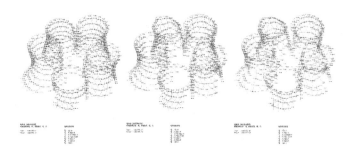

29

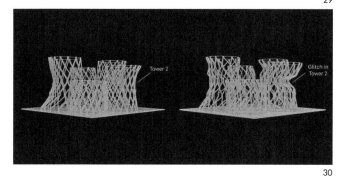

30

28 Evolution from original target geometry to actual sprayed.

29 Structural comparison (from left to right) original target, first iteration, third iteration. A color range from purple (stable) to red (unstable) indicates safety level.

30 3D printed models of original target geometry (left) and sprayed geometry (right).

31 Final pavilion.

Human Touch in Digital Fabrication López et al.

Catalog of Movements, Material Constraints and System Support

Although originally the catalog of movements was created to design the initial target geometry and to organize workers on site, the system's ability to register and incorporate variation proved effective, rendering the catalog obsolete for the specific pairing of sprayers at specific locations on site.

This was furthered by the material's drying time, which was not fast enough to instantaneously materialize an individual person's movements and tendencies, creating an inevitable difference between the movements performed during the spraying process and the actual solidified component.

Ultimately, this constraint highlighted the flexibility in the system, allowing any two people available on site to spray within the designated starting and ending points, which were provided and supported by the system.

CONCLUSION

The case study required a target final form prior to construction, however, future research could seek to forgo this step in favor of a more streamlined bottom-up process, using the GA and real-time feedback to produce solutions based on more local rules. Additionally, with a faster drying material, the catalog of movements (which was a closed and static source) could be substituted with a dynamic movement/pattern tracking and recognition algorithm, which could allow for emergent combinations and outcomes. Holistically, the implementation of 1)

real-time movement scanning, 2) machine learning, and 3) real-time feedback and guidance, would further the potential for an intuitive and non-deterministic approach to design and fabrication in an adaptive hybridized system—creating a multi-directional flow of information via a human-machine dialogue.

Moreover, many on-the-spot decisions were made during construction as a result of unforeseeable circumstances, demonstrating the accomplishments of the system. This was especially true for instances that could not have been premeditated prior to construction, such as 1) material malfunctions during spraying, which meant quickly swapping out defective cans of SPF with new ones, or 2) a mesh being cut too short, and thus being unable to reach the end-point, which necessitated the manual and timely repositioning of base and tripod jigs.

Construction also demonstrated that the feedback system successfully incorporated error, which yielded unexpected results while maintaining structural stability (Figure 31, 32, 33). Furthermore, despite the gap between drying time and human movement, the registration of variation in the final outcome produced a quality in tune with that of craftsmanship.

The human movement and materialization used in this research were the individual's arc and spray foam. However, the system as a methodological framework (Figure 18) is open ended, and could incorporate a number of other movements and materials in order to achieve a variation of outcomes in terms of process, form and function, demonstrating a potential in hybridized human-machine methods for the integration of humans in different kinds of on-site fabrication.

ACKNOWLEDGEMENTS

This research was conducted at the University of Tokyo as part of the Master's of Architecture course in Obuchi Laboratory, T_ADS (Advanced Design Studies). The research was realized under the guidance of staff and with first and second year master's course students.

Obuchi Lab:

Project Staff: Yusuke Obuchi, Kaz Yoneda, Kensuke Hotta, Kosuke Nagata, Toshikatsu Kiuchi

Second Year Students: Gilang Arenza; Hadin Charbel, Ann-Kristin Crusius, Lai Jiang, Samuel Lalo, Déborah López, Ratnar Sam, Pitchawut Virutamawongse, Jan Vranovsky, Lu Yuanfang, Ying Xu.

First Year Students: Akane Imai, Pan Kalin, Alric Lee, Nicky Li, Ittidej Lirapirom, Kenneth Lønning, Luca Marulli, Moritz Münzenmaier, Victor Wido, Christopher Wilkens, Isaac Yoo, Yihan Zhang.

Sato Lab:

Jun Sato (Professor), Mika Araki (Project Researcher)

Igarashi Lab:

Takeo Igarashi (Professor), Yosuke Takami (Student)

Thanks to Alisha Ivelich for editing and writing guidance.

REFERENCES

Bogue, Maxwell, Peter Dilworth, and Daniel Cowen. "3Doodler." Accessed May 6, 2016. http://the3doodler.com/

Carpo, Mario. 2011. *The Alphabet and the Algorithm*. Cambridge, MA: MIT Press.

Hight, Christopher. 2008. *Architectural Principles in the Age of Cybernetics*. New York: Routledge.

Hughes, Francesca. 2014. *The Architecture of Error: Matter, Measure, and the Misadventures of Precision*. Cambridge, MA: MIT Press.

Kilian, Axel. "Design Exploration through Bidirectional Modeling of Constraints." PhD diss., Massachusetts Institute of Technology, 2006.

Lagerkvist, Sofia, Charlotte von der Lancken, Anna Lindgren, and Katja Sävström. 2005. "Overview of Sketch Furniture Project." MoMA Multimedia, 2005. http://www.moma.org/explore/multimedia/audios/37/856

Leach, James, and Scott deLahunta. Forthcoming. "Dance 'Becoming' Knowledge." *Leonardo*.

Marble, Scott. 2010. "Imagining Risk." In *Building (in) the future: Recasting Labor in Architecture*, edited by Peggy Deamer and Phillip G. Bernstein. Princeton, NJ: Princeton Architectural Press. 38–43.

May, John. 2010. *Buildings Without Architects: A Global Guide to Everyday Architecture*. New York: Rizzoli.

Obuchi, Yusuke. 2015. "Arms Race." *Forty-Five*, August 9. Accessed April 20, 2016. http://forty-five.com/papers/33

Puentes, Pinochet, and Diego Ignacio. "Making Gestures: Design and Fabrication through Real Time Human Computer Interaction." PhD diss., Massachusetts Institute of Technology, 2015.

Pye, David. 1968. *The Nature and Art of Workmanship*. Cambridge: Cambridge University Press.

Rudofsky, Bernard. 1964. *Architecture without Architects: An Introduction to Non-Pedigreed Architecture*. New York: Museum of Modern Art.

Sato, Jun. Hogan, version 2. Windows. The University of Tokyo. 1993.

Tamaki, Emi, Takashi Miyaki, and Jun Rekimoto. 2011. "PossessedHand." *Proceedings of the 2011 Annual Conference on Human Factors in Computing Systems*, Vancouver, BC: CHI. 543–552.

Umetani, Nobuyuki, Yuki Koyama, Ryan Schmidt, and Takeo Igarashi. 2014. "Pteromys: Interactive Design and Optimization of Free-Formed Free-Flight Model Airplanes." *ACM Transactions on Graphics (TOG) - Proceedings of ACM SIGGRAPH 2014* 33 (4): 1–10.

Yoshida, Hironori, Takeo Igarashi, Yusuke Obuchi, Yosuke Takami, Jun Sato, Mika Araki, Masaaki Miki, Kosuke Nagata, Kazuhide Sakai, and Syunsuke Igarashi. 2015. "Architecture-Scale Human-Assisted Additive Manufacturing." *ACM Transactions on Graphics (TOG) - Proceedings of ACM SIGGRAPH 34* (4): 1–8.

Zoran, Amit, Roy Shilkrot, and Joseph Paradiso. 2013. "Human-Computer Interaction for Hybrid Carving." *Proceedings of the 26th Annual ACM Symposium on User Interface Software and Technology*, edited by Aaron John Quigley, Shahram Izadi, Ivan Poupyrev, and Takeo Igarashi. St. Andrews, UK: UIST. 433–440.

IMAGE CREDITS

Figures 1, 31: (Vranovský, 2015)
Figures 2–8, 10–22, 24–28, 30: (Obuchi Laboratory, University of Tokyo, 2015)
Figures 9, 23, 29: (Sato Laboratory, University of Tokyo, 2015)

Déborah López is completing her second Master's degree at Obuchi Lab at the University of Tokyo. She was awarded the *monbukagakusho* scholarship following the completion of a Bachelor of Arts and Master's of Architecture from the European University of Madrid. Her research currently examines the invisible forces that shape the city as well as the intersection of human and digital technologies.

Human Touch in Digital Fabrication López et al.

Hadin Charbel is a *monbukagakusho* scholar and is completing his Master's degree at Obuchi Lab at the University of Tokyo, following a Bachelor of Arts in Architecture from UCLA. His research investigates potential relationships between humans and computational tools in the fabrication process and he is currently investigating computer vision in the reading of the urban environment.

Yusuke Obuchi specializes in architectural design and computational design. After graduating from the University of Southern California and completing his studies at Princeton University, he served as Co-director of the Design Research Laboratory (DRL) in London. Since 2010, he has served as an Associate Professor in the Department of Architecture at the University of Tokyo. He has introduced world-leading computational design techniques to Japan, and explores digital fabrication processes to create models using programming and 3D modeling skills.

Takeo Igarashi is a Professor in the Department of Computer Science at the University of Tokyo. He is known for development of a sketch-based modeling system and a performance-driven animation authoring system. He has received several awards, including the IBM Science Prize, the JSPS Prize, the ACM SIGGRAPH 2006 Significant New Researcher Award, and the Katayanagi Prize in Computer Science.

Jun Sato is Principal at Jun Sato Structural Engineers Co., Ltd. / Associate Professor at University of Tokyo / Visiting Lecturer at Stanford University. He has researched and developed transparent / translucent structures, and lightweight & ductile structures through his collaborations with architects such as Kengo Kuma, Toyo Ito, Sou Fujimoto and Junya Ishigami, and through workshops with students. He received the Japan Structural Design Award in 2009.

Toshikatsu Kiuchi is an Assistant Professor in the Department of Architecture at the University of Tokyo's Graduate School of Engineering. He is a part of Advanced Design Studies, an advanced research and education laboratory, composed of a series of sub-units, including Digital Fabrication Lab (computation and fabrication), Sustainable Prototyping Lab (material sustainability in an urban context) and Design Think Tank (urban cybernetics and information), which he is currently leading.

Yosuke Takami is a PhD student at the Graduate School of Information Science and Technology at the University of Tokyo. He is a part of the Computer Graphics and User Interface Cluster. His research has focused on intuitive building methods for novice workers and developing guidance systems using augmented reality technologies.

Formeta:3D:
Posthuman Participant Historian

Interactive, Real-Time, Recursive Form Generation and Realization

Daniel Eisinger
Ball State University

Steven Putt
Design Collaborative

1

ABSTRACT

Formeta:3D is a project that engages the posthuman through the development of a machine that translates inputs from its surroundings into physical form in real-time. By responding to interaction with the inhabitants of its environs and incorporating the detected activity in the inflections of the produced form, it has an impact on the activity in the space, resulting in a recursive feedback loop that incorporates the digital, the physical, and the experiential. This paper presents the development of this project in detail, providing a methodology and toolchain for implementing real-time interaction with additive physical form derived from digital inputs and examining the results of an interactive installation set up to test the implementation.

1   Detail of one of the forms generated by Formeta:3D in real-time in response to activity observed by a webcam.

2   Previous Formeta:2D project developed by the authors as a means to explore real-time translation of inputs to a physical output.

3   One of the drawings produced by the Formeta:2D drawing machine.

INTRODUCTION

In a posthuman landscape, how is form generated? If form is autonomously designed, what are the inputs to the machine, when is the design realized, and how can the form continue to evolve? Formeta:3D is a project that explores these questions through the development of a machine that translates inputs from its surroundings into physical form in real-time, responding to interaction with the inhabitants of its environs and incorporating the detected activity in the inflections of the produced form. This in turn has an impact on the activity in the space, resulting in a recursive feedback loop that incorporates the digital, the physical, and the experiential.

Formeta:3D operates by translating motion detected in live video to 3D print toolpaths for immediate printing. In so doing, the system produces artifacts that capture in physical form the spatial activity at a given place and time. These objects are not merely records of activity, however, since they influence the human activity in a space by the fact of their evolving presence within it. The machine operates as both historian and collaborator, establishing itself as a player in human events.

This paper presents the development of this project in detail, providing a methodology and toolchain for implementing real-time interaction with additive physical form derived from digital inputs. It also examines the results of prints produced as a response to real-time inputs collected while the machine was set up as an interactive installation.

BACKGROUND

This work builds on a rich history of drawing machines that have been developed over the past half century as part of experimentation in art and architecture. In contrast to early examples in the 1950s and 1960s—including the creations of Jean Tinguely, which some reviewers would say regarded human and machine as diametrically opposed—more recent drawing machines have conceived of the machine as a creative partner, enabling a feedback loop between human and machine. In this view, "Where humans and machines are viewed as sharing patterns of organization, there can emerge the 'posthuman,' the splicing of human with the machine" (Hugunin 2000). Formeta:3D operates in this vein, machine and human acting both as observer and as participant in an iterative, non-deterministic process of form generation.

Precedents

One early project that sought to explore recursion or feedback within the CAM process was a project called "Shorting the Automation Circuit," developed by the sixteen*(makers) group in 2000. They created an instrument that would collect environmental data—light detected as the instrument was moved

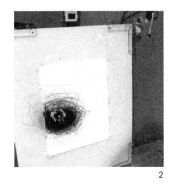

2   3

by wind currents (or as people passed by)—as well as a script that would translate this data into three-dimensional form. However, since this project was developed prior to the popular rise of 3D printing over the past decade, the group did not have direct access to a 3D printer. As a result, their system generated snapshots in time that did not reflect a real-time response to the forms as they were manufactured. The group speculated about the potential for an active feedback loop but did not have an opportunity to realize it or explore its ramifications (Ayres 2006).

Drawing machines such as the Artsbots of Leonel Moura have investigated other aspects of the machine, including artificial intelligence and swarm behaviors (Moura 2014). Moura's work frames the machine as an autonomous agent, dependent on human input primarily for its initial operating rules. While these rules may include responses to sensor input, he is more interested in the autonomous machine "taking the human out of the loop," than one that engages in a dialogue with humans.

Researchers at SCI-Arc have developed an advanced system for real-time human interaction and interfacing with the physical through robotics. Recent work is focused on bridging the digital and the physical in service of design through gesture recognition and environmental awareness enabled by 3D scanning. Live, the platform developed at SCI-Arc for real-time connectivity with their industrial robot arms, together with the various experiments developed on this platform, explore design ideation, blurred virtual and spatial communication, dynamic environments, tooling development, and autonomous decision-making in relation to a physical workpiece (Batliner, Newsum, and Rehm 2015). Many of the technologies employed and the associated toolchains are similar to those employed by Formeta:3D (in particular the Eyerobot project), and in fact many of the explorations could be framed similarly in the way that they engage the roles of the designer, the machine, and the user. Even so, Formeta:3D approaches real-time human-machine interaction from a different perspective, examining how formal and physical artifacts of human-machine interaction can passively record and yet actively influence human experiences and populate an evolving posthuman landscape.

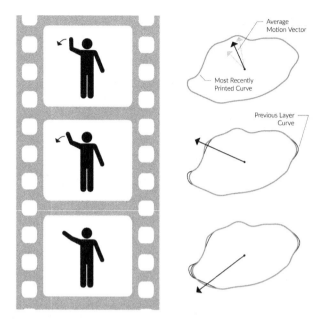

4   Response of Formeta:3D motion analysis to ongoing movement of human participant.

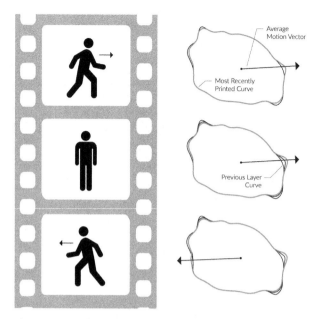

5   When a human participant stops, the algorithm retains the most recent motion vector until new motion is detected. This results in greater emphasis being placed on the earlier motion.

Previous Work

The machine presented in this paper is informed by and is an evolution of previously published work begun by the authors in two dimensions with the simple pen-on-paper drawing machine shown in Figure 2 (Wit, Eisinger, and Putt 2016). This drawing machine employed two LEGO Mindstorms motors positioned at the upper corners of a vertical drawing plane, controlling the movement of a marker using monofilament spooled and unspooled by the motors. The instructions for the motors were provided by a Raspberry Pi single-board computer, which generated drawing geometry in real-time based on image data collected via an attached camera. Multiple algorithms were developed as options for translating color data from the collected images to XY coordinates. The resulting drawings reflected the response to the predominant colors observed as well as to the extremes when intense colors were detected (Figure 3).

This earlier Formeta investigated how the machine can serve as a translator of inputs of one kind to outputs of another. It questioned the location of design authorship in relation to the machine, given that the drawings produced by the machine were influenced by a multitude of factors—including the human activity in the space, the lighting conditions, the physical imperfections of the machine, and the variety of technologies employed—in addition to the way in which the system components were assembled and coded by the designers. As with the drawing machines described by Hugunin, this machine was a creative partner, participating in the outcomes as much as the designers and those interacting with it.

METHODS

The realization of Formeta:3D required the development of a many-layered, low-latency toolchain from camera to printer as well as extensive mechanical and material calibration. This section discusses the main components and processes of the system as well as the final experiential setup.

Platform

The computational platform controlling Formeta:3D was grounded in Grasshopper, but included additional components located on intermediate and control devices. A Grasshopper definition performed the bulk of data collection and processing, relying on Firefly to obtain and pre-process the video feed and on Kangaroo to simulate forces acting on the geometry. Communication plugins, including gHowl and Slingshot, enabled recursion and the transmission of data to the intermediate control device, a Raspberry Pi. Python code on the Raspberry Pi processed coordinate data from Grasshopper and interfaced with the 3D printer, a kit-built, delta-style RepRap descendant with an Arduino/RAMPS controller and Marlin firmware, a Bowden extruder, an E3D hot end, and a heated bed.

Video Processing

The Firefly video stream feeding the Grasshopper definition for this project came from a standard HD webcam trained on the area in front of the 3D printer in order to facilitate interaction with the system and its product. Motion in that area was translated to 2D motion vectors produced by Firefly's color averaging component. These vectors, located at the XY plane origin in

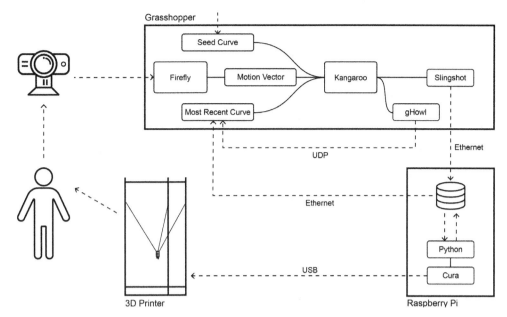

6   Overview of Formeta:3D platform and data flow. Human interaction influences the form produced by the 3D printer, which in turn influences the response of the human. Human and machine are both participants as the machine records the interaction on form.

Rhino, were smoothed, filtered, and averaged to produce a single primary vector (Figure 4), as well as to determine the location of an attractor point, which in turn influenced simulated forces acting on the evolving form.

Data Throughput and Recursion

Due to Firefly's ability to process the incoming video stream very quickly, it became necessary to throttle the output of the Grasshopper script to prevent slower downstream processes from being overwhelmed. This was accomplished by tracking a limited history of data for smoothing purposes and only allowing new geometry to be emitted every two seconds.

The data throttling also served to limit the speed of recursion in the definition, which was its own challenge. Grasshopper is not designed for recursion and in fact a workaround was required to circumvent Grasshopper's loop detection. The Slingshot and gHowl plugins eventually solved the problem (where Hoopsnake and others did not), enabling curve serialization and transmission through UDP network datagrams to the beginning of the definition.

Later versions of the definition used Slingshot exclusively, sending curve data to a MySQL database on the Raspberry Pi and retrieving that data at the beginning of the definition. This enabled the Python code to indicate to Grasshopper the success or failure of printing the received geometry, creating a closed loop process that reflected the actually printed layers in the recursion (Figure 6).

Geometric Constraints and Simulation of Forces

In order to facilitate the creation of a printable form that could express the effects of interaction over time, the generated toolpath geometry was based on a circular primitive, which would be pushed and pulled based on the location of the previously generated attractor point. While precision was not a goal, constraints were necessary to allow for the evolution of form in a fashion that would enable continued printing, providing a sufficient amount of material and structure to support each new layer.

The geometric constraints employed included limits on the toolpath deviation allowed from layer to layer and limits on the effects of simulated forces on the geometry. To accomplish this, the magnitudes of the motion vectors were mapped to a limited donut-shaped region around the previous layer curve. Further constraints were implemented as forces resisting change—Kangaroo "goals" set to pull curve division points to the previous layer curve and to minimize the change in distance between division points.

The forces compelling change to the geometry were also set up as Kangaroo goals. The primary goals affecting change were set up as loads pulling curve division points near the attractor point inward or outward on a gradient based on motion vector magnitude and distance from the attractor point. The effect of this configuration was similar to the capability of some software packages (particularly Maya) to allow "soft selection," where geometric operations on a given element are translated to surrounding elements on a gradient (Figure 7).

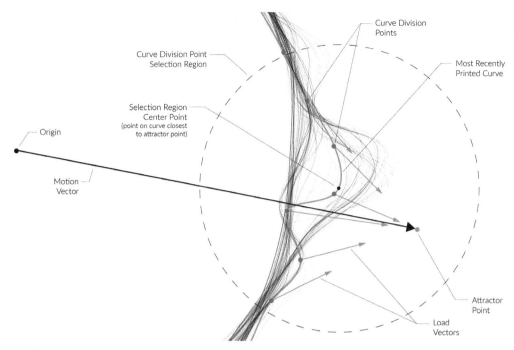

Labels on figure:
- Origin
- Motion Vector
- Curve Division Point Selection Region
- Selection Region Center Point (point on curve closest to attractor point)
- Curve Division Points
- Most Recently Printed Curve
- Attractor Point
- Load Vectors

7 Motion vector acting on toolpath. Motion detected by the webcam is translated to a vector emanating from the origin in virtual 3D space, which determines the position of an attractor point. Kangaroo forces pulling on curve division points in the vicinity pull the 3D print toolpath curve toward the attractor point to generate the next layer.

Interface with 3D Printer and G-Code Generation

The Marlin firmware controlling the 3D printer is a standard one for open source 3D printers, but its communication capabilities are relatively primitive. It uses a basic serial connection without built-in error-checking for USB communication, requiring a fair amount of code on the client side to ensure successful data transmission and synchronization. Rather than reinvent the wheel for Formeta:3D, open source Python code from the Cura project was leveraged to handle direct communication with the printer.

Unfortunately, using Cura's slicing logic for G-Code generation was not practical for this project. Cura's slicer is located in a separate module written in C code, which would have taken a significant amount of time to appropriate for the sake of this project. An alternative approach considered was generating 3D model files of each print layer for external slicing, but this proved too problematic in terms of performance and post-processing. In the end, custom G-Code generation routines were employed, even though this meant foregoing the benefits of the advanced acceleration and extrusion algorithms embedded in the slicer.

Physical Considerations and Calibration

As a first foray into the intricacies of 3D printing, this project required a great deal of learning related to the materials in use and their response to the controls provided by the printer. The single greatest challenge was in calibrating the extrusion rate of the PLA filament to ensure consistent printing. Initial tests attempted to push far more filament through than the hot end or the extruder could handle, resulting in inconsistent extrusion and filament jams. Other factors that required calibration over many iterations included the hot end temperature, print speed, and bed leveling.

Interactive Installation

Once adequately calibrated, it was possible to set up the Formeta:3D project in a public atrium space to allow the activity in that space to inform the generated form. The installation consisted of the 3D printer with a Raspberry Pi intermediate controller, a computer running the Grasshopper definition, and a webcam mounted on a six-foot-long dowel attached to the printer to give it a wider vantage point for tracking movement. The computer's monitor was mounted in such a way to allow people to see the video stream being captured, as well as the Rhino interface showing the generated geometry and the changing motion vectors being processed. The installation was active for hours at a time over the course of three days.

RESULTS AND REFLECTION

Realized over the course of a semester, Formeta:3D was completed in time for installation prior to Ball State University students leaving for the summer. The artifacts produced by the installation represent features of a new landscape shaped by machines for and in response to humans. No longer just a tool, the machine affects the (local) course of history with its own interpretation of events expressed in physical form.

Physical Outcomes

The print pictured in Figure 8 shows the longest running print of the public installation of Formeta:3D. It exhibits a number of features that exemplify the current behavior of the form generation logic, including how motion is expressed, how transitions between layers are handled, and various material effects resulting from how these elements of the algorithm were handled.

Motion is expressed on the pictured form in the wavy character of the walls as a result of the way in which the motion vectors pushed and pulled on the layer geometry. The changes in form that occur over many layers show times when the captured motion trended in a given direction. The moments where protrusions spilled over into loops of filament are due to the inability of underlying layers to support the magnitude of the diversion from the preceding layer geometry.

The moments where spillover happens also indicate a counterintuitive outcome of the form generation algorithm as it is currently implemented, since they represent the times where the least activity was taking place. The forces simulated in the Grasshopper definition do not have a significant cumulative effect unless those forces are continuously expressed in the same general direction. Since the definition currently retains the most recent motion vector until a new vector is captured (one that differs significantly enough from the last one to overcome the smoothing process), the result is that a lack of new activity translates to the greatest cumulative effect on the geometry (see diagram, Figure 5).

Transitions between layers are handled differently in different variants of the form generation algorithm. In Figure 8, the layer transition points are indicated by little balls of melted filament that accumulated where the print head paused while it waited to receive the next layer (see also Figure 1). At the moment, the location of those points is dictated only by the start and end points of the layer geometry as generated by Kangaroo. This represents an opportunity for further development, since the location of those points could be controlled by another input to provide an additional means for indexing the activity in the space.

Another approach for handling layer transitions is pictured in Figure 9, where a series of gaps are evident. In this approach, the filament and the print head were both retracted to avoid the seepage of melted plastic. This approach also kept the nozzle from sitting immersed in melted plastic, which can lead to burnt filament and a clogged nozzle. Burnt filament is evidenced at multiple points in Figure 8 in the form of dark brown streaks.

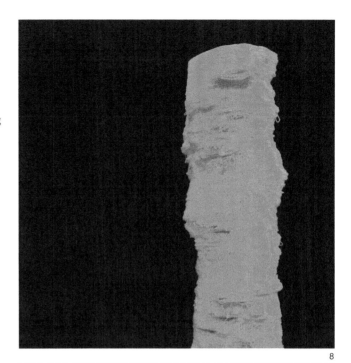

8

9

8 Extended print result. Balls of plastic on the surface indicate where the 3D printer hot end paused while waiting for the next 3D print layer to be generated.

9 An additional print result produced using an alternate layer transition approach that involved retracting the plastic filament while waiting for the next layer.

Formal Outcomes

The types of forms produced by the form generation algorithm as currently implemented are limited. As noted in the previous section, the wavy areas indicate varying motion expressed on form, but in general the form remains cylindrical without any major deviations. This is due in large part to the way in which the simulated forces behave as the geometry distorts itself to greater degrees. Beyond a certain point, Kangaroo gets overwhelmed, and the new toolpath layers lose all connection to the previous layers, gathering in a tangled mess to one side of the form and causing the printer to produce globs of material on that side (Figures 10–11).

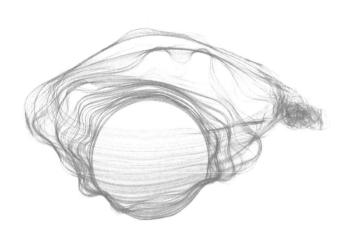

10 Rendering of toolpaths generated by Formeta:3D that produced the print result depicted in Figure 11.

11 3D print resulting from toolpaths shown in Figure 10. The globs of plastic on the top right side are the result of the Kangaroo simulation being overwhelmed by conflicting forces.

Due to this behavior, while each form produced is unique and represents a unique set of interactions at a specific time and place, at a distance the main differentiators between individual artifacts are the height of each print and/or the result of varying the seed geometry. On close inspection, the nature of the layer transitions or the presence of pronounced protrusions also differentiates each piece.

Beyond the programmed behavior of the machine, there are unexpected outcomes that differentiate outcomes and connect the results to the space by nature of the unexpectedness of the behavior. For example, a layer may fail to adhere to the previous one due to an imperfection in the machine or in the toolchain, resulting in bridges and other formations that make the printed form significantly different from its digital counterpart (Figures 10–11). Observing this happen during a print marks the experience due to its surprising behavior.

Human-Machine Interaction and Feedback

The public installation of Formeta:3D was successful at both recording interaction on and prompting interaction with the form being generated. As noted above, the machine produced unique prints, reflecting activity captured from its surroundings at a specific place and period of time. At a basic level, the machine was simply an historian, jotting down its interpretation of what it observed. But the fact that it was doing so in the moment, in full view of those around it, and in three-dimensional form created a stimulating feedback loop that led to activity that would not otherwise have taken place.

The initial draw toward interaction with the installation was the machine itself. The delta-style 3D printer does not have the familiar box-like shape of more common 3D printers like Makerbots, which prompted discussion about its function. Once aware of both the capabilities of the machine and its purpose in responding to activity in the space, people took notice of the camera and of themselves in the video feed displayed on the accompanying screen. The real point of understanding for those interacting with the machine was when they connected their own movement to the vector display on the screen, which provided immediate, highly responsive feedback. Then, they would begin to purposefully attempt to control the more subtle behavior of the machine with hand and body motions (Figure 12). The display of previously printed artifacts from earlier interactions allowed for discussion of individual features and speculation about what caused them.

CONCLUSION

Formeta:3D presents a methodology and a toolchain for implementing real-time interaction with additive physical form via digital inputs. While currently limited by the Grasshopper-based implementation of its core elements, this toolchain could be ported to a more flexible platform that would allow for more advanced decision making and machine learning, for example to amplify trends in form development for a given space. There is also the potential for more responsiveness in the system, since the current layer-by-layer approach for realizing form lacks the immediacy provided by machines, including the earlier Formeta:2D drawing machine and the SCI-Arc robot arms.

12 The Formeta:3D interactive installation as set up in the Ball State University College of Architecture and Planning main atrium space.

Additionally, while the creation of an occupiable space was never a goal for this research, a larger scale implementation would more than likely stimulate a greater degree of interaction with the machine and might more clearly express formal and spatial variation. Another iteration might focus less on stimulating interaction and observe human activity in a wider area, perhaps using facial recognition to generate multiple simultaneous motion vectors with differing origins to influence form generation.

The implementation of Formeta:3D adds a new element to the tradition of drawing machines by bringing a digital fabrication tool previously only used in non-real-time applications into the lexicon of real-time output tools. While real-time drawing machines that produce 3D elements are not new (Hugunin 2000), Formeta:3D realizes a feedback loop that operates in the digital, the physical, and the experiential realms simultaneously. This fact situates this project uniquely in relation to the work of Moura and SCI-Arc. Where Moura eschews human influence, Formeta:3D engages both autonomous form generation and human interaction. Where SCI-Arc's spatial explorations with the Eyerobot project are constantly in flux in relation to human interaction, Formeta:3D evolves a lasting artifact that records human experience additively. This situates the machine as historian, its products akin to pictorial histories, allowing for varied interpretations, conveyed from observer to observer. Here, this project embodies the posthuman, a machine recording human history autonomously while being inextricably tied to the involvement of human interaction.

ACKNOWLEDGEMENTS

Many thanks are due to the College of Architecture and Planning (CAP) at Ball State University, for the use of facilities for the Formeta:3D installation, and to CAP Interim Dean Phil Repp, for funding the purchase of the 3D printer for this project. The referenced previous work Formeta:2D was developed by Daniel Eisinger and Steven Putt with advising from Ball State faculty members Andrew John Wit and Joshua Coggeshall.

REFERENCES

Ayres, Phil. 2006. "Constructing the Specific." In *The Architecture Co-Laboratory: GameSetandMatch II : On Computer Games, Advanced Geometries, and Digital Technologies*, edited by Kas Oosterhuis and Lukas. Feireiss. Rotterdam: Episode Publishers. 314–321.

Batliner, Curime, Michael Jake Newsum, and Casey M. Rehm. 2015. "Live: Synchronous Computing in Robot Driven Design." In *Real Time: Proceedings of the 33rd eCAADe Conference*, vol. 2, edited by B. Martens, G. Wurzer, T. Grasl, W. E. Lorenz, and R. Schaffranek. Vienna, Austria: eCAADe. 277–286.

Hugunin, James. 2000. "in::FORMATION: The Aesthetic Use of Machinic Beings." *Leonardo* 33 (4): 249–261.

Moura, Leonel, and Henrique Garcia Pereira. 2014. "A New Kind of Art [Based on Autonomous Collective Robotics]." *Footprint* 15: 25–32.

Wit, Andrew, Daniel Eisinger, and Steven Putt. 2016. "Human Interaction-Oriented Robotic Form Generation." In *Robot 2015: Second Iberian Robotics Conference: Advances in Robotics*, vol. 2, edited by L. P. Reis, A. P. Moreira, P. U. Lima, L. Montano, and V. Muñoz-Martinez. Switzerland: Springer Verlag. 353–364.

IMAGE CREDITS

All images and diagrams produced by Daniel Eisinger. Figures 4, 5 and 6 include icons from the Noun Project created by Bradley Avison, Hoach Le Dinh, Irene Hoffman, Lastspark, Arthur Shlain, and Fernando Vasconcelos.

Daniel Eisinger is an instructor at the College of Architecture and Planning (CAP) at Ball State University (Indiana), where he also manages the college's digital fabrication and robotics labs. Eisinger holds a Bachelor of Science degree in Computer Science from Taylor University and spent a number of years in software development before obtaining his Master of Architecture degree from Ball State University.

Steven Putt is a Graduate Architect at Design Collaborative in Fort Wayne, Indiana, where he specializes in digital design and fabrication. As co-founder of the in-house research entity, dc.lab, Putt focuses on implementing computational and parametric design methods in day-to-day architectural practice. Putt holds a Master of Architecture degree from Ball State University.

Antithetical Colloquy

From operation to interaction in digital fabrication.

Diego Pinochet
DesignLab-School of Design
Adolfo Ibañez University

1

ABSTRACT

This paper, introduces a cybernetic approach to digital design and fabrication by embracing aspects of embodied interaction, behavior and communication between designers and machines. To do so, it proposes the use of body gestures, digital/tangible interfaces and Artificial Intelligence to create a more reciprocal way of making.

The goal is to present a model of designing and making as a 'conversation' instead a mere dialog from creator to executor of a predefined plan to represent an idea. In other words, this paper proposes a platform for interaction between two antithetical worlds—one binary/deterministic and the other perceptual/ambiguous—by focusing in the exploratory aspects of design and embracing aspects of improvisation, ambiguity, imprecision and discovery in the development of an idea.

1 'Making Gestures: An interactive personal fabrication system'. Diego Pinochet 2013–2015'.

INTRODUCTION

In the case of designers, architects, and artists, tools are part of a repertoire of cognitive, symbolic, and semiotic artifacts with which each explores and learns about design problems. Nonetheless, certain criticism has emerged about the use of digital tools in terms of the cognitive and creative aspects of the process of creating original work. The use of digital fabrication tools currently relies upon a model of 'tool operation' in which designers typically pause the ideation or 'creative' part of the process to produce a physical representation of an idea. Furthermore, in separating designing and prototyping from the development of that idea in a digital environment with a mouse or keyboard, and then 3d printing, laser cutting or CNC milling, the act of making and learning through exploration disappear. A designer cannot touch, feel or interact directly with the objects he or she creates; this insightful moment of sensing, feeling and discovering new things is lost.

This paper identifies three fundamental problems of digital design and fabrication processes, establishing a theoretical discussion about the use of digital tools in architecture at the early stages of the design process. Moreover, this paper inquires after possible implementations to engage architects in more perceptual digital design processes through the use of interactive fabrication machines. This paper shows the development of different interaction experiments with fabrication machines by the use of machine learning techniques to train and interact in a more personalized way with machines. In this paper, a model of interaction is proposed that seeks to transcend the 'hylomorphic' model (the imposition of form over matter) imperative in today's architectural design practice to a more reciprocal form of computational making. To do so, this paper seeks to reconcile design and making by exploring real-time interaction between mind, body, and digital fabrication tools. Furthermore, by using body gestures and imbuing fabrication machines with behavior, the goal is to establish an interaction model that embraces ambiguity and the unexpected, engaging the designer into more improvisational and insightful design processes. This presents a new perspective about digital fabrication where error and imprecision are incorporated into the unfolding process of making something to discover new things.

DIGITAL DESIGN AND MAKING, THREE PROBLEMS

By using digital tools, focusing on the representation of a design in the transition from mind, to model, to code, to machine, to material, it is possible to identify three fundamental problems about the use of the digital in design. The first one is the problem of black-boxed processes embedded into software and machines that might bias the design processes into more representational efforts instead of the creative/cognitive aspects of the design process. From this black-box concept, it is possible to identify the other two problems. The first is the use of generic operations—embedded into software and hardware—to impress a specific non-generic design idea. The final one, is the 'creative gap' that occurs as a result of using black-boxed generic operations in the transition from design idea to the fabrication of a prototype.

The 'Black-Box' Problem

Since the beginning of CAD during the 1960's, the concerns about the relationship between humans and technology in design were based upon the assumptions of a symmetrical symbiosis between human and machine. Moreover, the development of "intelligent" tools as "creative enhancers" was an enterprise that failed because of the naive assumptions of the human mind as an information processing machine and the simplistic view of human skill and expertise (Dreyfuss, 1986, p.12) that could be translated to combinatorial and discrete operations of data processing and analysis. Furthermore, the current use of digital tools somehow maintains this assumption by leaving 'the human' out of important and crucial moments of design, such as the physical manifestation of an idea, leaving the "tedious and time consuming tasks" such as drafting or calculating tool paths in a CAM software inside the computer as black-boxed opaque structures to the architect. In the case of digital fabrication techniques, the assumption that the digital is somehow a complete or sufficient representation of the real has ignored the phenomenological dimension of meaning (Perez-Gómez, 2002, p13), that involves aspects of materiality or human intervention as an active decision making actor. In the process of making through digital machines, the process follows a linearity of events from the generation emergence to the physical prototype.

This cause-effect relation between the idea and prototype, relies on the evaluation of results in order to modify the structure that originated that output. The translation from idea, to digital models, to G-code, to material, involves a series of translations and black boxed operations that happen without human intervention. Furthermore, these operations are dependent upon formal or codified knowledge about 'ways of representing' that don't take into account the perceptual and phenomenological contingency and interaction of the agents involved. This black boxed world, or what Perez-Gómez (2002, p13.) defines as "the between dimensions" (from idea to prototype), is a problem not present in analogue processes since they don't rely on predefined structures of codified knowledge. Conversely, these processes are based on a constant unfolding/evolving processes of tacit knowledge as situated action.

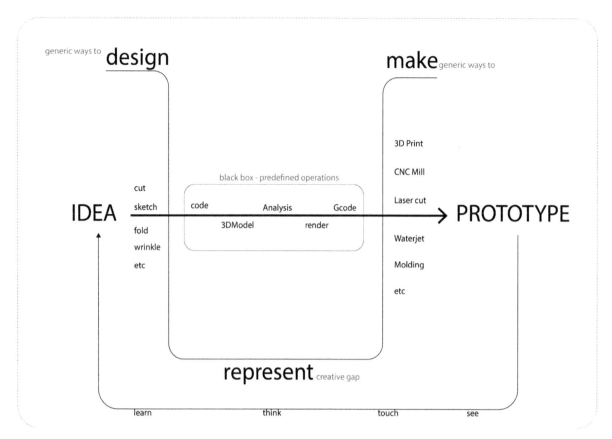

2   Design-make dichotomy.

Design Through The Generic.

If we consider that design is "something that we do" which is related to our unique human condition as creative individuals, one can argue that "design and making" is related to how we manifest and impress that uniqueness into our surrounding environment. Hence, it is valid to assert that after more than 50 years of CAD invention, the possibility to impress that uniqueness through the use of software and digital fabrication machines is limited. Moreover, because the machine is the one that determines the way something will be made according to predetermined structured procedures, the process of making, exploring and having feedback through seeing and doing is lost. Perceptual aspects of ideating and making something through a continuous unfolding of tools, material behavior and perceptual aspects of design, sometimes considered as 'tacit knowledge', 'personal knowledge' or even 'embedding,' are neglected.

Hence, why should designers accept that the physical manifestation of our ideas should be processed and expressed through this black-box using generic operations and constrained by predefined structures embedded in a software? Plotting a drawing, 3D printing or CNC milling are processes in which the software calculates the "optimal" or "average" operation to produce the physical manifestation of our ideas. Nonetheless, unlike digital processes that occur inside a software, designers don't 'make' according to calculations of optimal data. Moreover, the process of creation is the result of a continuous circulation between the interaction between the body and its senses (primarily vision and touch), tools (pen, knives, scissors) and matter (clay, paper, ink, wood) according to a constant evolving logic, perception and tacit knowledge about what happens at the moment of acting. If we consider cooking as an analogy, we can understand the difference between linear deterministic processes of making, such as 3D printing, and reciprocal perceptual processes such as drawing, painting or crafting. Cooking involves a predefined generic structure in the form of a recipe that indicates rules, procedures, and also quantities to be followed. Nonetheless, the act of cooking involves the perception and action of the chef by being aware of certain 'moments' impossible to quantify or code beforehand. Thickness, smell, taste, consistency, color, and so on are parameters often described with ambiguous sensorial instructions that are either visual—'bring to a boil' , 'until turns red' , 'until water is absorbed' – or physical – 'al dente', 'until it reaches hard consistency.'

The 'Creative Gap'.

To Dreyfus (1986), it is clear that computers are indispensable for some tasks due to some characteristics where they surpass human's capabilities such as precision or exhaustion. He asserts that computers are specifically useful in CAD applications due to their capacity to compute large amounts of information, improving efficiency by optimizing, drafting, analyzing and representing (p.xii).

Nonetheless, what happens if in the process of making a prototype the designer 'sees something else´ and needs to reformulate that design in real-time? How does this `black-boxing´—which rarely happens in analog design processes such as drawing or model making or even cooking—affect aspects of creativity and cognition of the design process? Furthermore, how can we reformulate the digital fabrication process in order to design and fabricate 'on the go' without worrying about elaborating deterministic and fixed structures? How can we enter the "space between dimensions" (Perez-Gomez, 2002, p.23) in order to enable a 'conversation' with fabrication tools to learn from error, imprecisions, see new things, be creative and find new meanings about what we do, on the making, not only before or after.

The design and fabrication practice through digital technology is constrained by a constant imposition of predetermined ideas over matter (Hylomorphism) by "a violent assault on a material prepared 'ad-hoc' to be informed with Stereotypes" (Flusser, 1991, p.43). Today's digital fabrication works by relying on a model that imposes a linearity of events in the transition from idea to prototype. The designer is forced to pause the creation, focusing on the representation of ideas neglecting the interaction between perception and action present in analog processes.

Hence, it is valid to ask how to bridge Design and Making through the use of technology engaging the designer into more creative processes? Moreover, how does the interaction between these antithetical worlds—the humans and machines- happen in order to generate more insightful and creative design processes?

TOOLS FOR REPRESENTATION VS. TOOLS FOR CREATIVITY

According to Schön (1987), design is a form of artistry and making, where learning about a specific topic or design emerges through actions (conscious and unconscious) and exploration (p. 29). The designer learns to design by knowing and "reflecting in action," reinterpreting and re-elaborating actions in the particular moment where the act of design takes place, producing new meanings and coherence. This suggests that every creative process is accompanied with a material representation as a by-product of a constant interaction with our surrounding objects. By interacting with our surrounding objects, we learn and produce meaning and therefore reason. As Robinson (2013) argues, reason, that is the power of the mind to think, understand, and form judgments by a process of logic, is actually a proprioceptive circulation of the relationship between mind, body, and things around us (p.60).

Digital tools (e.g. the computer, 3D printer, laser cutter, CNC milling machine) are commonly used to perform a task (a set of prescribed rules) which can be coded on an algorithm or inside a parametric model as a set of topological relations. In contrast, if design is considered an activity, moreover a cognitive one, it can be referred as "the way people actually realize their tasks on a cognitive level" (Visser, 2006, p. 28) by using knowledge, information and tools. Taking this into consideration, the problems identified in the previous chapter relate to the use of digital tools (CAD-CAM) as task performing machines instead of activity performing machines. In other words, we program machines to perform several tasks, however we don't interact with them to perform an activity. The little interaction between creator and executor (designer and machine) is constrained to an insufficient interface (clicking and typing) to grasp the main qualities of both as a fluid interaction (Figure 3)

Ingold (2008) asserts that in every creative endeavor "the role of the artist is not to reproduce a preconceived idea, novel or not, but to join and follow the forces and flows of material that bring the form of the work into being" (p.17). Robinson (2013) identifies this interaction (mind/tool) as 'Circulation,' in which tools become externalized mind, and cognition is internalized tools (p.35). This circulation is crucial to comprehend the role of digital fabrication tools in the creative part of design to discover the affinities and dissensions established by using tools as 'objects to sense' and 'objects to think' with. The mind-tool interface, is not static but one that implies a bi-directionality where humans – in this case designers – internalize tools as mind and externalize mind as tools as a "proprioceptive circulation"(Robinson, 2013, p.60). Thus, senses and action as gestures play a crucial role in this negotiation.

From Operation To Interaction

To Idhe (2003, p.91), the relationship between humans and technology is explained through "intentionality" and what he calls "middle ground" or "area of interaction/performance". Idhe asserts that the only way to define the relationship between humans and non-humans is through actional situations that happen in a specific time and place. This relates to real time interaction in which not only humans but also objects -which Idhe and Latour refer as the "non humans"- are redefined as a "Sociotechnical" Assemblage (Latour, 1994, p.64) that transforms both agents into

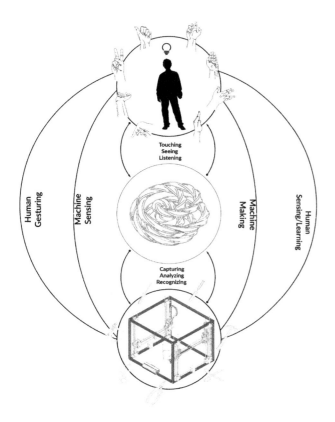

3 Interactive fabrication diagram. Design + Make on the go.

something else. Idhe (2003) argues that both human and objects enter into a dance of agencies "as the human with-intentions" (p.94) enters into the resistance and accommodation of mechanical agency provided by the object.

The actions and products derived from that interaction are possible neither by the human nor by the object, but by the relationship and actions enacted by their interactions.

Furthermore, in analog design processes, tools become almost invisible to us and act as mediated objects so the designer focuses in the specific action of 'making something'. Clark's (2004) interpretation about the relationship between humans and technologies according to degrees of transparency (p. 37), might be a useful perspective to understand why the problem of the generic, the creative gap, and black-box are relevant for today's design practice through technology. The difference between 'opaque' and 'transparent' technologies relies on the degrees of transparency according on how well technologies fit our individual characteristics as humans (Clark, 2004, p.37).

The more intricate and hard to use the technology is, the more opaque it is in relation to how it deviates the user from the

purpose of its use. Suffice it to say that the current model of digital fabrication is based in pure tool operation, neglecting real time interaction with tools, leaving important parts of cognitive processes of making aside. The many intricacies of digital tools—the black-box—lead the designer to pause creation and focus in the elaboration of plans for representation—the creative gap—that in many cases lead one to rely in software decisions—the generic—relegating the design act to an initial effort which is later rationalized by a fixed structure.

Therefore, the main problem this paper addresses is the linear communication—clicking and typing—established between human and machine at the different stages of the design process—from mind to software, from software to code and from code to material—which constrains the space for exploration, learning, and knowledge elicited about designs by neglecting the use of tools as 'objects to sense' and 'to think with.'

CREATOR/EXECUTOR INTERACTION: THREE EXPERIMENTS

In order to implement real time interaction between humans and machines, this paper explores the role of human gestures and the capability of a machine to discriminate, recognize, "understand" and act according to a specific gesture. To do so, three experiments are implemented: Gestural, Tangible and Collaborative interaction using a 2 Axis Drawing machine connected to different motion tracking sensors and computer algorithms in charge of search, object and gesture recognition (Figure 4).

The basic Machine setup, consisting of a 2-axis CNC machine controlled by GRBL, an open source firmware for Arduino capable of controlling up to 3 stepper motors using one UNO board. Python and C# were used to communicate hardware, software and the designer.

Gestural Interaction

Gestural Interaction (Figure 5), was implemented to test the use of Motion tracking sensors to interact with a CNC drawing machine, testing the use of gestures associated to specific operations. The goal was to recognize two types of movements, on one hand the free motion of the gestures in space as exploratory/improvisational gestures, and on the other hand the auxiliary movements as fixed operations to help the process of drawing.

To track gesture motion and get the data to perform gesture recognition, a LEAP motion sensor was used. This sensor, through the use of 2 infrared cameras is capable to track hands in space with high accuracy, with the position of hands represented by a 3D skeleton model. The data from the 3D skeleton model was used to perform gesture recognition using Gesture

Recognition Toolkit (GRT), a machine learning C++ library created by Nick Gillian to perform real time prediction from sensor data using one specific algorithm implemented by Gillian. By using the Adaptive Naive Bayes Classifier (ANBC), it was possible to perform gesture recognition using the LEAP motion sensor. Initially developed to recognize musical gestures, the ANBC proved to be very useful to distinguish other type of gestures due to its capacity to perform prediction from N-Dimensional signals, its capacity to weight specific dimensions of that signal (if one parameter is more important than others), and most of all, its capacity to be quickly trained using a small number of samples and the adaptation of the training model as the gestures are performed repetitively(Gillian, N., Knapp, R.B. and O'Modhrain, S. 2011, p.1).

To interact with the machine, hands' positions in space were used to perform free exploratory movements in XY dimensions controlling a pen. The data from the classified gestures were associated to specific auxiliary fixed operations such as drawing different geometrical shapes, or to start or stop motion from the machine. The implementation of gestural interaction was successful in order to enable the real time interaction that was intended with the machine. Moreover, the gesture recognition implementation allowed the generation of a workflow where the two types of movements (free and fixed) were successfully recognized and performed.

Tangible Interaction

The implementation of Tangible interaction (Figure 6) sought the development of a system to manipulate physical objects in space to interact with the machine. Through object recognition, by blob, edge, and color detection, it was possible to track objects' positions to control the CNC machine. By using computer vision algorithms, it was possible to recognize different objects extracting their positional data to recognize free and auxiliary movements.

This facilitated an indirect type of interaction, considering haptic as an important part of the interaction process. The user could have a different experience than gestural interaction by touching objects and producing machine movement. Combining the techniques developed from gestural interaction, using both a webcam and the Leap, it was possible to perform free movements of the objects in space in addition to the recognition of fixed gestures such as 'grabbing' an object to enable the motion of the machine while the object is manipulated.

Collaborative Interaction

If the implementation of the two the previous types of inter-action were concerned with capturing user intentions by using gestures and physical objects in space, the main goal of

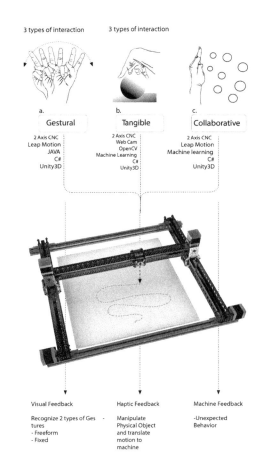

4   Human – machine interaction diagram.

collaborative interaction was the development of a dialog where the machine could produce unexpected output and behavior according to what the designer proposes by modifying a virtual environment using gesture recognition to get user input and Artificial intelligence as machine behavior (Figure 7). The goal was to engage designer and machine in a collaborative design environment where the human is not concerned with producing a design—e.g. drawing—but with setting up an environment that indirectly influences it by promoting the emergence of unexpected behavior from the machine to solve a problem. Furthermore, two different AI algorithms were implemented. First, Depth First Search (DFS) was used to create a random environment as a maze. Second, optimal search algorithm (A*) was used to find a solution to the proposed environment as the machine behavior. In this case, the human suggested a target inside the maze, and the machine tries to find the optimal path (using A*) to reach that target. The path calculated to the target is then translated to G Code. The software interface allows the user to specify in real-time a new target and/or modify the maze shape by moving objects in the virtual and physical environment. According to this, the machine can recalculate on the fly the new trajectory to the new target. In this case, it is the user who

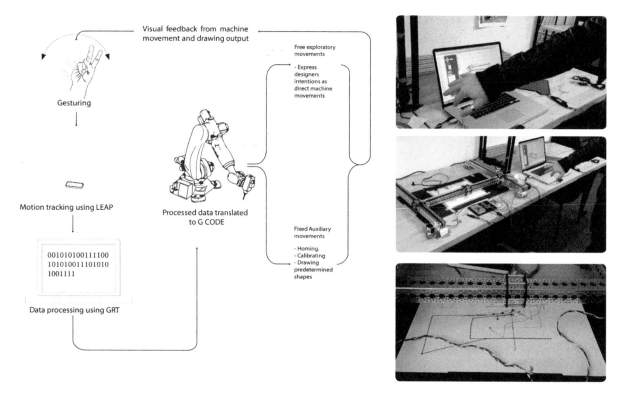

5 Gestural Interaction diagram.

generates indirect input so that the machine, through a constant recalculation of the maze solution from its current position to the target, generates motion expressed as drawings and unexpected behavior to designers.

CONCLUSIONS

By focusing on gestures, we can formulate mechanisms in which humans can communicate with machines establishing a fluid interaction with digital tools, capturing the moment where creativity and original work emerges. Moreover, this paper argues that creativity and originality don´t rely on pre-conceived ideas or elaboration of plans, but on the relationship of perception and action as creativity and knowledge promoters, in which the concept of real time interaction between humans, through gestures and machines, and through AI, is the key to use digital fabrication in more creative ways. At this point, one can assert that by the development of gestural machines, we can solve the three problems identified in the previous chapters—the black box, the creative gap and the generic. Moreover, in order to promote the improvisational aspects of design as exploration, we need not only machines emulating designer's movements but interactive machines that by the use of sensors and actuators can enhance our experience, establishing a constant circulation of action and perception from both sides. Nonetheless by stating this, it's not the intention of this paper to seek the development of 'Intelligent

cognitive design machines'. The proposal is that we can take advantage of current developments on robotics, artificial intelligence, CAD/CAM to grasp in a computational way the very aspects that make design an original enterprise.

The different types of interaction were successfully implemented in terms of the fluid response from the machine according to designer's gestures. The basic implementation of machine learning algorithms to discriminate fluid exploratory gestures from symbolic type of gestures (preprogrammed) served as the base to understand how the interaction with the development of a more sophisticated machine should take place. Furthermore, it was clear that simple act of translating hand motions into XYZ coordinates in the machine was insufficient in order to take advantage of its characteristics. Moreover, to achieve a seamless interaction between machine and designer, the interplay between the fixed gestures—associated to a machine´s pre-coded operations—and the free movements is crucial. The development of the three types of experiments as separate models for interaction worked as the basis for a robust software improvement at bigger and more complex scales by integrating different technologies in a circular loop of capturing, analyzing and expressing the analyzed human gestures as improvisation and consequently machine action(Figure 8)

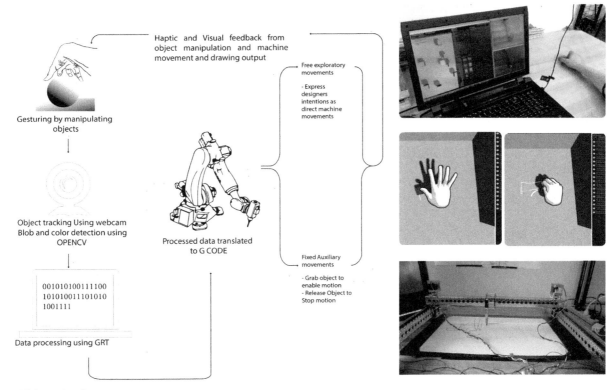

6   Tangible Interaction diagram.

By using fabrication machines imbued with behavior, performing in real time according to designer's improvisation, the problem of black-boxing can be solved. Moreover, by making tools capable of becoming internalized, designers can concentrate on design and making in a similar way analog tools are used. This paper showed experiments that did not focus on the manipulation of intermediate artifacts translating an intention into material, but rather on a process that participates seamlessly in a dance of agencies between designer, tools and materials. As a consequence, the problem of the 'creative gap' could be avoided. Because the system is interactive and dynamic, capturing and performing according to the specific resistances and contingencies of participants, tools and environment, there is no gap between idea and designed object. The system eliminates multiple translations present in digital fabrication processes so the designer focuses in creating by seeing and doing-making 'on the go'.

Finally, it proposes a solution to the problem of generic outcomes, allowing the impression of the self onto our material world, capturing designers' gestures and translating them as intentions in real time to materials. This process starts from the argument that the originality and the creativity emerge from the fact that no gesture is the same as any other, because of the ever-changing properties and characteristics of the participants (material, visual, psychological) of the interaction.

REFERENCES

Clark, Andy. 2004. *Natural-Born Cyborgs: Minds, Technologies, and the Future of Human Intelligence*. New York: Oxford University Press.

Dreyfus, Herbert L. 1986. *Mind Over Machine: The Power of Human Intuition and Expertise in the Era of the Computer*. New York: Free Press.

———. 1992. *What Computers Still Can't Do: A Critique of Artificial Reason*. Cambridge, MA: The MIT Press.

Flusser, Vilém. 2014. *Gestures*. Translated by Nancy A. Roth. Minneapolis, MN: University of Minnesota Press.

Gillian, Nicholas, R. Benjamin Knapp, and Sile O'Modhrain. 2011. "An Adaptive Classification Algorithm for Semiotic Musical Gestures." In *Proceedings of the 8th International Conference on Sound and Music Computing*. Padova, Italy: SMC.

Ihde, Don. 2002. *Bodies in Technology*. Minneapolis, MN: University of Minnesota Press.

Latour, Bruno. 1994. "On Technical Mediation—Philosophy, Sociology, Genealogy." *Common Knowledge* 3 (2): 29–64.

Maldonato, Mauro. 2014. *The Predictive Brain: Consciousness, Decision and Embodied Action*. Eastbourne, England: Sussex Academic Press.

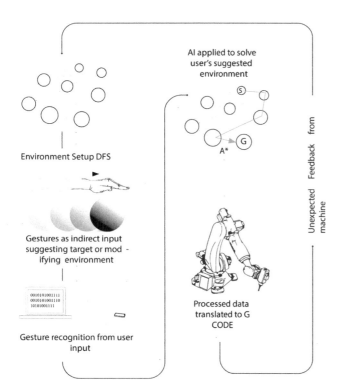

7 Collaborative Interaction diagram.

Nakamura, Fuyubi. 2008. "Creating or Performing Words? Observations on Contemporary Calligraphy." In *Creativity and Cultural Improvisation*, edited by Elizabeth Hallam and Tim Ingold.New York: Bloomsbury Academic. 79–98.

Pérez-Gómez, Alberto. 2012. "The Historical Context of Contemporary Architectural Representation. In *Persistent Modelling: Extending the Role of Architectural Representation*, edited by Phil Ayres. London: Routledge. 13–25.

IMAGE CREDITS

Figures 1, 3–4, 6, 8: Pinochet, 2015
Figures 2, 5, 7: Pinochet, 2014

Diego Pinochet Master of Science in Architecture Studies in Design and Computation, Massachusetts Institute of Technology. Master in Architecture and Architect, Catholic University of Chile. Computational research affiliate at MIT Group making. Diego is an expert in the development of tools for design, with extensive experience in design and computational geometry, manufacturing technologies and robotics. His research is mainly based on the interaction of humans and machines for design, architecture and construction. Since 2012 he works as an associate professor at the School of Design at the University Adolfo Ibáñez in the areas of design, digital fabrication and interaction through videogames

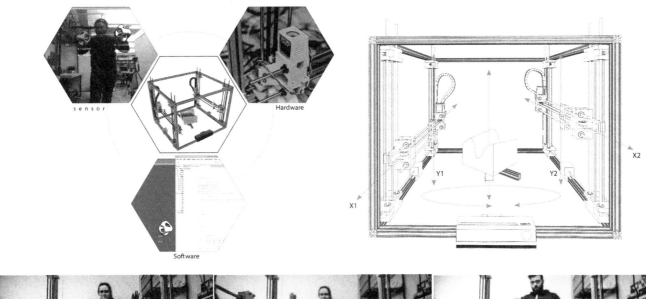

8   Making Gestures project implementation.

Material Frontiers

Kathy Velikov
University of Michigan, Taubman College of Architecture and Urban Planning

Within the past several years, architectural discourse has undergone a radical shift regarding questions of matter and materiality. If part of the posthumanist project has been to replace previously constructed boundaries among humans, technology, and the environment with intertwined hybrids, it is also possible to extend this thinking to the field of architecture and the breakdown of the Albertian (humanist) paradigm that separated acts of drawing and building. The disciplinary antagonism that had stood for many years between representation and materialization, between form and matter, has now given way to much more fluid and reciprocal methods of thinking and working through the interrelations among matter, form, structure, and notation.[1] Within this paradigm, which underlies the work in the *Programmable Matter*, *Generative Robotics*, and *Material Frontiers* sections, formational and performative agency is shared among designer, material, computational procedures, and often, environmental forces.

This section gathers papers that currently operate at the fringes of material experimentation within computational design, and anticipates the increased centrality of this research within the field as a whole. The following papers can be grouped into two distinct categories of investigation. The first group, which includes the papers by Tabbarah, Beaman, Clifford, and Twose and du Chantenier, focuses on questions of material agency that have only recently entered computational discourses. These include discussions around aesthetics, material ontology, and irregular formation that splinter dichotomous definitions of natural/non-natural, stable/unstable, familiar/unfamiliar, and planned/accidental, and that position computational design within concerns that extend beyond instrumentality and performance.

The papers that comprise the second group, including those by Estévez, Dade-Robertson, Sollazzo, Franzke, and Derme, are specifically engaged in experimentation with synthetic biologies intended toward architectural applications. While computational design has a fairly well established trajectory of exploration into the use of technologies that translate biological behaviors into morphogenetic codes and protocols for architectural formation, the synthesis of novel biological matter is a fairly recent development in the field. These papers are all engaged in developing working methods to crossbreed computational, biological, genetic, and electrochemical logics for new species of architectural and landscape materialization. These explorations look toward a future of architecture that goes beyond life-like behavior and intelligent interaction and actually incorporates living matter into the production of designed environments. With such work, the discourses of bioethics and biopolitics that have been central to posthumanist theory in relation to biomedicine will increasingly become relevant to architecture.

1. Achim Menges, "Material Systems, Computational Morphogenesis and Performative Capacity," in *Emergent Technologies and Design: Towards a Biological Paradigm for Architecture*, eds. Achim Menges, Michael Hansel, and Michael Weinstock (London: Routledge, 2010), 44.

Material Frontiers

Almost Natural Shelter:
Non-Linear Material Misbehavior

Faysal Tabbarah
American University of Sharjah

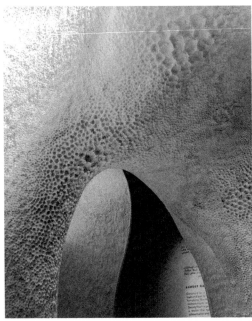

1

ABSTRACT

This paper critiques computational design and digital fabrication's obsession with both precision and images of natural patterns by describing a messy attitude towards digital and material computation that integrates and blurs between linear and non-linear fabrication, resulting in material formations and spatial affects that are beyond pattern and image and are almost natural. The motivation behind the body of work presented in the paper is to question the production of space and aesthetics in a post-human frontier as we embark on a new geological era that is emerging out of the unprecedented influence of the human race on the planet's ecological systems. The paper and the body of work posit that the blurring between the natural and the synthetic in the post-human frontier can materialize a conception of space that exhibits qualities that are both natural and synthetic.

The paper is organized in three parts. It begins by describing the theoretical framework that drives the body of work. Next, it describes early digital and material casting explorations that began to blur between linear and non-linear fabrication to produce almost natural objects. Finally, it describes the process of designing and making Almost Natural Shelter, a spatial installation that emerges from the integration of messy computational design methodologies and chemically volatile non-linear fabrication. In specific, High Density Foam is persuaded to chemically self-compute in an attempt at uncovering a shelter that has almost natural spatial qualities, such as non-linear textural differentiation and sudden migration between different texture types.

1   Photographs showing the final object in its context and its highly textured interior.

A NATURALLY IMPRECISE INTRODUCTION

"Your parents were wrong—messiness is a virtue"
Eric Schmidt and Jonathan Rosenberg (Schmidt 2014)

"Disciplined making has become a security blanket against the
realities, disruption and disorder of everyday life."
–Jeremy Till and Sarah Wigglesworth (Till 2001).

Almost Natural Shelter is a spatial installation that attempts to
respond to two prevailing tendencies that continue to proliferate within contemporary architectural work which falls under
the umbrellas of computation and digital fabrication. The first
tendency is the continuing false pursuit of ultimate precision in
architecture, which continues to make the production of the built
environment an unbearably slow endeavor. The second tendency
that *Almost Natural Shelter* interrogates is the fascination with
images of natural patterns which has been made easy and abundant by off-the-shelf computational design tools that continue
to adorn the computer screens of many. As such, *Almost Natural
Shelter*, a mostly synthetic object, presents a messy attitude
towards digital and material computation that integrates and
blurs between linear and non-linear operations, resulting in
material formations and spatial affects that are almost natural.

In her timely book, *The Architecture of Error: Matter, Measure, and
the Misadventures of Precision*, Francesca Hughes reveals the
fallacy of precision. She argues that the degrees of precision
that we hear about today are neither required for habitation
and construction, nor even possible to execute given the messy
reality of construction (Hughes 2014). Where the blame for
slowness falls is unclear. On the one hand, architects, who do not
wield as much control over construction as they desire, persist in
producing hopelessly precise solutions and drawings, and on the
other hand contractors who actually know the real impossibility
of absolute precision continue to demand it upfront.

Unsurprisingly, the pursuit of precision does not lurk in the
fringes of architectural history, it has always been part and parcel
of it, achieving legendary status with the spread of Modernism.
Having lost some of its conceptual stomping ground in early and
highly reactionary Post-Modernist projects, the contemporary
proliferation of computational design methodologies and digital
fabrication tools has given the pursuit of precision an undeniable
resurgence. This has allowed it a significant growth in momentum
within architectural practice, and with surprisingly more thrust, in
academia[1]. Amongst the many symptoms of this is that architects
are increasingly attempting to eradicate margins of error towards
the perfectly snap-fit, no matter the scale of the endeavor.
Excel sheets and performance analysis software reign supreme.
It sometimes seems that imprecision has become ethically

2

3

2 50 cm x 50 cm x 30 cm resin cast exhibiting misbehaved form and texture through non-linear casting.	**3** Detail of textures resulting from misbehaved casting operations.

4

5

objectionable, in much the same way that ornament was for Adolf Loos. All this, and I continue to see parts from fully digitally fabricated projects getting banged into place by rubber mallets or some other less benign tool. How, after watching Jacques Tati's *Mon Oncle* are architects not completely skeptical of the banality of space that is brought about by precision?

The same technological advances that are driving the pursuit of precision have helped accelerate the speed at which the built environment is produced. However, this acceleration is not nearly at an acceptable rate vis-à-vis other technological advances in other fields. There exists no Moore's Law for the built environment. One reason behind this is that architects operate spatially just as they did during Modernism, which results in excruciatingly similar space-types. Within all of this fetishization of precision, there exists a number of lost opportunities. More specifically, we might miss the window for experimentation with new space conceptions[2] that truly engage the zeitgeist (Giedion 1941).

These issues are not without response within the contemporary discipline. David Ruy decries the recent shift away from architectural objects and champions an architectural embrace of strangeness towards the production of new objects (Ruy 2013). The development of precise and complex networks and systemic relations within architecture that continues to amplify the pursuit of material precision lies in opposition to the pursuit of strangeness that Ruy champions. A more applied response to this critique is found in the material and intellectual work of the Possible Mediums Project, where its founders continue to explore unconventional 2D and 3D materialities.

The motivation behind the work in this paper is to question the production of space and aesthetics in a post-human frontier as we embark on a new geological era that is emerging out of the enormous influence of the human race on the planet's ecology. More specifically, it attempts to present the contribution that a conscious and designed blurring between the natural and the synthetic in the post-human frontier can materialize a space conception that exhibits qualities that are both natural and synthetic. *Almost Natural Shelter* describes a faster and messier architectural world-view that embraces material misbehavior through integrating linear and nonlinear fabrication techniques, resulting in a highly naturalized space conception and aesthetic.

LETTING GO AND MASTERING THE MESSY
Letting go of control when working with materials implies a disciplined and collaborative relationship between all agents in a design process. In the case of the examples presented in this paper, there exists a three-part collaboration between the author as designer and human agent, digital information and

6

4   Early digital studies showing amplified textural readings within compression-only structures.

5   3D-printed prototypes of the early digital studies.

6   Process image showing a resin cast during the de-molding process.

7-8 Resin cast chairs that exhibit textural surface qualities.

material behavior (and misbehavior). This collaborative relationship is amplified when using materials and material processes that are less prone to submission. *Figure 2 + Figure 3* show an early exploration into this attitude towards messy making. These explorations are essentially a resin casting workflow that explores the relationship between the moldable casting material (epoxy resin) and a formwork (Expanded Polystyrene – EPS) that reacts in a volatile manner when exposed to the exothermic process of curing resin. This process results in an object that deviates to varying controllable degrees from its intended form and is characterized by a high resolution differentiated texture. These early explorations succeeded in responding to the two overarching critiques set out at the beginning of the paper, namely the creation of objects through an imprecise fabrication method that results in an almost natural aesthetic that shies away from linear patterns in favor of differentiated textures.

There remains a resistance towards this attitude in the production of architecture in general. Jeremy Till and Sarah Wigglesworth lay the blame squarely at Mies' feet. (Perhaps the blame should go to Miesian disciples, and not to Mies himself.) Through his cult of personality, the Miesian mythical attention and expression of detail continues to give credence to the pursuit of precision. Teachers of architecture continue to speak of that corner detail with extreme reverence. David Spaeth has gone as far as referring to him as "the architectural conscience of this age," (Spaeth 1988) (Till and Wigglesworth 2001).Today, it is not the spatial discipline of Mies that is at work, it is the fetishization of the detail, and ultimately of precision. Thus, the

potential of developing new space conceptions, as Mies had masterfully contributed to through his oeuvre, goes untouched. Miesian ideas such as openness, lightness, transparency and movement are still driving much of architecture. These had their moment, but this must pass.

Ultimately, Till and Wigglesworth put it aptly when they write that "the project to provide society's salvation through recourse to architectural honesty, truth, economy of means and precise tectonics appears deeply flawed and delusional" (Till and Wigglesworth 2001). Precision is a delusion; it is high time architects just let go of it. Thus, mastering a messy collaboration with volatile materiality is needed. There is an unusual video floating around the internet where Richard Long, acclaimed English land artist and winner of the 1989 Turner Prize, paints a relatively large and rather intense wall mural with a specific mixture of mud in less than 18 minutes. Long's oeuvre is abundant with mud paintings at different scales and in different configurations, giving him a unique experience with his medium of choice. On top of portraying his technical mastery and control over the material, the video shows a quick production of a work of art if there ever was one. However, while the mural is painterly on the surface, it straddles a very particular material and conceptual space between the messy and the precise, the orderly and the misbehaved. This is evident by watching the artist produce the mural; one can easily note that his strokes do not lack intentionality. Long embraces the volatility and precariousness of his medium, allowing it a very long leash to express its misgivings.

9

10

11

12

9 Photograph showing the 5-axis routing process of the different blocks prior to gluing.

10 Photograph showing the 5-axis routing process of the different blocks prior to gluing.

11 Photograph showing full object under gluing. Epoxy resin and fiberglass were applied on the outer surface to increase object bonding and lateral stability.

12 Photograph showing full object under gluing. Epoxy resin and fiberglass were applied on the outer surface to increase object bonding and lateral stability.

13 Photograph showing some of the analog tools used to create the textures including a hand-held wire cutter and a hot gun.

14 Photograph showing full object while the outer timber skin was being installed.

13

14

15 Photograph showing the interior of the object during the process of texturing.

Long's attitude towards the production of art is antithetical to the production of most of contemporary architecture in general, and work that falls under the umbrellas of computational design and digital fabrication in specific. He posits a relationship between material, artist and product that is volatile and collaborative as opposed to submissive. Ultimately, the artist exerts control, but it is a radically different kind of control than the one that have proliferated with the advent of digital fabrication techniques. His kind of control results in an aesthetic of exuberant naturalism characterized by material misbehavior and messiness. Revisiting his work and others of a similar kind through the lens of computational design methodologies and digital fabrication techniques has the potential to illuminate alternative modes of contemporary architectural production that result in a new space conception and aesthetic discourse at a time when most architecture relying on contemporary tools ironically looks homogeneous in its pursuit of surface differentiation. A faster, messier and highly energetic form of spatial practice can emerge through an oblique and deviant adoption of technological advancements.

MISBEHAVED DIGITAL THINGS

What follows is a description of a series of experiments that are influenced by the ideas and critique described above and whose overarching aesthetic is strongly driven by digital and material misbehavior. The relatively small scale results have paved the way for the design and construction of a 3 m x 3 m x 3 m (approx. 10 ft x 10 ft x 10 ft) installation that will be described below.

Initial digital studies (*Figure 4*) relied on a custom-built computational system that produces results that amplify textural readings within compression-only structures. Historically, compression

only structures, masonry construction such as columns and domes, exhibited high degrees of part-to-whole relationships. Sometimes, as with examples of rustication, mass begins to breakdown in favor of textural readings. Thus, this made an appropriate starting point and test condition. Here, the process attempts to transform discrete readings of column grids (discrete lines) and domes (discrete surfaces) towards non-discrete textural object readings that blur part-to-whole relationships. In the initial stages of the exploration, these studies were continuously fabricated utilizing additive manufacturing techniques. This was doubly preferred given that these techniques prefer compression only structures and perform much weaker in tension.

Digital misbehavior in these models comes in two forms, the first structural and the second organizational. The structural misbehavior occurs through allowing a high margin of error for the above set goal of compression-only solutions. For example, the approach allows portions of the system to refuse to optimize towards compression and remain under tension. Some of these conditions translate into failed 3D prints, as seen in *Figure 5*, which becomes a form of evaluation criteria (i.e. how much structural misbehavior is allowed).

Organizationally, the explorations are deployed to structure themselves in one of two rigid organizational patterns, a radial (dome) and a bi-axial (column) grid. The misbehavior occurs through allowing a high degree of flexibility for the elements within the form to exist simultaneously and in different degrees between the two patterns. The central image in *Figure 4* and *Figure 5* shows an organizationally misbehaved condition.

Almost Natural Shelter: Non-Linear Material Misbehavior Tabbarah

16 Photograph showing the interior of the object after the process of texturing.

17 Photograph showing the full object on exhibition.

18 Photograph showing an interior vertical funnel.

17

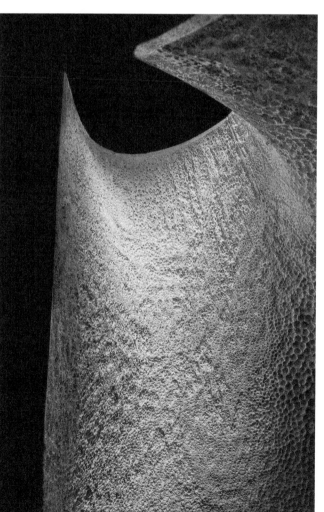

18

MISBEHAVED MATERIAL THINGS

The studies were scaled up through developing a resin casting workflow that explores the relationship between the moldable casting material (epoxy resin) and a formwork (Expanded Polystyrene – EPS) that reacts in a volatile manner when exposed to the exothermic process of curing resin. As a scale test, these models aimed to produce highly textural things that exhibit the ability to perform as chairs, as seen in *Figure 7 + Figure 8*.

The choice of exploring misbehavior through casting comes from a desire to re-question the operation's contemporary history. Casting is typically a very precise type of construction and easily reproducible. Since the advent of reinforced concrete in the early 20th Century, procedures have been developed to ensure quality control. Thus, in trying to relinquish *some* control at the beginning of the process, the question arises: How can one make casting a little less precise?

The solution lies in rethinking the relationship between the casting material and the formwork (*Figure 6*). In typical casting processes, the formwork is primary in that it directs the flow and form of the casting material and allows it to cure with a little temperature control. Within this body of work, resin was chosen as a casting material and was paired with a specific form of polystyrene. Here, the relationship between the two flattens since the exothermic reaction between the cast material and the formwork produces deep textures (*Figure 3*). Basically, the heat from the curing resin burns some of the closest layers in the formwork. Making the formwork deep enough allows the process to cool over time and permits some form of misbehaved formation to emerge. The resultant textures penetrate the surface into the core of the chairs to truly allow for a reading of a textural formation and blurs any reading of pattern or unit-to-whole relationship.

PRECISE FABRICATION

In March 2016, an opportunity presented itself to conduct the above research on a spatial scale with a given space of 3 m x 5 m. This was funded by 1971 Design Space in Sharjah, U.A.E, to be exhibited at Design Days Dubai 2016. Conceived as a hollow object, the interior of *Almost Natural Shelter* rejected traditional part-to-whole relationships through the design of deep and spatial textural formations. This shelter emerges from the integration of messy computational design methodologies and chemically volatile non-linear fabrication. In specific, high density EPS is persuaded to chemically self-compute in an attempt at uncovering a shelter that has almost natural spatial qualities such as non-linear textural differentiation and interplay between line, surface and mass. The blurring between the synthetic and the natural is also evident in the manufacturing processes at play within this shelter. A juxtaposition of slow and precise manufacturing techniques, such as cutting and gluing, with messy and fast operations such as melting. This results in a determinate and synthetic exterior and an interior that exhibits an amplification of texture that is almost natural.

The first response was to linearly scale up the chair casting process to create a spatial object at 3 m x 3 m x 3 m. While the space available was larger, the 3 meter cube was a function of the largest single object that can be transported on a flat-bed truck without police escort. The casting process proved materially unscalable because the amount of heat generated from the exothermic reaction at that size exceeded the amount generated from casting the chairs exponentially, while the EPS's ability to withstand that heat does not scale up equally. This resulted in the epoxy resin completely melting the EPS prior to curing.

The above failure lead to rethinking and scaling down the chemical reaction that acts on EPS. The phasing and manufacturing of *Almost Natural Shelter* mimics the two-step process that takes place when manufacturing the chairs. The first, linear and highly controlled, is the 3-axis milling of the negative form in 60 cm x 60 cm x 10 cm blocks of EPS that were eventually stacked and glued. The second, non-linear and slightly volatile, is the exothermic resin casting into the negative spaces of the

EPS. At the larger scale, a 5-axis mill was utilized to cut out a predetermined form. The overall form was produced in 3 large sections and eventually glued into its final composition as seen in the milling process in *Figure 9 + Figure 10*. Epoxy resin and fiberglass were applied on the outer surface to increase EPS bonding and lateral stability (*Figure 11 + Figure 12*), resulting in a single transportable volume that weighs less than 500 Kilograms; a temporary base was constructed to allow a small forklift to position it in its exact location.

IMPRECISE FABRICATION

The second phase in the process utilizes the volume produced via the 5-axis mill as a canvas on which major guidelines for a differentiated pattern are drawn using chalk by the 5-axis arm. This is then transformed into a heterogeneous textural formation in a 2-step process. The first step in the process is the cutting out of the pattern using a hand-held hot wire cutter (*Figure 13*). The tool used allows for a quick adjustment of profile and size, and while manual, far exceeds the speed it would have taken to create the cavities in using the 5-axis mill, which has a double effect if one factors in the cost of machine time. It was also conceptually desirable given the lack of interest in deploying a precise pattern onto the surface, especially as it will completely transform into something other and far more differentiated in character.

After the cavities are fully made (*Figure 15*), the second step in the process includes using an off-the-shelf white enamel spray paint. Using a solvent-based spray will react to the porous surface of the EPS and induce a chemical reaction whereby the integrity of the cavities will be dissolved. To achieve a highly differentiated condition, the spray was applied in different speeds and at different distances from the EPS surface as these variables change the strength at which the EPS dissolves.

The final layer is the assembly of a series of timber panels on the outside face of the object. 3 m x 8 cm timber panels were glued using silicone onto the fiberglass surface with a 2 mm gap. The introduction of this skin was meant as a hint at the blurring between the artificial and the natural that takes place on the interior surface. Thus, the exterior is the only surface where precision gets introduced to identify to observers that they are entering a fully artificial condition, even though the resulting textural formation might indicate otherwise. This was further put to test during the exhibition, where visitors invariably showed various expressions of surprise at being told that this was neither coral nor any other purely natural material, but simply EPS.

CONCLUSION

Further explorations within this umbrella of research fall under two main strands. The first is the development of a more streamlined process where the messiness is machine automated in a loop from bits to materials. The second strand might require moving away from the discrete object reading where the material formation can be deployed in conjunction with other material systems on a fully functional architectural space. For example, this includes designing installation techniques as well as ensuring that the material passes code such as being fire resistant. This will also push these explorations to become truly space-making tools and generators of space conception in the post-human frontier.

ACKNOWLEDGEMENTS

Almost Natural Shelter was funded by 1971 Design Space in Sharjah, United Arab Emirates and was first displayed in Design Days Dubai 2016, in the United Arab Emirates.

This work was done with Khawla Al Hashimi and Nada Taryam.

NOTES

1. Graduate programs at leading institutions continue to develop degree courses that aim to explore the space of ultimate precision.
2. The use of the idea of space conception is based on its appearance in Sigfried Giedion's *Space, Time and Architecture*. Giedion describes the emergence of a new space conception in Modernism in relation to older space conceptions. The first space conception is defined by volumes and objects in space, and the second by questions of interiority brought about by the vaulting problem. According to Giedion, the space conception developed in Modernism is concerned with the simultaneity of volume and interiority as space generating and defining.

REFERENCES

Giedion, Sigfriend. 1941. *Space, Time and Architecture: The Growth of a New Tradition*. Cambridge: Harvard University Press.

Hughes, Francesca. 2014 *The Architecture of Error: Matter, Measure, and the Misadventures of Precision*. Cambridge, Ma: MIT Press. p 5.

Ruy, David. 2013. "Returning to (Strange) Objects," in *Adaptive Ecologies: Correlated Systems of Living*. Edited by T. Spyropoulos. London: AA Publications. p. 268–277.

Schmidt, Eric, Jonathan Rosenberg, Alan Eagle. 2014. *How Google Works*. New York: Grand Central Publishing. p 38.

Spaeth, David. 1988. 'Ludwig Mies van der Rohe: A Biographical Essay," in *Mies Reconstructed*. Edited by J. Zukowsky. New York: Rizzoli International Publications. p 33.

Till, Jeremy and Sarah Wigglesworth. 2001. "The Future is Hairy" in *Architecture: The Subject is Matter*. Edited by J. Hill. London: Routledge. p 15.

IMAGE CREDITS

Figures 1–2: Pachica, 2016
Figures 3–4: Tabbarah, 2015
Figures 5–6: Tabbarah, 2014
Figures 7–8: Abu Shakra, 2015
Figures 9–15: Tabbarah, 2016
Figure 16: Roldan, 2016
Figures 17–18: Kalo, 2016

Faysal Tabbarah is an architect and educator, currently Assistant Professor of Architecture at the American University of Sharjah, U.A.E. He is also co-founder of Architecture + Other Things, a collaborative and interdisciplinary platform exploring alternative models of architectural practice and design to produce work at multiple scales and within multiple disciplines. He received his Master's in Architecture and Urbanism at the Architectural Association School of Architecture's Design Research Lab (AADRL) in London and his B.Arch from AUS. He has held professional positions in architecture and design in London and the U.A.E.

Experimental Material Research — Digital Chocolate

Simon Twose
Rosa du Chatenier
Victoria University of Wellington

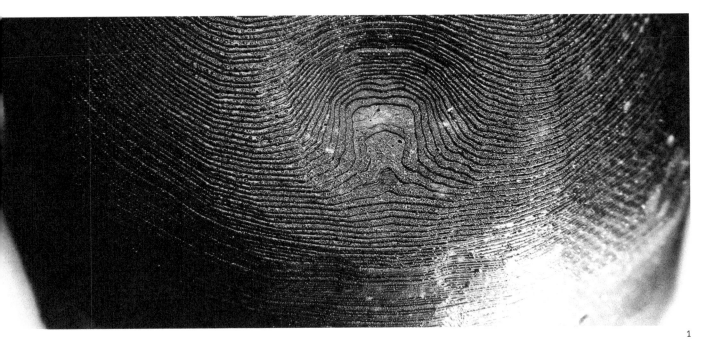

1

ABSTRACT

This research investigates the aesthetics of a shared agency between humans, computation and physical material. 'Chocolate' is manipulated in physical and virtual space simultaneously to extract aesthetic conditions that are a sum of human and non-human relations. This is an attempt to further the knowledge of designing, giving physical and digital materials force in determining their own aesthetics. The research springs from work in speculative aesthetics, particularly N. Katherine Hayles's OOI (object-oriented inquiry) and Graham Harman's OOO (object-oriented ontology) and explores how these ideas impact contemporary computational architectural design.

To study this, a simple material has been chosen, chocolate, and used as a vehicle to investigate the dynamics of physical and digital materials and their shared/differing 'resistances to human manipulation' (Pickering 1995). Digital chocolate is 'melted' through virtual heat, and the results printed and cast in real chocolate, to be further manipulated in real space. The resistances and feedback of physical and digital chocolate to human 'prodding' (Hayles 2014) are analyzed in terms of a material's qualities and tendencies in digital space versus those in physical space. Observations from this process are used to speculate on an aesthetics where humans, computation and physical material are mutually agential. This research is a pilot for a larger study taking on more complex conditions, such as building and cities, with a view to broadening how aesthetics is understood in architectural design. The contribution of this research to the field of architectural computation is thus in areas of aesthetic speculation and human/non-human architectural authorship.

1   Chocolate cast 1 – chocolate model formed through a process of 'melting through touch" in digital space.

INTRODUCTION

Aesthetics traditionally has human perception at its centre. Recent thinking challenges this, arguing that non-humans contribute, mysteriously, to aesthetic viewpoints (Hayles, 2014). In this research, dynamic performances of matter are given aesthetic priority; they are 'forceful' in drawing their own form. This disrupts our imaginative and aesthetic presumptions, particularly when designing built space, prompting unexpected solutions to surface. This research attempts to apply ideas in speculative aesthetics, object-oriented ontology (OOO) and object-oriented inquiry (OOI), to the field of computational architectural design, through a study of comparative physical/digital materialities. Through a series of experiments in forming digital 'chocolate', the research pursues a design ontology inclusive of the aesthetic agency of objects and computation. As such, the objectives of this research are to identify and analyse the forcefulness of non-human agents in architectural design in order to speculate as to what their influence might be on architectural aesthetics.

Key contextualizing texts in this area are Hayles' "Speculative Aesthetics and Object-Oriented Inquiry," in which Hayles outlines an approach to understanding the world through an aesthetic agency shared between humans, objects and artificial intelligences (Hayles 2014); Brian R. Johnson's "Virtuality and Place," which argues for a blending of physicality and virtuality by claiming that they are inherently both based on experience (Johnson 2002); Graham Harman's notions of the 'allure' objects have to one another (Harman 2002) and work such as Sean Cubitt's Digital Aesthetics, which speaks of the aesthetics of the digital and their effect on culture and society. These point to a literary context which is extended by the digital design research in this paper.

A key precedent for the research is the installation *Skulls* by Robert Lazzarini, first exhibited at the BitStreams exhibition in 2001 (Hansen 2003). *Skulls* is a series of sculptures created by 3D scanning a skull ,which is then distorted in digital software and the distorted results prototyped in bone. This creates a set of strange, hybrid objects that allow a human appreciation of aesthetics internal to digital 'materiality', confronting viewers with the plasticity of digital space in physical form (Hansen 2003). In essence, the research in this paper investigates physical/digital material hybridity, in which design feedback shuttles between analogue and virtual conditions, allowing the aesthetic decisions in designing to be a sum of more than human factors.

2

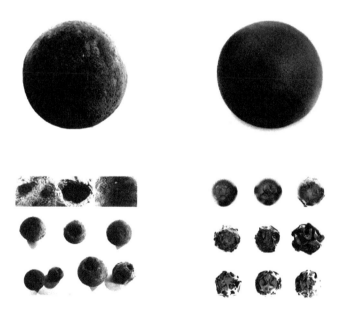

3

2 *Skulls* – Robert Lazzarini's *Skulls* installation, first exhibited at BitStreams exhibition in 2001.

3 Formal chocolate explorations – melting through touch, in physical space on the left and in digital space on the right and below.

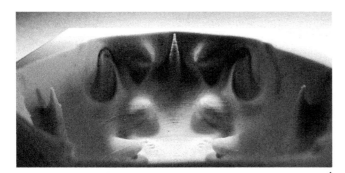

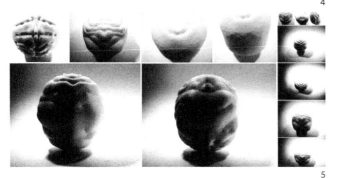

4

5

6

METHOD

This research tested the resistances, affordances and feedback of computational and physical material through a series of 'chocolate' experiments; chocolate was deformed virtually and the results cast in real chocolate using 3D prototyped and silicon molds. These experiments are the first steps in testing how designing might be inflected by multiple aesthetic agents. This research will be developed further, encompassing progressively more complex architectural conditions, and go on to inform the design for a public building.

Material Characteristics

Chocolate was chosen for its physical characteristics, its ability to melt and be cast into free-form shape, as well as its sensory potential, giving feedback of smell, touch and taste. Chocolate's propensity to melt at the touch was chosen as a way to merge human actions with the digital realm. Melting was performed in RealFlow software after experiments simulated manual application of the cursor on virtual chocolate spheres. Successive experiments in melting the sphere were performed through digital equivalents to physical touch to achieve unusual or unruly aesthetic results.

Digital/Physical Modelling Process

A sphere of real chocolate was cast by making a plaster mould from a glass sphere, which was then used to cast molten chocolate. A similar sphere was created in RealFlow by assigning particles in a spherical volume with particle properties resembling chocolate, such as physical qualities of viscosity, particle

attraction, melting point and surface tension. A heat emitter was created in the software to transfer heat energy to the particles in the virtual 'chocolate' to simulate the melting of a warm finger probing the chocolate. The effects of this on the form of the sphere were animated and rendered to create a virtual version of someone melting the chocolate with their fingers. One thousand iterations were made of this virtual melting, which fine-tuned the digital version of the physical and sensory event of melting chocolate with the fingers.

Casting

The next stage was to transpose the digitally 'designed' chocolate into real chocolate. In order to allow the 'trace of the digital' (Dixon 2007) to come through into the physically cast chocolate, the unruly meltings were repeated around an axis of symmetry. This allowed the aleatory results of melting through warm virtual fingers to be tied to computation, as this symmetrical duplication was not possible in the real material. This was achieved through two virtual heat emitters, set up at exactly opposing sides of the chocolate sphere, giving an artificial formal logic to performances in formlessness.

In order to cast the digital meltings, the symmetrical chocolate distortions were initially inversely 3D printed to make a mold using Rhinoceros software. The resulting series of chocolate objects was a physical record of formal manipulations through virtual senses of touch. The objects carry the imprint of human play, yet due to their symmetry possess a shape not possible if the play was performed in the physical world. They are

4   Molds 1 – close up of 3D printed mold, shaped by the formal manipulations of chocolate melting in digital space.

5   3D prints of formal explorations – of the chocolate in digital space, to be cast to create silicone molds, to which the physical chocolate will be cast.

6   Melting physical chocolate process – close up of the process of melting the physical chocolate.

7   Molds 2 – 3D printed molds, symmetrical along one axis. Formally interesting but difficult to cast into, due to double-buttress intrusions in the surface. Material limits of the chocolate were stretched to the point in which digital form was no longer discernible.

7

somewhat normal as abstract forms when viewed on a screen yet have a strange physical presence as objects, particularly when placed in relation to one another, rotated in the hand, smelt, melted, or tasted.

The difficulties in removing the chocolate from the 3D printed molds led to interesting textures derived from the resistance of the chocolate to their rigid formwork. The surface only approximated the spheres, and the accidental inclusions added another layer of unruly material forming, becoming a physical approximation of a virtual approximation of human interaction in the real. These inclusions led to a surface texture that bordered on the abject from a human perspective, yet was integral to the aesthetics of the materials and the performances in their making.

The next casting test was to 3D print the melted volumes and create flexible silicone molds around them. This experiment was more successful in casting accurate versions of the digital chocolate, yet had less aesthetic intrusion into the process. The models created with this technique had a striated surface appearance, replicating the 3D printer's way of making objects from layers of fine filaments. This trace of the manufacturing, as well as the symmetry evident in the form of the chocolate, connected the object to aesthetics inherent to the digital realm, though the object has a singular physical presence in the real. The objects' strong aroma and presence as possibly edible things —slightly abject confections—deflected them from being dumb physical prints of digital forms, and they became forceful in strange, sensory ways. This ties them to similarly strange hybrid

conditions described in the *Skulls* sculptures, curiously conflating virtual and palpable space.

RESULTS

This iterative design research produced hybrid objects with the trace of both physical and virtual relations. Human actions, such as melting, were simulated virtually in frictionless digital space which produced smooth forms. These forms were impacted by printing and casting, and the sensory objects that resulted. The process provided successive iterations that bound non-human agency in an aesthetic dialogue with human design actions. The result was a test in generating a hybrid aesthetic, part-way between computation, material and sensorially-driven human input.

Digital/Material Agency

The digital's capacity to simulate physical properties and simplified, human 'design' actions produced gooey, unpredictable formal results, and by having the computer duplicate these gooey forms across an axis of symmetry, as they unfolded the forms spoke of the frictionless, accurate and plastic aesthetics of the digital. They also spoke of a certain smoothness and lack of resistance to the movements of the human operating the actions. The making of these gooey virtual objects in physical chocolate created a second order of variation, bringing in resistances inherent to physical material. Through melting, molding and unmolding, small imperfections, bumps, bubbles, and variations of tone compromised the perfect symmetry from the digital, and gave the objects a fine surface detail which contributed to an aesthetic unpredictability. Their potential as confectionery added

8   Chocolate cast 2 – close up of chocolate cast, created through silicone-molding process of 3D printed form, which physical chocolate was then cast into.

9   Chocolate cast 3.

10 Chocolate cast 4.

a curious sensorial feedback. This dialogue between digital and physical aesthetic dynamics bound together unpredictable forces in the forming of the object and rendered an aesthetic agency in designing.

Human Agency

This project is an experiment in deflecting human aesthetic agency, so, despite setting the process into play, there has been an openness to aesthetic feedback from multiple non-human sources. The thinking surrounding OOO and OOI suggests all agency as ontologically equivalent; the smooth plasticity of the digital somehow combines with the recalcitrant ruptures and meltings of the real and the embodied actions of the human, directed by contingent sensorial motives. This is proposed as an

intensification of the architectural design process where feedback traditionally comes from multiple processes and materials.

Iterative Process

This research is a simplified version of a typical iterative design process, accentuating the shared agency of humans with their non-human collaborators in design-decision making. Materials, both real and virtual, push back and impress themselves on the design process, creating dynamic relations that are forceful authors of aesthetics.

CONCLUSION

Such hybrid aesthetic influences are being developed in the design for a public building, which follows on from this early work, with the hope that new aesthetic understanding will emerge as architectural complexity is fed into the process. The future work tests in more detail how a design methodology emphasizing human, computational and material collaboration can affect the design of architectural space, be it a building or perhaps, if encompassing larger materialities, a city. The first stage of this has been to melt the urban context for the public building as if it were a found material. A mix of photogrammetry, prototyping, casting, chocolate molding, heating, touching and tasting has shifted the site from a known spatial phenomenon to a complex blend of aesthetic agency, shared between human, computer and material. Other key architectural aspects of the design are being developed in the same way, such as form, tectonics, surface, materiality and programme. The project is a test bed for an architecture of ontological equivalency between real non-us, digital non-us and us.

Feedback loops in the aesthetic development of the design become strange and tightly circular in this mode. Abject forms thrown up by corrupt photogrammetry files or distorted molds are ameliorated by their sensorial richness as melted chocolate, making them candidates as architectural forms whereas they otherwise might be ruled out. This points to a shuttling between the aesthetic forcefulness of computation, things and human that departs from a solely human centered mode where visual perception is dominant. Aesthetics internal to real materials, their relations to each other and their relations to their digital doppelgangers are allowed to emerge through iterative designing and merge with human sensibilities, albeit in ways which are not entirely comfortable. Strangeness is perhaps an inevitable architectural condition of such hybrid sensorial feedback.

The success of this research is that it identifies curious aesthetic conditions at work in hybrid digital/ physical design operations that point to further study, both of design methodology and the architecture that might emerge from it. The failures of

Digital Chocolate Twose, du Chatenier

11 Experimental architectural investigation in chocolate – furthur experiments, as an inquiry into the future applications of speculative aesthetics and OOI to architecture.

this research are that, at present, it has a small scope and small collection of data to back up observations about shared aesthetic agency and feedback. This is an ongoing project and work is continuing on these difficult questions.

This research adds to the field of computational design in architecture through discovery of aesthetic conditions gained from a hybrid, ontological methodology, blending human sensory capacities with those of non-humans: objects, materials, computation. It wonders about what architecture might be, if composed of multiple desires and tastes, spanning human and non-human sensibilities.

IMAGE CREDITS

Figures 1–12: du Chatenier and Twose, 2016
Figure 2: Robert Lazzarini, 2000 ©

REFERENCES

Cubitt, Sean. 1998. *Theory, Culture and Society: Digital Aesthetics*. London: Sage Publishing.

Dixon, Steve. 2007. *Digital Performance*. Cambridge: The MIT Press.

Hansen, Mark B.N. 2004. *New Philosophy for New Media*. Cambridge: The MIT Press.

Harman, Graham. 2002. *Tool-Being: Heidegger and the Metaphysics of Objects*. New York: Open Court.

Hayles, N. Katherine. 2014. Speculative Aesthetics and Object-Oriented Inquiry (OOI). *Speculations: A Journal of Speculative Realism* V:158–179.

Johnson, Brian R. 2002. "Virtuality and Place". *Thresholds - Design, Research, Education and Practice in the Space Between the Physical and the Virtual: Proceedings of the 2002 Annual Conference of the Association for Computer Aided Design in Architecture*, Edited by G. Proctor. Pomona: ACADIA.

Pickering, Andrew. 1995. *The Mangle of Practice: Time, Agency and Science*. Chicago: University of Chicago Press.

Simon Twose is an architect and Senior Lecturer at the School of Architecture, VUW New Zealand. His work focuses on design research, looking particularly at the crossings and transferences between drawing and built space. Twose has contributed work to four Venice Biennales, Adam Art Gallery and the Prague Quadrennial, PQ15.

Rosa du Chatenier is a Masters of Architecture student in her final year at Victoria University (NZ), and is interested in the crossing of physical/digital architectural design approaches to inspire new design methodologies. After her studies are complete, she is looking for work!

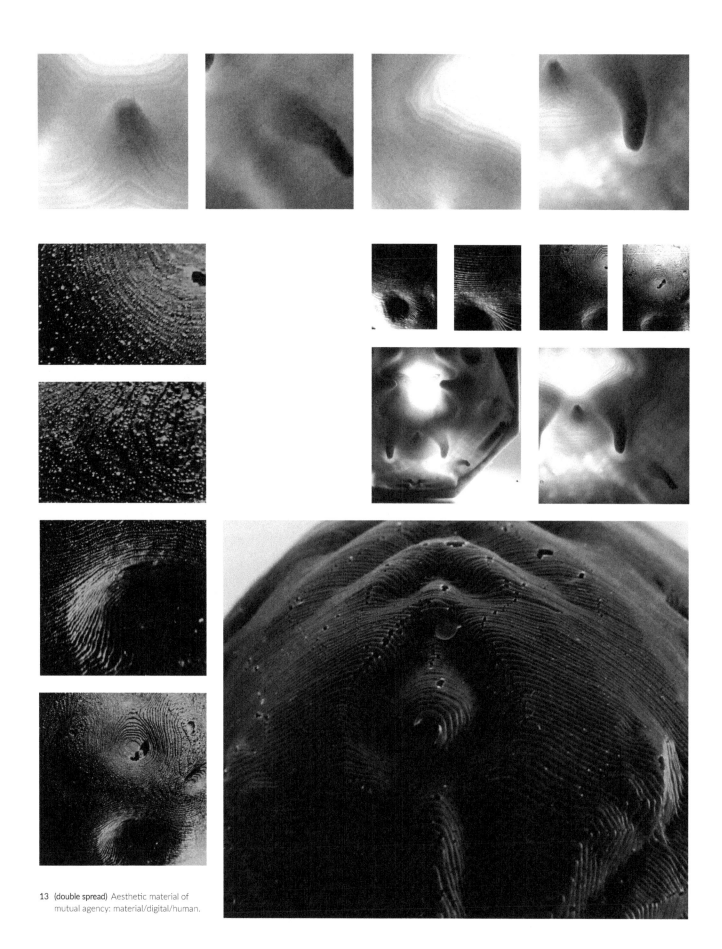

13 (double spread) Aesthetic material of mutual agency: material/digital/human.

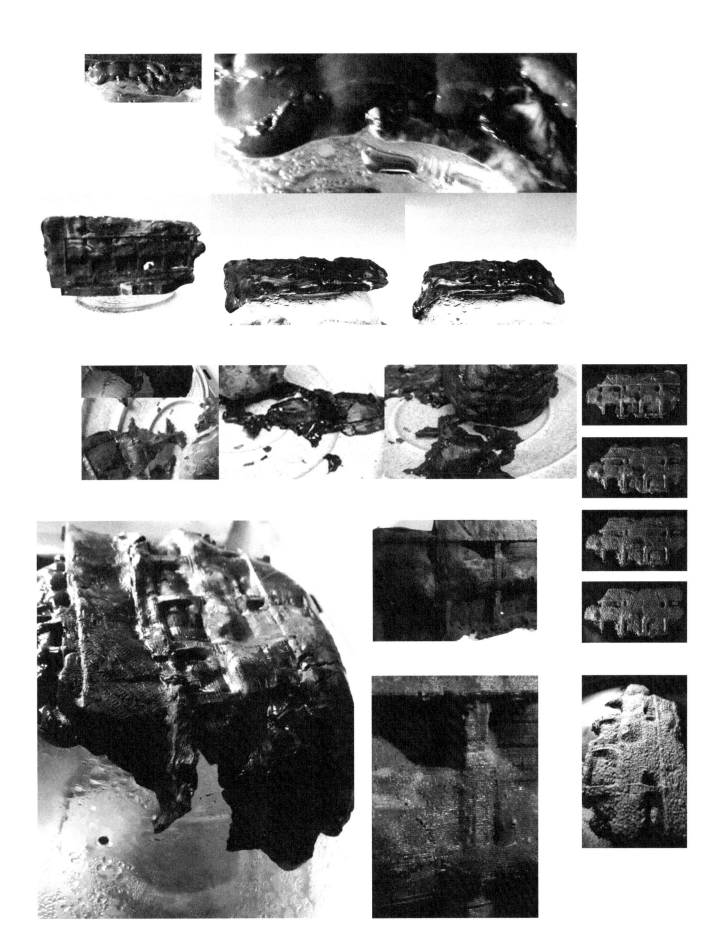

Landscapes After The Bifurcation of Nature

Models for Speculative Landformations

Michael Leighton Beaman
Rhode Island School of Design
Beta-field

1

ABSTRACT

Landformations have not historically been the purview of design production or intervention. Whether it is the spatial extensions in which they emerge, the temporal extensions in which they operate, the complexities of their generative and sustaining processes, or a cultural and institutional deference to a notion of *natural* processes, designers as individuals or design as a discipline has not treated landformation as an area of design inquiry. But the inability to grasp nature fully has not stopped geological-scale manipulation by humans. In fact, anthropogenic activity is responsible for the re-formation of more of the Earth's surface than all other agents combined.

And yet as designers we often disregard this transformation as a design problem, precisely because it eludes the artifices of information visualization employed by designers. This paper examines ongoing research into the generation of *speculative landformations* through an analysis of underlying geological and anthropogenic processes as the quantitative basis for creating generative computational models (figure 1). The Speculative Landformations Project posits human geological-scale activity as a design problem by expanding the operability and agency of environmental design practice through hybrid human/digital computations.

1   3D surface articulation of existing topography from USGS, Digital Elevation Model (DEM).

GEOLOGICAL AGENCY
Anthropogenic by Design

Through a systematic manipulation of the landscape, humans account for the fastest geological transformation of the Earth's surface in its 4.54 billion year history (Wilkinson 2005). Anthropogenic activity is responsible for the re-formation of more of the Earth's surface than all other mechanisms combined. Agricultural and industrial practices impact the majority of this change. When, combined with formations and programs traditionally held within the domain of design and architectural practice—spaces of habitation, occupation, protection, and labor—humans have steadily increased the depth of the Earth's anthropocentric event layer to span over 2,000 meters (Sanderson et al. 2002).

The goal of this planetary re-formation project has been to domesticate our environment through the mitigation or exploitation of its material processes and effects. Human intervention, however, has been unable to target specific environmental processes, and cannot generate isolated effects. The interconnectedness of constituent agents, pressures, and materials in any ecological/environmental system undermines the effect of creative and projective spatial design practices such as architecture, landscape architecture, and urban design.

The *Speculative Landformations Project* (SLP) is an ongoing body of research that re-frames the question of anthropogenic landformation as a design problem. Through the creation of a hybrid analytical model which combines human and non-human computational methods, the SLP reverses the predictive goal of analysis to the speculative goal of design. This reversal focuses on two aspects of landformation: *phenomena* —the events that are produced as a result of information exchange—and *artifacts*—the formal results of those events (Beaman & Hong 2016). In this early set of iterations, the SLP is centered on examining landformation morphology (figures 2 & 3) and process typology as vehicles for examining existing ground conditions and generating potential configurations.

Post-Human Propositions

Landformations have not historically been the purview of design production or intervention. Whether it is the spatial extensions in which they emerge, the temporal extensions in which they operate, the complexities of their generative and sustaining processes, or a cultural and institutional deference to a notion of natural processes, neither designers as individuals nor design as a discipline has treated landformation as an area of design inquiry (Smith 1994).[1] This perception of a distinction between what exists and what is within human capacity to know, and thus intervene in—what Alfred North Whitehead called the " bifurcation of nature"

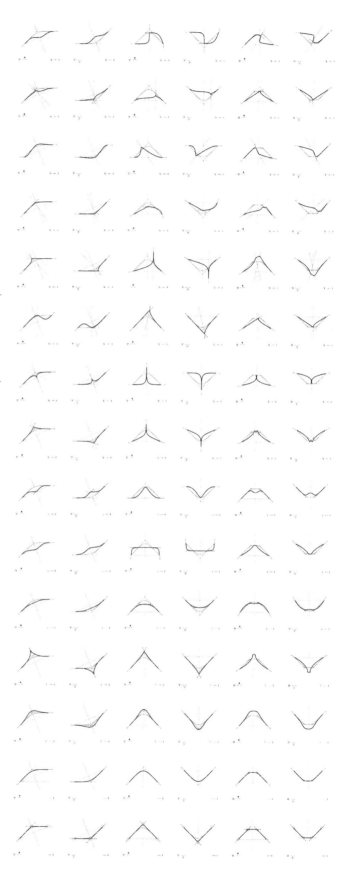

2   Landformation Morphology—*abridged*: Profile Analysis.

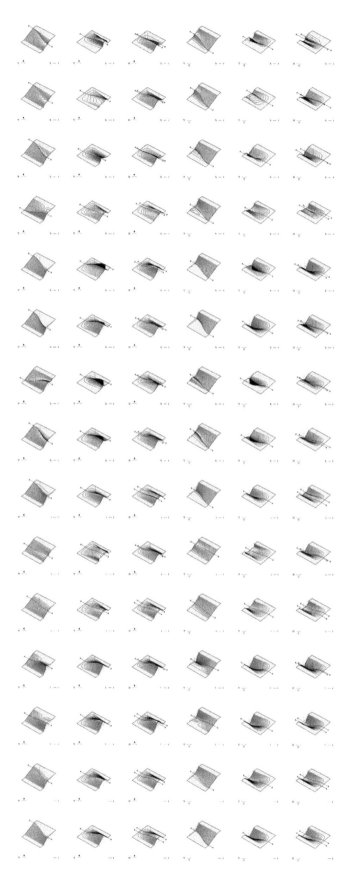

—undoubtedly colors the tendency to address landformations as entities which lie outside design practice (Whitehead 1920, passim). But the inability to grasp nature fully has not stopped geological-scale manipulation by humans.

This was first achieved through direct manipulation, relying on heuristic knowledge bases such as cultural traditions and practices (Goudie 2009). The scale of intervention widened as representational models of landformation systems developed. These "interpretive structures" allowed scientists and engineers to investigate both exogenic and endogenic land forming processes by mitigating the amount of information needed to describe an observed process or condition (Weisberg 2013, 15).[2] Computational models operate much in the same way as conventional representational models in that they reduce the complexity of an actual environment by creating a corollary between it and a computational model in an effort to study actual phenomena and artifacts through abstracted ones contingent on less information .[3]

Computational methodologies in landscape design, much like their counterparts in architectural design, assist in the design process by magnifying the ability to test a set of procedures carried out within a domain of variables, generating multiple solutions through an iterative process. Solutions are constrained to the structure of the model which means that the methodology and evaluation of model-to-target correlations determine the validity of computational (and indeed all) models (Weisberg 2013). However, for design, correlation to actual environments is only half the problem. The other half includes the creation of future configurations of current environments or wholly new ones that replace, or exist parallel to those problematized. This marks a departure from the analytical or predictive capacity needed in the sciences to the projective or speculative capacity required in design. How can these two capabilities be bridged?

Whitehead again offers an alternative framework through his use of the term "propositions"—a way of suspending the desire toward judgment of truth in favor of opening up discovery. (Shaviro 2009, 3). All representational models have latent biases which preclude certain sets of information in an effort to affect others. While these have come to be known in part as a "simple fictions" (Weisberg 2013), "purely fictional entities are not constituents of propositions" (Walton 1990, 32).[4] Proposition leverages the strengths of the logical (model structure), and the strengths of the analogical (intuitions, resemblances, or affinities assigned or imbued within the model). The combination resides somewhere between the fictional and the factual. Its advantages become apparent in computational models where both analogical and logical structures combined can exceed the representational capacity of either alone. It is in this hybridized position that speculation is viable.

3   Landformation Morphology — *abridged:* Surface Analysis.

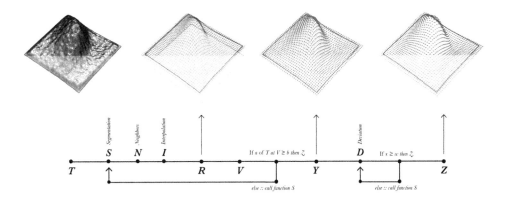

T = Target Surface (NURBS Surface, Mesh, or Point Collection)
R = Representational Surface (sub-srf)
V = Control Vector (vector)
Y = Output: Subsurface Configuration (subsrf)
Z = Output: Target Surface
a = the slope of a surface along the control vector
b = the average slope of a target complex surface along the control vector
x = the percentage of allowable deviation
w = the percentage of allowable deviation

4 Landformation Analysis-to-Proposition Procedure (simplified).

It is within this conceptual framework that the SLP examines how propositional structures within computational models might open up design inquiry in a way that bridges analysis and projection (figure 4).

SPECULATIVE LANDFORMATIONS
Analysis to Projection

Computational proposition can open landformation to design inquiry. The SLP seeks a process for landform exploration and production that extends the ability of designers to understand and communicate the operational, material, and technological history of constructed landscapes as well as their potential material, phenomenological, and technological futures. To build a model that worked in this way, we looked toward other disciplines that explore this general approach. Though there is a long history of analytical and predictive modeling of landforms and formation processes , only those instrumental to the SLP's early stages are covered here.

We began by understanding topographic characteristics as formal manifestations of dynamic systems in either stable or perturbed states. But to be able to utilize these formations in a design process, the ability to project them as discrete properties of future conditions was needed. In geomorphology, the practice of delimitation is a question defining morphometric properties. Landformations that exhibit similar formal features or geometric qualities are delimited in similar ways (Jasiewicz & Stepinski 2013). Within this regime, cycles and periods, end conditions, aspect transitions, material and geometric discontinuities, and qualitative thresholds are used to establish a provisional landformation ontology. The classification procedure can be described computationally and performed recursively or serially.

Landforms are differentiations in the Earth's surface. The process of identifying individual landformations within this continuous surface field, spanning both subaerial and subaqueous environments, falls into two approaches: 1) differentiation by formal analysis (paradigmatic), or 2) differentiation by process analysis (syntagmatic). Categorization by form relies on identifying surface features as having unique yet iterative physical qualities. Categorization by process focuses on the relationship between formation process and formal results. In both cases, landform differentiation can emerge at different scales and with varying degrees of consistency.

The SLP began with an analysis of landformations in two categories: *Morphologies* and *Process Typologies*. Within these categories we then identified boundary conditions which elicited two formation tendencies in relation to Earth's surface features: *protrusions* and *depressions*. Delineation of these two can be further determined by the pattern of field qualities (material intensity and distribution) and flow qualities (movement, direction, and speed) that each formation establishes.

- *Morphologies*: A morphometric analysis of landformations examines surface attributes at face-Value with no regard for how they have come to be. These were developed through point-based global variances within two contexts: slopes and inflections. Variances include elevational differences, slope angle, concavity/convexity, and curvature. Profile derivatives were generated through localized variations to global forms through push-pull operations applied either symmetrically or asymmetrically. These profiles correspond to both anthropogenic and non-anthropogenic physiographic features.

- *Process Typologies:* A process analysis of landformations examines patterns of material fields and flows, and form ontogeny. This analysis approach defines landformations as belonging to a linear program of activity: coastal, eolian, fluvial, glacial, igneous, lacustrine, and tectonic and in what ontogenetic state of ablation or superposition each surface is defined. Processes can be analyzed through scalar recursion, identifying features at various scales.

The first iteration of morphological analysis was generated from USGS Digital Elevation Models (DEM) of existing land surfaces, which could be reduced to a simple point cloud. To further reduce the amount of information used to describe each landform, and to evaluate it within a computational analysis schema, each target landformation was described using a representational surface with lower resolution. This was done using three steps: segmentation, neighbor finding, and interpolation (Agarwal, et.al. 2006). Next, a set of if/then questions were posed between a "representational surface" and a target surfaces to test the representational surface's fidelity through geometric similarity. [5]

Once a catalogue of surface morphological attributes was established, the evaluative process could be reversed so that any set of new attributes applied to an existing or constructed base surface generated a new singular or collection of formations. When produced in series, secondary evaluative processes such as landformation programming can be either assessed or applied using a similar reversible methodology as outlined above (figure 5).

Similarly, process analysis yielded residual patterns which could be reversed to create new landformation features. These can generally be subdivided into either additive or subtractive formation processes—features could be classified as being in a state of material subtraction (e.g. erosion or deposition) or addition (e.g. accretion of sedimentation). Since land forms are in a persistent state of formation, process analysis allows for a visualizations of past and future configurations.

To test the viability of this process, six propositional landformation hybrids were created: Cryostatic Buttes, Parasitic Pyroclastic Cones, Trellised Escarpments, River Ridges, Branched Basins, and Thermokarstic Ranges. From these models, one instantiation was printed (figure 6–11). For each series, landformations were generated by selecting attributes, defining their relative influence in the computational assemblage, and assigning them to a randomly

5   Global Surface Geometry-to-Local Surface Features: Speculative Landformations
    Series 02.

Landscapes After the Bifurcation of Nature Beaman

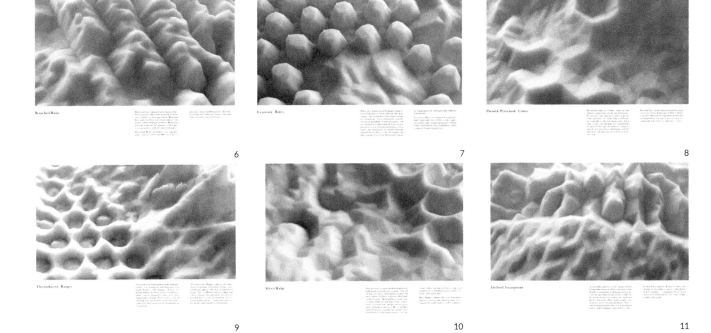

6

7

8

9

10

11

selected initiation surface. Each instantiated landform was defined through point locations at two scales (global and local) along with an articulation resolution, inverting the analytical process.

Since the primary concern was establishing representational fidelity within a highly restrictive scope as a means of revealing affinities, and subsequently categories, of landformations, dynamical fidelity (the match between modeled and actual forms) was initially downplayed. Representational fidelity was determined by how speculative surfaces compared against known and modeled phenomena with special interest to geometric similarity—its surface structure and articulation as much as it surface affect or resemblance.

FICTIONAL FACTUALS
Conclusion

While results to this point have been limited to abstract artifacts (Weisberg, 2013), landformations generated as fictional, instantiated surface conditions, future explorations will focus on speculations that correspond to existing historically defined typologies and established systems of *artificial* landform production. This phase will examine relationships of scale, shape, resolution, and occurrence of speculative formation typologies, the expectations associated with each typology, and the operative techniques used to produce them at full-scale.

In reversing the analytical function of landformation computational models to projective ones, discrepancies, incongruencies, and infidelities inherent in representational frameworks become

6   Branched Basins | Speculative Landformation Series 01.

7   Cryostatic Buttes | Speculative Landformation Series 01.

8   Parasitic Pyroclastic Cones | Speculative Landformation Series 01.

9   Thermokarstic Ranges | Speculative Landformation Series 01.

10 River Ridges | Speculative Landformation Series 01.

11 Trellised Escarpments | Speculative Landformation Series 01.

characteristics of the actual geometric composition of the speculative formations. Correspondence between the subject of the model and the model itself can be reinterpreted as the foundational compositional logic of a now generative process (figure 12).

Both architecture and landscape architecture commonly relies on the well-established techniques of abstraction and substitution (the removal or displacement of specific information to adhere and/or conform to a representational system) to reduce the complexity and data-scale of a given condition to a discrete and edited set of information. This effort equalizes the disparate types of information contained in an environment into a common vision-based, effects-oriented framework. This framework in turn has become ingrained in how designers work within the design disciplines. These systems are highly coded and ripe for procedural exploration. The connection between representations that focus on analysis and those that are projective, or rather the designer's transition from one area of focus to another, invites a high degree of interpretation. And, while this can not be avoided, considering computational models as being both inside

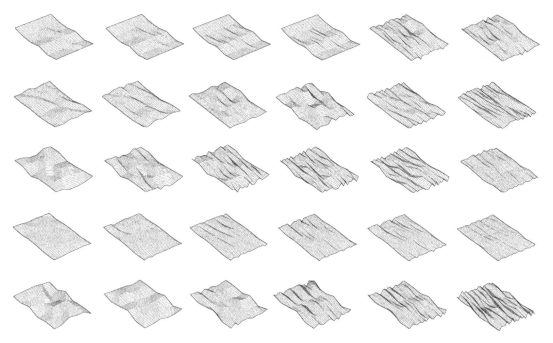

12 Sample Landformation Families.

and outside of factuality (i.e. propositions) helps intertwine both domains in ways that ground speculation and expand agency.

ACKNOWLEDGEMENTS
Speculative Landformations Project:

- Research + Design Team: Michael Leighton Beaman and Zaneta Hong.
- Student Assistants: Joshua Jow and Foad Vahidi

NOTES

1. Barry Smith defines objects as "extended in space" and processes as "extended in time".
2. See Weisberg on the definition of models in science which have here been extended into design.
3. In this case I am using the term "abstract" in the general sense, but I would argue that this applies to the specific use of abstract, as in "abstract artifacts" or "abstract structures" just as well. See Weisberg 2013, 18—19.
4. Kendal Walton offers a view of propositions as a type of constructive and ruled "fiction" . See also Godfrey-Smith 2009.
5. For more on geometric similarity and model-to-target fidelity, see Weisberg 2013, 40—45.

REFERENCES

Agarwal, Pankaj. Lars Arge and Andrew Danner. 2006. "From Point Cloud to Grid DEM: A Scalable Approach." in *Progress in Spatial Data Handling: 12th International Symposium on Spatial Data Handling*. edited by A. Riedl, W. Kainz and G. Elmes. Berlin: Springer. 771—788.

Beaman, Michael and Zaneta Hong. 2016. "Material Resonance" in *Innovations in Landscape Architecture*. Edited by J. Anderson and D. Ortega. Routledge New York. 127—128

Godfrey-Smith, Peter. 2009. "Models and Fictions in Science" in *Philosophical Studies*, vol 143: 101—116

Goudie, Andrew. 2009. *The Human Impact On the Natural Environment: Past, Present and Future. Seventh ed.* Chichester, West Sussex: Wiley-Blackwell.

Jasiewicz, Jaroslaw and Tomasz Stepinski. 2013. "Geomorphons — A Pattern Recognition Approach to Classification and Mapping of Landforms" in *Geomorphology*, vol 182: 147—156.

Sanderson, Eric W. and Jaiteh Malanding, Marc A. Levy, Kent H. Redford, Antoinette V. Wannebo, Gillian Woolmer. 2002. "The Human Footprint and the Last of the Wild," in *Bioscience*, vol 52 (10): 891—904.

Smith, Barry. 1994. "Fiat Objects" in *Parts and Wholes: Conceptual Part—Whole Relationships and Formal Mereology, 11th European Confrence on Artificial Intelligence, Amerstaerdam, 8 Aug. 1994:* European Coordinating Commitee for Artificial Intelligence. 15—23.

Shaviro, Steven. 2009. *Without Criteria: Kant, Whitehead, Deleuze, and Aesthetics.* Cambridge: MIT Press.

Taylor, Timothy. 2010. *The Artificial Ape — How Technology Changed the Course of Human Evolution.* New York: Palgrave McMillian.

Walton, Kendall. 1990. *Mimesis as Make-Believe.* Cambridge: Harvard University Press.

Weisberg, Michael. 2013. *Simulation and Similarity — Using Models to Understand the World.* New York: Oxford University Press.

Whitehead, Alfred North. 1920. *The Concept of Nature, Tarner Lectures Delivered in Trinity College, November, 1919.* Middleton.

Wilkinson, Bruce H. 2003. "Humans as Geologic Agents: A Deep-time Perspective," in *Geology.* vol 33 (3): 161—164.

IMAGE CREDITS

Figure 1: Surface Articulations (© Beaman & Hong, 2016).
Figure 2: Landformation Profile Morphology — *abridged* (© Beaman & Hong, 2016).
Figure 3: Landformation Surface Morphology — *abridged* (© Beaman & Hong, 2016).
Figure 4: Landform Propositions 01 — *abridged* (© Beaman & Hong, 2016).
Figure 5: Landformation Global Geometry & Local Features (© Beaman & Hong, 2016).
Figures 6—11: Speculative Landformations: Series 01 — *various* (© Beaman & Hong, 2016).
Figure 12 : Sample Landformation Families (© Beaman & Hong, 2016).

Michael Leighton Beaman is Founder + Principal of Beta-field, Co-founder + Principal of GA Collaborative, Design and Technology writer for Architectural Record, and currently teaches at the Rhode Island School of Design. His research and writing focuses on the theory and application of technology in Architecture and Landscape Architecture and its implications for cultural, environmental and socially conscious design practices. Beaman was named an AIA Emerging Practitioner, MacDowell Fellow, and UVa Teaching Fellow. Prior to RISD, he was an Assistant Professor of Architecture at the University of Texas, Austin and has taught at Harvard University, NC State University, and Kigali Institute of Science and Technology.

The McKnelly Megalith

A Method of Organic Modeling Feedback

Brandon Clifford
Massachusetts Institute of
Technology / Matter Design

1

ABSTRACT

Megalithic civilizations held tremendous knowledge surrounding the deceivingly simple task of moving heavy objects. Much of this knowledge has been lost to us from the past. This paper mines, extracts, and experiments with this knowledge to test what applications and resonance it holds with contemporary digital practice. As an experiment, a sixteen-foot tall megalith is designed, computed, and constructed to walk horizontally and stand vertically with little effort. Testing this prototype raises many questions about the relationship between form and physics. In addition, it projects practical application of such reciprocity between architectural desires and the computation of an object's center of mass. This research contributes to ongoing efforts around the integration of physics-based solvers into the design process. It goes beyond the assumption of statics as a solution in order to ask questions about what potentials mass can contribute to the assembly and erecting of architectures to come. It engages a megalithic way of thinking which requires an intimate relationship between designer and center of mass. In doing so, it questions conventional disciplinary notions of stasis and efficiency.

1 Erecting, *McKnelly Megalith*, Killian Court: MIT, 2015.

INTRODUCTION

With carving starting around 1100 A.D., the Moai of Rapa Nui[1] weigh up to eighty tons apiece. Since the Dutch discovery of this Pacific island in 1722, visitors have wondered at these megalithic figures, asking the inhabitants how their ancestors possibly moved the statues from quarry to site. The Rapa Nui claim their ancestors never moved the Moai; rather, the Moai walked themselves. For centuries, this anthropomorphic explanation was considered superstitious poppycock by all but the islanders. It is this mystery that has fostered book titles by such as *The Mystery of Easter Island* (Routledge 1919) and *Aku-Aku: the Secret of Easter Island* (Heyerdahl 1958) by early researchers to the island. It wasn't until 2012 that Archaeologists Carl Lipo and Terry Hunt were able to prove the Moai were in fact transported in a vertical position (Lipo 2012;Hunt 2011). In a similar manner to how one might shimmy a refrigerator into place, the Moai were pulled back and forth by ropes employing momentum to transport these unwieldy megaliths. This (re)discovery brings new meaning to the assumed folklore that the statues 'walked themselves' from the quarry to the Ahu[2].

As with most megalithic sites, a feat of strength is challenged in order to produce a mystical spectacle. The challenge the Rapa Nui posed to themselves was to carve megaliths from a quarry in a horizontal position, dislodge these figures from the rock face and stand them vertically, then walk the Moai a considerable distance with relatively little energy. Once in position, they continued to carve and refine the statues in order to shift the center of mass back into a stable standing position. While the most common image of this ritual is the Moai lined up nicely on the Ahu, some (Hamilton 2008) have posited that the landscape of the quarry is equally calibrated, suggesting this wonder is larger than the objects. This ritual was a living process, not a design intended to result in a conclusion. In fact, a number of 'Road Moai'[3] are in a state of suspended animation. Carved to perform the act of walking, these Moai fell during transport and are still resting alongside the roads. The act of walking these massive Moai from quarry to Ahu must have been quite a spectacle.

In the digital era, architects and designers are employing computation to better inform their creations with concepts, principles, and constraints previously considered extra-disciplinary—structure, environmental conditions, physics, and material behaviors to name a few. When calibrating the center of mass to perform the movement behavior from quarry to site, the Rapa Nui were solving a multi-variable problem that bears similarities with the types of computation designers are working with today; however, the Moai are not the result of a problem-solving approach (engineering), nor an exclusively aesthetic concern (sculpture). Moai are a cultivated result of a conflated design practice that engenders practical concerns with cultural performance, which results in a marvel. This paper presents research into translating this ancient knowledge into contemporary computation methods. The result is a physical spectacle.

BACKGROUND

This research illuminates and exercises challenges of the megalithic era that become acutely problematic and hold resonance with contemporary practice. In proposing the act of transporting a large stone a distance and standing it vertically, the designer is required to become in tune with the center of mass relative to the desired actions and motions of the stone. This is not necessarily the case with heavy machinery, where one can overpower an uninformed geometry with a crane. In the world of the megalith, a failure to calibrate is a catastrophic event, but produces a different value set. Though commonly considered to be a hindrance today, mass is a tool for the presumably primitive civilizations that elected to work with megaliths. These stones are erected because of their mass, not in spite of their mass. This way of thinking is best described by Carolyn Dean in her influential book, *A Culture of Stone* (Dean 2010), which examines what is possible when a culture believes stone to be alive. While this challenge could be perceived irrelevant now that heavy equipment exists, there is a value in embedding topics such as mass into the design process in order to more intelligently assembly and erect architectures.

In developing reciprocity between center of mass and the figure's geometry, embodied computation allows the artifact to work for itself. One proof of this concept is the Gömböc, conceived of my mathematician Vladimir Arnold and later proven by Gábor Domokos and Péter Várkonyi (Domokos 2007). It has two states of equilibrium, one stable and the other unstable, explaining the mathematics behind how turtles right themselves. This delightful kinetic action builds a relationship between physics and form; a relationship that has entranced architects and engineers. For instance, recent work by John Ochsendorf (Ochsendorf 2010), Philip Block (Andriaenssens 2014), and others into compression-only structures is a contemporary expression of a time-honored tradition that includes Felix Candela, Antoni Gaudí, and Frei Otto (Rasch 1995). These works also deal with physics and the relationship with mass and form; however, they work under an assumption of singularity form-finding, or finding a form for a given constraint, often gravity. This motivation extends deeper through the compression-only structures of Gothic Master Masons, back to the megalithic eras where the immediate relationship is purposefully elusive. Why is this relationship largely absent from standard practice today? Perhaps the false assumption that architects should be relegated to producing representations of architectural intent has restricted the

2   Resting Position – In this position, the center of mass is directly over the resting pivot, resulting in a stable equilibrium. Two verification geometries can be drawn from this position, the foot geometry is vertical to the ground and the chin is parallel.

3   Weighted Step Position – With the additional mass of one person to the eyehole geometry, the combined center of mass is disturbed, pulled closer to the head. This shifting of the COM results in a moment relative to the resting pivot rolling the megalith forward on the belly. Ultimately it finds a new stable position when the upper verification geometry strikes a horizontal.

4   Standing Position – Once in the standing position, the original center of mass is above the 'C' shaped foot geometry, resulting in a stable condition.

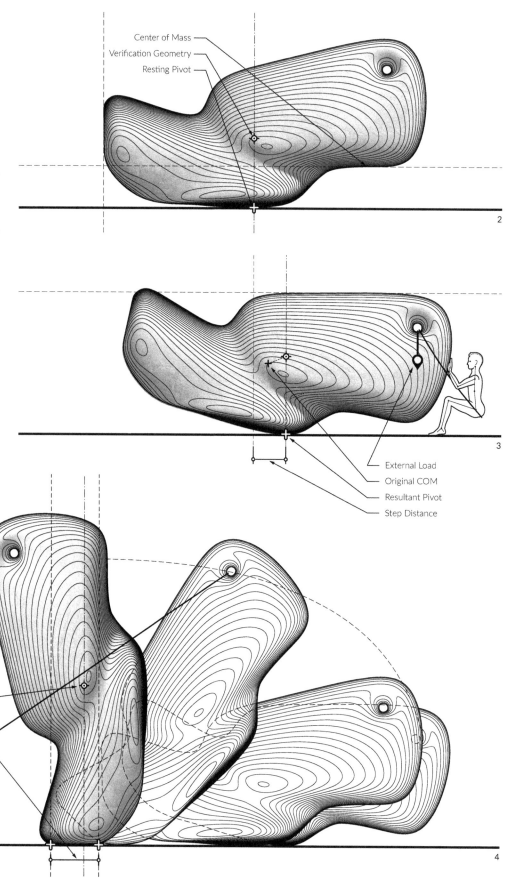

Center of Mass
Verification Geometry
Resting Pivot

2

External Load
Original COM
Resultant Pivot
Step Distance

3

Stable Center of Mass
Footprint

4

profession to drawings, often uninformed by physical limitations, giving rise to the consultancy of architectural engineer. In spite of this false assumption, new research is developing design tools to re-integrate gravity with design, such as a paper by Axel Kilian on particle-spring systems for structural form-finding (Kilian 2005) and the recent work by Daniel Piker (Piker 2016) into building a physics engine titled 'Kangaroo.' As computation re-entangles our relationship with the physical world, topics from megalithic era knowledge are able to re-insert themselves.

A retired carpenter named Wally Wallington is dedicating his retirement to the problem of moving heavy things with little effort (Wallington 2016). While illuminating, his work operates under a similar assumption to those moving an Egyptian Obelisk (Dibner 1950)—that the form of the stone does not change, rather, the mechanism used to move it could be designed to be minimal. The research of this paper builds upon some of those findings, engaging them in a computation strategy for formal design, akin to the Moai of Rapa Nui.

The challenge of megalithic construction is to safely move a heavy object with little force. A vast majority of megalithic constructions evoke the mystery of transportation from quarry to site, and then the mystery around the erecting or assembly of the artifacts. These two mystical motivations exist in megalithic constructions, regardless of location or time-period—from Egyptian Obelisks, to Cairns and Henges of northern Europe, down to the Inca stonework, and over to Rapa Nui. The challenge proposed for this research is to design an artifact that intelligently transports horizontally and stands vertically. In doing so, it challenges the designer to create a single figure that resolves multiple stable conditions while simultaneously ensuring the forces required to transition between stable conditions can be safely accomplished within human abilities. The following paper illuminates this embodied computation process.

PROPOSED ACTIONS

Three positions are required to perform the action of walking horizontally and standing vertically. These three positions include the resting position, the weighted step position, and the standing position. While it is possible to design an artifact to perform any one of these positions, it is significantly more difficult to accommodate all three in a single object.

Resting Position

The resting position is a horizontal position where the megalith rests on the belly geometry. This position needs a base that is able to pivot, and therefore is required to balance on a point. In this position, the megalith needs to be able to spin, but should resist rolling over.

Weighted Step Position

By adding the mass of a single person to one end, the center of mass is shifted, allowing a new equilibrium position to be found. This new position should be far enough from the resting position to consider it a step. This weighted step position maintains the same program and constraints of the resting position, as the megalith will rotate 180 degrees in plan to release it back to the resting position—a second step has occurred.

Standing Position

The standing position is ninety degrees in elevation relative to the resting position. The challenge of reaching this standing position is significantly greater than the weighted step position, but only with respect to force. It does not have to deal with the pivot constraint. The participants should be able to erect the megalith from a 'safe' distance; defined by a radius around the object equal to height. Once in the standing position, the megalith should achieve a stable resting position able to resist wind-loads up to sixty miles per hour.

METHODS

The foundation of this method involves determining the location of the center of mass of an object and the effects it has on the object's form throughout the design process. This relatively simple calculation achieves its complication by adjusting the model to drive the center of mass where it is needed in order to resolve the three different scenarios. This method is broken down into geometries, computation, prototypes, and material methods.

Constant and Variable Geometries

The megalith is designed with a series of different conditions, some of which are fixed, and some variable. These conditions include the belly, rails, and feet, which determine how the megalith touches the ground. These also include an eyehole, which is a certain distance away from the resting point on the ground for maximum leverage. In addition, there are a number of verification and variable geometries.

Belly Geometry – The belly geometry is the lower surface that interacts with the ground in both the resting horizontal position and the weighted step position. This geometry needs to be able to allow the transition between the two positions as well as the ability for both to spin about a point. A perfect sphere would perform both of these tasks easily; however, the added constraint of stability is required. A cylinder would allow rolling in one direction and stability in the other, but does not allow rotation. To resolve this constraint, a bi-directional calculus geometry defines a variable curvature surface. This surface maintains an even curve between the two resting positions to ease this transition. The curvature in the opposite direction ensures that

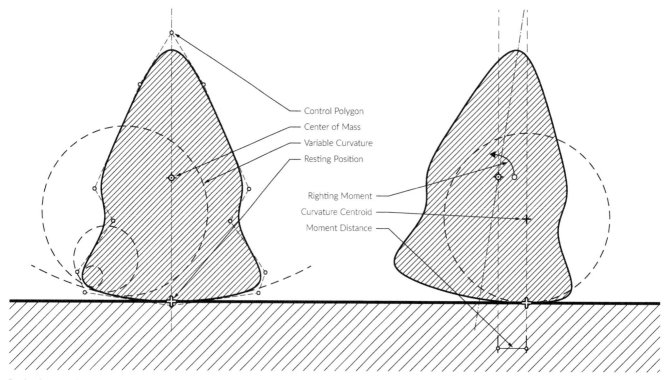

Control Polygon
Center of Mass
Variable Curvature
Resting Position

Righting Moment
Curvature Centroid
Moment Distance

5   Section describing the relationship between the curvature of the belly geometry and the height of the center of mass. When tipped to a side, a righting moment is produced resulting in a stable positioning on a single point to allow for rotation.

the megalith rests on a single point while resisting overturning. A constant curvature such as a circle allows a horizontal force to roll an object eternally because the center of mass is also the center of the radius. The geometry of the megalith has a large curvature at the resting position where the center of the geometry is well above the center of mass. This condition allows the object to roll easily, but the variable curvature rapidly transitions the radius from large to small, thus inverting this condition and transitioning the object from an unstable to a stable position. The result of this transition is a righting moment, which trusts from the resting pivot back to the center of mass until the object returns to the original resting position.

Rails Geometry – When the force is pulled down on the tail to erect the megalith into the standing position, the desire for being able to spin dissipates in favor of a stable and directional rolling up. Rail geometries emerge from the belly to provide the stability in the cross-axis, while the curvature of these rails attempt to alleviate the strain of standing the piece vertically. There is an unique moment of transition from a positive curvature of the resting belly to a negative curvature with two rails. This transition occurs seamlessly with curvature continuity derived from the calculus-based geometry.

Foot Geometry – Once in the standing position, the two rails transition into a 'C' shaped geometry where the rails are

co-planar, providing a wide base for the megalith to stand upon. The center of mass is calibrated to stand within this 'C' shape, therefore ensuring stability. This 'C' geometry was favored over a flat base due to uncertainties about the flatness of the ground the megalith was performing on.

Eyehole Geometry – This aperture is employed twice, the first for the loaded step position, and the second to aid in the erection. For that reason, both of those forces gain the most leverage from being as far horizontal and vertical away from the center of mass. The eye is designed for a human arm to fit inside for the stepping position, where the most force is being placed. In this action, the entire mass of a human is pulling on this point, so a significant amount of material is placed under the hole to resist pullout. In the erection process, this concern is lessened as the ropes are mainly serving as stabilizers.

Verification Geometries – Because of difficulties in knowing the proper thickness of the Glass Fiber Reinforced Concrete (GFRC) coat, it was determined to provide a series of verification geometries in the megalith to verify whether it was performing as calculated when in these positions. For example, the chin is a curvature continuous with a horizontal line when in the resting position. This way, one might know right away if that is not striking a horizontal, whether something went wrong in the calculation, or in the application. The other verification geometry

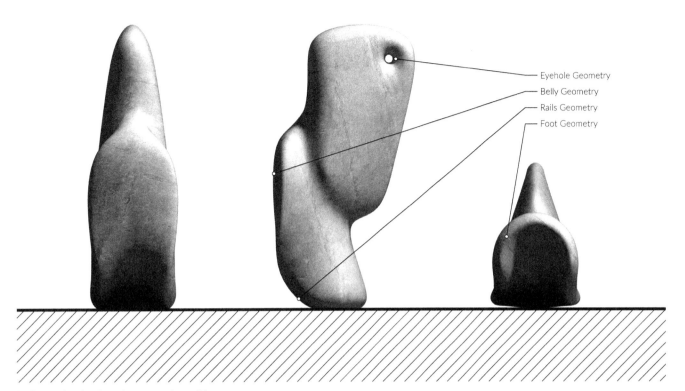

Eyehole Geometry
Belly Geometry
Rails Geometry
Foot Geometry

6   Drawing locating the various geometric conditions.

is in the angle of the face. This angle is coincident with the angle of the weighted step to verify how far to pull the head down to make a proper step. In this project, these verification geometries are fixed as a design strategy; however, in future work, they could become dependent upon other constraints. For example, the angle of the face verification geometry could be derived from the distance of the eyehole from the pivot point. The further the distance, the easier it would be to pull the megalith into the step position, but the lesser the angle one could pull down before the chin hits the ground, thus relocating the eye hole and the face angle verification geometry.

Variable Geometries – While the previous geometries are variable throughout the design process, during the computation portion, those geometries become fixed and informed with parameters. Inside the computation loop, the variable geometry, which allows the center of mass to be driven where it is desired to be, are the cheek geometries. The cheeks are allowed to inflate or deflate to make the head heavier or lighter than the base, and they are also allowed to do this in the vertical dimension to lift or lower the center of mass, giving complete control over the location.

Computation
In order to properly compute this multi-variable problem, the computation is broken down in to deriving the center of mass of the object, operating on the object through recursive loops, and

simulating the resting positions in order to validate those goals.

Center of Mass – In order to perform these three tasks, two different center of mass locations are required. The resting position and the standing position each share the same center of mass location. As there are no other forces acting on the megalith at those times, they are simply starting in two different positions and find different resting positions—horizontal and vertical. The loaded step position is not shared with the other two center of mass locations. It increases the mass of the mega-lith by the weight of a human. Given the form is produced of one lb/ft³ density Expanded Polystyrene Foam (EPS) Foam and a half inch thick shell of GFRC (see materials), the calculation accepts a given geometry as input and parses these two materials by a volume and area calculation. The volume is assigned to the EPS foam, and the area to the GFRC. In previous iterations, the surface of the geometry offset the dimensional thickness of the proposed shell and calculated with that volume; however, the difference between the two was less than 0.001 inch, well under the tolerance threshold for this project. As a result, the area calculation was favored in order to increase the calculation speed. After the area is calculated, it is multiplied by the thick-ness of the coat to receive the proper mass of the GFRC shell. The resulting center of mass fulfills the requirements of the resting and standing position. In order to calculate the loaded step position, a point load of 160 lbs is added to the eyehole geometry to represent the mass of a human. This pulls the global center of mass closer to the head, producing a new resting position on the belly geometry and thus making a twenty-four inch step. In calculating these centers of mass, the global mass without the point load needs to rest perpendicular above the belly in the horizontal position and needs to be low enough to maintain stability from rolling over. It also needs to be high enough to land within the footprint of the vertical position. This is a negotiation. The loaded step center of mass needs to pull forward enough to provide a step, but not so far forward that the head become top heavy, thus making the condition unstable. While the center of mass calculation can tell you where the COM is, it cannot drive the center of mass to where you would like it to be.

$$x_{cm} = \frac{\sum_i m_i x_i}{\sum_i m_i}$$

Variable Recursion – The input geometry is derived from a polygonal control point rig to produce a calculus-based curvature continuous surface with T-Splines in Rhinoceros. While a number of the control points are dedicated to creating the fixed geometries as described previously, within three points of those moments, a series of free points are variables to allow the center of mass to drive to the desired location. This calculation measures the distance between the current location of the center of mass and the desired location of the center of mass. As a fraction of this dimension, the variable points in the rig are scaled a proportional distance from the central axis, thus re-distributing the center of mass. This calculation is repeated until the COM distance difference is less than the tolerance (0.001 inches). As a result, the calculation adjusts the form to drive the center of mass to the desired location of the designer.

Resting Positions – While a significant amount of feedback is produced to ensure stability from curvature continuous surfaces, the simulation of resting positions is helpful for the designer to visualize what effects are being had when slight changes to the geometry are produced. For this calculation, a solver cuts a section through the model along the symmetric axis. Since this is a primarily directional calculation, the resting positions can be solved in section. From the center of mass of the object, the sectional curve through the artifact searches for perpendicular conditions between the curve and the center of mass. It then solves for the most likely resting position scenario by relaxing these angles relative to the given horizontal. These simulations do not animate the figure; they simply locate the likely resting position, or the point where the megalith interacts with the ground. These positions are then verified with prototypes and the verification geometries.

Prototypes

While the computation verifies the location of the center of mass relative to a geometry, it does not simulate the kinetic actions. In order to supplement this gap in the computation, physical prototypes are produced to validate claims. Powder rapid-prototypes served as an ideal form of prototyping. The final geometry is offset to produce a shell geometry, and the interior is hollowed out to re-locate the center of mass correctly. This is because a solid geometry volume centroid is not the same location as a surface area centroid. One can design a sphere with many wrinkles on one half and the area centroid would be pulled to that wrinkled half, while the volume centroid might not move. Once the ideal centroid matches the prototype centroid, the prints are tested. These prototypes served as incredible knowledge tests as they could withstand attempts to overturn, flip, re-right, and worst case scenarios before the final prototype was performed. They also served as devices to verify that geometries rock and roll at certain intervals when applied with forces from different angles, but return to the resting position in a repeatable manner. While an infinite number of geometries could balance the artifact in the computation, the physical prototypes allowed those geometries to become refined to better perform. For instance, the belly geometry is a difficult one to visualize in the computation, but in physicality, certain aspects become clear.

Materials and Methods

The megalith is composed of a solid core of Expanded Polystyrene Foam (EPS) and coated in a shell of Glass Fiber Reinforced Concrete (GFRC). This material method allowed the project to be constructed as a positive object coated in concrete to reduce the weight enough for safety and liability concerns, while providing a substantially heavy artifact to test. The difference in material density is appropriately calculated (see computation).

EPS foam core – The EPS foam core is subtractive machined on a 3-axis computer numerically controlled (CNC) machine in six-inch depths. These sub-units are then assembled with polyurethane glue in two halves. Once the two sub-assemblies have bonded, they are bonded to each other. Before this moment, the glue-up can be done on a flat surface with weights holding them in place, but when this bi-lateral glue-up happens, the two halves are ratchet clamped together.

GFRC Shell – The EPS foam positive is then covered in a woven mesh of glass fiber and turned on its side to receive a half-inch layer of GFRC. In order to ensure that the proper mass of concrete is applied, scales are used to give a live feedback and judge how many batches are required. Especially because the GFRC is sprayed onto the piece, some of the GFRC is atomized,

7 Resting Position, *McKnelly Megalith*, Killian Court: MIT, 2015. While perceived to be flat on the ground, this is the least stable of the three figures on this page. The belly geometry allows the megalith to teeter.

8 In the process of erection, *McKnelly Megalith*, Killian Court: MIT, 2015. From this position, one can see that a majority of the force employed to erect the megalith is forced down from the tail to a pulley which re-directs the rope. This allows the force to be at a forty-five degree angle at the height of the force, but also ensures once standing, the Megalith will not stand on top of it's own rigging.

9 Standing Position, *McKnelly Megalith*, Killian Court: MIT, 2015. This is a stable condition, with ropes running through the eyehole to stabilize the megalith during the erection process.

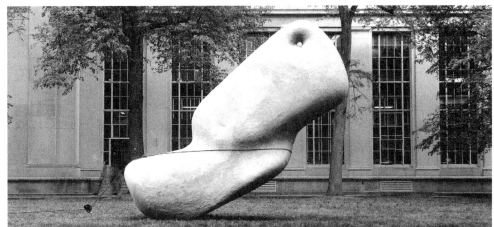

10   Image of the Belly Geometry while the *McKnelly Megalith* is in the standing position. Swirl stains from the grass can be seen from walking the megalith across Killian Court.

11   A bias view of the lower foot geometry demonstrating the 'C' shape which produces a flat foot.

12   A bias view from the front of the megalith, showing the tapper from the top of the head down to the chin, a result of the solver while trying to pull the center of mass lower.

or over-sprayed, making it difficult to judge based on the number of batches. This step is crucial, because the actual mass is important to know in order to determine whether humans can easily move it. Attempts were made to judge how thick the coats are with stand-offs, or depth gauges; however, these attempts fell short of their goal, as the EPS foam is softer than the concrete coat, making depth gauges unsure where they bottom out, and stand-off gauges got in the way and often lost in the process of spraying. The scales served as the primary judge of mass. Unfortunately, it is not possible to know where the shell might be thicker or thinner. Using this method, one may know the actual mass of the piece, but not the exact location of the center of mass. It isn't until the performance that we guarantee this with verification geometries.

The next step in the process is to complete the coat on the other side of the megalith. This process reveals two problems—the first being a cold-joint in the concrete, and the second being the complication of rolling this mass over. The cold-joint is resolved with the woven fiber mesh. Half of this mesh is embedded in the previous cast, and the other half is loose, awaiting the second coat. The problem of rolling the megalith over onto its other side is an interesting one, as it is not considered in the calculations. All the calculations consider a completed shell, but until the second half is coated, the center of mass is no longer down the central axis, but pulled to the side of the first coat. This means that in the act of rolling it over, the megalith rolled from its right side, up to its assumed 'stable' position of the final resting design, and continued over to land on its left side. As expected, but not as calculated, the megalith landed on spare blocks of foam to support it for the second coat. Once completed, the megalith is righted in a similar manner to the previous flip. This time, however, since the center of mass is re-distributed to the central axis, it did not flip completely over, but instead righted itself into the resting position. While these actions were not designed into the computation, future research could investigate these moments as part of the problematic.

RESULTS

The resulting *McKnelly Megalith* is sixteen feet long (or high in the standing position) by six feet of depth and eight feet of height. The EPS occupies 237 cubic feet, weighing 237 lbs. The half-inch thick GFRC shell occupies 268 square feet and weighs 1,763 pounds. In total the megalith weighs 2,000 pounds (1 Ton). While this mass is significantly lighter than the inspiration Moai, it is significantly heavier than one would expect a human to be able to maneuver. Each step travels two feet, allowing it to transport at 300 feet per hour.

CONCLUSIONS

This paper demonstrates a process whereby organic modeling can interact with physics-based information modeling in order to perform megalithic actions. It proves the ability to prototype, test, and manufacture at a large scale; however, this project asks more questions of the role of the architect and designer than it answers, thus a number of future works could test these questions. These future interventions could include dealing with significantly heavier masses, perhaps through the solid casting of concrete, or the carving of stone. It could also include other stability calculations such as buoyancy, which could incorporate counter masses and dynamic conditions. It could also be informed by wind-loads, and a host of other physics-based information. This method could inform architecture with information to aid in the erection and assembly of parts, or it could inform the stability of perceived unstable artifacts.

ACKNOWLEDGEMENTS

This research was conducted in the graduate options architecture studio titled 'Megalithic Robotics' at the *Massachusetts Institute of Technology* in the Spring of 2015 co-taught by Brandon Clifford and Mark Jarzombek (professors) with Carrie Lee McKnelly (Assistant Instructor). Students include Sam Ghantous, Anastasia Hiller, Karen Kitayama, Dan Li, Hui Li, Patrick Evan Little, Tengjia Liu, Ryan McLaughlin, Kaining Peng, Alexis Sablone, and Luisel Zayas. The computation employs *T-Splines* (www.tsplines.com) as the organic modeler to inform *Grasshopper* (www.grasshopper3d.com), a plugin developed by David Rutten for *Rhinoceros* (www.rhino3d.com), a program developed by Robert McNeil. The *McKnelly Megalith* is dedicated to the memory of Steve and Rendy McKnelly.

NOTES

1. Rapa Nui is commonly known as Easter Island.
2. Ahu /'ahú / *n*. Stone platforms upon which the Moai stand. There are over 300 known Ahu on Rapa Nui.
3. The term 'Road Moai' is coined in 'The Statues that Walked' (Hunt 2011)

REFERENCES

Andriaenssens, Sigrid, Philippe Block, Diederik Veenendaal, and Chris Williams, eds. 2014. *Shell Structures for Architecture: Form Finding and Optimization*. London: Routledge.

Dean, Carolyn. 2010. *A Culture of Stone: Inka Perspectives on Rock*. Durham, NC: Duke University Press.

Dibner, Bern. 1950. *Moving the Obelisks*. Norkwalk CT: Burndy Library.

Domokos, Gabor, Peter Varkonyi, and Vladimir Arnold. 2007. "Turtles, Eggs and the 'Gömböc.'" *Hungarian Quarterly* 48 (187): 43–47.

Hamilton, Sue, Susana N. Arellano, Colin Richards, and Francisco Torres. 2008. "Quarried Away: Thinking about Landscapes of Megalithic Construction on Rapa Nui (Easter Island)." In *Handbook of Landscape Archaeology*, edited by Bruno David and Julian Thomas. Walnut Creek: Left Coast Press. 176–186.

Heyerdahl, Thor. 1958. *Aku-Aku, the Secret of Easter Island*. Chicago: Rand McNally.

Hunt, Terry L., and Carl P. Lipo. 2011. *The Statues That Walked: Unraveling the Mystery of Easter Island*. New York: Free Press.

Kilian, Axel and John Ochsendorf. 2005. "Particle-Spring Systems for Structural Form Finding." *Journal of the International Association for Shell and Spatial Structures* 46 (2): 77–84.

Lipo, Carl P., Terry L. Hunt, and Sergio Rau Haoa. 2012. "The 'Walking' Megalithic Statues (Moai) of Easter Island." *Journal of Archaeological Science* 40 (6): 2859–2866.

Ochsendorf, John and Michael Freeman. 2010. *Guastavino Vaulting: The Art of Structural Tile*. Princeton: Princeton Architectural Press.

Piker, Daniel. "Kangaroo." Accessed July 6, 2016. www.grasshopper3d.com/group/kangaroo

Rasch, Bodo and Frei Otto. 1995. *Finding Form: Towards an Architecture of the Minimal*. Stuttgart: Edition Axel Menges.

Routledge, Katherine. 1919. *The Mystery of Easter Island*. London: Sifton, Praed.

Wallington, Wally. "The Forgotten Technology." Accessed July 6, 2016. www.theforgottentechnology.com

IMAGE CREDITS

Brandon Clifford is an Assistant Professor at the Massachusetts Institute of Technology and Principal at Matter Design. Brandon received his Master of Architecture from Princeton University in 2011 and Bachelor of Science in Architecture from the Georgia Tech in 2006. He worked as project manager at Office dA from 2006–09, LeFevre Fellow at OSU from 2011–12, and Belluschi Lecturer at MIT from 2012–16. Brandon has been awarded the Design Biennial Boston Award, the Architectural League Prize, as well as the prestigious SOM Prize. Brandon's translation of past knowledge into contemporary practice continues to provoke new directions for digital design.

Towards Genetic Posthuman Frontiers in Architecture & Design

Alberto T. Estévez
ESARQ, School of Architecture
UIC Barcelona
Universitat Internacional de
Catalunya

1

ABSTRACT

This paper includes a brief history about the beginning of the practical application of real genetics to architecture and design. Genetics introduces a privileged point-of-view for both biology and the digital realm, and these two are the main characters (the protagonists) in our posthuman society. With all of its positive and negative aspects, the study of genetics is becoming the cornerstone of our posthuman future precisely because it is at the intersection of both fields, nature and computation, and because it is a science that can command both of them from within—one practically and the other one theoretically.

1   *A Strange Alive Planet:* collage with scanning electron microscope photo of a *Malvaceae* pollen grain at 6000x, with photos of bioluminescent *Ctenophores* and fireworks (photos taken by the author).

Meanwhile, through genetics and biodigital architecture and design, we are searching at the frontiers of knowledge for planetary benefit. In order to enlighten us about these issues, the hero image (Figure 1) has been created within the framework of scanning electron microscope (SEM) research on the genesic level, where masses of cells organize themselves into primigenic structures. Microscope study was carried out at the same time as the aforementioned genetic research in order to find structures and to learn typologies that could be of interest for architecture, here illustrated as an alternative landscape of the future. Behind this hero image is the laboratory's first effort to begin the real application of genetics to architecture, thereby fighting for the sustainability of our entire planet and a better world.

2   *Genetic Barcelona Project:* The magic light of the GFP lemon trees. Center, image of a possible world (image of *Casa Milà* by Antoni Gaudí). Right, real comparison between a lemon tree leaf with GFP and another without GFP from the same tree type: above photo taken with conventional reflex camera, and below photo taken with special UV camera (author's images and photos).

INTRODUCTION

We know that the answer is in nature and that nature is the answer. The more that science advances, the more we know of what we call nature and the more we understand that nature is the answer. But, "if Nature is the answer, what was the question?" (Wagensberg 2007). We are exploring and interrogating "the question" through interdisciplinary endeavours involving fields such as material science, biology, genetics, art, architecture, civil engineering, design, computer graphics, and human-computer interaction. We are exploring posthuman frontiers. One such frontier can be explored at the intersection where genetics meets biology and the digital, and can be applied to architecture (and in other contexts, to art, civil engineering, and design). This is the intersection at which we find ourselves, and the one that this paper explores.

In order to pursue this line of inquiry, the Genetic Architectures Research Group & Office and the Biodigital Architecture Master Program was founded at the ESARQ, the School of Architecture of UIC Barcelona (Universitat Internacional de Catalunya). This is where we work to create architecture and design, with geneticists focused on architectural objectives, and architects researching the fusion of biological and digital techniques.

A BIT OF (GENETIC) HISTORY

It might now make a little sense to give a little history on the subject. Between December 1999 and January 2000, a snippet of information went viral, with the media publishing increasingly more information on it and constantly "infecting" each other. In this case, it was a story about genetics: press, radio, and television were very quickly inundated by news reports on this subject.

Then, watching how genetics offered such a huge field in the world of health and nutrition, I wondered about the application of genetics to architecture and design. For that reason, in 2000 we created the aforementioned research group and Master's degree program, and also connected it to a PhD program.

At the same time, without us knowing about it yet, Eduardo Kac was working on *Alba*, a transgenic artwork that attempted to create bioluminescence in a rabbit through the use of green fluorescent protein (GFP). As is it known, the GFP gene is widely used in genetics as a marker; an indicator that allows for easy verification of genetic transformation success. And though it is now supplied to genetic laboratories without problem, the natural source of this gene was originally a jellyfish called *Aequorea victoria*, from the Northeast Pacific, in which GFP glows in the dark. After the transformation of the cells that have been required for each case, the gene synthesizes the protein, which allows the cells to emit a bright green color when exposed to blue or black light. However, GFP is also present in hundreds of sea species, with green, orange, and red colors, as in sharks, eels, seahorses, fish, coral, etc. This discovery has recently given rise to *fluo diving*, night diving in fluorescent underwater marine life, as if one were floating in the Avatar movie.

One day in January of 2003, while talking with the geneticists in our group about the use of GFP in research, Dr. Miquel-Àngel Serra asked "what else can the GFP be used for other than being an indicator?" As an architect it was clear to me: "for illuminating architectural spaces!" At that moment we began research for getting trees to work as "lamps" illuminating streets, plants illuminating homes, vegetation illuminating the roadsides without

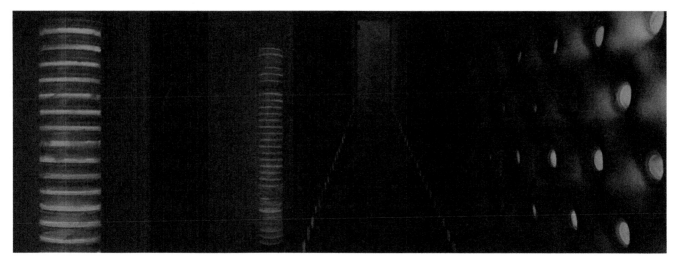

3 "Biolamps": the first systematically fully illuminated apartment with living light (human eye view: photos by the author, taken with a conventional reflex camera).

electricity: the creation of plants with natural light by genetic transformation for urban and domestic use had emerged.

So, in October of 2005, thanks to our geneticist Dr. Agustí Fontarnau along with Dr. Leandro Peña, we successfully obtained the first seven lemon trees with luminescent leaves (Figure 2) provided by GFP. These modified lemon trees get their green fluorescent protein through the expression of the GFP gene. That gene was transferred to the lemon tree cells through an in vitro culture experiment using a DNA vector containing GFP genes (the DNA containing the GFP gene was not spliced; it was inserted into the lemon tree genome and kept intact inside of it). Some of the transformed cells regenerated a new plant, with cells expressing GFP, knowing that the glowing properties can be seen by microscope from the beginning of its cellular transformation. In two months, the trees were approximately 30 cm high, so that we were able to directly see the bioluminescent properties with our own eyes. We then took photos with blue light and a conventional reflex camera, or along with Dr. Josep Clotet, also from our University, taking photos with our special UV camera.

We started with GFP, since it is one of the most studied genes, as geneticists use it as a common cellular marker. The functionality of our project was clear: the trees in this project were made with the objective of being of architectonic and urban use. It was the first time in architectural history that geneticists had worked for an architect. In 2005, we also presented this research under the name of the *Genetic Barcelona Project* to the mayor of Barcelona.

Regarding durability, the results were good: today, more than 10 years later, the leaves have the same luminescence, and the initial little lemon trees continue to grow depending on soil availability. They can also be also multiplied by planting their branches, becoming non-manufactured "lamps," for free! But from the beginning, the lightning efficiency was very poor and needed special light inputs in order to achieve enough brightness.

Through a second phase of this project, to make more efficient and useful bioluminescent vegetation, we arrived at "Biolamps" (Figure 3). In 2007, we started to research bacterial bioluminescence for urban and domestic use. We were also involved with research into how to achieve bioluminescent plants with a bacterial gene group responsible for bioluminescence at the same time.

In this phase, in 2008 we began to create "Biolamps," a kind of "battery" with bioluminescent bacteria that are originally found in abyssal fish. With them, we created the first fully illuminated living light apartment without electricity. For the first time in architectural history, a whole home was illuminated using bioluminescence, without any electrical installation.

The digital design and manufacturing of the *Biodigital Lamps Series*, and its use as "Biolamps" had also begun (Figure 4). These lamps are based on an analysis of radiolarian structures and pollen. This analysis was applied to the digital development of architecture and design, first by using an SEM, which we have used from 2008 to now. This continues along the lines established through the idea of "bio-learning," which offers the benefits of the structural, formal, and processual efficiency that we can learn from nature. Using CAD-CAM technology, once we believe that the drawings have reached the desired result, will proceed to its digital manufacturing, directly on a scale of 1:1. In this case, we can take advantage of different parts or levels where this technology allows for interscalarity. This allows us to easily change the scale of the jewelery and lamps to that of the pavilion. Research starts by choosing a system, and a structural,

4   *Biodigital Lamps Series*, being used as "Biolamps" digitally 3D printed (author's images and photos with the collaboration of Diego Navarro).

architectonic, and design idea using geometry in order to draw it. Finally, in order to manufacture this, research with digital machines needs to be carried out with the confidence that "what can be drawn can be built" (Figure 5 and Figure 9).

Paradoxically, unlike the GFP lemon trees, the second phase of this bioluminescence research was very effective for lighting, but problematic in terms of durability: every 10 days, the "bio-batteries" needed to be changed. The other option was fabricating a lamp that could guarantee the required air-tightness, oxygen, and food. However, it was determined that the lamp was too complicated to manufacture compared with a simple bioluminescent plant or tree.

We are now in the third phase, trying to introduce the genes responsible for bioluminescence in ornamental plants. First, we obtained two stable lines using bioballistics, with the plasmid pLDLux integrated into the genome of the *Nicotiana tabacum* W38 chloroplast, and we can assert that the expression of the bacterial operon luxCDABE is correct and stable. We have also done the same using bioballistics with different species of ornamental flowers, such as *Begonia semperflorens*, *Codariocalyx motorious*, *Mathiola incana*, and *Dianthus caryophyllus* (Figure 6). However, due to the low bioluminescence provided by the pLDLux vector (probably due to a lack of a LuxG gene whose mission is to participate in the turn-over of FMN, flavin mononucleotide), our efforts are now being focused on finalizing a vector of chloroplast transformation possessing LuxCDABEG genes. A work in progress!

After its presentation at various congresses and publications in 2005, 2006, 2007, etc., we can say that the diffusion of this research has been a success (Estévez 2005; Estévez 2006;

Dollens and Estévez 2007). In 2010, the American "Bioglow" company took our idea for producing plants which can illuminate human spaces. Soon thereafter, in 2012, the "Glowing Plant Project" began to search for the same genes (but not without controversy). "Bioglow," led by a geneticist, has also seen this as a potentially powerful niche. Its second project, the "Glowing Plant Project," was led by a businessman who also saw this as a great business opportunity. Since then, there have even been different cases when the forgetful mass media has occasionally come out with the "amazing" news about the "novel" idea of illuminating trees illustrated using Photoshop (i.e.: Swain 2010; Rincon 2013; Brooks 2014).

GENETIC POSTHUMAN FRONTIERS

We can see the enormous potential that nature offers us in order to assure a better future for our planet. This is the path towards genetic posthuman frontiers. After fifteen years of work, a big difference remains between what we can imagine and what we can achieve, because everything depends on getting money for research.

For example, we have already explored three ways of using bioluminescence in order to drastically reduce energy consumption for night lighting and the pollution it produces. We are now at the threshold of a fourth method, which might be more effective, using bioluminescent fungi. We are preparing the identification of the responsible genes for bioluminescence of *Mycena*, *Gerronema* and *Armillaria*. A fifth possibility is researching bioluminescent yeast, which is more experimental and perhaps more spectacular.

This is thereby the beginning of a revolutionary change in the cultural posthuman understanding of light, city and architecture. This is also applicable to heat and habitat. What is at the end of

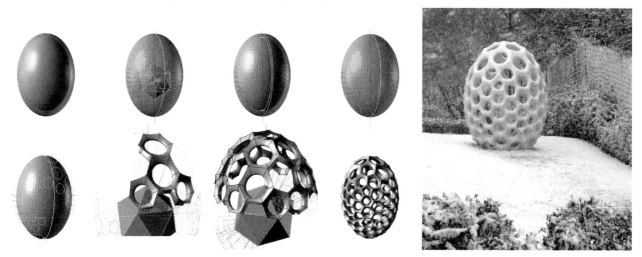

5   Some images of the drawing process, and digitally manufactured pavilion (left: drawings with the collaboration of Daniel Wunsch; right: author's photo of the *Biodigital Barcelona Pavilion* and previous).

the road? The satisfaction of meeting three of the most basic humans needs solved in the most natural and sustainable way: natural light, heat, and habitat, and living without consuming energy and producing pollution, fuelled by the power offered by natural processes. Trees and plants offer biolight and bioheat naturally in streets and homes; there are even vegetable genes responsible for providing warmth. Imagine living biohouses consisting of trees or mushrooms with inhabitable conditions that can be purchased in malls; seeds which can be planted in the ground and grow alone; this all opens up an infinite unexplored posthuman horizon.

How can we visualize posthuman cities and future houses? As "soft and furry (hairy)" architecture, living cities, and houses (Figure 7). The city of the future will be 50% biological technology and 50% digital technology (100% biodigital). Living houses that grow alone, trees that give light at night, plants that provide warmth in the winter: a city that is more like a forest than a landscape of shipping containers on the port. After all, where do we prefer to live, in boxes or in trees? Our cities are destroying nature wherever they grow. We need to assure that every human footprint becomes a creator of life. We need to change our reality with life!

(GEN)ETHICS
Genetic research for architecture also requires precautions, like avoiding accidents and contamination, as in conventional medical research, or in simple heart surgery. Science requires responsibility and we are establishing strict procedures for testing in hermetic environments, breeding plants without pollen, or by acting in chloroplast to avoid pollination problems. Our team includes philosophers dealing with bioethical matters, like Dr. Josep Corcó and Dr. Xavier Escribano. We hyphenate the word

"gen-etics" (meaning "ethics") in our research when the need for planetary sustainability justifies our work.

Nonetheless, from an objective point of view, there is no ethical difference between acting on "the surface of things" and acting at the intramolecular level. Once we accept the organic and fluid configuration of nature, there is, ethically speaking, not much difference between the production of a Japanese bonsai and a fluorescent rabbit. Bonsais are socially accepted even though they are the result of "tormented" living matter, while a fluorescent rabbit is not less happy than a black or white one.

What's more, the most extreme action would be eating a living being because we simply don't kill it but instead we make it disappear into our own cells. However, nobody is put in jail for eating a chicken sandwich. Since this applies to even the most extreme action—eating—it automatically follows that any other less drastic action is permissible, excluding ill treatment.

But of course, if we work with genetic material, we must accept our responsibility as illustrated by the "domino effect" that takes place in time and space, and that has been explained using the example of a butterfly. In spatial terms, there is the "butterfly effect," where the beating of a butterfly's wings in China is said to be able to trigger a storm in US. In terms of time, we can refer the dramatic book *A Sound of Thunder* (Bradbury 1953), where a prehistoric butterfly is accidentally killed by a traveller from the future thereby changing life millions of years later.

Precisely, it's not only our actions on genetic material, but all of our actions that have a corresponding domino effect millions of years later. At least everything is part of nature, but with genetics, "a new and vast territory is removed from the realm

6   Author's photos of the current research with different species of ornamental flowers already genetically transformed, but with too low bioluminescence.

of randomness and enters into the realm of morality. We are captives of our own competence, of our own capabilities, by which we recreate what we only wanted to represent, or we transgress the natural order that we only pretended to repair" (Rubert De Ventós 2015).

KAC'S AFFAIR

Furthermore, the way Eduardo Kac explains his work should be approached in terms of ethics. He defined Transgenic art as "a new art form based on the use of genetic engineering techniques to transfer synthetic genes to an organism or to transfer natural genetic material from one species into another, to create unique living beings" (Kac 1998). The only wrong and confusing aspect of this definition is the word "unique," like when he claims to be some kind of God-Creator and oversteps the definitions that humans have agreed on. This transgression does not do science any favors. The account that he likes to offer in public (like the following excerpt from an interview) does more harm than good:

Kac: "it took—seven years!—of work on the Edunia petunia before I managed to introduce my own DNA into it. [...] I put my DNA into its 'veins,' and now it is producing my human proteins. The green phosphorescent rabbit and the "plantimal" aren't nature... I created them! [...] With Alba (2000) and the plantimal Edunia (2003), I also relieve God of his status as a creator-myth and turn him into a lab worker, a technician working in a trans-genic workshop."

Amiguet: "You don't seem very humble."

Kac: "I don't copy reality: I create it" (Amiguet and Kac 2012).

It has a negative effect that somebody with a strong presence in

the media speaks without any scientific accuracy, and demon-strates terminological and ideological confusion. In addition, the necessary clarifications and criticisms below are not arguments taken from the authors' subjective point of view, because they have a background that is substantiated by the previously mentioned scientists, geneticists, and philosophers in our research group:

It is not true that he spent seven years on this project, it simply took seven years to happen.

It is not true that he inserted his DNA into the plant, it was more like having a "microbrick" inserted into an enormous set of many thousands of "microbricks." In any case, this "microbrick" is identical to the ones that we all have and it is not in any sense specifically or uniquely "his."

It is not true that the resulting plant produces "his" human proteins. Rather, it produces human proteins that are chemically identical to those of any human.

It is not true that by inserting a gene taken from an animal into a plant, it becomes a "plantimal." Just like a virus can mutate our cell's DNA and cause a tumour, this does not make us a "humanirus."

It is not true that the rabbit and the plant in question "are not nature."

It is not true that he created this rabbit and this plant.

It is not true that he relieves God of his status as a creator-myth and turns him into a lab worker, a technician in a transgenic

7   Author's images of biolamps and bioceilings based on research with SEM.

workshop, because the definition of "God" includes he who "creates from nothingness." Genetic manipulation simply involves repositioning existing "microbricks."

It is not true that he "creates reality," because the gene that he integrates into an enormous pre-existing genetic structure already existed before. Therefore, he doesn't create a single gene, he simply changes its position.

Basically, by inserting a gene from another being into the rabbit and the plant, these did not cease to be "natural," nor did they cease to be nature. This gene "repositioning" has been carried out anonymously by the pharmaceutical and agricultural food industries long before Kac's projects, with more complexity and implications, and on a large scale.

DECONSTRUCTING NATURE

However, this emergent character of life is what humanity has to take advantage of, and this is why we are interested in investigating how genetics can be applied to architecture. The idea is to take advantage of nature's capacity for self-organisation, growth, and reproducibility "for free." Therefore, we look for plants that emit light or heat and will help find the energy-saving mechanisms that our world needs, and that will be usable as construction materials and even as entire habitats. We can begin to imagine, in a not so distant reality, "streetlights," "heaters," and even entire houses that grow on their own.

Given that this research also focuses on the use of genetics, we can also consider possible architectural uses at the level at which undefined cellular masses emerge and self-organise, as the first structural step. We can study this with an SEM, which has an extremely high resolution, allowing to us see images magnified

thousands of times. This opens up a little-known dimension of reality, which, depending on how the images are read or interpreted, can lead to a fascinating level of surreality (Figure 8). As a result of research carried out in this framework, it was possible to create strange and surprising new images: "altered" photographs of natural structures at their most Genesis-like and primitive level. Artistic works and architectural plans based on biotechnological work that have an enigmatic evocative power.

CONCLUSION

To conclude, describing the future development of our work and providing a reasonable projection of the research into future applications, it can be said that the equilibrium of our planet—for our own survival—needs several things: accurate and precise reset of our behaviour, education, basic habits, food, ability to manage waste, and consumer goods. For example, we have to get used to the idea that we don't need so much light at night (our eyes have a wide range capacity that we don't use), just as we put on a sweater if it is cold.

Consumption of energy must be radically reduced: a middle-sized European city of only 100 km² spends 10 million Euros annually just on the maintenance of its street lights (new lamps, repairs, repainting), in addition to electricity consumption. If multiplied by all the cities on the five continents, the figure is absolutely astronomical. Therefore, bioluminescence will substitute artificial lighting, at some levels—there is no doubt about that. Nature is always teaching us, in this case with in many bioluminescent ways, from bacteria and plankton to algae, fungi, insects, etc. It would be like suicide for the next generations if we didn't learn about it.

And this is only a fraction of the possible scope of the application of genetics for architecture and design. When architects stop

Towards Genetic Posthuman Frontiers in Architecture & Design Estévez

8 *Living city:* the enigmatic evocative power of SEM images.

needing conventional construction industries and start working with geneticists, who are the bricklayers of the posthuman future, we may begin moving towards science, architecture, and design collaborations where genetics becomes integral to architectural research and production, with infinite possibilities.

The research into the architectural application of cutting edge biological and digital techniques—with the benefits that come from the inclusion of genetics, like efficiency, economy, renewable use, and self-replication—is crucial, relevant, and urgent: it must be pursued before it is too late for our planet, which has reached the limits of its sustainability. "We have, because human, an inalienable prerogative of responsibility which we cannot devolve." (Sherrington 1940).

ACKNOWLEDGEMENTS
I would like to express my very great gratitude to the geneticists Dr. Miquel-Àngel Serra, Dr. Agustí Fontarnau, Dr. Leandro Peña, Dr. Josep Clotet, and Dr. Aranzazu Balfagón for his valuable collaboration on the planning and development of this research. The same can been said of our philosophers Dr. Josep Corcó and Dr. Xavier Escribano, and the entire staff of the UIC Barcelona for its support, especially all the other research group members, like Mr. Diego Navarro and Mr. Daniel Wunsch, among others.

REFERENCES
Amiguet, Lluís and Eduardo Kac. 2012. "Eduardo Kac, creador del conejo fosforescente." *La Vanguardia*, February 4.64 (obc). Also at http://www.lavanguardia.com/lacontra/20120204/54248805338/eduardo-kac-el-perro-es-una-tecnologia.html

Bradbury, Ray. 1953. "A Sound of Thunder." In *The Golden Apples of the Sun*, Ray Bradbury. New York: Doubleday & Company.

Brooks, Katherine. 2014. "In The Not So Distant Future, Glow-In-The-Dark Trees Could Replace Street Lights." *Huffington Post*, March 30. http://www.huffingtonpost.com/2014/03/30/daan-roosegaarde_n_5044578.html

Estévez, Alberto T. 2005. "Genetic Barcelona Project." *Metalocus* 17: 162–165.

———. 2006. "Genetic Barcelona Project: Cultural and lighting implications". In *Urban Nightscape 2006, Proceedings of the International Commission on Illumination*. Athens: CIE. 86–88.

———. 2007. "The Genetic Creation of Bioluminescent Plants for Urban and Domestic Use." Translated by Dennis Dollens. *Leonardo* 40 (1): 18, 46.

Kac, Eduardo. December 1998. "Transgenic Art." Originally published in *Leonardo Electronic Almanac*, 6 (11). Cambridge, MA: MIT Press. Also at http://www.ekac.org/transgenic.html

Rincon, Paul. 2013. "The Light Fantastic: Harnessing Nature's Glow." *BBC News*, January 24. http://www.bbc.com/news/science-environment-21144766

Rubert De Ventós, Xavier. 2015. In *¿Humanos o posthumanos?: Singularidad tecnológica y mejoramiento humano*, edited by Albert Cortina and Miquel-Àngel Serra. Barcelona: Fragmenta. 257–261, 161–165.

Sherrington, Charles. 1940. *Man on his Nature*. Cambridge: Cambridge University Press.

9 *Biodigital Furniture Series*, with the application of "Biolamps", digitally designed for digital manufacturing.

Towards Genetic Posthuman Frontiers in Architecture & Design Estévez

Swain, Frank. 2010. "Glowing Trees Could Light Up City Streets."
New Scientist, November 24. https://www.newscientist.com/article/
mg20827885.000-glowing-trees-could-light-up-city-streets/

Wagensberg, Jorge. 2008 (2002). *Si la naturaleza es la respuesta, ¿cuál era
la pregunta?* Barcelona: Tusquets.

IMAGE CREDITS

Figures 1, 2, 3, 6, 7, 8: Alberto T. Estévez.
Figures 4, 9: Alberto T. Estévez and Diego Navarro.
Figure 5: Alberto T. Estévez and Daniel Wunsch.

Alberto T. Estévez (b. Barcelona, 1960) holds degrees in Architecture
(Universitat Politècnica de Catalunya, 1983), Ph.D. in Architecture
(Universitat Politècnica de Catalunya, 1990), Art History (Universitat
de Barcelona, 1994), and Ph.D. in Art History (Universitat de Barcelona,
2008). With a professional office of architecture and design (Barcelona,
1983-today), he also taught at different universities before becoming
founding Director at the ESARQ School of Architecture (Universitat
Internacional de Catalunya, Barcelona, 1996). He also founded
the Genetic Architectures Research Group & Office, the Biodigital
Architecture Master (2000–today), and he is also Director of the UIC
Architecture Ph.D. Program. He has written more than 100 articles and
books, and has participated in a large number of exhibitions and conferences around Europe, America and Asia.

Thinking Soils

A synthetic biology approach to material-based design computation

Martyn Dade-Robertson
Javier Rodriguez Corral
School of Architecture,
Newcastle University

Helen Mitrani
School of Civil Engineering
and Geosciences,
Newcastle University

Meng Zhang
Department of Applied Sciences,
Northumbria University

Anil Wipat
School of Computing Science,
Newcastle University

Carolina Ramirez-Figueroa
Luis Hernan
School of Architecture,
Newcastle University

1

ABSTRACT

The paper details the computational modelling work to define a new type of responsive material system based on genetically engineered bacteria cells. We introduce the discipline of synthetic biology and show how it may be possible to program a cell to respond genetically to inputs from its environment. We propose a system of synthetic biocementing, where engineered cells, living within a soil matrix, respond to pore pressure changes in their environment when the soil is loaded by synthesising new material and strengthening the soil. We develop a prototype CAD system which maps genetic responses of individual bacteria cells to geotechnical models of stress and pore pressure. We show different gene promoter sensitivities may make substantial changes to patterns of consolidation. We conclude by indicating future research in this area which combines both *in vivo* and *in silico* work.

1 Artists impression of a bio-based
self constructing foundation.

INTRODUCTION

In this paper we will introduce a computer model for a novel type of material-based design computation. Our model is based on bacteria which are engineered to sense pressure changes in their environment. Our project involves *in vivo* (within the living) experiments to identify genes which respond to pressure by regulating their expression, but our paper will focus on the *in silico* (in the computer) component of the project to map values of gene expression into a geophysical context. The project has the broader aim of developing a system of intelligent material synthesis, where bacteria, growing within soils or soil-like matrices, could respond to mechanical change by producing materials to improve soil resistance. Such a system could be implemented without the need for soil excavations since the bacteria could be seeded and would grow throughout the soil matrix. Advances in computational processes have enabled modelling and simulation to be conducted much earlier in the design process and computational models are used to *synthesise* design outcomes. For example, processes based on *material design computation* use models of material performance to find the optimal form for a structure based on the most efficient distribution of materials given a set of design requirements (Oxman and Rosenberg 2007). We propose a new design process based on systems in which both modelling and manufacture are combined into an engineered biological system. In this case, the designer does not define the material form, but rather designs a system where the material is synthesised in direct response to an environmental context.

While we will highlight the *in vivo* factors in designing such a system, this paper focuses on the *in silico* part of our process through the development of a modelling and editing tool which bridges the design of individual DNA molecules and the large-scale physical modelling of volumes of soil. The project is conducted using knowledge from synthetic biology, which will be introduced here. By describing the demonstrator application and the results obtained from it, we will show how it may be possible to design material forms through genetic manipulation of living cells, thus extending the work of material-based design computation into a new sort of *in vivo* material computation. Our starting point is to consider a design scenario where a heavy load is placed on a weak soil. We imagine an engineered bacteria, living in the soil, that is capable of sensing pore pressure changes within the soil and responding by producing material to solidify the soil matrix. Such a system would create a synthetic foundation beneath the structure. However, designing such a system is not straightforward, and as we shall show here, the computational modelling illustrates some unforeseen emergent properties of such a system.

BACKGROUND
Synthetic biology

Before describing the application, it is worth briefly introducing the aims of synthetic biology and some basic biological understanding. The aim of synthetic biology, according to the Royal Society of Engineers is to "design and engineer biologically based parts, novel devices and systems as well as redesigning existing, natural biological systems" (Voigt 2012).

All living cells contain chromosomes which are long chains of deoxyribonucleic acid (DNA) molecules made of smaller molecules know as nucleotides of adenine, thymine, guanine and cytosine (abbreviated as A, T, G and C). Defined sequences of nucleotides can be grouped into genes. Genes are the molecular units of heredity and are transcribed by an enzyme called RNA polymerase into messenger ribonucleic acid (mRNA). The mRNA is then read by molecular machines know as ribosomes and translated into proteins (long chains of amino acids) according to the codes carried by the genes. Proteins, in turn, provide the structural parts of a living cell and drive the cell's metabolism (see Figure 2). Importantly for synthetic biology, the expression of genes is regulated. In many cases, genes are not simply being expressed constitutively, but are turned on or off depending on whether the cell needs their product at a given time. The regulation of gene expression, it is proposed, can be harnessed by building gene circuits. Regions of DNA can be broken into 'parts,' which not only encode for the proteins (genes), but also have promoter regions which, through their interaction with other molecules in the cell, can either inhibit or promote the transcription of a gene. Other parts include areas for the ribosome to bind when translation is initiated (ribosomebinding sites) and terminators which indicate the transcription of a gene. These different parts can be assembled into 'devices,' as shown in Figure 3. Promoters can be used to control a single gene or a gene cluster and different promoters are sensitive to different chemical or physical conditions within the cell. By recombining promoters that are sensitive to a specific condition with genes that express a protein or proteins that we want to produce, we can create new genetic circuits.

An important factor in this project is that a gene can rarely be said to have been turned on or off, but rather their expression is up- or downregulated. Expression profiles can be mapped for specific genes showing the genetic response (in terms of the amount of a gene product produced) against a given input. In our case, we are interested in potentially pressure-sensitive genes; i.e. the expression of the genes that will be regulated by changing pressures in their environment.

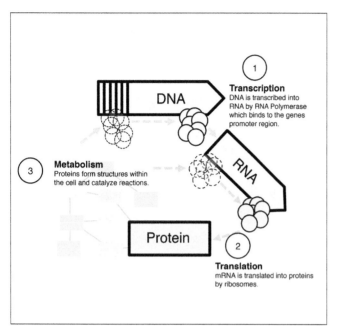

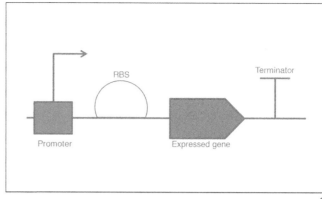

2   Schematic illustration of the 'central dogma' in biology, the processes of translation and transcription from DNA to proteins.

3   Diagram of a genetic 'device' consisting of a promoter which regulates the genes transcription into mRNA, a ribosome binding site (RBS) which recruits the ribosome which initiates the process of translation and a gene which encodes for a protein. The diagram is produced using SBOL (Synthetic Biology Open Language) visual.

Design basis in biology

In our project, we hypothesised the existence of a pressure-sensitive gene or genes in the bacteria *Escherichia coli (E. coli)*. Pressure sensing systems have been discovered in bacteria, including *E.coli* (Welch et al. 1993). We performed transcriptome shotgun sequencing (RNA-seq) to annotate and quantify all mRNAs presented in *E. coli* cells under higher pressure and compare them with the control, that is, *E. coli* cells under normal atmospheric pressure. Using this technique, we have identified 122 genes which are upregulated at pressures of 10atm, and 16 which are downregulated (defined by genes which have a more than 3 fold change in their expression when compared to normal atmospheric pressure).

In our system, we want to locate one or more pressure-sensitive gene promoters. Using this pressure-sensitive promoter, we can then build a gene circuit which controls the synthesis of extra cellular materials. In this way, soils would be cemented where pressures were highest.

Dynamic behaviour of saturated soils

The proposed system requires an understanding of the behaviour of granular materials (like soils) under load and the relationship between this behaviour and the environment of the bacteria cell itself. In this paper, we are interested in saturated soils which would carry bacteria and the nutrients that enable them to survive.

In saturated soils, the pores between the grains are filled with water. In unloaded sediments, the pressures in the pore water are hydrostatic over the depth of the soil layer, but as saturated sediments are loaded, pore pressures can locally increase, which exerts a counteracting force in the material. Because water is incompressible, a load will cause a localised increase in pore pressure before the water is allowed to flow away from the pores and pressure is equalized throughout the system. This implies that fluid flow through the material must be slow or stagnant, thus allowing pockets of saturated sediment to build different pressures (Bredehoeft and Hanshaw 1968).

Pore pressure is defined as "the pressure of the fluid in the voids (pores) between the individual grains comprising a soil's matrix" (Strout and Tjelta 2005). Pore pressure is important because it is a factor in the fundamental geotechnical concept of effective stress, where actual stress is calculated as total stress minus pore pressure. Pore pressure is therefore a function of:

- The permeability of the soil.
- The amount of excess pressure
- The length of the drainage path through the soil.

The restructuring process of a saturated soil under loading is known as consolidation.

DEVELOPING THE *IN SILICO* MODEL

An *in silico* model was developed to examine the vertical stress and pore pressure encountered by soils under load and map soil mechanics over time. Specifically, the model looks at the initial vertical stress developed in a soil under load and the transition from pore pressure to stress within the soil skeleton. This is then used to predict how bacteria in different locations within the soil

Thinking Soils Dade-Robertson et al.

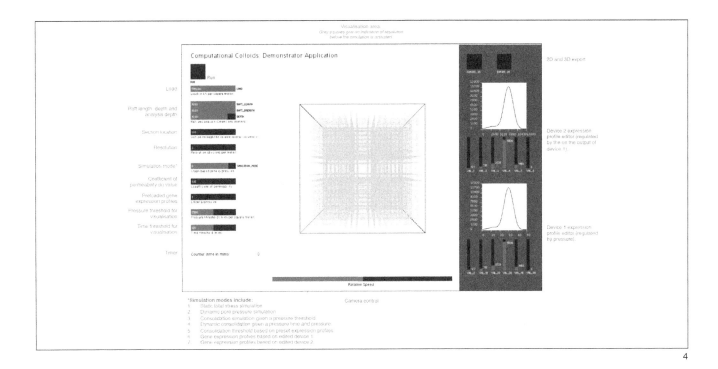

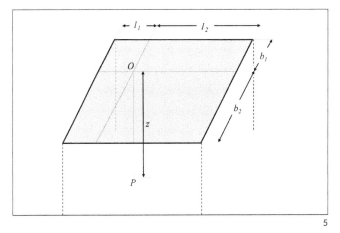

4   Annotated screenshot of simulation
    software developed for the
    geotechnical simulations.

5   Diagram to illustrate the method of
    calculation of stress underneath a
    rectangular load.

might respond in terms of the regulation of a pressure-sensitive gene promoter. A graphical user interface was built to allow the model to be explored interactively and for the results to be visualized dynamically (see Figure 4).

Modeling vertical stress

The vertical stress model is based on the Bossinesq equation (Bossinesq 1871) (initially adapted for point loads and based on the simulation of wave propagation), which assumes a saturated, homogeneous, isotropic, elastic material. Whilst this is not strictly true for soils, it has proven to be sufficiently accurate for most geotechnical contexts, although its limitation will be discussed later. The Bossinesq equation is then coupled with an equation which calculates the influence of a load distributed over a rectangular foundation. While influence of a load on a rectangular foundation will also occur outside the limits of the foundation plate itself, this mode only simulates the distribution of stress directly underneath a foundation (where the most significant stress would be expected). The process, as defined by (Tomlinson 2001), is in two stages. For each point of interest beneath the foundation, two values are required:

- $m = b/z$
- $n = l/z$

where b is the breadth of the foundation, l is the length of the foundation and z is the depth at the point of measurement. In the case of a rectangular foundation, for each point of interest the foundation is split into four rectangles by lines intersecting at point O as seen in Figure 5.

For each rectangle, an influence value is calculated using values for length and breadth (11 or 12 and b1 or b2) and the depth of the point of interest (z). Using values obtained for m and n, in each case the following equations are used to calculate the influence (I) of each rectangle using equations taken from (Newmark 1935):

$$I = \frac{1}{4\pi} \left(\frac{2mn\sqrt{m^2+n^2+1}}{m^2+n^2+m^2n^2+1} \cdot \frac{m^2+n^2+2}{m^2+n^2+1} \right.$$

$$\left. + \tan^{-1} \frac{2mn\sqrt{m^2+n^2+1}}{m^2+n^2+1-m^2n^2} \right) \tag{1}$$

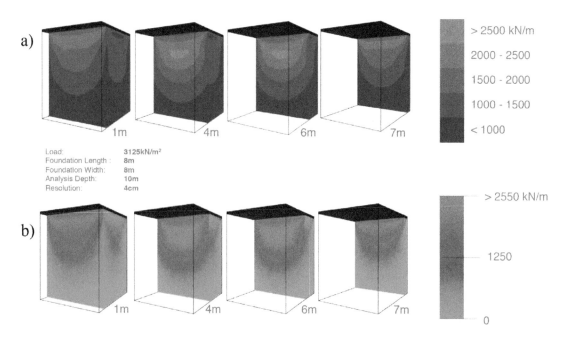

a)

	> 2500 kN/m
	2000 - 2500
	1500 - 2000
	1000 - 1500
	< 1000

1m   4m   6m   7m

Load: **3125kN/m²**
Foundation Length : **8m**
Foundation Width: **8m**
Analysis Depth: **10m**
Resolution: **4cm**

b)

> 2550 kN/m

1250

0

1m   4m   6m   7m

6 Simulations used to show the magnitude of vertical stresses under an 8m x8m raft foundation with a total stress of 200,000 kN. The images show two different visualisations based on (a) contours and (b) a continuous gradient.

When $m^2 + n^2 + 1$ is larger than m^2n^2 the value for I ends up as a negative so this equation is used instead:

$$I = \frac{1}{4\pi}\left[\frac{2mn\sqrt{m^2+n^2+1}}{m^2+n^2+m^2n^2+1}\cdot\frac{m^2+n^2+2}{m^2+n^2+1}\right. \quad (2)$$
$$\left. + \tan^{-1}\left(\pi - \frac{2mn\sqrt{m^2+n^2+1}}{m^2+n^2+1-m^2n^2}\right)\right]$$

The influence is calculated for each of the four rectangles and the stress for point P (σ) can then be calculated using the equation:

$$\sigma_z = q(I_1 + I_2 + I_3 + I_4) \quad (3)$$

where q is the load per square meter over the foundation.

To build up a profile of stresses underneath the foundation, these equations were implemented in software code to systematically iterate over a matrix of points beneath the foundation and return a stress value for each point. The code was implemented in *Processing* (v. 2.2.1), a programming environment developed at MIT and used predominantly for visualisation. The *Processing* language is based on Java and uses the same syntax and structure, but has the advantages of built-in libraries for graphics processing and presentation and a simplified programming

environment. This makes it a useful language for generating quick software 'sketches' and quickly developing complex visualisations which may require many more lines of code or the extensive use of external libraries in other programming languages. The code implements a type of finite element analysis, where the area underneath the loaded foundation is split into voxels. The values for each point in the voxel matrix was calculated using the equations described above for each position in the grid. The output of the Influence equation was validated using lookup table from Tomlinson (2001).

Figure 6 shows the typical output from a rendering that maps vertical stresses through a volume of soil 8x8x10 loaded with 200,000 kN. In this case, points at 3 cm intervals are analysed through the soil volume. The resulting visualisations show two types of rendering, where each voxel is given a colour value where either red indicates higher stress and green indicates low stress or isobars depict blocks of red shades indicating different stress ranges. The visualisations are also shown with sections taken at different points through the soil volume.

Modeling pore pressure dissipation with time
The vertical stress through a material given a constant load does not change. However, in the context of soils, stress will be transferred from pore pressure to the soil skeleton over time. In saturated soils, it is assumed that the initial load will be taken entirely by pore pressure, but that as water flows out of the pores, the water pressure in the material will dissipate and

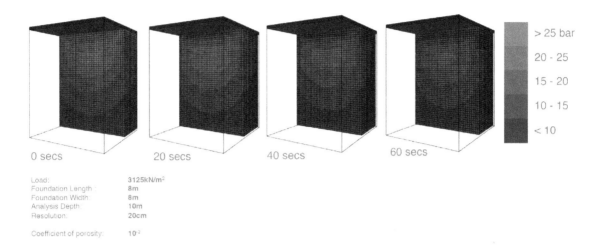

Load: 3125kN/m²
Foundation Length: 8m
Foundation Width: 8m
Analysis Depth: 10m
Resolution: 20cm

Coefficient of porosity: 10⁻²

7 Example simulation of pore pressure dissipation over time.

the load will eventually be taken entirely by the soil skeleton. The speed and spatial distribution of this is important for this study, as the proposed system depends on the soil maintaining high pore pressures long enough for the bacteria to detect and respond to the higher pressure levels.

In civil engineering, it is rarely necessary to do detailed calculations of pore pressure changes underneath a foundation. The process of consolidation leads to the soil underneath a load being compressed, and civil engineers are required to understand the maximum bearing load for a given soil and the time and magnitude of settlement. To this end, civil engineers tend to use a one-dimensional model of consolidation which assumes that pressure is even across the underneath of a foundation and that water flows upwards to the lowest areas of pressure on top of the soil. There are two and three-dimensional models of pore pressure, but these are complex and computationally expensive to implement. In this case, a one-dimensional method based on Darcy equations of flow (Atkinson 1993) has been chosen. The equation—which assumes that the flow of water through the soil is upwards as pore pressure rises—calculates the velocity of flow (V) based on the known coefficients of permeability for different soils (k), the unit weight of water (γ_w), the excess pore pressure ($\delta\bar{u}$) in the location and the depth of point of interest in the soil (δz).

$$V = \frac{k}{\gamma_w} \cdot \frac{\delta\bar{u}}{\delta z} \qquad (4)$$

The velocity is used to calculate the rate at which the pore pressure will subside. The effect shown by running simulations over time is that the areas closest to the top of the soil surface (the underside of the foundation raft) tend to dissipate quickly, while deeper points tend to show slower flow velocities.

Modeling integration with gene data

The different models were implemented in the same code and a graphical user interface was developed that allowed the different simulation modes to be accessed and run and for the variables to be edited using sliders. The interface allowed for different simulations to be run quickly and without accessing the main code and for the results to be output as images or as 3D models for further analysis. In addition, the dynamic pressure values were referenced against hypothetical gene promoter profiles that simulate gene expression as measured in terms of enzyme activity (measured as U mg⁻¹). The units in this instance, however, are less important than the relative expression, which are indicated as promoter profiles on editable graphs on the left hand side of the interface. These promoter profiles could be adjusted and the test rerun. Two scenarios were modelled in this context. In the first scenario, a pressure-sensing promoter is placed on the same gene as a hypothetical material synthesis gene. In other words, both the sensing and material synthesis are done as part of the same genetic device in the same organisms (see Figure 6). In the second scenario, two devices are modelled—imagining two different organisms with two different genetic circuits. In cell type 1 genes, the pressure-sensing promoter is connected to a

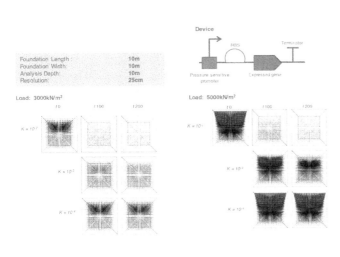

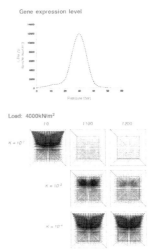

8   Annotated screenshot of simulation software developed for the geotechnical simulations.

9   Visualisations to show gene expression levels in a three dimensional volume of soil given the interaction of two gene circuits where the product of pressure regulated Device 1 regulates the expression of Device 2. The simulations are based on different response profiles for Device 2. The size of the cubes in each matrix indicates the relative level of expression and the colors indicate levels of stress in the soil.

8

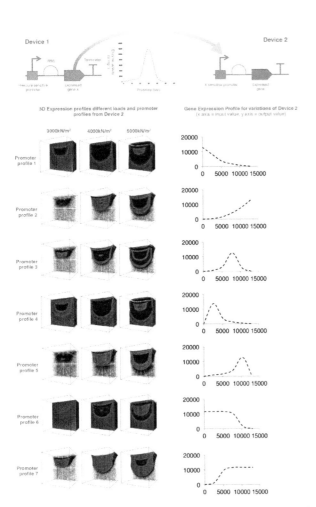

gene which codes for a signalling molecule. A signalling molecule is a substance that can be exported outside the cell and detected by other cells. The second cell contains a device with a promoter that is sensitive to the signalling molecule expressed by Device 1. The promoter belonging to Device 2 is used to control the material synthesis gene. Both scenarios were run with different loads, different soil permeability and scenario two was run for different promoter profiles for Device 2. In this case, the cube size at each point in the matrix is proportional to the level of gene expression from the product of Device 2. A section is also cut through the soil volume to show the internal profile of the soil matrix (see Figure 8). The computational model was run for various loading scenarios on a 10 m x 10 m raft foundation with an analysis depth of 10 meters.

RESULTS

Spatial distribution of high pore pressure

The stress models shown in Figure 6 illustrate the uneven nature of stress transfer through the soil with comparatively low pressure directly underneath the foundation and a zone of high pressure some distance below. However, these results need to be treated with caution. The method for calculating vertical stress assumes that the material being analysed is an elastic solid. Whilst this model (based on the wave theory of Bossenesq) is often used in soil mechanics for modelling basic consolidation, more sophisticated models exist. The stress characteristics are likely to vary for different soil types. It is anticipated that further research will show that it is unlikely that stress distributions in many types of soil would appear as they do in Figure 6, and that the highest

9

Thinking Soils Dade-Robertson et al.

stress should be felt directly below the foundation plate (see for example (Powrie 2014)).

While this vertical stress model needs more development, the model of pore pressure changes with time, which shows that an area of high pressure persisting deep into the soil volume (see figure 7) is more reasonable as the water deeper into the volume of soil has further to travel. This is, however, a theoretical case and should also be investigated in more realistic geotechnical contexts where, for example, there is the presence of bedrock or soils with variable porosities. Any boundary between a porous material and a non- or less-porous material would usually be considered as a means of water escape and many standard geotechnical models assume water escaping vertically and horizontally. To cope with these more complex situations, the model would have to be developed to a full model for 3 dimensional consolidation.

Gene expression in the soil volume

In the absence of detailed data on pressure sensitivity at the magnitudes of pore pressure highlighted in this study, this aspect of the research is largely hypothetical. However, the application of hypothetical data has proven to be valuable and has highlighted some design challenges and potentials of our proposed system. Using hypothetical data has also enabled us to build a framework on which real data can be interpreted.

The model has been developed while assuming a pressure-sensitive promoter with a similar profile but which is an order of magnitude more sensitive, thus showing activity between 1MPa and 5MPa. Running the simulation with different loads shows that, for loads that exceed the peak promoter sensitivity, zones of low gene activity exist within the highest pressure areas of the soil volume. The situation becomes more complicated if the model simulates a device containing a pressure-sensitive promoter that signals to another device with its own sensitivity to the signalling product of Device 1 as shown in Figure 9. The simulations show gene expression levels 20 minutes after loading for a range of different profiles for Device 2, including profiles with peaks of activity and situations where the promoter is positively and negatively regulated by the product of the first genetic device. The results show a wide range of expression patterns with some of the most interesting results demonstrating an interference-like effect resulting in two bands of high or low expression.

These results have potential implications for design. In a synthetic consolidation system, the appearance of low areas of gene expression within high-pressure areas of the soil matrix could cause problems of instability and their effects may need to be mitigated. For example, using soils with comparatively higher

porosity would mean that areas of highest pressure would diminish quickly, thus bringing the pressurized volume of soil within the range of the pressure-sensitive promoter. More than one type of promoter sensitive to different pressure ranges and sensitivities to the signal of Device 1 might also be tried.

These results might, however, also offer design potential. By combing promoters with different sensitivities through a signalling system, the designer might have tight controls over the morphology of the consolidated soils. While the immediate applications for such a process are not clear, it might be possible to imagine a process whereby underground structures are created using consolidated soils, where the more friable material is excavated, thereby leaving caverns and holes in the ground. When combined with different patterns of loading, complex structures could be created in this way.

To explore this further, a better understanding of how synthetic consolidation would occur would need to be developed with reference to threshold values for meaningful levels of consolidation—i.e. when the resulting material can be considered substantially more load-bearing than the material around it. The current model also doesn't take into account the fact that the process of consolidation itself will change the dynamics of pore pressure and prevent the flow of water from leaving the pores. This may mean that high pressure areas will persist in the pores permanently, and will in turn change the dynamics of consolidation.

CONCLUSION
Limitations of this study

There are limitations to this analysis. Currently these broad values for pore pressure are obtained from standard formulas based on soils which are saturated with water. A bacteria mix is likely to be much thicker than water and the liquid flow through the soils will be slower, thus maintaining higher pressures in the soil for each condition longer.

The narrow range of loading values should also be expanded. The current values of loading are orders of magnitude higher than the bearing capacity of most soils because much of the research conducted so far on pressure sensing in bacteria is concerned with responses of high pressures. Based on our in vivo work, we believe that there are promoters which are sensitive to much smaller pressure changes and the model needs to be developed to reflect this.

The model also does not take into account many of the microbiological parameters which would need to be considered. It is likely, for example, that the bacteria would not distribute themselves

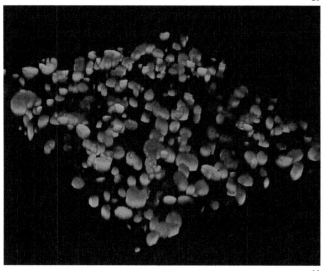

10 Compression test being performed on a column of agarose gel.

11 Confocal (3D) microscope images of bacteria illuminated with GFP.

in meters. The sensitivity of the promoter parts, shown here as editable graphs, is hypothetical. But as our in vivo work develops, we should be able to plot promoter sensitivity for a selection of our known pressure sensitive genes, and in the future, we may be able to edit promoters to give us the desired sensitivity profile—sculpting material responses to force by altering sequences of DNA and through the interaction of many different genetic devices and engineered organisms. This work also hints at the possibility of extending our application domain. In addition to formally characterizing the profiles of pressure-sensing promoters, therefore, we have begun to extend the experimental work to growing the bacteria in hydrogels. Hydrogels act for us as a surrogate soil with some similarities with weak clay-type soils. The bacteria can grow through the 3D matrix of the hydrogel material and we can visualize their growth and activity in three dimensions. Our aim will be to build an engineered bacterium using our newly discovered pressure-sensing promoter. The pressure-sensing promoter will be connected to a signaling molecule (such as green fluorescent protein [GFP]) so that we can visualize the promoter activity in a volume of hydrogel. Figures 10 and 11 show us beginning to test the mechanical properties of hydrogels and visualizing bacteria which glow as they express GFP production. Such a demonstrator would indicate the first steps towards a new type of responsive material which might have much broader architectural applications.

ACKNOWLEDGEMENTS

We would like to thank the EPSRC for their support for the project: *Computational Colloids: Engineered bacteria as computational agents in the design and manufacture of new materials and structures*. Project number: EP/N005791/1.

REFERENCES

Atkinson, J. 1993. *The Mechanics of Soils and Foundations*. Maidenhead: McGraw-Hill.

Bossinesq, J. 1871. "Théorie de L'intumescence Liquide, Appelée Onde Solitaire Ou de Translation, Se Propageant Dans Un Canal Rectangulaire." *Comptes Rendus de l'Academie Des Sciences* 72: 755–759.

Bredehoeft, J., and B. Hanshaw. 1968. "On the Maintenance of Anomalous Fluid Pressures. I Thick Sedimentary Sequences." *Geological Society of America Bulletin* 79 (9): 1097–1106.

Newmark, Nathan. 1935. "Simplified Computation of Vertical Pressures in Elastic Foundations." *Engineering Experiment Station* 33 (24).

Oxman, Neri, and Jesse Louis Rosenberg. 2007. "Material-Based Design Computation: An Inquiry into Digital Simulation of Physical Material Properties as Design Generators." *International Journal of Architectural Computing* 5 (1): 26–44. doi:10.1260/147807707780912985.

evenly through the soil; rather, they would migrate to the surface where oxygen levels are highest. The models would need to take these factors into consideration.

Implications and future work

While there are limitations in the model described here, the editing interface hints at a new type of CAD application which connects the behavior of cells, defined at the molecular level (at scales of nanometers) with consolidation patterns measured

Thinking Soils Dade-Robertson et al.

Powrie, W. 2014. *Soil Mechanics: Concepts and Applications*. London: CRC Press.

Strout, James Michael, and Tor Inge Tjelta. 2005. "In Situ Pore Pressures: What Is Their Significance and How Can They Be Reliably Measured?" *Marine and Petroleum Geology* 22 (1–2): 275–285. doi:10.1016/j.marpetgeo.2004.10.024.

Tomlinson, M.J. 2001. *Foundation Design and Construction*. Harlow: Prentice Hall.

Voigt, Christopher. 2012. "Synthetic Biology Scope, Applications and Implications." *Synthetic Biology* 1 (1): 1–2. doi:10.1021/sb300001c.

Welch, Timothy J, Anne Farewell, Frederick C Neidhardt, and Douglas H Bartlett. 1993. "Stress Response of Escherichia Coli to Elevated Hydrostatic Pressure." *Journal of Bacteriology* 175 (22): 7170–7177.

IMAGE CREDITS

Figure 1: Carolina Remirez-Figueroa, Luis Hernan, Martyn Dade-Robertson, 2016

Figures 2–9: Martyn Dade-Robertson, 2016

Figure 10: Javier Rodriguez Corral, 2016

Figure 11: Javier Rodriguez Corall, Meng Zhang, 2016).

Dr. Martyn Dade-Robertson is a Reader in Design Computation at Newcastle University where he is Co-Director of the Architectural Research Collaborative (ARC), Director of the MSc in Experimental Architecture and leads a research group on Synthetic Biology and Design (www.synbio.construction). Martyn holds an MPhil and Ph.D. from Cambridge University in Architecture and Computing. He also holds a BA in Architectural Design and an MSc in Synthetic Biology from Newcastle University. Along with over 25 peer reviewed publications Martyn published the book *The Architecture of Information* with Routledge in 2011. Email: martyn.dade-robertson@newcastle.ac.uk

Javier Rodriguez Corral is a PhD in the School of Architecture, Planning and Landscape at Newcastle University. Javier graduated from Cardiff University in 2015 (MSc Civil Enginnering) and from Universitat Politècnica de Catalunya in 2014 (BEng Construction Engineering). His current research focuses on the interaction between microorganisms and soils, the effect on the mechanical properties and an investigation on potential bio-mediated soil reinforcement techniques. Email: j.rodriguez-corral2@newcastle.ac.uk

Dr. Helen Mitrani obtained her PhD in geotechnics in 2006 from the University of Cambridge. This research focussed on novel liquefaction remediation methods for existing buildings, including soil cementation. Following this, Mitrani worked for 5.5 years as a consultant engineer for Arup and achieved Chartered Engineer status with the Institution of Civil Engineers in 2011. Mitrani joined Newcastle University as a Lecturer in Structural Engineering in 2012. Her current research interests are at the interface of structural and geotechnical engineering and include ground improvement, dynamic soil structure interaction and the fatigue performance of soils and structures. Email: helen.mitrani@newcastle.ac.uk

Dr. Meng Zhang is a Senior Lecturer at Northumbria University. Meng Zhang studied for her B.Sc. in Biotechnology at ShanDong Agriculture University, China. In the UK she was then rewarded Overseas Research Scholarship by UK Universities and studied for a Ph.D. in Proteomics in Northumbria University. Meng is currently a lecturer at Northumbria University where her research interests are on Proteomics and Metabolomics study of pathogens and Synthetic Biology on Biocatalysis. Meng has published widely on Molecular Biology and Microbiology and is currently investigating wider applications of Synthetic Biology. Email: meng.zhang@unn.ac.uk

Prof. Anil Wipat is Professor of Bioinformatics in the School of Computing Science, co-directs the Interdisciplinary Computing and Complex Biology Systems Group (ICOS - http://ico2s.org/) and founder and associate-director of the Newcastle Centre for Synthetic Biology and Bioeconomy (CSBB). AW trained first as a molecular microbiologist and subsequently as a computer scientist. Recent work has focussed on design in Synthetic Biology. Anil is also very active in the development of standards for data exchange in Synthetic Biology and is the Chair of the International SBOL steering committee. He has over 100 refereed publications many of which are interdisciplinary in nature. Email: anil.wipat@newcastle.ac.uk

Carolina Ramirez-Figueroa is a PhD researcher and designer based in the UK. Her work addresses the intersection between design and synthetic biology, especially in the way living challenges existing models of design and material production. In Synthetic Morphologies (www.syntheticmorphologies.com), she develops a number of design interventions which helps her understand the impact, background and potential of living systems in the fabrication of the built environment. Carolina holds an MSc in Digital Architecture, and trained originally in Architecture. Email: p.c.ramirez-figueroa@newcastle.ac.uk

Luis Hernan is a designer and doctoral researcher based in Newcastle University. Initially trained as an architect, Luis pursued a master's degree in Digital Architecture before embarking in his doctoral research, which explores the role of wireless infrastructure in contemporary experience of architectural space (www.digitalethereal.com). His work also extends to living technologies, investigating the interface between digital, physical and living computation. Email: j.l.hernandez-hernandez@newcastle.ac.uk

Symbiotic Associations

Aldo Sollazzo
Efilena Baseta
Noumena

Angelos Chronis
Institute for Advanced
Architecture, Catalonia /
Noumena

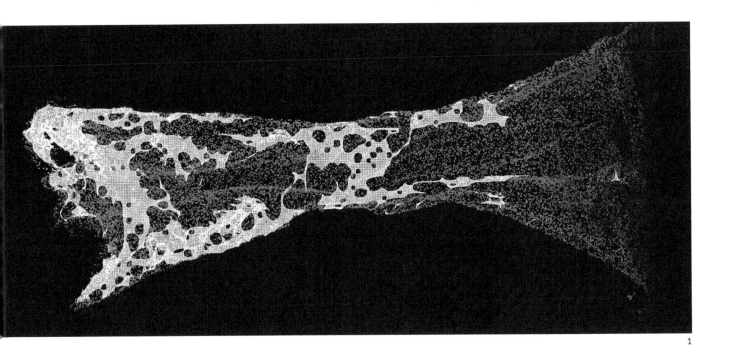

1

ABSTRACT

Soil contamination poses a series of important health issues, following years of neglect, constant industrialization, and unsustainable agriculture. It is estimated that 30% of the total cultivated soil in the world will convert to degraded land by 2020 (Rashid et al. 2016). Finding suitable treatment technologies to clean up contaminated water and soil is not trivial, and although technological solutions are sought, many are both resource-expensive and potentially equally unsustainable in long term. Bacteria and fungi have proved efficient in contributing to the bioavailability of nutrients and in aggregating formation in degraded soils (Rashid et al. 2016).

Our research aims to explore the possible implementation of physical computing, computational analysis, and digital fabrication techniques in the design and optimization of an efficient soil remediation strategy using mycelium. The study presented here is a first step towards an overarching methodology for the development of an automated soil decontamination process, using an optimized bio-cell fungus seed that can be remotely populated using aerial transportation. The presented study focuses on the development of a methodology for capturing and modeling the growth of the mycelium fungus using photogrammetry-based 3D scanning and computational analysis techniques.

1   Point cloud analysis.

2

INTRODUCTION

Soil contamination, also referred to as land pollution, is the degradation or destruction of the earth's surface, specifically soil, through xenobiotic (non-biological) chemicals or other substances, typically as a byproduct of human activities. Soil, being a porous and permeable mixture of unconsolidated mineral and rock fragments, acts naturally as a filter that allows the free flow of water through the pores and spaces between its particles. Land pollution thus poses a series of important health issues that have become prominent in recent years, following years of neglect, constant industrialization, and unsustainable agriculture. It is estimated that 30% of the total cultivated soil in the world will convert to degraded land by 2020 (Rashid et al. 2016). It is argued that the impacts of land degradation not only pose a serious challenge to sustainable development, but may also amplify the underlying social and political weaknesses which can contribute to global security threats (Barbut 2016).

Therefore, there is an evident need for proactive solutions for the reduction of land degradation, such as the promotion of sustainable land management and the restoration of degraded soils. Finding suitable treatment technologies to clean up contaminated water and soil is not trivial, and although technological solutions are sought many are both resource-expensive and potentially equally unsustainable in long term.

Bacteria and fungi have proved efficient in contributing to the bioavailability of nutrients and in aggregating formation in degraded soils (Rashid et al. 2016). Bacterial and fungal inocula have the potential to reinstate the degraded land's fertility by mobilizing key nutrients to the crop plants while remediating the soil's structure and improving its aggregation and stability.

In our effort to develop cost-effective procedures for cleaning up contaminated water and soil, our research focuses on studying the potential processes of using decontaminating agents—such as the mycelium fungus—to reinstate the degraded soil. Overall, our aim is to explore the implementation of physical computing, computational analysis, and digital fabrication for the design and optimization of an efficient soil remediation strategy using mycelium. The study presented in this paper is a first step towards an overarching methodology for the development of an automated soil decontamination process, using an optimized bio-cell fungus seed that can be remotely populated using aerial transportation. The study focuses solely on developing a methodology for capturing and modeling the growth of the mycelium fungus using the aforementioned technologies.

Many mathematical models of fungal growth and function have been and are currently being developed across different disciplines, including biotechnology (Ashley et al. 1990), soil science (Boddy et al. 2000), chemistry (Galli et al. 2008), mycology (Davidson et al. 2011) and biology (Lin et al. 2016), among others. We have found these models difficult to implement in our research and not focused on the aspect in we are most interested; namely, capturing the three-dimensional growth patterns of fungal growth. Our study is structured around data collection from either single or multiple apparatuses to detect

3   Point cloud construction after 3D scanning process.

4   Data extraction. Parsing point cloud results.

5   Apparatus 01. Where strings are parsed according to their length.

3

the mycelium's growth in 3D. The digitally fabricated apparatuses, which contain a substratum and the inoculated mycelium fungus, aim to provide an initial experimentation test-bed for the further development of a bio-cell, which is the basis for the automated soil decontamination process.

To detect the growth pattern of mycelium, we have developed three-dimensional imaging and computational techniques to capture and analyze the growth process. In order to acquire the three-dimensional form of the mycelium growth, we have used photogrammetry-based 3D scanning. Different 3D scanning techniques have been previously used for detecting the growth of fungi (Senthilkumar 2016) and other biological organisms (Vilena 2010). In our study, we focus on affordable and accessible technology with an adequate level of accuracy. Further to the 3D imaging acquisition, computational analysis of the three-dimensional data is employed to iteratively detect the growth of mycelium.

Previous research on biological growth analysis in architecture has mainly focused on the development of design methodologies and morphogenetic computation. Growth modeling has been employed for structural optimization (Klemmt 2014, Itsuka 2014), morphological strategies (Klemmt and Bollinger 2015), and more commonly, for developing design methodologies and frameworks for generative design (Biloria and Chang 2012; Ahmar et al. 2013; Teixeira 2015). Biological fabrication techniques, similar to the ones envisioned under the overarching research goals of this study, have also been investigated (Araya et al. 2012), albeit in a completely different context and with a focus on producing physical design components. Thus, we can assume that this study

is breaking new ground in the development of a fungal growth model under the scope of an automated soil decontamination process and within the context of architectural research.

METHODS

The study is structured in different parts, organized among analog and computational operations. The first part is focused on the design of a digitally fabricated apparatus for the observation of the mycelium growth. The next step is the generation of a 3D point cloud of this apparatus from a series of captured images using photogrammetry-based 3D scanning software (Agisoft PhotoScan). Following the 3D reconstruction process, the analysis of the 3D point cloud is done by parsing the data based on a specified color range in Rhinoceros 3D using Grasshopper. The computational analysis is then concluded by applying an octree subdivision to the parsed cloud in order to extract the main vectors that define the growing directions of the mycelium. The following part of the paper focuses on a detailed description of each part mentioned above. In order to establish an optimal process for the study, we tested different apparatuses with the intention to define the best method for data collection.

Apparatus

During the study, we developed several types of apparatus, each based on a combination of digital fabrication techniques and approaches to overcome the various challenges. The first apparatus was a 3D printed hexaedra structure with four faces capped with laser-cut metacrilatic plates. Between those faces, a grid of thin strings was inserted to provide a continuous surface, distributed in multiple directions where the mycelium could possibly grow. This first apparatus posed several difficulties for

Symbiotic Associations Sollazzo, Baseta, Chronis

3D scanning, as the spacing was very narrow and the strong symmetry of the structure generated significant noise in the point cloud. This led to the development of a few iterations of more linearly distributed grids. Overall, during the whole study, we designed a number of apparatuses, each with a specific observation focus, and tested different substrates and grid structures.

The main substrate consisted of straw, which was infused with a mix of water and honey to add complex sugars as a nutrient base for the mycelium. This substrate was inoculated with grain spawn of *Pleurotus ostreatus* (oyster mushroom), at a ratio of about 20% of the total substrate mass. The inoculation took place around a Bunsen burner gas flame to ensure a sterile working environment. Colonization of the different substrates was complete after 2 to 3 weeks, and produced mushrooms 5 weeks after inoculation. In some cases, we added plastic from 3D printing material in order to observe the possible decomposition of the material by the mycelium. For the digital fabrication process, the machines used were commercial 3D printers and laser cutters.

3D Scanning

With regards to the photogrammetry process, the collection of images was performed using a Nikon D3200 with a resolution of 25 megapixels and a fixed prime 50 mm lens. The analysis and reconstruction of the 3D point cloud was performed solely within PhotoScan, photogrammetry software which processes digital images to generate 3D point clouds.

The images used can be arbitrary, both in terms of positions as well as exposure or focal length—although a constant focal length was proven to give better results—thus the software is efficient in both controlled and uncontrolled conditions, which was ideal for our study. Other scanning frameworks were tested, mainly based on infrared depth sensors, but due to both the scanning inaccuracy as well as their outdoor performance limitations, they did not prove adequate for the purposes of the study. Image alignment and 3D point cloud reconstruction, which are fully automated processes in PhotoScan, were implemented in the study with adequately accurate results in both outdoor and indoor conditions. In order to get precise results for the level of accuracy needed for our analysis, around 30 to 50 pictures were used for every reconstruction iteration. Between building dense clouds and generating meshes, the processing time of reconstruction was on average 30 minutes, Once the cloud was generated, the entire process shifted to Rhinoceros 3D / Grasshopper.

Growth Analysis

In Grasshopper, the main tool used was the Volvox library, developed as a part of the DURAARK project at the Center for

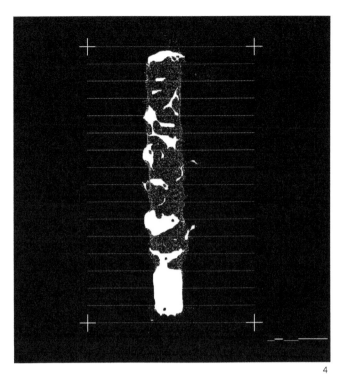

4

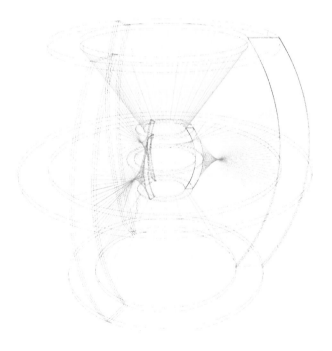

5

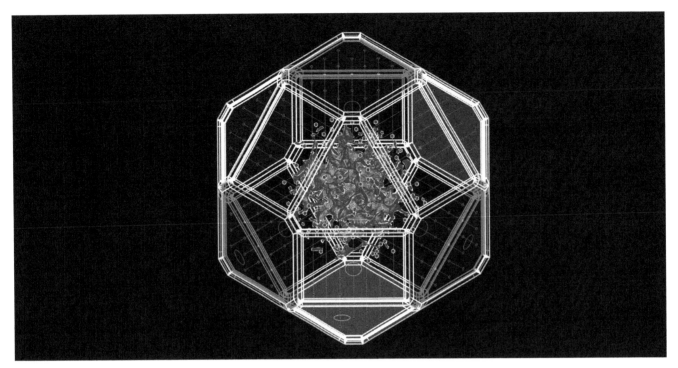

6   Apparatus 00. Designed to host different cameras to detect bio growth.

Information Technology and Architecture (CITA) by Henrik Leander Evers and Mateusz Zwierzycki. Volvox is a point cloud library which enables the creation, editing, and analysis of point cloud data. The library allows direct import of the formats which are exported from PhotoScan. Volvox allowed us to work with a large amount of points and perform multiple operations regardless of the quantity of data stored in each file. To detect the mycelium growth, a color range that corresponds to the mycelium was defined in Grasshopper. The point cloud is then clustered in 3D space by converting each band from RGB values to XYZ coordinates, and the range converted to a bounding box to select the growth points from the entire list. The final output is a list of boolean values which are used to cull the data from the point cloud. The entire point cloud list is also used in an octree subdivision, which organizes all points in the main eight quads of the 3D space recursively, by subdividing according to a minimum set of points that each cube can contain. Finally, from the overall amount of cubes generated for each quad in 3D space, we extract the main vectors representing the main growing direction of the mycelium.

The intention of this study is to perform this operation iteratively in order to generate a time-based growth model for the mycelium. At each time step, the growth can be scanned, reconstructed, and analyzed, thus creating an overlapping series of growth snapshots. This series will then be used to draw conclusions about the growth patterns of mycelium and to produce a growth model for our purposes.

RESULTS AND FUTURE WORK
Data acquisition
The study went through several apparatuses and different data collection methods, which resulted in a large amount of processed data. This data collection demonstrated substantial incremental improvements for each of the methods employed. From each single observation, we collected more than 50 pictures, generating a parse cloud of 56.000 points in PhotoScan, which was converted to a dense point cloud of more than 1.6 million points. Using the Volvox library, this point cloud was processed in Grasshopper, where we were able to parse the number of points associated with mycelium, translating percentage and growth direction into vectors. Collecting this information is crucial for our future work, which aims to generate a growth model that will allow us to predict the behavioral patterns for the organisms and which can be used to inform multiple design strategies.

Although our iterative experimentation has proven our initial hypothesis of detecting a growth pattern through our methodology, our experiments faced several difficulties in data collection, mainly related to the structure and accessibility of the camera in the internal part, and the ability of PhotoScan to reconstruct 3D point clouds in narrow and symmetric environments. One of the main pitfalls of the software is that it relies heavily on the pixel count of the acquired images. This makes the automation of the process cumbersome, as it relies on costly and difficult-to-automate acquisition methods, such as high-definition DSLR cameras, to collect the image data.

Future work

At this stage, this initial study shows the potential of the method, but it has yet to be optimized by automating many of the steps, which are currently seen as separate entities. Specifically, the image acquisition could be automated with a robotic arm rotating around the apparatus. The 3D point cloud reconstruction could also be optimized by culling all information related to parts of the picture outside the apparatus. Although this operation is possible in PhotoScan, the method allowed in the software is not automatable, since it relies on manual correction of the pictures. For the computational methods implemented in Grasshopper, an interface needs to be developed which can store and allow access to all the information generated during the process. This interface can be open and accessible online to promote collaboration with other researchers developing similar studies. Automated image uploading and a data visualization service would enable an initial study of the overlapping analyses generated.

CONCLUSIONS

In this study, we presented a first step towards a methodology for the development of an automated soil decontamination process. The overarching research aims to produce an optimized bio-cell fungus seed that can be remotely populated using aerial transportation. This first step has proven that with the employment of appropriate technologies and computational techniques, we are able to get initial results of the growth pattern detection of mycelium. Although technical challenges still remain, we consider the initial results promising for future work and aim to continue developing this methodology further.

Significant further steps need to be taken, including the automation of the capturing and analysis process, and the development of accompanying appropriate interfaces for the dissemination of the work to a wider research audience. Nevertheless, we expect future developments to yield more interesting results and a more robust growth detection process, which will be the basis for the further development of the overall research goals. Moreover, this methodology can also be used to inform other studies associated not only with mycelium, but to similar image-based observations of living organisms.

The end goal of our research is to test the impact of mycelium growth on contaminated soil. Our next generation of apparatuses will focus on monitoring mycoremediation by adding data from pH sensors, humidity sensors, and thermal cameras to the dataset presented in this paper. The future work of this study is part of an overarching research effort to merge computational methods and traditional apparatuses for the observation of biological reactions. The results obtained by the overall process aims to converge to the development of a series of bio-cells—small hosts that can host both the mycelium and the substratum.

7   Preparation of the substrate.

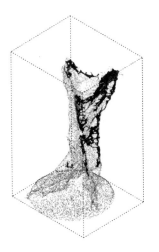 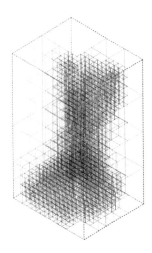 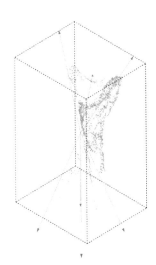

8 Parsing point cloud, recursive sampling and growth orientation.

These bio-cells can then be involved in a wider aerial robotic research project involving drones, and focused on empowering reforestation and food production on dry or contaminated lands

REFERENCES

Aynsley, M., A. C. Ward, and A. R. Wright. 1990. "A Mathematical Model for the Growth of Mycelial Fungi in Submerged Culture." *Biotechnology and Bioengineering* 35 (8): 820–830. doi:10.1002/bit.260350810.

Barbut, Monique and Sasha Alexander. "Land Degradation as a Security Threat Amplifier: The New Global Frontline." In *Land Restoration: Reclaiming Landscapes for a Sustainable Future*, edited by Martin Frick and Jennifer Helgeson, 3–12. Boston: Academic Press.

Biloria, Nimish and Jia-Rey Chang. 2012. "HyperCell: A Bio-Inspired Information Design Framework for Real-Time Adaptive Spatial Components." In *Digital Physicality: Proceedings of the 30th eCAADe Conference*, vol. 2, edited by Henri Achten, Jiri Pavlicek, Jaroslav Hulin, and Dana Matejovska. Prague, Czech Republic: eCAADe. 573–583.

Boddy, Lynne, John M. Wells, Claire Culshaw, and Damian P. Donnelly. "Fractal Analysis in Studies of Mycelium in Soil." In *Fractals in Soil Science*, edited by Yakov A. Pachepsky, John W. Crawford, and Walter J. Rawls. Developments in Soil Science 27. London: Elsevier. 211–238.

Davidson, Fordyce A., Graeme P. Boswell, Mark W. F. Fischer, Luke Heaton, Daniel Hofstadler, and Marcus Roper. 2011. "Mathematical Modelling of Fungal Growth and Function." *IMA Fungus* 2 (1): 33–37.

doi:10.5598/imafungus.2011.02.01.06.

El Ahmar, Salma, Antonio Fioravanti, and Mohamed Hanafi. 2013. "A Methodology for Computational Architectural Design Based on Biological Principles." In *Computation and Performance: Proceedings of the 31st eCAADe Conference*, vol. 1, edited by Rudi Stouffs and Sevil Sriyildiz. Delft, the Netherlands: eCAADe. 539–548.

Fischer, Thomas, Torben Fischer, and Cristiano Ceccato. 2002. "Distributed Agents for Morphologic and Behavioral Expression in Cellular Design Systems." In *Thresholds - Design, Research, Education and Practice in the Space Between the Physical and the Virtual: Proceedings of the 2002 Annual Conference of the Association for Computer Aided Design in Architecture*, edited by George Proctor. Pomona, CA: ACADIA. 111–121.

Galli, E., E. Brancaleoni, F. Di Mario, E. Donati, M. Frattoni, C. M. Polcaro, and P. Rapanà. 2008. "Mycelium Growth and Degradation of Creosote-Treated Wood by Basydiomycetes." *Chemosphere* 72 (7): 1069–72. doi:10.1016/j.chemosphere.2008.04.014.

Iitsuka, Mayum. 2014. "Biological Data-Mining and Optimization: In the Case of Immunorium Project." In *ACADIA 14: Design Agency—Projects of the 34th Annual Conference of the Association for Computer Aided Design in Architecture (ACADIA)*, edited by David Gerber, Alvin Huang, and Jose Sanchez. Los Angeles: ACADIA. 115–118.

Klemmt, Chirstoph. 2014. "Compression Based Growth Modelling." In *Acadia 2014: Design Agency—Proceedings of the 34th Annual Conference of the Association for Computer Aided Design in Architecture*, edited by David Gerber, Alvin Huang, and Jose Sanchez. Los Angeles: ACADIA. 565–572.

Klemmt, Christoph and Klaus Bollinger. 2015. "Cell-Based Venation Systems." In *Real Time: Proceedings of the 33rd eCAADe Conference* vol. 2, edited by Bob Martens. Vienna, Austria: eCAADe. 573–580.

Lin, Xiao, Gabriel Terejanu, Sajan Shrestha, Sourav Banerjee, and Anindya Chanda. 2016. "Bayesian Model Selection Framework for Identifying Growth Patterns in Filamentous Fungi." *Journal of Theoretical Biology* 398: 85–95. doi:10.1016/j.jtbi.2016.03.021.

Rashid, Muhammad Imtiaz, Liyakat Hamid Mujawar, Tanvir Shahzad, Talal Almeelbi, Iqbal M. I. Ismail, and Mohammad Oves. 2016. "Bacteria and Fungi Can Contribute to Nutrients Bioavailability and Aggregate Formation in Degraded Soils." *Microbiological Research* 183: 26–41. doi:10.1016/j.micres.2015.11.007.

Senthilkumar, Thiruppathi, Digvir S. Jayas, Noel D. G. White, Paul G. Fields, and Tom Gräfenhan. 2016. "Detection of Fungal Infection and Ochratoxin A Contamination in Stored Barley Using Near-Infrared Hyperspectral Imaging." *Biosystems Engineering* 147: 162–73. doi:10.1016/j.biosystemseng.2016.03.010.

Teixeira, Frederico Fialho. 2015. "Biology, Real Time and Multimodal Design: Cell-Signaling as a Realtime Principle in Multimodal Design." In *Real Time: Proceedings of the 33rd eCAADe Conference* vol. 2, edited by Bob Martens. Vienna, Austria: eCAADe. 551–562.

Villena, G. K., T. Fujikawa, S. Tsuyumu, and M. Gutiérrez-Correa. 2010. "Structural Analysis of Biofilms and Pellets of *Aspergillus Niger* by Confocal Laser Scanning Microscopy and Cryo Scanning Electron Microscopy." *Bioresource Technology* 101 (6): 1920–1926. doi:10.1016/j.biortech.2009.10.036.

IMAGE CREDITS
Figures 1,3,4,5: (Sollazzo, 2016)
Figure 6: (Klein-Agbo-Ola, 2016)
Figure 2,7: (Postma. 2016)
Figure 8: (Sollazzo-Baseta, 2016)

Aldo Sollazzo is an architect and researcher. With a Master in Architectonic Design in 2007, Master in Advanced Architecture at IAAC (Institute for Advanced Architecture of Catalunya) in 2012, and a Fab Academy diploma in 2014 at the Fab Lab Barcelona, he is an expert in computational design and digital fabrication. Since 2011, Aldo is the manager of Noumena. He is also a founder of Fab Lab Frosinone and the Director of Reshape – digital craft community. Since 2015, he is Head of IaaC Visiting Programs.

Efilena Baseta is an architectural engineer, who studied at the National Technical University of Athens (NTUA), with a Master's degree in Advanced Architecture from the Institute for Advanced Architecture of Catalonia (IAAC). Her current interest lies in exploring material behaviors, physically and digitally, in order to create real-time responsive architectural structures.

Angelos Chronis is a PhD Candidate at the Institute of Advanced Architecture of Catalonia in Barcelona, under the Innochain network. He teaches at IaaC and at the Bartlett School of Architecture at University College London. He has previously worked as an Associate for the Applied Research + Development group at Foster + Partners. He is also actively involved in scientific committees as an author, reviewer and organizer as well as participating in lectures, workshops and architecture crits internationally.

Fluid Morphologies

Hydroactive Polymers for Responsive Architecture

Luke Franzke

Dino Rossi

Prof. Dr. Karmen Franinović
IAD, Zurich University of the Arts

1

ABSTRACT

This paper describes Hydroactive Polymers (HAPs), a novel way of combining shape-changing Electroactive Polymers (EAPs) and water for potential design and architectural explorations. We present a number of experiments together with the Fluid Morphologies installation, which demonstrated the materials through an interactive and sensory experience. We frame our research within the context of both material science and design/architecture projects that engage the unique material properties of EAPs. A detailed description of the design and fabrication process is given, followed by a discussion of material limitations and potential for improving robustness and production. We demonstrate fluid manipulation of light and shadow that would be impossible to achieve with traditional electromechanical actuators. Through the development of this new actuator, we have attempted to advance the accessibility of programmable materials for designers and architects to conduct hands-on experiments and prototypes. We thus conclude that the HAP modules hold a previously unexplored yet promising potential for a new kind of shape-changing, liquid-based architecture.

1   Details of HAP device in the Fluid Morphologies installation.

INTRODUCTION

One of the most exciting areas in design and architecture in recent years has come from practitioners engaging in new developments from material science (Rossi, Nagy, and Schlueter 2014; Decker 2015). In particular, a lot of attention has been given to shape-changing materials such as EAPs, which have been explored since the early 1990s in various material science labs (Biggs et al. 2013). It wasn't until the late 2000s that architects and designers began to work with them, typically to actuate structural elements. However, there is great potential for experimentation beyond simple kinetic activation, as we have demonstrated through our hands-on approach to EAP research.

We believe that this engagement of designers and architects with material science is best approached through an embodied, hands-on creative process, especially with regard to the unintuitive nature of elastomer deformations. We frame this paper and our research within the context of the Active Material approach, which explores the active properties of materials and sets out to uncover related aesthetic potentials (Franinović and Franzke 2015). The active materials we work with include those that can change their states and properties when exposed to certain stimuli such as humidity, light, or electricity. Hydroactive Polymers (HAPs), as described in this paper, work not only with the active properties of EAPs, but also with the active properties of water. Through the electrical actuation of HAPs, water oscillates and accentuates the movement of the module. In this interplay between electrical charges, the polymer response, and the dynamics of water, a complex material activity emerges and further involves people and the surrounding environment in interaction.

BACKGROUND

EAPs in Architecture/Design

EAPs have found their way into several architecture and design projects over the last several years. ShapeShift sought to imagine EAPs at an architectural scale through a speculative, adaptive facade prototype (Kretzer and Rossi 2012). The Homeostatic Facade System developed an adaptable solar shading system that took advantage of both the flexibility and reflectivity of an industrially produced EAP (Decker 2013). Reef implemented EAPs in a "living" ceiling (Mossé, Gauthier, and Guggi 2011). A Degree of Freedom integrated EAPs with rigid elements to create a dynamic structure (Rhoné and Genet 2014). Sound to Polymer proposed a dynamic wall surface which adapted in response to its acoustic environment (Joucka 2015). There has also been EAP Architecture research related to adaptive facades carried out through computer simulation and modeling (Krietemeyer and Dyson 2011; Stouffs 2013).

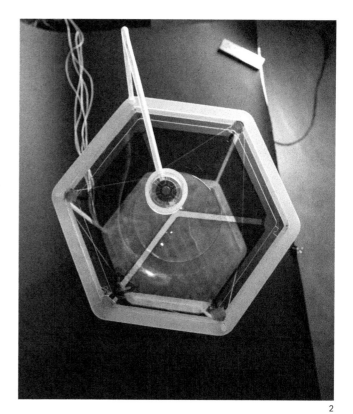

2

3

2   Details of installation construction showing the LED light source.

3   Video stills from the first experiment using water as an EAP electrode. See note 3.

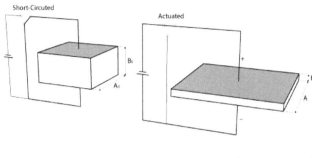

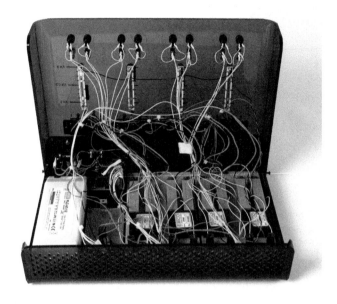

4

5

6

So far, there have been few design-oriented projects that explore the optical properties of EAPs. Most notably, Kaleidoscope demonstrated variable color control through the contraction and expansion of tinted areas on the EAP membrane (Luna 2014). Similar effects have also been demonstrated in science literature, with EAPs used to emulate color-changing chromatophore cells in octopuses (Fishman, Rossiter, and Homer 2015).

EAP Functionality

The type of EAP technology used in this project is commonly referred to as a Dielectric Elastomer (Bar-Cohen, Sherrit, and Lih 2001). This type of EAP consists of two electrodes separated by an elastic dielectric material (Pelrine, Kornbluh, and Kofod 2000). When there is an electrical potential between the two electrodes, an electrostatic force compresses the elastomer, forcing it to expand laterally (Figure 4). When the two electrodes are short circuited, the elastomer returns to its original state.

EAPs provide a number of novel actuation possibilities that are difficult or impossible to achieve with traditional mechanical actuators. One of the most interesting aspects for designers is the organic movement that EAPs produce, an aesthetic counterpoint to the mechanical nature of traditional actuators.

DIY EAP Fabrication Method

Our EAP fabrication method is based on techniques developed in the Laboratory for Mechanical Engineering Systems at the Swiss Federal Laboratories for Materials Science and Technology (EMPA). We learned the basics of this approach

through hands-on instruction at the EMPA laboratories in Dübendorf. The process includes the use of a motorized biaxial stretching device capable of pre-tensioning thin membranes of VHB, a common double-sided adhesive tape produced by 3M (Lochmatter and Kovacs 2008).

Inspired by EMPA's approach, we built a manual biaxial stretching mechanism with opposing scissor-hinges (Figure 5). With this mechanism we were able to produce EAP actuators up to approximately 420 x 420mm from 1 mm 3M VHB Type 4910, although the same approach can work on a larger scale. After stretching to 250–350% of the original size, the elastomer was hand impregnated on both sides with Carbon Black (Ketchenblack EC300 from AkzoNobel), using a cellular foam applicator. We created detailed and repeatable patterns for electrically active areas with laser-cut stencils made from 3M backing foil. We demonstrated this technique in numerous workshops to enable other designers and artists to experiment and work with EAP actuators.[1] We developed a custom-built EAP driver for easy and programmable actuation of these DIY EAPs (Figure 6).

Liquids, Shape Control, and Tunable Lenses

Some examples of liquids combined with EAPs exist in material science literature, demonstrating applications for tunable lenses (Carpi et al. 2011; Maffli et al. 2015). Similar technology has also been used for tactile interfaces (Frediani et al. 2014; Carpi, Frediani, and Rossi 2010). However, these examples primarily function with enclosed volumes for tiny mechanisms, making them suitable for imaging applications and consumer electronics but

Fluid Morphologies Franzke, Rossi, Franinović

7

not for an architectural scale. The technique we developed shows that an open-construction water EAP, what we refer to as HAPs, can act as a large-scale programmable lens and shape-changing element. We also demonstrate that water can act not only as a compliant material, but also as an active part of HAP modules.

EXPERIMENTS WITH LIQUIDS

Our first experiments with liquids and EAPs began during the Contraction Expansion workshop run in collaboration with Liquid Things art research in Vienna, Austria (Franinović, Wille, and Korschner 2012). In preparation for the workshop, we experimented and prepared homemade electrorheological fluids (ERF) which harden when exposed to electricity. We used them in the workshop to connect two EAPs and to replace carbon black as an electrode. Both tests exhibited exciting possibilities for haptic sensations and interactions due to the hardening of the ERF.[2] During the workshop we decided to repeat these experiments with water due to the difficulty of preparing well functioning electrorheological fluids.

Vertical Movement

Water was tested as an electrode on a pre-stretched elastomer, with a carbon electrode on the underside. A wire was hung from above, contacting the water to provide electrical current. When activated, the EAP exhibited strong unidirectional deformation under the weight of the water (Figure 3).[3] In addition to impressive deformation under the weight of water, we discovered another benefit of using a dynamic electrode. When holes appeared in regions of the polymer, we moved the water away

from that location and the undamaged part of a polymer would continue to activate. These early HAP experiments demonstrated the potential for exceptionally strong activation.

Lateral Movement and Interaction

In 2014, Karmen Franinović conducted a residency at the Liquid Things project where she further explored the phenomenon (Moñivas 2017). We experimented with two independent carbon black electrodes on the underside of the VHB with water as the electrode on the topside. When the electrodes were activated simultaneously, the movement of the water was vertical, similar to previous experiments. However, when they were activated independently, an oscillating movement of the water could be generated. The motion grew stronger as the pendulum effect gained momentum over the course of interaction. Changing the timing of the actuation modulated the rhythm of this movement. While pressing the buttons of the driver, we tended to begin synchronising our body movements with the movement of the water. Such interaction provided a sense of continuous physical engagement, despite a discrete button interface. Furthermore, curious visual effects were generated from the moving, irregularly-shaped water lenses (Figure 7).

Asymmetrical Electrodes

With the goal of finding the most dynamic movement possible, we tested asymmetrically shaped electrodes, which produced an irregular motion. In a further step, we combined several asymmetrical electrodes, resulting in a large deformation and dynamic

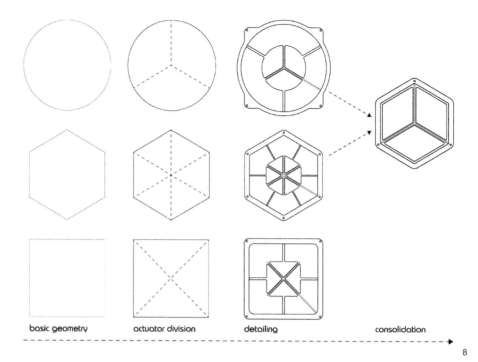

8   HAP geometry development. To reduce high stress regions the angles of the hexagon were filleted. Three actuation zones were chosen as the minimal number allowing a full range of motion.

9   From left: a) triangular tiling, b) rectangular tiling, c) hexagonal tiling, and d) rhombille tiling.

10   Detail of silicone touchpad interface.

11   Construction of silicone touchpad interface.

basic geometry   actuator division   detailing   consolidation

8

movement of the water. The motion of the HAP varied based on the rhythm of the inputs provided by the user, the size and the shape of the electrodes, and the amount of water. Impressed by the range of motions and the engaging and complex interaction, we set out to develop a demonstration of these qualities.

Acqua Alta

The research developed over the course of the Liquid Things residency was directed towards an installation for the Venice Biennale of Architecture in 2014. The Acqua Alta project proposed an integration of dynamic HAP elements into a venetian beam ceiling. The shape-changing membranes were to perceptually destabilize the old wooden structure. They would move in relation to the real-time changes in water tides and the presence of visitors. The installation was intended to reflect the powers of water over the city of Venice and the human influences on the lagoon. Although the project was not constructed, the concept opened up new potential for ceilings, integrating dynamic elements and liquids into existing structures.

FLUID MORPHOLOGIES INSTALLATION

The Fluid Morphologies installation was developed for the Swiss Spirit exhibition at the Phillips Collection in Washington DC. The goal of our exhibit was to present the state of our research, but also to engage the visitors to interact with the material. Our installation combined EAP, water, and light into an interactive sculpture. The HAP module was divided into three separately actuatable zones on top of which water was suspended. An LED light source illuminated the water from above, creating shadow play below. A silicone-based touch interface mirrored the actuatable EAP zones and allowed visitors to control the movement of the EAP and water, and thereby distort the light.[4]

Movement and Interaction

We began by testing possibilities beyond the vertical up/down, and back and forth pendulum movement. We found that three electrodes enabled a circular motion of the HAPs, in addition to the up/down and pendulum motions. With careful activation of the three zones, it was possible to build up momentum for ever greater circular motion. The three active areas provided an optimal amount of shape control while still demanding focus and anticipation from the user to maintain control. Based on these experiments, we decided to develop a sculptural element with an interface that would allow the visitors to trigger and play with the various surface deformations.

Support Structure

We created a lightweight tensile structure to display the HAP modules, while allowing them to be easily replaced (Figures 1 and 2). The HAP modules connected to electrical contacts on the structure through self-weight in order to provide power for the three individual activation zones. The rigid elements were made from laser-cut, sandblasted plexiglas and the tensile elements from transparent nylon thread. Electrical wiring was discretely piped through channels and tubes. Additionally, the installation was designed to be disassembled and packed into a standard suitcase and reassembled on site.

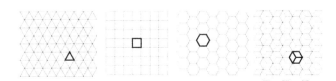

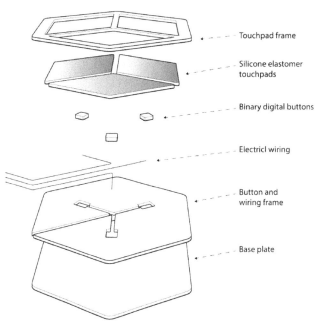

Touchpad frame

Silicone elastomer
touchpads

Binary digital buttons

Electricl wiring

Button and
wiring frame

Base plate

9

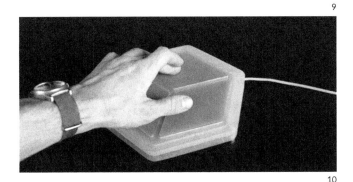

10

11

Electrode Design

In our final design, the EAP electrode was composed of three individual activation zones within a hexagonal frame. This geometry was arrived at through a series of physical experiments that ranged in shape from triangles and squares to hexagons and circles. Our goal was to develop a module that could be easily combined into a larger surface and thus we explored tiling patterns. Triangles, squares, and hexagons were the three regular tiling (or monohedral) geometries we considered (Grünbaum and Shephard 1987) (Figure 9 a, b, c). The final geometry we chose can be tessellated through rhombille tiling (Figure 9d).

The hexagon geometry, due to its the wide angles, was closest to a circle. The latter, while not a regular tiling geometry, allowed for an even deformation of the thin EAP film element when actuated. Sharp corners cause regions of high stress in the thin film and can lead to material failure. The wide angles of the hexagon in combination with filleted corners helped us work around this issue while still enabling the regular tiling (Figure 8).

Interface

The installation was controlled through our EAP driver, together with a tactile interface designed to resemble the form and material qualities of the HAPs (Figure 10 and 11). A soft feel silicone controller was created, which allowed the user to activate specific areas of the EAP. Although the push-button interaction was discrete, the nature of the silicone touchpad lent itself to gradual application of pressure. In combination with the muscle-like

movement of the HAP, this reinforced an illusion of continuous interaction.

New EAP Electrode Connection Method

One inherent challenge when working with EAPs is the interface between rigid and compliant elements. The VHB film is thinly stretched and at high risk of puncture when combined with rigid wires. Similar issues exist when working with other soft or unconventional electronic materials, such as e-textile (Perner-Wilson, Buechley, and Satomi 2010).

We overcame these issues by fabricating soft electrical connections to act as intermediaries between conventional wiring and the EAP (Figure 12 and 13). The connectors were assembled by stenciling traces of carbon black onto strips of unstretched VHB film. These strips then bridged between the wire terminals and the EAP membrane, so that the thinly stretched active areas would never be in direct contact with non-elastic electrical connections.

RESULTS

The results showed that different electrode constellations, combined with the water electrode, make the EAPs respond in a complex, fluid, and dynamic manner to gravity and user interaction.

Audience Response

The installation (Figure 14 and 15) was presented with two identical modules in order to allow several participants to control

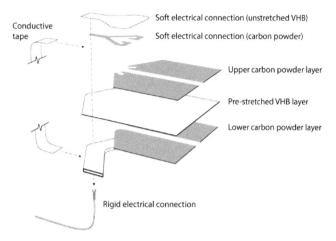

Conductive tape

Soft electrical connection (unstretched VHB)

Soft electrical connection (carbon powder)

Upper carbon powder layer

Pre-stretched VHB layer

Lower carbon powder layer

Rigid electrical connection

12

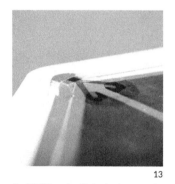 

13 14

8   VHB based soft electrode connector.

9   Detail of VHB based soft electrode connector.

10 Visitors interacting with the two modules in the exhibition space.

the motion of the HAPs simultaneously. Due to the oversized nature of the touchpad, it was possible for a number of people to interact with the interface at once. During the event, we gauged visitors' impressions through direct observation of interaction and casual discussion.

It became obvious that the installation acted as an eye opener to the potential of such technology, not only in terms of architecture but also in robotics, design, and fashion. While visitors speculated at future applications of such technologies, they were most enthusiastic about aesthetic aspects of this dynamic material experience. These preliminary observations showed the importance of sensual interactions, as well as emotional and social aspects related to EAPs, which deserve to be examined in the future.

Material Response

One known limitation of do-it-yourself EAPs is the longevity of the extremely thin and delicate membranes. We had a certain

degree of success in addressing this issue through our development of elastomer-based electrical connections, which reduced failure in this critical area. This new method greatly reduced issues related to puncturing of the VHB at electrical connections. This could also be suitable for applications beyond EAP, where soft electronic components must interface with rigid ones, as in e-textile fabrication.

In our experiments, the VHB elastomer underwent rapid material fatigue. This appeared independent of the number of activations occurring, signaling that it is the weight of the water that strained the VHB elastomer and led to its deterioration. To overcome this constraint, we designed the structure for rapid exchange of EAP membranes via replaceable inserts. This was an intermediate solution, and other options for improving material fatigue remain to be explored. Silicone-based EAPs, for example, may offer greater robustness and potential new avenues for fabrication. Silicone-based fabrication methods demonstrated in material science have involved printing techniques and tools familiar to many designers (Rosset and Shea 2015), offering greater potential for knowledge transfer into design disciplines.

CONCLUSIONS AND FUTURE DIRECTIONS

In this paper, we described a novel HAP actuator that combines water and EAP to create a dynamic, shape-changing actuator. We explored the potential of these actuators through a number of experiments and demonstrated our outcomes with an interactive installation.

In order to improve the robustness of HAPs beyond our success with soft electrical connections, our future research will focus on experimenting with silicone elastomers to replace the more fragile VHB used in this project. Once these issues of longevity are addressed, we will be able to move our focus to a more architectural scale. Improvements to lifespan would open up possibilities for further design explorations of this programmable material. Future enquiries should also include a quantifiable analysis of the material performance, which was beyond the scope of this paper. Finally, the interaction with such a system on different scales must be explored in a structured way.

These improvements will enable us to develop a more robust dynamic and interactive ceiling system at an architectural scale (Figures 16 and 17). Such an installation would continue the line of research established by projects such as Stratus and Resonant Chamber, which explored the potential of responsive ceiling installations to deal with air quality and sound, respectively (Thün et al. 2012). Whether driven directly by people in the space, algorithmically, or by environmental parameters, an array of these dynamic lenses would create an immersive environment of rippling light.

15 Visitors interacting with the two modules in the exhibition space.

ACKNOWLEDGEMENTS

We would like to thank the Swiss Embassy Washington, Phillips Collection, Roman Kirschner and EMPA. Special thanks to Florian Wille, who as a member of the Enactive Environments research group between 2011 and 2013, was a key contributor to the material explored in this paper.

NOTES

1. These results can also be reproduced by following our online video tutorials, at https://www.youtube.com/watch?v=uw8FLgiXsmk and https://www.youtube.com/watch?v=PgNKeqOCOKE
2. Video demonstration available at https://vimeo.com/53367582
3. Video demonstration can be seen at https://vimeo.com/161240281
4. Video at https://vimeo.com/163376887

REFERENCES

Bar-Cohen, Yoseph, Stewart Sherrit, and Shyh-Shiuh Lih. 2001. "Characterization of the Electromechanical Properties of EAP Materials." *Proceedings SPIE 4329, Smart Structures and Materials: Electroactive Polymer Actuators and Devices:* 319–327 doi:10.1117/12.432663

Biggs, James, Karsten Danielmeier, Julia Hitzbleck, Jens Krause, Tom Kridl, Stephan Nowak, Enrico Orselli, Xina Quan, Dirk Schapeler, Will Sutherland, and Joachim Wagner. 2013. "Electroactive Polymers: Developments of and Perspectives for Dielectric Elastomers" *Angewandte Chemie International Edition* 52 (36): 9409–9421.

Carpi, Federico, Gabriele Frediani, and Danilo De Rossi. 2010.

"Hydrostatically Coupled Dielectric Elastomer Actuators." *IEEE/ASME Transactions on Mechatronics* 15 (2): 308–315.

Carpi, Federico, Gabriele Frediani, Simona Turco, and Danilo De Rossi. 2011. "Bioinspired Tunable Lens with Muscle-Like Electroactive Elastomers." *Advanced Functional Materials* 21 (21): 4152–4158.

Decker, Martina. 2013. "Emergent Futures: Nanotechnology and Emergent Materials in Architecture." At *4th Conference of the Building Technology Educators' Society: Techtonics of Teaching.* Bristol, RI: BTES.

———. 2015. "Soft Robotics and Emergent Materials in Architecture." In *Real Time: Proceedings of the 33rd eCAADe Conference,* vol. 2, edited by B. Martens, G. Wurzer, T. Grasl, W. E. Lorenz, and R. Schaffranek. Vienna, Austria: eCAADe. 409–416.

Fishman, Aaron, Jonathan Rossiter, and Martin Homer. 2015. "Hiding the Squid: Patterns in Artificial Cephalopod Skin." *Journal of The Royal Society Interface* 12 (108): 20150281.

Franinović, Karmen, and Luke Franzke. 2015. "Luminous Matter: Electroluminescent Paper as an Active Material." In *Proceedings of the 9th International Conference on Design and Semantics of Form and Movement.* Milan, Italy: DeSForM. 37–47.

Franinović, Karmen, Florian Wille, and Roman Kirschner. 2012. Contraction-Expansion." *Liquid Things* workshop, Vienna. Accessed April 19, 2016. http://www.liquidthings.net/?portfolio= retackling-contraction -and-expansion

16 Close-up visualization of a ceiling installation of an array of HAPs.

Frediani, Gabriele, Daniele Mazzei, Danilo Emilio De Rossi, and Federico Carpi. 2014. "Wearable Wireless Tactile Display for Virtual Interactions With Soft Bodies." *Frontiers in Bioengineering and Biotechnology* 2: 31 doi: 10.3389/fbioe.2014.00031

Grunbaum, Branko. and Geoffrey C. Shephard. 1987. *Tilings and Patterns*. New York: W. H. Freeman and Co.

Joucka, Hashem. 2015. "Sound to Polymer 1.0." *Iaac Blog*, June 14. http://www.iaacblog.com/programs/sound-to-polymer/ Accessed April 19, 2016.

Kolodziej, Przemyslaw, and Josef Rak. 2013. "Responsive Building Envelope as a Material System of Autonomous Agents." In *Proceedings of the 19th Conference on Computer-Aided Architectural Design Research in Asia*, edited by R. Stouffs, P. Janssen, S. Roudavski, and B. Tunçer. Singapore: CAADRIA. 945–954.

Kretzer, Manuel, and Dino Rossi. 2012. "Shapeshift." *Leonardo* 45 (5): 480–481.

Krietemeyer, Elizabeth A., and Anna H. Dyson. 2011. "Electropolymeric Technology for Dynamic Building Envelopes." In *Parametricism (SPC) ACADIA Regional Conference Proceedings*, edited by Janghwan Cheon, Steve Hardy, Timothy Hemsath. Lincoln, NE: ACADIA. 75–84.

Lochmatter, Patrick, and Gabor Kovacs. 2008. "Design and Characterization of an Active Hinge Segment Based on Soft Dielectric EAPs." *Sensors and Actuators A: Physical* 141 (2): 577–587.

Luna, Dulce. 2014. "Kaleidoscope." *Iaac Blog*, March 15. http://www.iaac-blog.com/programs/kaleidoscope/ Accessed April 19, 2016.

Maffli, Luc, Samuel Rosset, Michele Ghilardi, Federico Carpi, and Herbert Shea. 2015. "Ultrafast All-Polymer Electrically Tunable Silicone Lenses." *Advanced Functional Materials* 25 (11): 1656–1665.

Moñivas Mayor, Esther. 2017. "Observing From Inside the Liquid Things Drift: The Studio as a Flux Condenser." In *Raw Flow*, edited by R. Kirschner. Berlin: De Gruyter. Forthcoming.

Mossé, Aurellie, David Gauthier, and Kofod Guggi. 2012. "Towards Interconnectivity: Appropriation of Responsive Minimum Energy Structures in an Architectural Context." *Studies in Material Thinking* 7: Article 06.

Perner-Wilson, Hannah, Leah Buechley. and Mika Satomi. 2010. "Handcrafting Textile Interfaces From a Kit-of-No-Parts." In *Proceedings of the Fifth International Conference on Tangible, Embedded, and Embodied Interaction*. Funchal, Portugal: TEI. 61–68.

Pelrine, Ron, Roy Kornbluh, and Guggi Kofod. 2000. "High-Strain Actuator Materials Based on Dielectric Elastomers." *Advanced Materials* 12 (16): 1223–1225.

Rhoné, Jim, and Martin Genet. 2014. "A Degree Of Freedom." Master's thesis, École normale supérieure. Accessed April 19, 2016. https://issuu.com/jimrhone/docs/rhone_jim_r10

Rosset, Samuel, and Herbert R. Shea. 2015. "Towards Fast, Reliable, and Manufacturable DEAs: Miniaturized Motor and Rupert the Rolling Robot." In *Proceedings SPIE 9430, Electroactive Polymer Actuators and Devices (EAPAD)*. 943009. doi:10.1117/12.2085279

17 Visualization of a ceiling installation of an array of HAPs.

Rossi, Dino, Zoltán Nagy, and Arno Schlueter. 2014. "Soft Robotics for Architects: Integrating Soft Robotics Education in an Architectural Context." *Soft Robotics* 1 (2): 147–153. doi:10.1089/soro.2014.0006.

Thün, Geoffrey, Kathy Velikov, Mary O'Malley, and Lisa Sauvé. 2012. "The Agency of Responsive Envelopes: Interaction, Politics and Interconnected Systems." *International Journal of Architectural Computing* 10 (3): 377–400.

IMAGE CREDITS

Figure 6: Photo by Florian Wille.
Figure 15: Photo by Alexander Morozov, courtesy of the Phillips Collection.
All other figures by the authors.

Luke Franzke is a design researcher and teacher at ZHdK. Luke completed a BA in Multimedia at Victoria University, Melbourne in 2006 and has several years experience as a UI designer and developer. Through the MA in Interaction Design program at ZHdK, he joined the Enactive Environments research group, where he developed new forms of interfaces from transient materials. Luke teaches both product design and programming basic courses, and does research in the area of emerging material technology in Interaction Design. luke.franzke@zhdk.ch

Dino Rossi holds degrees in Environmental Studies (BA, University of California at Santa Cruz, USA), Environment & Energy in Architecture (MA, Architectural Association, UK), and Computer Aided Architecture Design (MAS, ETH Zürich, CH). He has experience as a designer, fabricator, and researcher, and his work has been published in several conferences, journals and books. His research interests include the development and implementation of novel soft actuators in speculative design contexts. dino.rossi@zhdk.ch

Prof. Dr. Karmen Franinović is head of Interaction Design research and education at ZHdK. Her research focus is on experiences that engage bodily and spatial knowledge in interaction. She leads research projects on sonic interaction, movement rehabilitation and responsive architectures. Karmen's practical and theoretical work challenges established interaction paradigms to foster more critical and playful uses of technology in everyday life. She is the founder of Enactive Environments group and Zero-Th Studio. karmen.franinovic@zhdk.ch

Growth Based Fabrication Techniques for Bacterial Cellulose

Three-Dimensional Grown Membranes and Scaffolding Design for Biological Polymers

Tiziano Derme
Daniela Mitterberger
Independent researcher/MäID

Umberto Di Tanna
Independent researcher

1

ABSTRACT

Self-assembling manufacturing for natural polymers is still in its infancy, despite the urgent need for alternatives to fuel-based products. Non-fuel based products, specifically bio-polymers, possess exceptional mechanical properties and biodegradability. Bacterial cellulose has proven to be a remarkably versatile bio-polymer, gaining attention in a wide variety of applied scientific applications such as electronics, biomedical devices, and tissue-engineering. In order to introduce bacterial cellulose as a building material, it is important to develop bio-fabrication methodologies linked to material-informed computational modeling and material science. This paper emphasizes the development of three-dimensionally grown bacterial cellulose (BC) membranes for large-scale applications, and introduces new manufacturing technologies that combine the fields of bio-materials science, digital fabrication, and material-informed computational modeling. This paper demonstrates a novel method for bacterial cellulose bio-synthesis as well as in-situ self-assembly fabrication and scaffolding techniques that are able to control three-dimensional shapes and material behavior of BC. Furthermore, it clarifies the factors affecting the bio-synthetic pathway of bacterial cellulose—such as bacteria, environmental conditions, nutrients, and growth medium—by altering the mechanical properties, tensile strength, and thickness of bacterial cellulose. The transformation of the bio-synthesis of bacterial cellulose into BC-based bio-composite leads to the creation of new materials with additional functionality and properties. Potential applications range from small architectural components to large structures, thus linking formation and materialization, and achieving a material with specified ranges and gradient conditions, such as hydrophobic or hydrophilic capacity, graded mechanical properties over time, material responsiveness, and biodegradability.

1   Bacterial cellulose differential growing. Physical model of the membrane and representation of the funicular system.

INTRODUCTION

Contemporary digital fabrication tools are able to produce geometrically complex objects and structures, yet most of the constructs are not generally sustainable nor energy efficient (Oxman et al. 2012). In contrast, organic self-assembling processes produce little to no waste, using small amounts of energy to produce multi-functional and adaptable systems (Vincent 2012). Despite the recognized capabilities of natural processes to generate complex structures of organic and inorganic multi-functional composites (shells, corals, teeth, wood, silk, horn, collagen, and muscle fibers), the use of bio-materials for large scale architectural and engineering applications is still underdeveloped (Benyus 1997; Vincent 2012). Their structural and functional diversity, especially that of bio-polymers, underlines their capacity to replace existing synthetic polymers and to provide new methods of bio-fabrication as well as new applications and structures.

Cellulose, one of these bio-polymers, is one of the most abundant biodegradable materials in nature, and has been the topic of wide investigations in macromolecular chemistry (Mohite et al. 2014). The high water content of bio-synthetic cellulose (99%) and its mechanical properties make it a versatile material that can be manufactured in various sizes and shapes. Bacterial cellulose has many unique properties, including high purity, high water retention, and a hydrophilic nature, tensile strength, thermal stability, and biodegradability.

Because of these unique properties, it is an attractive candidate for a wide range of applications, including within architecture and engineering (i.e.: water retaining structures, architectural components, etc.), but due to the lack of suitable fabrication methods and digital design tools, cellulose is still disregarded as a building material. Although recent developments within biochemistry and microelectronic engineering have improved knowledge of biological materials, it is still not possible to produce bacterial cellulose on an industrial scale and control the three-dimensional (3D) outcomes through standard manufacturing and digital techniques (Fernandez et al. 2013). Current approaches towards virtual and physical prototyping with non-fuel-based materials also lack the capacity to model and fabricate with continuously varying material properties (Oxman 2011). In order to introduce cellulose as a building material, it is therefore important to develop bio-fabrication methodologies linked to materially informed computational modeling.

In nature, morphogenesis is a biological process describing the formation of a shape of an organism, inseparably linking formation and materialization (Menges 2007). By contrast, architecture is characterized by prioritizing form-generation over inherent material logic. The integration of morphogenesis within the field

of architecture suggests a bottom-up, material-driven design with specified ranges and gradient conditions (Soldevila 2015). This paper emphasizes the development of new manufacturing technologies for the production of 3D-grown bacterial cellulose (BC) membranes.

In this paper, this claim is substantiated and a solution is offered, making the following contributions:

1. *Investigation of bio-inspired fabrication methodologies and virtual and physical prototyping. The factors affecting cellulose bio-synthesis are demonstrated—mainly growth medium, environmental conditions, and the formation of derivatives (§1.1 and §2.1). The design of the culture medium is a key influence for the growth of microorganisms, and therefore in stimulating the formation of three-dimensional membranes.*

2. *The transformation of the bio-synthesis of bacterial cellulose into BC based bio-composite is shown, leading to the creation of new materials with additional functionalities and properties. Furthermore, we clarify the factors that affect the bio-synthetic pathway of bacterial cellulose, such as bacteria, nutrients, and medium culture properties. Particular focus is given to the creation of a natural polymer which could grow to any thickness, shape, and robust structure (§1.2 and §2.3, 2.4).*

3. *In-situ self-assembly fabrication techniques and scaffolding techniques for bacteria cellulose are presented and show the creation of a bio-composite via the fermentation of bacteria strain A. Xylinum along the surface of natural fibers (§1.3 and § 2.2, 2.5).*

1.1 Bio-Inspired Fabrication Methodologies and Virtual and Physical Prototyping

Through the combination of non-fuel-based materials with

2   Consolidation through drying. Bacterial cellulose three-dimensional morphology, with variable thickness.

3 Bio-synthesis of bacterial cellulose: strikers for preparation of *Acetobacter xylinum* culture.

4 Bio-synthesis of bacterial cellulose: Microfluidics system to provide continuous nutrient to the culture.

5 Bio-synthesis of bacterial cellulose: bacterial cellulose growing after 5 days.

3

4

5

advances in biological sciences, genetics, and bio-engineering of bacteria, a new set of possible bio-fabrication technologies can be developed.

Current design practice is mostly characterized by the domination of shape over matter, consequently prioritizing virtual shape-defining parameters over physical material and fabrication constraints, leading to a geometric-centric design phase (Menges 2007; Oxman 2011). Nevertheless, some recent developments in direct digital manufacturing enable a shift towards a material-centric design practice, such as water-based fabrication techniques (Oxman 2011). Additive manufacture (AM) technologies for rapid prototyping employ virtual, computer-aided designed models, and translate them into thin horizontal successive cross-sections to define three-dimensional physical objects (Sachs et al. 1993).

AM technologies have become an efficient and common means to deliver geometrically precise functional prototypes in relatively short periods of time (Oxman 2012). At the same time, there is a need to expand manufacturing processes towards bio-fabrication-based approaches, borrowing techniques from biological science and tissue engineering. In contrast to AM technologies, which relate to a specific controlled output, bio-fabrication techniques in a water medium aim to create dynamic feedback or reciprocity within a specific context. This process certainly escapes the pitfalls of bio-mimicry in favor of bio-synthesis, as a substance forms through interaction with a living organism. As a result, this will proliferate methods to fabricate synthetic

composites by generating novel morphogenetic mechanisms linked to bacteria and different states of matter. This process is clearly described within §2.3, focusing on the different levels of oxygen within the water medium. As a result, 3D growth of cellulose can be directly manipulated.

1.2 Bio-Nanocomposite and Bacteria Engineering

Engineering the bio-synthesis of bacterial cellulose (BC) into BC-based nanocomposites leads to better mechanical and thermal properties, or additional functionalities that are useful in many applications and fields. They could be categorized, for instance, as high-strength materials, plant-mimicking materials, electrically-conductive materials, catalytic materials, antimicrobial materials, thermo-responsive materials, and many others. More explicitly, the process describes a structure with growth-induced material properties reacting to external stimuli and resulting in hierarchically structured forms (Soldevila 2015). Bacterial cellulose production depends heavily on several factors, including culture medium and environmental conditions. The culture medium (Figure 3) contains a carbon source, nitrogen source, and other nutrients required for the bacteria to grow. In normal static and aerobic conditions, the bacteria will form a pellicle (flake) (Figure 5) at the surface of the culture medium. This pellicle will grow in thickness slightly, but the thickness is limited by the supply of oxygen and nutrients to the bacteria. Recent studies found it possible to optimize the thickness and strength of BC by creating continuous systems of sub-ministration of nutrients (Figure 4). This creates a potential condition where the natural polymer could grow to any thickness and shape to form a

Growth-based fabrication techniques for bacterial cellulose
Derme, Mitterberger, Di Tanna

structure (Gateholm et al. 2012). The variation of culture medium components can alter the metabolism of the bacteria (Figures 13 and 14, and §2.3) and consequently the mechanical properties and bio-synthesis pathways. Such 3D bacterial cellulose-based structures and membranes can be prepared with a target topological condition, thickness, and strength. In this approach, the distribution of material properties is informed by structural and environmental performance criteria contributing to the internal physical makeup of the membrane. It thus requires a set of virtual and physical prototyping tools and methods to support variable fabrication approaches and modeling (Oxman 2011).

It is precisely this complex and dynamic exchange between organism, environment, and functionality that makes synthetic life valuable for architecture.

1.3 Self-Assembly Fabrication Techniques

Self-assembly fabrication of 3D structures and membranes can be described as a multi-step fermentation process in the presence of natural fibers or other polymers. This can furthermore lead to the formation of BC-based hybrids or nanocomposites (Qiu and Netravali 2014). Ultimately, advances in self-assembly fabrication techniques will lead to different manufacturing methods, implying a condition of growing and morphogenesis. Taking as reference the biosynthetic process of bacterial cellulose, these methods are able to control three-dimensional shapes and material behavior. Self–assembly fabrication can be achieved by providing a scaffold to support and guide the process of growing to define precise target geometry (Figures 6–9 and §2.2).

Cellulose is a polymer which forms the cell wall of eukaryotic plants and algae and can also be found as the major constituent of the cell wall of fungi. Despite this, a few bacteria can also secrete cellulose. A notable example is *Acetobacter xylinum*, which is known to secrete cellulose as part of its metabolism of glucose and other carbohydrates. The bacteria strains of *A. Xylinum* have been found to grow preferentially on the surface of natural fibers (Figures 17 and 18) or certain polymer molecules, rather than freely in the medium. Therefore, starch, soy resin, or polyvinyl alcohol (Figures 6 and 7) provide ideal substrates for the bacteria to grow on and can therefore lead to the formation of BC-based hybrids or bio-composites. Bio-composites are composite materials reinforced via natural plant-based fibers or certain polymers (Qiu and Netravali, 2014). Biosynthesizing and engineering the original BC into-BC based composites could potentially find several applications in the fields of architecture, engineering, and product design, opening up a substantial change in how products and buildings are form-found, designed, and fabricated (Oxman 2011). A variety of methods, digital and analog, have been utilized for achieving this purpose. Polyvinyl

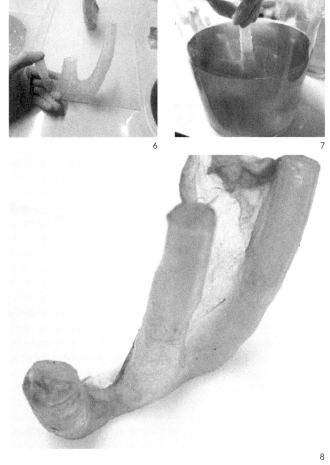

6

7

8

9

6   BC-induced growth over permanent scaffold. PLA scaffolding and PVA immersion.

7   BC-induced growth over permanent scaffold. Immersion of scaffold into culture medium.

8   BC-induced growth over permanent scaffold. Image of the membrane during the drying process.

9   BC-induced growth over permanent scaffold. Detail of the resulting membrane.

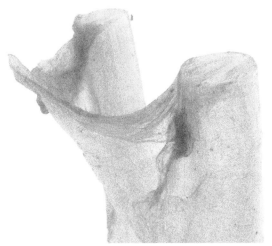

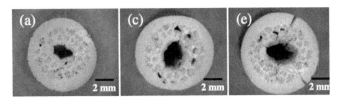

10

11

12

alcohol hardening (Figures 6–9 and §2.2) and sisal fiber deposition (Figures 17 and 18 and §2.5) can be used as excellent in-situ bio-fabrication techniques.

A key point for potential future development is the augmentation of the mechanical properties of the polymer, including its strength and stiffness. In all cases, consolidation is usually achieved by drying, sintering, or solidification techniques. BC in particular has the capacity to be calcified (Figure 10). This solidification method is currently used in biomedical applications for tissue and bone engineering using hydroxyapatite, chitosan, alginate, and agarose. In this case, it may be possible to generate a material with properties similar to those of bones if we introduce an agent in its mineral phase into the bacteria culture medium during the formation of BC.

METHODS
This process is intended to show novel BC fermentation methods and scaffolding techniques for controlling shape, thickness, and material-driven computing experiments.

2.1 Preparation of *Acetobacter xylinum* (AXy) Culture (Figures 3–5)
Preparation of the AXy culture (Figure 3): The bacteria was grown in broth and subsequently streaked on agar with an incubation period of 24 hours. A single colony of AXy was inoculated in a flask containing mature coconut water, which was sterilized at 121 C° for 15 minutes. The flask was left at 25 C° or room temperature for 2–3 days.

Preparation of the culture medium (Figure 4): 5% sugar and 0.5% of ammonium sulphate were added to mature coconut water, and the mixture was brought to boil for 5 minutes. 1% acetic acid (99.85%) was added and mixed. A micro-fluid system (Figure 5) was applied to provide a continuous flow of nutrients in order to favor an unlimited growth of BC (Gateholm et al. 2012).

2.2 In Situ Self-Assembling (Figures 6–9, 11, 12)
We prepared different scaffolding methods to induce BC growth over a predetermined shape. Two strategies of scaffolding were introduced: permanent and bio-degradable. Figures 6–9 and 12 show the creation of shells and membranes using BC as a binding agent. The scaffold remains embedded within the structure. Bio-degradability (dissolving after the bio-synthesis process) of a scaffold is shown in figure 11, using a sodium-alginate scaffold.

The culture medium was prepared in a glass beaker (2 liters) and 10% of AXy culture was added. The flask was left at 25 C° for 7 days without disturbance. Oxygen was supplied during the photosynthetic period to direct the growth on the scaffolding.

2.3 Adhesion Growing (Figures 13 and 14)
This methodology showed a potential to grow three-dimensional morphology with a target geometry. Static and anaerobic conditions caused the inversion of the metabolism of the bacteria, producing a growing which was no longer superficial, but instead adhered to the morphology of the containing element.

Growth-based fabrication techniques for bacterial cellulose
Derme, Mitterberger, Di Tanna

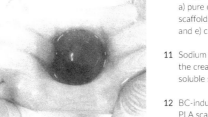

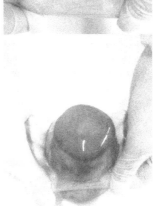

10 Images of cellulose-based scaffolds a) pure cellulose and composite scaffolds of c) cellulose agarose, and e) chitosan–alginate.

11 Sodium alginate medium led to the creation of result of a bio-film soluble scaffolding.

12 BC-induced growth over permanent PLA scaffold led to the creation of a growth shell structure.

13 Three-dimensional growth morphology under static conditions: resulting geometry.

14 Three-dimensional growth morphology under static conditions,removal operations from the containing boundary.

13

14

The culture medium was prepared in a latex flask (1.5 liter) and 10% of AXy culture was added. The flask was left at 25 C° for 10 days without disturbance.

2.4 Bacterial Cellulose with Differential Growing Patterns (Figures 1, 15, 16)

We prepared BC membranes with different growing patterns and different thicknesses, and used a particle-spring system as a "tension-only" hypothetical funicular model to simulate the behavior of the membrane (Figure 16).

The culture medium was prepared in a glass beaker (2 liter) and 10% of AXy culture was added. The flask was left at 25 C° for 4 days without shaking and nutrients were continuously added. The flask was then left at 25 C° for 5 days. Drying method: freezing (4 hours) and drying (4 days at 10 C°) .

2.5 Scaffolding Technique Using Sisal Fiber (Figures 17, 18)

We prepared a scaffolding technique with BC-modified sisal fibers (produced in an incubating shaker), using 60 cm long sterilized sisal fibers. These were added to the culture media and sterilized. AXy was inoculated into the culture media and BC-modified sisal fibers were extracted after 3 days culture. Preliminary studies were done by simulating the material behavior according to the pattern deposition of the fibers.

RESULTS

This approach demonstrated the first steps in the design of a complementary technology able to expand the potential of

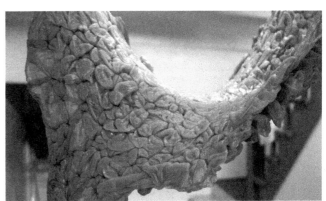

15

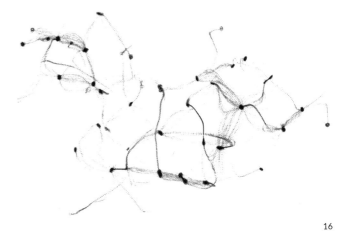

16

15 BC differential growing. Physical model of the membrane and representation of the funicular system.

16 Particle spring simulation, "tension only" funicular system of the membrane.

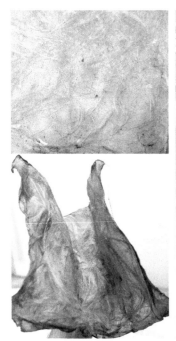

prototyping processes for bio-synthesis. Specifically, experiments 2.1 and 2.4 showed that BC is moldable in cultivation. By meticulous control of the addition of fermentation media through the use of a microfluidic system, BC can grow to a potentially unlimited thickness and shape. Moreover, other experiments confirmed the hypothesis that BC grows preferentially on natural fibres such as sisal. It has been noted that the creation of BC-based bio-composites increase the mechanical properties of the bio-polymer, offering the potential to create structural and graded properties. Finally it has been successfully proven that it is possible to grow BC around a predefined shape; the addition of oxygen during the fermentation drove the cellulose to follow and adhere to specific surface conditions. Despite the results, all the experiments are still far away from a real application on a large scale and the accuracy of material-driven computing methodologies, relative to prototypes as grown, needs to be further explored.

CONCLUSIONS

This paper outlined novel bio-fabrication and scaffolding techniques for the control of 3D membranes and morphologies of bacterial cellulose, positioning the results in the field of bio-materials science, digital fabrication, and material-informed computational modeling. Future developments of the present research include:

- Developing a range of structural BC bio-composites using hydroxyapatite, chitosan, or lining as agents to calcify the 3D membranes.
- Exploring the conditions for target-geometry-based growing. Define the role of the scaffold compared to the growing pattern of BC.
- Investigating BC as a matrix for growth and cultivation of photosynthetic microorganisms such as algae and cyanobacteria in order to generate a material able to perform photosynthesis and react to photo stimuli.

ACKNOWLEDGEMENTS

We thank our colleagues from University of Tokyo and Yusuke Obuchi and Jutamat Klinsoda from the Department of Applied Microbiology Institute of Food Research and Product Development, Kasetsart University, Thailand who provided insight and expertise that greatly assisted the research.

17

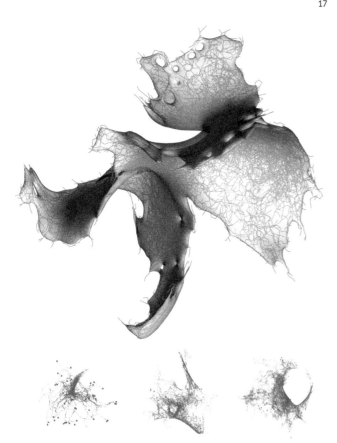

18

17 Result of BC bio-composite using sisal fibers. Bacterial cellulose as binding material has been observed to grow preferably around natural fibers.

18 Recursive system for sisal fiber deposition and scaffolding creation. The aggregation of the fibers relative to a hypothetical growing ratio of BC.

REFERENCES

Benyus, Janine M., 1997. *Biomimicry: Innovation Inspired by Nature*. New York: Quill.

Growth-based fabrication techniques for bacterial cellulose
Derme, Mitterberger, Di Tanna

Vincent, Julian. 2012. *Structural Biomaterials*. Princeton, NJ: Princeton University Press.

Fernandez, Javier G. and Donald E. Ingber. 2013. "Bioinspired Chitinous Material Solutions for Environmental Sustainability and Medicine." *Advanced Functional Materials* 23 (36): 4454–4466.

Mohite, Bhavna V. and Satish V. Patil. 2014. "A Novel Bio Material: Bacterial Cellulose and its New Era Applications" *Biotechnology and Applied Biochemistry* 61 (2): 101–110.

Oxman, Neri, Elizabeth Tsai, and Michal Firstenberg. 2012. "Digital Anisotropy: A Variable Elasticity Rapid Prototyping Platform." *Virtual and Physical Prototyping* 7 (4): 261–274.

Oxman, Neri. 2011. "Variable Property Rapid Prototyping." *Virtual and Physical Prototyping* 6 (1): 3–31.

Qiu, Kaiyan and Anil N. Netravali. 2014. "Fabrication and Applications of Bacterial Cellulose Based Nanocomposites" *Polymer Reviews* 54 (4): 598–626.

Sachs, Emanuel, Michael Cima, James Cornie, David Brancazio, Jim Bredt, Alain Curodeau, Tailin Fan, Satbir Khanuja, Alan Lauder, John Lee, and Steve Michaels. 1993. "Three-Dimensional Printing: The Physics and Implications of Additive Manufacturing." *CIRP Annals: Manufacturing Technology* 42 (1): 257–260

Menges, Achim. 2007. "Computational Morphogenesis: Integral Form Generation and Materialization Processes." In *Proceedings of the 3rd International ASCAAD Conference on Embodying Virtual Architecture*, edited by Ahmad Okeil, Aghlab Al-Attili, and Zaki Mallasi. Alexandria, Egypt: ASCAAD. 725–744.

Gateholm, Paul, Henrik Backdahl, Theodore Jon Tzavaras, Rafael V. Dovalos, and Michael B. Sano. 2012. Three Dimensional Bioprinting of Biosynthetic Cellulose (BC) Implants and Scaffolds for Tissue Engineering. US Patent 0190078A, filed September 28, 2010, and issued July 26, 2012.

Soldevila, Laia Mogas. 2015. "Water-based Digital Design and Fabrication" MS Thesis, Massachusetts Institute of Technology.

IMAGE CREDITS
Figures 1–9, 11–18: Derme, Mitterberger, Di Tanna, 2015
Figure 10: Jang et al., 2012

Tiziano Derme is teaching at University of Melbourne a design and research studio. Tiziano received his Architectural Masters degree with honors at the University of Rome "La Sapienza ". He worked several years for "New-Territories" with Francois Roche and in 2015 he was researcher and tutor assistant at University of Tokyo at T(ADS) Advanced Design Studies. His body of work ranges between Sustainable Prototyping, Design Computing and synthetic biology. Since 2016 Tiziano is also director and co-founder of multidisciplinary design practice *MäID FutureRetrospectiveNarrative*.

Daniela Mitterberger is currently teaching at University of Melbourne a design and research studio. Daniela received her Architectural Masters degree with honors at the Academy of Fine Arts Vienna and was nominated for the big price of the Academy of Fine Arts and the Hunter Douglas Award 2015. Her work ranges between human body as architecture and the definition of technology and tools within the field of architecture. In 2015 Daniela worked for "New-Territories" with Francois Roche and since 2016 Daniela is director and co-founder of the multidisciplinary design practice *MäID FutureRetrospectiveNarrative*.

Umberto Di Tanna received his architectural Master degree at the University of Rome "La Sapienza". In 2012 he was researcher for CRR(Rome Research Centre) on a spin-off regarding open source manufacturing systems in architecture. His work ranges between sustainable prototyping, biomaterials and scenography. In 2013 Umberto worked for SHSH Architecture and scenography and in 2015 he was co-founder of Quyala_synthetic Environments, an independent research cluster focused on biologically inspired fabrication, based in Bangkok.

ACADIA 2016 Credits

Conference Chairs

Geoffrey Thün
SITE CHAIR

Associate Dean for Research and Creative Practice, Associate Professor, University of Michigan Taubman College

Founding Principal, RVTR

Geoffrey Thün holds a Masters of Urban Design from the University of Toronto, and a Bachelor of Architecture and Bachelor of Environmental Studies from the University of Waterloo. He is currently an Associate Dean of Research and Associate Professor at the Taubman College of Architecture and Urban Planning at the University of Michigan where he teaches design studios, courses in urban systems, site operations, and material systems.

Thün is a founding partner of the research-based practice RVTR, which serves as a platform for exploration and experimentation in the agency of architecture and urban design within the context of dynamic ecological systems, infrastructures, materially and technologically mediated environments, and emerging social organizations. He exhibited at the 2016 Venice Architecture Biennale and his academic research and creative practice has collected many awards and accolades which most recently includes an R+D Honorable Mention (2016) and an R+D Award (2011) from *Architect Magazine*, *Journal of Architectural Education* Best Design as Research Article (2013), The Architizer A+ Award Program's Architecture + Sound Jury Award (2013), and an ACSA Faculty Design Award (2012).

Kathy Velikov
TECHNICAL COMMITTEE CO-CHAIR

Associate Professor, University of Michigan Taubman College
Founding Principal, RVTR

Sean Ahlquist
TECHNICAL COMMITTEE CO-CHAIR

Assistant Professor, University of Michigan Taubman College
Member, Cluster in Computational Media and Interactive Systems

Kathy Velikov holds a Master of Arts from the University of Toronto and a Bachelor of Architecture and Bachelor of Environmental Studies from the University of Waterloo. As an Associate Professor at Taubman College of Architecture and Urban Planning at the University of Michigan, Velikov teaches design studio, ecology, technology, and urbanism courses in the Architecture program, as well as courses in the Master of Science in Material Systems program.

Velikov is a registered architect and founding partner of the award winning research-based practice RVTR, which serves as a platform for exploration and experimentation in the agency of architecture and urban design within the context of dynamic ecological systems, infrastructures, materially and technologically mediated environments, and emerging social organizations. She exhibited at the 2016 Venice Architecture Biennale and is co-author of *Infra Eco Logi Urbanism* (Park Books, 2015), has been published in *IJAC*, *Leonardo*, *JAE*, *New Geographies*, *eVolo*, *Volume*, *[bracket] Goes Soft*, and *MONU*, and has work in Brownell and Swackhamer's *Hypernatural*, Ng and Patel's *Performative Materials in Architecture*, Gerber and Ibanez's *Paradigms in Computing*, and Trubiano's *High Performance Homes*.

Sean Ahlquist holds a Master of Architecture from the Emergent Design and Technologies Program at the Architectural Association in London. As a part of the Cluster in Computational Media and Interactive Systems, his work connects Architecture with the fields of Material Science, Computer Science, Art & Design, and Music, with particular focus on material computation.

Ahlquist's research formulates computational design frameworks where materiality functions as an a priori agent in the organization of architectural systems and their spatial tectonics. As an Assistant Professor at Taubman College of Architecture and Urban Planning at the University of Michigan, Ahlquist teaches at all levels, including ongoing involvement with the Master of Science in Material Systems program. Ahlquist continues to publish widely on the topic of computational design, including a reader entitled *Computational Design Thinking*, co-edited with Achim Menges, which collects and reflects up seminal texts formulating a systems and material based approach to architecture and design methodologies.

Matias del Campo
TECHNICAL COMMITTEE CO-CHAIR
Associate Professor, University of Michigan Taubman College
Co-founder, SPAN Architects

Wes McGee
WORKSHOP CO-CHAIR
Assistant Professor, University of Michigan Taubman College
Founding Partner, Matter Design

Chilean born and Austrian native, Matias del Campo graduated with distinction from the University of Applied Arts in Vienna, Austria. del Campo co-founded SPAN Architects in 2003 with Sandra Manninger and the globally-acting practice has become known for its sophisticated application of contemporary technologies in architectural production. The practice investigates the architectural qualities of sensorial and spatial conditions informed by Baroque geometries, romantic atmospheres, and biological systems, as they are combined with the manifold qualities of contemporary, algorithm-driven approaches.

As Associate Professor of Architecture at Taubman College of Architecture and Urban Planning at the University of Michigan, del Campo explores robotic fabrication processes as they relate to formal and tectonic architectural tendencies. del Campo's work has garnered wide recognition, including first place competition entry for the Austrian Pavilion at the Shanghai World Expo in 2010, first place winning competition entry for the new Brancusi Museum in Paris, France in 2008, and the exhibition of the work of SPAN at the Venice Architecture Biennale in 2012.

Wes McGee is an Assistant Professor in Architecture and the Director of the FABLab at the University of Michigan Taubman College of Architecture and Urban Planning. His work revolves around the interrogation of machinic craft and material performance, with a research and teaching agenda focused on developing new connections between design, engineering, materials, and process as they relate to the built environment through the creation of customized software and hardware tools. With the goal of seamlessly integrating fabrication constraints with design intent, the work spans multiple realms, including algorithmic design, computational feedback of material properties, and the development of novel production processes which utilize industrial robots as bespoke machines of architectural production.

As a founding partner and senior designer in the studio Matter Design, his work spans a broad range of scales and materials, always dedicated to re-imagining the role of the designer in the digital era. In 2013 Matter Design was awarded the Architectural League Prize for Young Architects & Designers. Wes frequently presents work at national and international conferences on design and fabrication, and the work of Matter Design was recently featured in the book *PostDigital Artisans* by Jonathan Openshaw (Frame publishers), as well as "Next Progressives" in *Architect Magazine*. In 2014 he was the co-chair of the Robots in Architecture Conference, hosted at the University of Michigan.

Catie Newell
WORKSHOP CO-CHAIR
Assistant Professor, University of Michigan Taubman College
Founding Principal, Alibi Studio

Sandra Manninger
PROJECT/EXHIBITION CHAIR
Assistant Professor, University of Michigan Taubman College
Co-founder, SPAN Architects

Catie Newell is an Assistant Professor of Architecture at the University of Michigan's Taubman College of Architecture and Urban Planning. Newell joined the faculty in 2009 as the Oberdick Fellow. She received her Masters of Architecture from Rice University and a Bachelor of Science in Architecture from Georgia Tech. Newell teaches design and fabrication courses and is heavily involved in the Masters of Science in Material Systems concentration.

Newell is also the founding principal of Alibi Studio. Newell's creative practice has been widely recognized for exploring design construction and materiality in relationship to the specificity of location and geography and cultural contingencies. Newell has won the SOM Prize for Design, Architecture and Urban Design, the Architectural League Prize for Young Architects and Designers, and the ArtPrize Best Use of Urban Space Juried Award. Alibi Studio has exhibited at the 2012 Architecture Venice Biennale and at the 2015 Lille3000 Triennial. In 2013 Newell was awarded the Cynthia Hazen Polsky and Leon Polsky Rome Prize in Architecture. She is a Fellow of the American Academy in Rome.

Sandra Manninger began her training at the Federal Higher Technical Institute for Education and Experimentation in Graz, Austria. She then joined the Technical University in Vienna where she graduated as an architect under the auspices of Bart Lootsma. Manninger is currently Assistant Professor of Practice at Taubman College of Architecture and Urban Planning at the University of Michigan. Manninger is Co-founder of SPAN Architects together with Matias del Campo. The practice focuses on the integration of advanced design and building techniques that folds, nature, culture, and technology into one design ecology.

Sandra Manninger received a Research Fellowship, awarded from the University for Applied Arts Vienna, the Schindler Scholarship, awarded by the Federal Ministry for Education and Art together with the Museum for Applied Art Vienna, and most recently a Research Fellowship at CERN, granted by the Federal Chancellery Austria.

She has received numerous awards such as the Ephemeral Structures Competition in Athens, the Architecture and Digital Fabrication competition, the competition for the new Media Center of the University for Music and Performance Art, Vienna, or the Experimental Tendencies in Architecture award granted by the Republic of Austria. She has won numerous architecture competitions such as the Brancusi Museum in Paris, the PUGA (a mobile Public Gallery), the bidding for the HAWK Headquarters, and the Austrian Pavilion for the Shanghai Expo.

Her work is part of the permanent collection of the FRAC Centre Orleans, the Luciano Benetton Collection, the MAK Museum of Applied Arts, the Albertina in Vienna, and the Pinakothek in Munich and she has been published in numerous magazines and books. She has written and presented papers at numerous conferences and considers herself lucky to travel the world to counsel and lecture at various institutions.

Session Moderators

Ellie Abrons
Assistant Professor, University of Michigan Taubman College
Principal, EADO
Member, T+E+A+M

Ellie Abrons is a designer, educator, and the principal of EADO. She is an Assistant Professor at the University of Michigan's Taubman College of Architecture and Urban Planning, where she was the A. Alfred Taubman Fellow in 2009–2010. Her work focuses on material experimentation and reuse, digital fabrication, and explorations of formal allusion. Ellie received her M.Arch from the University of California Los Angeles, where she graduated with distinction, and her BA in art history and gender studies from New York University. Ellie is the recipient of a residency fellowship at the Akademie Schloss Solitude in Stuttgart, Germany and her work has been exhibited at the Venice Biennale, Storefront for Art and Architecture, A+D Gallery, and the Architectural Association. An exhibition of Ellie's work, entitled Inside Things, was recently shown at SCI-Arc and she is a contributor (as part of T+E+A+M) to the U.S. Pavilion at the 2016 Venice Architecture Biennale.

Dana Cupkova
Caste Assistant Professor, Track Chair MSSD, Carnegie Mellon University, School of Architecture

Neil Leach
Architect, Curator, Writer

Dana Cupkova is a founding partner and principal of Epiphyte Lab, an interdisciplinary design and research practice. She currently holds the Lucian & Rita Caste Assistant Professorship at Carnegie Mellon University's School of Architecture; and serves as a Track Chair for the school's MSSD program.

Dana's design work studies the built environment at the intersection of ecology, computational processes, and systems analysis. In her research, she interrogates the relationship between design-space and ecology as it engages computational methods, thermodynamic processes, and experimentation with geometrically-driven performance logic.

Her current design-research in collaboration with manufacturing industry has been supported by the NYSCA Grant, the Center for Architecture Foundation Arnold W. Brunner Grant, the Cornell University Faculty Innovation Grant, Architectural League of NY, AIA Urban and Regional Solution Grant, the Pennsylvania Infrastructure Technology Alliance Grant, etc; presented and published internationally at venues and conferences such as ACADIA, *IJAC*, Design Modeling Symposium and others.

Neil Leach teaches at the European Graduate School, Tongji University and Florida International University, and is a member of the Academia Europaea. He has been working on a NASA funded research project to develop robotic fabrication technologies to print structures on the Moon and Mars. He has published 27 books including *Designing for a Digital World* (Wiley, 2002), *Digital Tectonics* (Wiley, 2004), *Digital Cities* (Wiley, 2009), *Machinic Processes* (CABP, 2010), *Fabricating the Future* (Tongji UP, 2012), *Scripting the Future* (Tongji UP, 2012), *Robotic Futures* (Tongji UP, 2015) and *Digital Factory: Advanced Computational Research* (CABP, 2016).

Paul Nicholas
Associate Professor, Centre for Information Technology and
Architecture, Royal Academy of Fine Arts, School of Architecture

Gilles Retsin
Founder, Gilles Retsin Architecture

Teaching Fellow, UCL the Bartlett School of Architecture

Senior Lecturer, University of East London

Paul Nicholas holds a PhD in Architecture from RMIT University, Melbourne Australia. After co-founding the design firm Mesne Design Studio, and practicing with Arup Consulting Engineers from 2005 and AECOM/Edaw from 2009, Paul joined the Centre for Information Technology and Architecture (CITA), Copenhagen Denmark in 2011. Paul is currently head of the masters programme CITAstudio: Computation in Architecture. Paul's particular interest is the development of innovative computational approaches that establish new bridges between design, structure, and materiality. His recent research explores sensor enabled robotic fabrication, multiscale modeling, and the idea that designed materials such as composites necessitate new relationships between material, representation, simulation and making.

Gilles Retsin is the founder of Gilles Retsin Architecture, a young award-winning London based architecture and design practice, investigating new architectural models which engage with the potential of increased computational power and fabrication to generate buildings and objects with a previously unseen structure, detail and materiality. He graduated from the Architectural Association Design Research Lab in London. Alongside his practice, Gilles directs a Research Cluster at UCL - the Bartlett school of Architecture investigating robotic manufacturing and large-scale 3D printing, and he is a senior lecturer at UEL.

Jenny Sabin
Arthur L. and Isabel B. Wiesenberger Assistant Professor,
Director of Graduate Studies, Cornell University

Principal, Jenny Sabin Studio

Jose Sanchez
Assistant Professor, University of Southern California

Director of Plethora Project

Jenny E. Sabin is an architectural designer whose work is at the forefront of a new direction for 21st century architectural practice—one that investigates the intersections of architecture and science, and applies insights and theories from biology and mathematics to the design of material structures. She is principal of the award-winning practice, Jenny Sabin Studio, an experimental architectural design studio based in Ithaca and Director of the Sabin Design Lab at Cornell AAP, a trans-disciplinary design research lab with specialization in computational design, data visualization and digital fabrication. Sabin is the Arthur L. and Isabel B. Wiesenberger Assistant Professor in the area of design and emerging technologies and Director of Graduate Studies in the Department of Architecture at Cornell University. Sabin's collaborative research and design including bioinspired adaptive materials and 3D geometric assemblies has been funded substantially by the National Science Foundation with applied projects commissioned by diverse clients including Nike Inc., Autodesk, the Cooper Hewitt, the FRAC, the American Philosophical Society Museum, the Museum of Craft and Design, the Philadelphia Redevelopment Authority and the Exploratorium.

Jose Sanchez is an Architect / Programmer / Game Designer based in Los Angeles, California. He is the director of the Plethora Project, a research and learning project investing in the future of on-line open-source knowledge. He is also the creator of Block'hood, an award-winning city building video game exploring notions of crowdsourced urbanism named by the Guardian one of the most anticipated games of 2016.

He has taught and guest lectured in several renowned institutions across the world, including the Architectural Association in London, the University of Applied Arts (Angewandte) in Vienna, ETH Zurich, The Bartlett School of Architecture, University College London, and the Ecole Nationale Supérieure D'Architecture in Paris.

Today, he is an Assistant Professor at USC School of Architecture in Los Angeles. His research 'Gamescapes', explores generative interfaces in the form of video games, speculating in modes of intelligence augmentation, combinatorics and open systems as a design medium.

Kyle Steinfeld
Assistant Professor of Architecture, University of California, Berkeley

Dr. Philip Yuan
Professor of Digital Design Research Center (DDRC), the College of Architecture and Urban Planning (CAUP), Tongji University

Founding partner of ARCHI-UNION and FAB-UNION

Council Member of Architecture Society of China (ASC) - Co-founder of Digital Architecture Design Association (DADA) in ASC

Kyle Steinfeld, Assistant Professor specializing in digital design technologies, is the author of the forthcoming "Geometric Computation: Foundations for Design", a text that seeks to demystify computational geometry for an audience of architecture students and design professionals, and is the creator of Decod.es, a platform-agnostic geometry library intended to promote computational literacy in creative design. He has been the recipient of a number of fellowships for research in design technology, recently serving as an IDEA fellow at Autodesk in 2014 and as a Hellman Fellow in 2012. His broad research interests include collaborative design technology platforms, design computation pedagogy, and bioclimatic design visualization. Professionally, he has worked with and consulted for a number of firms, including SOM, Acconci Studio, KPF, Höweler+Yoon, and Diller+Scofidio. He teaches design studios, core courses in representation, and advanced seminars in digital modeling and visualization.

Philip F. Yuan is dedicating in the combination of research, pedagogy and design experiments. The major research field is on the methodology of computational design and fabrication. The research consists of the fabrication of computational material research, robotic fabrication technology, the prototype of future digital factory, etc. Furthermore, the research also contains the Integration of Physical Wind Tunnel and CFD Simulation Technology as well as Acoustics and Theater Design. Philip F. Yuan has published books including *A Tectonic Reality*, *Fabricating the Future*, *Scripting the Future*, *Theater Design*, *Robotic Futures*, *From Diagrammatic Thinking to Digital Fabrication* and *Digital Workshop in China*. Philip F. Yuan has been honored the Youth Architects Prize of ASC, the Nomination Award in 2014 ARC SIA, Feng Jizhong Architectural Education Award, Public prize of Shenzhen Biennia, etc. And he has also been shortlisted for WA China Architectural Award of World Architecture and the 2014 Wienerberger Brick Award.

ACADIA Organization

The Association for Computer Aided Design in Architecture (ACADIA) is an international network of digital design researchers and professionals that facilitates critical investigations into the role of computation in architecture, planning, and building science, encouraging innovation in design creativity, sustainability and education.

ACADIA was founded in 1981 by some of the pioneers in the field of design computation including Bill Mitchell, Chuck Eastman, and Chris Yessios. Since then, ACADIA has hosted over 30 conferences across North America and has grown into a strong network of academics and professionals in the design computation field.

Incorporated in the state of Delaware as a not-for-profit corporation, ACADIA is an all-volunteer organization governed by elected officers, an elected Board of Directors, and appointed ex-officio officers.

PRESIDENT
Jason Kelly Johnson California College of the Arts
pres@acadia.org

VICE-PRESIDENT
Michael Fox Cal Poly Pomona
vp@acadia.org

SECRETARY
Gregory Luhan University of Kentucky
secretary@acadia.org

TREASURER
Mike Christenson North Dakota State University
treasurer@acadia.org

MEMBERSHIP OFFICER
Phillip Anzalone New York City College of Technology
membership@acadia.org

TECHNOLOGY OFFICER
Andrew Kudless California College of the Arts
webmaster@acadia.org

DEVELOPMENT OFFICER
Shane Burger Woods Bagot, Smartgeometry
development@acadia.org

COMMUNICATION OFFICER
Adam Marcus California College of the Arts
communications@acadia.org

BOARD OF DIRECTORS

January 2016–December 2017
Kory Bieg The University of Texas at Austin
Shane Burger Woods Bagot, Smartgeometry
David Jason Gerber University of Southern California
Alvin Huang University of Southern California
Kathy Velikov University of Michigan
Joshua Bard *(alternate)* Carnegie Mellon University
Gregory Luhan *(alternate)* University of Kentucky
Adam Marcus *(alternate)* California College of the Arts
Chris Perry *(alternate)* Rensselaer Polytechnic Institute

January 2015–December 2016
Phillip Anzalone New York City College of Technology
Mike Christenson North Dakota State University
Dana Cupkova Carnegie Mellon University
Gregory Luhan University of Kentucky
Nathan Miller Proving Ground LLC
Chandler Ahrens *(alternate)* Washington University of St. Louis
Duks Koschitz *(alternate)* Pratt Institute
Kyle Steinfeld *(alternate)* University of California–Berkeley

Conference Management & Production Credits

SITE CHAIR
Geoffrey Thün Associate Dean, Associate Professor, University of Michigan Taubman College | Founding Principal, RVTR

TECHNICAL CO-CHAIRS
Kathy Velikov Associate Professor, University of Michigan Taubman College | Founding Principal, RVTR
Sean Ahlquist Assistant Professor, University of Michigan Taubman College | Cluster Member, Comp. Media & Interactive Systems
Matias del Campo Associate Professor, University of Michigan Taubman College | Co-founder, SPAN Architects

WORKSHOP CO-CHAIRS
Wes McGee Assistant Professor, FABLab Director, University of Michigan Taubman College | Founding Partner, Matter Design
Catie Newell Assistant Professor, University of Michigan Taubman College | Founding Principal, Alibi Studio

PROJECTS/EXHIBITION CHAIR
Sandra Manninger Assistant Professor, University of Michigan Taubman College | Co-founder, SPAN Architects

PROJECT MANAGEMENT
Kate Grandfield Administrative Specialist, University of Michigan Taubman College

PLANNING AND ORGANIZATION
Katee Cole Conferences and Events Manager, University of Michigan Taubman College
Deniz McGee Public Relations Associate, University of Michigan Taubman College
Amber LaCroix Senior Director, Marketing Communications, University of Michigan Taubman College
Sandra Patton Business and Finance Manager, University of Michigan Taubman College

WEBSITE AND IT SUPPORT
Bryan Ranallo Senior Web Designer, University of Michigan Taubman College
Bill Manspeaker Information Technology Manager, University of Michigan Taubman College

EXHIBITION PRODUCTION
Maryann Wilkinson Exhibition Director, University of Michigan Taubman College
Dustin Brugmann FAB Lab Research Associate, University of Michigan Taubman College
Asa Peller FAB Lab Research Associate, University of Michigan Taubman College
Kallie Sternburgh Research Associate, University of Michigan Taubman College and RVTR
Daniel Tish Research Associate, University of Michigan Taubman College and RVTR

GRAPHIC DESIGN FOR PROMOTIONAL MATERIALS
Liz Momblanco Graphic Designer, University of Michigan Taubman College

COVER AND SECTION IMAGERY
Matias del Campo / SPAN Architects

COPY EDITING FOR PUBLICATIONS
Pascal Massinon, Ph.D.
Mary O'Malley

GRAPHIC DESIGN AND LAYOUT FOR PUBLICATIONS
Rebekka Kuhn Principal, Rebekka Kuhn Communications Design

Peer Review Committee

Henri Achten Associate Professor, Czech Technical University in Prague

Sean Ahlquist Assistant Professor, University of Michigan

Chandler Ahrens Assistant Professor, Washington University in St. Louis

Gil Akos Mode Lab

T. Jason Anderson Associate Professor, California College of the Arts (CA)

Phillip Anzalone Assistant Professor, New York City College of Technology

Phil Ayres Associate Professor, CITA / KADK

Brad Bell Associate Professor, University of Texas Arlington

Shajay Bhooshan Professional/Practitioner Zaha Hadid Architects

Kory Bieg Assistant Professor, The University of Texas at Austin

Nimish Biloria Assistant Professor, Delft University of Technology

Johannes Braumann Robots in Architecture

Sigrid Brell-Cokcan Full Professor, Association for Robots in Architecture/ RWTH Aachen University

Danelle Briscoe Associate Professor, The University of Texas at Austin

Nicholas Bruscia Assistant Professor, State University of New York at Buffalo

Shane Burger Principal and Director of Technical Innovation Woods Bagot

Jane Burry Associate Professor, RMIT University

Bradley Cantrell Associate Professor, Harvard University

Joseph Choma Assistant Professor, Clemson University

Mike Christenson Associate Professor, North Dakota State University

Angelos Chronis PhD Candidate Institute of Advanced Architecture of Catalonia

Brandon Clifford Assistant Professor, MIT

Kristof Crolla Assistant Professor, The Chinese University of Hong Kong

Jason Crow Assistant Professor, Louisiana State University

Dana Cupkova Assistant Professor, Carnegie Mellon University

Martyn Dade-Robertson Associate Professor, Newcastle University

Daniel Davis Director of Spaces and Cities Research WeWork

Matias del Campo Associate Professor, University of Michigan

Anders Holden Deleuran PhD Fellow CITA / KADK

Christian Derix Principal and Director of Woods Bagot SUPERSPACE Woods Bagot

Mark Donohue Associate Professor, California College of the Arts

Jefferson Ellinger Associate Professor, UNC Charlotte

Thom Faulders Full Professor, California College of the Arts

Jelle Feringa Professional/Practitioner Odico formwork robotics

Laura Forlano Assistant Professor, Illinois Institute of Technology

Michael Fox Full Professor, Cal Poly Pomona

Ursula Frick PhD Researcher, ETH Zurich

Pia Fricker Lecturer, ETH Zurich

Nataly Gattegno Associate Professor, California College of the Arts

Jordan Geiger Assistant Professor, State University of New York at Buffalo

David Gerber Assistant Professor, University of Southern California

Marcelyn Gow Associate Professor, Southern California Institute of Architecture

Yasha J. Grobman Assistant Professor, Technion - Israel Institute of Technology

John Haymaker Professor, Researcher, Technologist, Perkins+Will

Alvin Huang Assistant Professor, University of Southern California

Jason Johnson Associate Professor, University of Calgary

Jason Kelly Johnson Full Professor, California College of the Arts

Joshua Joshua Assistant Professor, Carnegie Mellon University

Lars Junghans Assistant Professor, University of Michigan

Sawako Kaijima Assistant Professor, Singapore University of Technology and Design

Jyoti Kapur Doctoral Student, University of Borås

Karen Kensek Adjunct Professor, University of Southern California

Ted Kesik Full Professor, University of Toronto

Omar Khan Associate Professor, University at Buffalo

Axel Kilian Assistant Professor, Princeton University

Simon Kim Assistant Professor, University of Pennsylvania

Kevin Klinger Associate Professor, Ball State University

Chris Knapp Assistant Professor, Bond University

Branko Kolarevic Full Professor, University of Calgary

Manuel Kretzer Visiting Professor, University of Arts Braunschweig

Oliver David Krieg Assistant Professor, University of Stuttgart

Andrew Kudless Associate Professor, California College of the Arts

Riccardo La Magna Postdoctoral Scholar, University of Stuttgart

Charles Lee Professional/Practitioner Corgan

Julian Lienhard Visiting Professor, HCU Hamburg / str.ucture

Russell Loveridge Managing Director of the NCCR Digital Fabrication ETH Zurich

Adam Marcus Assistant Professor, California College of the Arts

Tyrone Marshall Professional/Practitioner

Bob Martens Associate Professor, TU Wien

Malcolm McCullough Full Professor, University of Michigan

Wes Mcgee Assistant Professor, University of Michigan

Achim Menges Full Professor, University of Stuttgart

AnnaLisa Meyboom Associate Professor, University of British Columbia

Nathan Miller Founder Proving, Ground LLC

Volker Mueller Research Director, Bentley Systems, Incorporated

Tsz Yan Ng Assistant Professor, University of Michigan

Paul Nicholas Assistant Professor, CITA / KADK

Rivka Oxman Visiting Professor, Technion - Israel Institute of Technology

Guvenc Ozel Lecturer, UCLA

Dimitris Papanikolaou Postdoctoral Scholar, Harvard University

Vera Parlac Assistant Professor, University of Calgary

Andrew Payne Principal, Research Engineer, Autodesk

Santiago Perez Assistant Professor, University of Arkansas

Chris Perry Assistant Professor, Rensselaer Polytechnic Institute

Brady Peters Assistant Professor, University of Toronto

Dave Pigram Course Director, Senior Lecturer supermanoeuvre | ETH Zurich | UTS

Nicholas Pisca Professional/Practitioner 0001D LLC

Maya Przybylski Assistant Professor, University of Waterloo

Nick Puckett Assistant Professor, OCAD University

Mette Ramsgaard Thomsen Full Professor, CITA / KADK

Matthias Rippmann Postdoctoral Scholar ETH Zurich

Ronnie Parsons Partner Mode Lab

David Ruy Full Professor, Pratt Institute

Jose Sanchez Assistant Professor, University of Southern California

Blair Satterfield Associate Professor, University of British Columbia

Anton Savov Research Associate TU Darmstadt

Simon Schleicher Assistant Professor, University of California - Berkeley

Tobias Schwinn Lecturer, University of Stuttgart

Mark Shepard Associate Professor, University at Buffalo

Shane Ida Smith Assistant Professor, University of Arizona

Ian Smith Full Professor, EPFL

Aaron Sprecher Associate Professor, McGill University

Kyle Steinfeld Assistant Professor, UC Berkeley

Robert Stuart-Smith Assistant Professor, Architectural Association School of Architecture, UCL-Computer Science

Marc Swackhamer Associate Professor, University of Minnesota - Twin Cities

Martin Tamke Associate Professor, The Royal Danish Academy of Fine Arts

Josh Taron Associate Professor, University of Calgary

Geoffrey Thün Associate Professor, University of Michigan

Skylar Tibbits Research Scientist, MIT

Bige Tuncer Associate Professor, Singapore University of Technology and Design

Kathy Velikov Associate Professor, University of Michigan

Joshua Vermillion Assistant Professor, University of Nevada-Las Vegas

Peter von Buelow Full Professor, University of Michigan

Michael Weinstock Director, Emergent Technologies and Design AA School of Architecture

Andrew Wit Assistant Professor, Temple University

Andrew Witt Assistant Professor, Harvard University

Robert Woodbury Distinguished Professor/Endowed Chair, Simon Fraser University

Wei Yan Associate Professor, Texas A&M University

Shai Yeshayahu Assistant Professor, University of Nevada-Las Vegas

ACADIA 2016 Sponsors

PLATINUM SPONSOR + AWARD

ACADIA would also like to acknowledge the generosity of Autodesk in their support of additional scholarships and awards including the Autodesk ACADIA Student Conference Travel Scholarships, and the Autodesk ACADIA Research Excellence Awards to support outstanding peer-reviewed papers and projects.

SILVER SPONSOR

PERKINS + WILL

BRONZE SPONSOR

HKS LINE

FLUX HDR

SPONSOR

 VECTORWORKS
A NEMETSCHEK COMPANY

ARKTURA

Oasys
YOUR IDEAS BROUGHT TO LIFE

 **BLOOM**
GENERAL CONTRACTING, INC.

CONFERENCE MEDIA PARTNER

THE
ARCHITECTS
NEWSPAPER

WORKSHOPS MEDIA PARTNER

 **Archinect**